Lecture Notes in Compute  528

Edited by G. Goos and J. Hartmanis

Advisory Board: W. Brauer   D. Gri

J. Maluszyński   M. Wirsing (Eds.)

Programming Language Implementation and Logic Programming

3rd International Symposium, PLILP '91
Passau, Germany, August 26-28, 1991
Proceedings

Springer-Verlag
Berlin Heidelberg New York
London Paris Tokyo
Hong Kong Barcelona
Budapest

Series Editors

Gerhard Goos
GMD Forschungsstelle
Universität Karlsruhe
Vincenz-Priessnitz-Straße 1
W-7500 Karlsruhe, FRG

Juris Hartmanis
Department of Computer Science
Cornell University
Upson Hall
Ithaca, NY 14853, USA

Volume Editors

Jan Maluszyński
Department of Computer and Information Science
Linköping University, S-581 83 Linköping, Sweden

Martin Wirsing
Fakultät für Mathematik und Informatik, Universität Passau
Postfach 25 40, W-8390 Passau, FRG

CR Subject Classification (1991): F.4.1-2, D.3.1, D.3.4, F.3.3, I.2.3

ISBN 3-540-54444-5 Springer-Verlag Berlin Heidelberg New York
ISBN 0-387-54444-5 Springer-Verlag New York Berlin Heidelberg

Typesetting: Camera ready by author
Printing and binding: Druckhaus Beltz, Hemsbach/Bergstr.
2145/3140-543210 - Printed on acid-free paper

Preface

This volume contains the papers which have been accepted for presentation at the Third International Symposium on Programming Language Implementation and Logic Programming (PLILP '91) held in Passau, Germany, August 26-28, 1991. The Symposium was preceded by two workshops which took place in Orléans, France, May 16-18, 1988 and in Linköping, Sweden, August 20-22, 1990 (their proceedings were published as LNCS volumes 348 and 456 respectively).

The aim of the Symposium was to explore new declarative concepts, methods and techniques relevant for the implementation of all kinds of programming languages, whether algorithmic or declarative ones. The intention was to gather researchers from the fields of algorithmic programming languages as well as logic, functional and object-oriented programming.

In response to the call for papers, 96 papers were submitted. The Program Committee met on April 30 and selected 32 papers, chosen on the basis of their scientific quality and relevance to the Symposium. At the Symposium, two invited talks were given by Hassan Aït-Kaci and David B. MacQueen. Several software systems were presented, showing new developments in the implementation of programming languages and logic programming.

This volume contains the two invited presentations, the selected papers and abstracts of the system demonstrations.

On behalf of the Program Committee the Program Chairmen would like to thank all those who submitted papers and the referees who helped to evaluate the papers.

The support of

> ALP (Association of Logic Programming),
> FORWISS (Bayer. Forschungszentrum für Wissensbasierte Systeme),
> INRIA (Institut National de Recherche en Informatique et en Automatique),
> University of Passau

is gratefully acknowledged. Andy Mück, Margit Berger, Heidrun Walker and several other members of the department provided invaluable help throughout the preparation and organization of the Symposium. We also would like to thank Springer-Verlag for their excellent cooperation concerning the publication of this volume.

June 1991
Linköping
Passau

Jan Maluszyński
Martin Wirsing

Conference Chairmen

Martin Wirsing, Univ. of Passau (FRG)
Jan Maluszyński, Linköping University (Sweden)

Program Committee

Maurice Bruynooghe, Katholieke Univ. Leuven (Belgium)
Pierre Deransart, INRIA, Rocquencourt (France)
Seif Haridi, SICS, Stockholm (Sweden)
Stefan Jähnichen, GMD, Univ. of Karlsruhe (FRG)
Claude Kirchner, INRIA Lorraine, CRIN, Nancy (France)
Bernard Lang, INRIA, Rocquencourt (France)
Giorgio Levi, Univ. of Pisa (Italy)
Gary Lindstrom, Univ. of Utah, Salt Lake City (USA)
Heikki Mannila, Univ. of Helsinki (Finland)
Torben Mogensen, Univ. Copenhagen (Denmark)
Jaan Penjam, Estonian Academy of Sciences, Tallin (USSR)
Masataka Sassa, Univ. of Tsukuba (Japan)
Péter Szeredi, SzKI Budapest (Hungary)
Reinhard Wilhelm, Univ. of Saarbrücken (FRG)

List of Referees

Many other referees helped the Program Committee in evaluating papers. Their assistance is gratefully acknowledged.

V. Akella	B. Brüderlin	V. Dumortier
K. Ali	J. Burghardt	T. Ehler
M. Alpuente	M. Carlsson	C. Eisenbeis
M. Alt	D. Caromel	H. Emmelmann
N. Andersen	V. Cengarle	M. Falaschi
U. Assmann	P.H. Cheong	C. Fecht
E. Astesiano	H. Christiansen	A. Felty
K. Balogh	S. Cluet	A. Feng
R. Barbuti	M. Codish	C. Ferdinand
E. Bertino	P. Codognet	G. Ferrand
B. Bertolino	D. Craeynest	D. Fortin
A. Blikle	F. Dederichs	U. Fraus
R. Bol	E. De La Clergerie	L. Fribourg
P. Bonatti	D. De Schreye	Y. Futamura
A. Bondorf	B. Demoen	M. Gabbrielli
S. Bonnier	M. Denecker	R. Gabriel
A. Bossi	L. Dewez	H. Ganzinger
J.-L. Bouquard	Y. Deville	M. Gengenbach
M. Breu	R. Dietrich	L. George
A. Brogi	W. Drabent	A. Geser

R. Giacobazzi
P.-Y. Gloess
G. Gonthier
J. Goossenaerts
N. Graham
R. Grosu
S. Grumbach
M. Hanus
A. Haraldsson
B. Hausman
R. Heckmann
F. Henglein
R. Hennicker
J. Henno
A. Hui Bon Hoa
H. Hußmann
T. Ida
J.M. Jacquet
R. Jagadeesan
S. Janson
G. Janssens
B. Jayaraman
M. Jourdan
R. Kaivola
U. Kastens
R. Keßler
P. Kilpeläinen
H. Kirchner
Y. Kiyoki
F. Kluzniak
G. Koch
G. Kock
L.B. Kovács
S. Krisher
K. Kuchcinski
K. Kuse
A. Lalouet

T. Langer
J. Leszczylowski
H. Lock
J. Loeckx
A. Lomp
P. Mancarella
L. Mandel
S. Mantha
L. Maranget
A. Mariën
A. Márkus
B. Martens
M. Martelli
M. Matskin
M. Mazaud
A. Megrelis
P. Mello
D. Mery
B. Möller
D. Montesi
R. Moolenaar
T. Moore
A. Mück
F. Müller
A. Nakamura
H. Nakamura
D. Nazareth
F. Nickl
U. Nilsson
K. Ohmaki
S. Okui
J. Paakki
P. Padawitz
D. Parigot
D. Pedreschi
V. Pollara
C. Queinnec

A. Quéré
M. Raber
T. Rapcsák
F. Regensburger
P. Réty
B. Reus
M. Rosendahl
F. Rossi
F. Rouaix
M. Rusinowitch
K. Sakai
D. Saktin
G. Sander
B. Schieder
R. Schott
H. Seidl
R. Stabl
T. Streicher
M. Suarez
D. Subramanian
M. Szöts
O.-M. Tammepuu
T. Tammet
N. Tamura
J. Tarhio
J. Tepandi
T. Tokuda
F. Turini
E. Ukkonen
A. Van Acker
K. Verschaetse
P. Viry
J. Vollmer
K. Wada
P. Weemeeuw
D. Yves
J. Zachary

Table of Contents

Standard ML of New Jersey

Andrew W. Appel*
Princeton University

David B. MacQueen
AT&T Bell Laboratories

Abstract

The Standard ML of New Jersey compiler has been under development for five years now. We have developed a robust and complete environment for Standard ML that supports the implementation of large software systems and generates efficient code. The compiler has also served as a laboratory for developing novel implementation techniques for a sophisticated type and module system, continuation based code generation, efficient pattern matching, and concurrent programming features.

1 Introduction

Standard ML of New Jersey is a compiler and programming environment for the Standard ML language[26] that has been continuously developed since early 1986. Our initial goal was to produce a working ML front end and interpreter for programming language research, but the scope of the project has expanded considerably. We believe that Standard ML may be the best general-purpose programming language yet developed; to demonstrate this, we must provide high-quality, robust, and efficient tools for software engineering.

Along the way we have learned many useful things about the design and implementation of "modern" programming languages. There were some unexpected interactions between the module system, type system, code generator, debugger, garbage collector, runtime data format, and hardware; and some things were much easier than expected. We wrote an early description of the compiler in the spring of 1987[7], but almost every component of the compiler has since been redesigned and reimplemented at least once, so it is worthwhile to provide an updated overview of the system and our implementation experience.

Our compiler is structured in a rather conventional way: the input stream is broken into tokens by a lexical analyzer, parsed according to a context-free grammar, semantically analyzed into an annotated abstract syntax tree, type-checked, and

*Supported in part by NSF grant CCR-9002786.

translated into a lower-level intermediate language. This is the "front end" of the compiler. Then the intermediate language—*Continuation-Passing Style*—is "optimized," closures are introduced to implement lexical scoping, registers are allocated, target-machine instructions are generated, and (on RISC machines) instructions are scheduled to avoid pipeline delays; these together constitute the "back end."

2 Parsing

Early in the development of the compiler we used a hand-written lexical analyzer and a recursive-descent parser. In both of these components the code for semantic analysis was intermixed with the parsing code. This made error recovery difficult, and it was difficult to understand the syntax or semantics individually. We now have excellent tools[8, 32] for the automatic generation of lexical analyzers and error-correcting parsers. Syntactic error recovery is handled automatically by the parser generator, and semantic actions are only evaluated on correct (or corrected) parses. This has greatly improved both the quality of the error messages and the robustness of the compiler on incorrect inputs. We remark that it would have been helpful if the definition of Standard ML[26] had included an LR(1) grammar for the language.

There are two places in the ML grammar that appear not to be context free. One is the treatment of data constructors: according to the definition, constructor names are in a different lexical class than variable names, even though the distinction depends on the semantic analysis of previous `datatype` definitions. However, by putting constructors and variables into the same class of lexical tokens, and the same name space, parsing can be done correctly and the difference resolved in semantic analysis.

The other context-dependent aspect of syntax is the parsing of `infix` identifiers. ML allows the programmer to specify any identifier as infix, with an operator precedence ranging from 0 to 9. Our solution to this problem is to completely ignore operator precedence in writing our LALR(1) grammar; the expression $a + b * c$ is parsed into the list $[a, +, b, *, c]$ and the semantic analysis routines include a simple operator precedence parser (35 lines of ML).

Each production of our grammar is annotated by a semantic action, roughly in the style made popular by YACC[16]. Our semantic actions are written like a denotational semantics or attribute grammar, where each fragment is a function that takes inherited attributes as parameters and returns synthesized attributes as results. Within the actions there are occasional side-effects; e.g. when the typechecker performs unification by the modification of ref-cells.

A complete parse yields a function p parameterized by a static environment e (of identifiers defined in previous compilation units, etc.). No side-effects occur until p is applied to e, at which point e is distributed by further function calls to many levels of the parse tree. In essence, before p is applied to e it is a tree of closures (one pointing to the other) that is isomorphic to the *concrete* parse tree of the program. Yet we have not had to introduce a myriad of data constructors to describe concrete parse trees!

Delaying the semantic actions is useful to the error-correcting parser. If an error in the parse occurs, the parser might want to correct it at a point 10 tokens previous; this means discarding the last few semantic actions. Since the actions have had no side-effects, it is easy to discard them. Then, when a complete correct parse is constructed, its semantic value can be applied to the environment e and all the side-effects will go off in the right order.

Finally, the treatment of mutually-recursive definitions is easier with delayed semantic actions; the newly-defined identifiers can be entered into the environment before the right-hand-sides are processed.

There is one disadvantage to this arrangement. It turns out that the closure representation of the concrete parse tree is much larger than the annotated parse tree that results from performing the semantic actions. Thus, if we had used a more conventional style in which the actions are performed as the input is parsed, the compiler would use less memory.

Our parser-generator provides, for each nonterminal in the input, the line number (and position within the line) of the beginning and end of the program fragment corresponding to that nonterminal. These are used to add accurate locality information to error messages. Furthermore, these line numbers are sprinkled into the annotated abstract syntax tree so that the type checker, match compiler, and debugger can also give good diagnostics.

3 Semantic analysis

A *static environment* maps each variable of the program to a *binding* containing its *type* and its runtime *access* information. The type is used for compile-time type checking, and is not used at runtime. The access information is (typically) the name of a low-level λ-calculus variable that will be manipulated by the code generator. Static environments also map other kinds of identifiers—data constructors, type constructors, structure names, etc.—to other kinds of bindings.

Our initial implementation treated environments imperatively: the operations on environments were to add a new binding to the global environment; to "mark" (save) the state of the environment; to revert back to a previous mark; and, for implementation of the module system, to encapsulate into a special table everything added since a particular mark. We did this even though we knew better—denotational semantics or attribute grammars would have us treat environments as pure values, to be combined to yield larger environments—because we thought that imperative environments would be faster.

We have recently changed to a pure functional style of environments, in which the operations are to create an environment with a single binding, and to layer one environment on top of another nondestructively, yielding a new environment. The implementation of this abstract data type has side effects, as sufficiently large environment-values are represented as hash tables, etc. We made this change to accommodate the new debugger, which must allow the user to be in several environments simultaneously; and to allow the implementation of "make" programs, which need explicit control over the static environments of the programs being compiled. Though we were willing to suffer a performance degradation in exchange for this flexibility, we found "pure" environments to be just as fast as imperative ones.

This illustrates a more general principle that we have noticed in ML program development. Many parts of the compiler that we initially implemented in an imperative style have been rewritten piecemeal in a cleaner functional style. This is one of the advantages of ML: programs (and programmers) can migrate gradually to "functional" programming.

Type checking

The main type-checking algorithm has changed relatively little since our earlier description[7]. The representations of types, type constructors, and type variables have been cleaned up in various ways, but the basic algorithm for type checking is still based on a straightforward unification algorithm.

The most complex part of the type-checking algorithm deals with *weak* polymorphism, a restricted form of polymorphism required to handle mutable values (references and arrays), exception transmission, and communication (in extensions like Concurrent ML[28]). Standard ML of New Jersey implements a generalization of the imperative type variables described in the Definition[26, 34]. In our scheme, *imperative* type variables are replaced by *weak* type variables that have an associated degree of weakness: a nonnegative integer. A type variable must be weak if it is involved in the type of an expression denoting a reference, and its degree of weakness roughly measures the number of function applications that must take place before the reference value is actually created. A weakness degree of zero is disallowed at top level, which insures that top-level reference values (*i.e.* those existing within values in the top level environment) have monomorphic types. The type-checking algorithm uses an abstract type *occ* to keep track of the "applicative context" of expression occurrences, which is approximately the balance of function abstractions over function applications surrounding the expression, and the *occ* value at a variable occurrence determines the weakness degree of generic type variables introduced by that occurrence. The *occ* value at a let binding is also used to determine which type variables can be generalized.

The weak typing scheme is fairly subtle and has been prone to bugs, so it is important that it be formalized and proven sound (as the Tofte scheme has been [Tofte-thesis]). There are several people currently working on formalizing the treatment used in the compiler[17, 38].

The weak polymorphism scheme currently used in Standard ML of New Jersey is not regarded as the final word on polymorphism and references. It shares with the imperative type variable scheme the fault that weak polymorphism propagates more widely than necessary. Even purely internal and temporary uses of references in a function definition will often "poison" the function, giving it a weak type. An example is the definition

```
fun f x = !(ref x)
```

in which f has the type $1\alpha \rightarrow 1\alpha$, but *ought* to have the strong polymorphic type $\alpha \rightarrow \alpha$. This inessential weak polymorphism is particularly annoying when it interferes with the matching of a signature specification merely because of the use of an imperative style within a function's definition. Such implementation choices should be invisible in the type.

Research continues on this problem[17, 22, 38], but there is no satisfactory solution yet.

The interface between the type checker and the parser is quite simple in most respects. There is only one entry point to the type checker, a function that is called to type-check each value declaration at top level and within a structure. However, the interface between type checking and the parser is complicated by the problem of determining the scope or binding point of explicit type variables that appear in a program. The rather subtle scoping rules for these type variables[26, Section 4.6][25, Section 4.4] force the parser to pass sets of type variables both upward and downward (as both synthesized and inherited attributes of phrases). Once determined, the set of explicit type variables to be bound at a definition is stored in the abstract syntax representation of the definition to make it available to the typechecker.

4 Modules

The implementation of modules in SML of NJ has evolved through three different designs. The main innovation of the second version factored signatures into a symbol table shared among all instances, and a small instantiation environment for each instance[23]. Experience with this version revealed problems that led to the third implementation developed in collaboration with Georges Gonthier and Damien Doligez.

Representations

At the heart of the module system are the internal representations of signatures, structures, and functors. Based on these representations, the following principal procedures must be implemented:

1. signature creation—static evaluation of signature expressions;

2. structure creation—static evaluation of structure expressions;

3. signature matching between a signature and a structure, creating an *instance* of the signature, and a *view* of the structure;

4. definition of functors—abstraction of the functor body expression with respect to the formal parameter;

5. functor application—instantiation of the formal parameter by matching against the actual parameter, followed by instantiation of the functor body.

It is clear that instantiation of structure templates (i.e. signatures and functor bodies) is a critical process in the module system. It is also a process prone to consume excessive space and time if implemented naively. Our implementation has achieved reasonable efficiency by separating the *volatile* part of a template, that which changes with each instance, from the *stable* part that is common to all instances and whose representation may therefore be shared by all instances. The volatile components are stored in an instantiation environment and they are referred to indirectly in the bindings in the shared symbol table (or static environment) using indices or paths into the instantiation environment. The instantiation environment is represented as a pair of arrays, one for type constructor components, the other for substructures.

The static representation of a structure is essentially an environment (i.e., symbol table) containing bindings of types, variables, etc., and an identifying stamp[26, 33, 23]. In the second implementation a signature was represented as a "dummy" instance that differs from an ordinary structure in that its volatile components contain dummy or *bound* stamps and it carries some additional information specifying sharing constraints. The volatile components with their bound stamps are replaced, or *instantiated*, during signature matching by corresponding components from the structure being matched. Similarly, a functor body is represented as a structure with dummy stamps that are replaced by newly generated stamps when the functor is applied.

The problem with representing signatures (and functor bodies) as dummy structures with bound stamps is the need to do alpha-conversion at various points to avoid confusing bound stamps. To minimize this problem the previous implementation insures that the sets of bound stamps for each signature and functor body are disjoint. But there is still a problem with signatures and functors that are separately compiled and then imported into a new context; here alpha-conversion of bound stamps is required to maintain the disjointness property. Managing bound stamps was a source of complexity and bugs in the module system.

The usual way of avoiding the complications of bound variables is to replace them with an indexing scheme, as is done with deBruijn indices in the lambda calculus[13]. Since in the symbol table part we already used indices into instantiation arrays to refer to volatile components, we can avoid the bound stamps by using this *relativized* symbol table alone to represent signatures.

To drop the instantiation environment part of the signature representation, leaving only the symbol table part, we need to revise the details of how environments are represented. Formerly a substructure specification would be represented in the symbol table by a binding like

$$A \mapsto \text{INDstr } i$$

indicating that A is the ith substructure, and the rest of the specification of A (in the form of a dummy structure) would be found in the ith slot of the instantiation environment. Since we are dropping the dummy instantiation environment we must have all the information specifying A in the binding. Thus the new implementation uses

$$A \mapsto \text{FORMALstrb}\{\text{pos} = i, \text{spec} = sig_A\}$$

as the binding of A. This makes the substructure signature specification available immediately in the symbol table without having to access it indirectly through an instantiation environment.

Another improvement in the representation of signatures (and their instantiations) has to do with the *scope* of instantiation environments. In the old implementation each substructure had its own instantiation environment. But one substructure may contain relative references to a component of another substructure, as in the following example

```
signature S1 =
sig
  structure A : sig type t end
  structure B : sig val x : A.t end
end
```

Here the type of B.x refers to the first type component t of A. This would be represented from the standpoint of B as a relative path [*parent, first substructure, first type constructor*]. To accommodate these cross-structure references when each structure has a local instantiation environment, the first structure slot in the instantiation environment contains a pointer to the parent signature or structure. Defining and maintaining these parent pointers was another source of complexity, since it made the representation highly cyclical.

The new representation avoids this problem by having a single instantiation environment shared by the top level signature and all its *embedded* signatures. An embedded signature is one that is written "in-line" like the signatures of A and B in the example above. In the above example, the new representation of A.t within B is [*first type constructor*] since A.t will occupy the first type constructor slot in the shared instantiation environment.

A nonembedded signature is one that is defined at top level and referred to by name. The signature S0 in the following example is a nonembedded signature.

```
signature S0 = sig type t end
signature S1 =
sig
  structure A : S0
  structure B : sig val x : A.t end
end
```

In this case the type A.t of x uses the indirect reference [*first substructure, first type constructor*] meaning the first type constructor in the local instantiation environment of A, which is the first structure component in the instantiation environment of S1. S1 and B share a common instantiation environment because B is embedded in S1. But S0, the signature of A, is nonembedded because it was defined externally to S1. It therefore can contain no references to other components of S1 and so it is given its own private instantiation environment having the configuration appropriate to S0.

Signature Matching

The goal of the representation of signatures is to make it easy to instantiate them via signature matching. A signature is a template for structures, and a structure can be obtained from the signature by adding an appropriate instantiation environment (and recursively instantiating any substructures with nonembedded signature specifications).

The signature matching process involves the following steps: (1) Create an empty instantiation environment of a size specified in the signature representation. (2) For each component of the signature, in the order they were specified, check that there is a corresponding component in the structure and that this component satisfies the specification. When this check succeeds it may result in an instance of a volatile component (e.g. a type constructor) that is entered into the new instantiation environment. (3) Finally, having created the instantiation structure, any sharing constraints in the signature are verified by "inspection."

Functors

The key idea is to process a functor definition to isolate volatile components of the result (those deriving from the parameter and those arising from generative declarations in the body) in an instantiation environment. Then the body's symbol table is *relativized* to this instantiation environment by replacing direct references by indirect paths. As in the case of signature matching, this minimizes the effort required to create an instance of the body when the functor is applied, because the symbol table information is inherited unchanged by the instance.

Defining a functor is done in three steps: (1) The formal parameter signature is *instantiated* to create a dummy parameter structure. (2) This dummy structure is bound to the formal parameter name in the current environment and the resulting environment is used to parse and type-check the functor body expression. If a result signature is specified the functor body is matched against it. (3) The resulting body structure is scanned for *volatile* components, identified by having stamps belonging to the dummy parameter or generated within the body, and references to these volatile components are replaced by indirect positional references into an instantiation environment.

The instantiation of the parameter signature must produce a structure that is *free* modulo the sharing constraints contained in the signature. In other words, it must satisfy the explicit sharing constraints in the signature and all implicit sharing constraints implied by them, but there must by no extraneous sharing. The algorithm used for this instantiation process is mainly due to George Gonthier and is vaguely related to linear unification algorithms. This instantiation process is also used to create structures declared as *abstractions* using the abstraction declaration of Standard ML of New Jersey (a nonstandard extension of the language).

Given this processing of the functor definition, functor application is now a fairly straightforward process. The actual parameter is matched with the formal parameter signature yielding an instantiation environment relative to the parameter signature. This is combined with a new instantiation environment generated for the functor body using freshly generated stamps in new volatile components.

5 Translation to λ-language

During the semantic analysis phase, all static program errors are detected; the result is an abstract parse tree annotated with type information. This is then translated into a *strict* lambda calculus augmented with data constructors, numeric and string constants, n-tuples, mutually-recursive functions; and various primitive operators for arithmetic, manipulation of refs, numeric comparisons, etc. The translation into λ-language is the phase of our compiler that has changed least over the years.

Though the λ language has data constructors, it does not have pattern-matches. Instead, there is a very simple case statement that determines which constructor has been applied at top level in a given value. The pattern-matches of ML must be translated into discriminations on individual constructors. This is done as described in our previous paper[7], though Bruce Duba has revised the details of the algorithm.

The dynamic semantics of structures and functors are represented using the same lambda-language operators as for the records and functions of the core language. This means that the code generator and runtime system don't need to know anything about the module system, which is a great convenience.

Also in this phase we handle equality tests. ML allows any hereditarily nonabstract, nonfunctional values of the same type to be tested for equality; even if the values have polymorphic types. In most cases, however, the types can be determined at compile time. For equality on atomic types (like integer and real), we substitute an efficient, type-specific primitive operator for the generic equality function. When constructed datatypes are tested for equality, we automatically construct a set of mutually-recursive functions for the specific instance of the datatype; these are compiled into the code for the user's program. Only when the type is truly polymorphic—not known at compile time—is the general polymorphic equality function invoked. This function interprets the tags of objects at runtime to recursively compare bit-patterns without knowing the full types of the objects it is testing for equality.

Standard ML's polymorphic equality seriously complicates the compiler. In the front end, there are special "equality type variables" to indicate polymorphic types that are required to admit equality, and signatures have an eqtype keyword to denote exported types that admit equality. The eqtype property must be propagated among all types and structures that *share* in a functor definition. We estimate that about 7% of the code in the front end of the compiler is there to implement polymorphic equality.

The effect on the back end and runtime system is just as pernicious. Because ML is a statically-typed language, it should not be necessary to have type tags and descriptors on every runtime object (as Lisp does). The only reasons to have these tags are for the garbage collector (so it can understand how to traverse pointers and records) and for the polymorphic equality function. But it's possible to give the garbage collector a map of the type system[1], so that it can figure out the types of runtime objects without tags and descriptors. Yet the polymorphic equality function also uses these tags, so even with a sophisticated garbage collector they can't be done away with. (One alternative is to pass an equality-test function along with every value of an equality type, but this is also quite costly[36].)

Finally, the treatment of equality types in Standard ML is irregular and incomplete[15]. The *Definition* categorizes type constructors as either "equality" or "nonequality" type constructors; but a more refined classification would more accurately specify the effects of the ref operator. Some types that structurally support equality are classified as nonequality types by the *Definition*.

6 Conversion to CPS

The λ-language is converted into *continuation-passing style* (CPS) before optimization and code generation. CPS is used because it has clean semantic properties (like λ-calculus), but it also matches the execution model of a von Neumann register machine: variables of the CPS correspond closely to registers on the machine, which leads to very efficient code generaton[18].

In the λ-language (with side-effecting operators) we must specify a call-by-value (strict) order of evaluation to really pin down the meaning of the program; this means that we can't simply do arbitrary β-reductions (etc.) to partially evaluate and optimize the program. In the conversion to CPS, all order-of-evaluation information is encoded in the chaining of function calls, and it doesn't matter whether we consider the CPS to be strict or nonstrict. Thus, β-reductions and other optimizations become much easier to specify and implement.

The CPS notation[30] and our representation of it[5] are described elsewhere, as is a detailed description of optimization techniques and runtime representations for CPS[4]. We will just summarize the important points here.

In continuation-passing style, each function can have several arguments (in contrast to ML, in which functions formally have only one parameter). Each of the actual parameters to a function must be *atomic*—a constant or a variable. The operands of an arithmetic operator must also be atomic; the result of the operation is bound to a newly-defined variable. There is no provision for binding the result of a function call to a variable; "functions never return."

To use CPS for compiling a programming language—in which functions are usually allowed to return results, and expressions can have nontrivial sub-expressions—it is necessary to use *continuations*. Instead of saying that a function call $f(x)$ returns a value a, we can make a function $k(a)$ that expresses what "the rest of the program" would do with the result a, and then call $f_{cps}(x, k)$. Then f_{cps}, instead of returning, will call k with its result a.

After CPS-conversion, a source-language function call looks just like a source-language function return—they both look like calls in the CPS. This means it is easy to β-reduce the call without reducing the return, or vice versa; this kind of flexibility is very useful in reasoning about (and optimizing) tail-recursion, etc.

In a strict λ-calculus, β-reduction is problematical. If the actual parameters to a function have side effects, or do not terminate, then they cannot be safely substituted for the formal parameters throughout the body of the function. Any actual parameter expression could contain a call to an unknown (at compile time) function, and in this case it is impossible to tell whether it does have a side effect. But in CPS, the actual parameters to a function are always atomic expressions, which have no side effects and always terminate; so it's safe and easy to perform β-reduction and other kinds of substitutions.

In our optimizer, we take great advantage of a unique property of ML: records, n-tuples, constructors, etc., are *immutable*. That is, except for ref cells and arrays (which are identifiable at compile time through the type system), once a record is created it cannot be modified. This means that a *fetch* from a record will always yield the same result, even if the compiler arranges for it to be performed earlier or later than specified in the program. This allows much greater freedom in the partial evaluation of fetches (e.g. from pattern-matches), in constant-folding, in instruction scheduling, and common subexpression elimination than most compilers are permitted. (One would think that in a pure functional language like Haskell this *immutable record* property would be similarly useful, but such languages are usually *lazy* so that fetches from a lazy cell will yield different results the first and second times.)

A similar property of ML is that immutable records are not distinguishable by address. That is, if two records contain the same values, they are "the same;" the expressions

    [(x,y), (x,y)]
    let val a = (x,y) in [a,a] end

are indistinguishable in any context. This is not the case in most programming languages, where the different pairs (x,y) in the first list would have different addresses and could be distinguished by a pointer-equality test.

This means that the compiler is free to perform common sub-expression elimination on record expressions (i.e. convert the first expression above to the second); the garbage collector is free to make several copies of a record (possibly useful for concurrent collection), or to merge several copies into one (a kind of "delayed hash-consing"); a distributed

implementation is free to keep separate copies of a record on different machines, etc. We have not really exploited most of these opportunities yet, however.

7 Closure conversion

The conversion of λ-calculus to CPS makes the control flow of the program much more explicit, which is useful when performing optimizations. The next phase of our compiler, *closure conversion*, makes explicit the access to nonlocal variables (using lexical scope). In ML (and Scheme, Smalltalk, and other languages), function definitions may be nested inside each other; and an inner function can have *free variables* that are bound in an outer function. Therefore, the representation of a function-value (at runtime) must include some way to access the values of these free variables. The *closure* data structure allows a function to be represented by a pointer to a record containing

1. The address of the machine-code entry-point for the body of the function.

2. The values of free variables of the function.

The code pointer (item 1) must be kept in a standardized location in all closures; for when a function f is passed as an argument to another function g, then g must be able to extract the address of f in order to jump to f. But it's not necessary to keep the free variables (item 2) in any standard order; instead, g will simply pass f's closure-pointer as an extra argument to f, which will know how to extract its own free variables.

This mechanism is quite old[19] and reasonably efficient. However, the introduction of closures is usually performed as part of machine-code generation; we have made it a separate phase that rewrites the CPS representation of the program to include closure records. Thus, the output of the closure-conversion phase is a CPS expression in which it is guaranteed that no function has free variables; this expression has explicit record-creation operators to build closures, and explicit fetch operators to extract code-pointers and free variables from them.

Since closure-introduction is not bundled together with other aspects of code generation, it is easier to introduce sophisticated closure techniques without breaking the rest of the compiler. In general, we have found that structuring our compiler with so many phases—each with a clean and well-defined interface—has proven very successful in allowing work to proceed independently on different parts of the compiler.

Initially, we considered variations on two different closure representations, which we call *flat* and *linked*. A *flat* closure for a function f is a record containing the code-pointer for f and the values of each of f's free variables. A *linked* closure for f contains the code pointer, the value of each free variable *bound by the enclosing function*, and a pointer to the enclosing function's closure. Variables free in the enclosing function can be found by traversing the linked list of closures starting from f; this is just like the method of *access links* used in implementing static scope in Pascal.

It would seem that linked closures are cheaper to build (because a single pointer to the enclosing scope can be used instead of all the free variables from that scope) but costlier to access (getting a free variable requires traversing a linked list). In fact, we investigated many different representational tricks on the spectrum between flat and linked closures[6], including tricks where we use the *same* closure record for several different functions *with several different code-pointers*[5, 4].

In a "traditional" compiler, these tricks make a significant difference. But in the CPS representation, it appears that the pattern of functions and variable access narrows the effective difference between these techniques, so that closure representation is not usually too important.

There are two aspects of closures that are important, however. We have recently shown that using linked or merged closures can cause a compiled program to use much more memory[4]. For example, a program compiled with flat closures might use $O(N)$ memory (i.e. simultaneous live data) on an input of size N, and the same program compiled with linked closures might use $O(N^2)$. Though this may happen rarely, we believe it is unacceptable (especially since the programmer will have no way to understand what is going on). We are therefore re-examining our closure representations to ensure "safety" of memory usage; this essentially means sticking to flat closures.

We have also introduced the notion of "callee-save registers."[9, 4] Normally, when an "unknown" function (e.g. one from another compilation unit) is called in a compiler using CPS, all the registers (variables) that will be needed "after the call" are *free variables of the continuation*. As such, they are stored into the continuation closure, and fetched back after the continuation is invoked. In a conventional compiler, the *caller* of a function might similarly save registers into the stack frame, and fetch them back after the call.

But some conventional compilers also have "callee-save" registers. It is the responsibility of each function to leave these registers undisturbed;

if they are needed during execution of the function, they must be saved and restored by the *callee*.

We can represent callee-save variables in the original CPS language, without changing the code-generation interface. We will represent a continuation not as one argument but as $N + 1$ arguments $k_0, k_1, k_2, \ldots, k_N$. Then, when the continuation k_0 is invoked with "return-value" a, the variables $k_1, \ldots, k_N$ will also be passed as arguments to the continuation.

Since our code generator keeps all CPS variables in registers—including function parameters—the variables $k_1, \ldots, k_N$ are, in effect, callee-save registers. We have found that $N = 3$ is sufficient to obtain a significant (7%) improvement in performance.

8 Final code generation

The operators of the CPS notation—especially after closure-conversion—are similar to the instructions of a simple register/memory von Neumann machine. The recent trend towards RISC machines with large register sets makes CPS-based code generation very attractive. It is a relatively simple matter to translate the closure-converted CPS into simple abstract-machine instructions; these are then translated into native machine code for the MIPS, Sparc, VAX, or MC68020. The latter two machines are not RISC machines, and to do a really good job in code generation for them we would have to add a final peephole-optimization or instruction-selection phase. On the RISC machines, we have a final instruction-scheduling phase to minimize delays from run-time pipeline interlocks.

One interesting aspect of the final abstract-machine code generation is the register allocation. After closure-conversion and before code generation we have a *spill* phase that rewrites the CPS expression to limit the number of free variables of any subexpression to less than the number of registers on the target machine[5, 4]. It turns out that very few functions require any such rewriting, especially on modern machines with 32 registers; five spills in 40,000 lines of code is typical.

Because the free variables of any expression are guaranteed to fit in registers, register allocation is a very simple matter: when each variable is bound, only K other variables are live (i.e. free in the continuation of the operation that binds the variable), where $K < N$, the number of registers. Thus, any of the remaining $N - K$ registers can be chosen to hold the new value.

The only place that a register-register move is ever required is at a procedure call, when the actual parameters must be shuffled into the locations required for the formal parameters. For those functions whose call sites are all evident to the compiler (i.e. those functions that are not passed as parameters or stored into data structures), we can choose the register-bindings for formal parameters to eliminate any moves in at least one of the calls[18]. By clever choices of which register to use for the bindings described in the last paragraph, we can almost eliminate any remaining register-register moves that might be required for the other procedure calls.

9 The runtime system

The absence of function returns means that a runtime stack is not formally required to execute programs. Although most CPS-based compilers introduce a runtime stack anyway[30, 18], we do not. Instead, we keep all closures (i.e. activation records) on the garbage-collected heap. This not only simplifies some aspects of our runtime system, but makes the use of first-class continuations (call-with-current-continuation) very efficient.

Because all closures are put on the heap, however, SML/NJ allocates garbage-collected storage at a furious rate: one 32-bit word of storage for every five instructions executed, approximately[4]. This means that the most important requirement for the runtime system is that it support fast storage allocation and fast garbage collection.

To make heap allocations cheap, we use a generational copying garbage collector[2] and we keep the format of our runtime data simple[3]. Copying collection is attractive because the collector touches only the live data, and not the garbage; we can arrange that almost all of a particular region of memory is garbage, then just a few operations can reclaim a very large amount of storage. Another advantage of copying collection is that the free area (in which to allocate new records) is a contiguous block of memory; it is easier to grab the first few words of this block than it would be to manage a "free list" of different-sized records.

Indeed, we keep pointers to the beginning and end of the free area in registers for fast access. Allocation and initialization of an n-word record requires n store instructions at different offsets from the free-space register, followed by the addition of a constant (the size of the new record) to the register. We perform allocations in-line (without a procedure call), and we use just one test for free storage exhaustion to cover all the allocations in a procedure (remember that in CPS, procedures don't have internal loops). Furthermore, we can perform

this test in one single-cycle instruction by clever use of the overflow interrupt to initiate garbage collection[4].

Overall, garbage-collection overhead in Standard ML of New Jersey (with memory size equal to 5 times the amount of live data) is usually between 5 and 20%; this means that for each word of memory allocated, the amortized cost of collecting it is about 1/4 to 1 instruction. Thus, copying a data structure (reading it and writing a new copy) takes only two or three times as long as traversing it (examining all the fields). This encourages a more side-effect-free, functional style of programming.

In addition to the garbage collector, the runtime system provides an interface to operating system calls[3]. Higher-level services like buffered I/O are provided by a "standard library" written in Standard ML. There are also many C-language functions in the runtime system callable from ML; but we have not yet provided an easy interface for users to link their own foreign-language functions to be called from ML. Since the overhead for calling a C function is rather high, we have implemented half a dozen frequently-used functions (e.g. allocation of an array or a string) in assembly language.

There is also an ML interface to operating system signals[27] that uses the call/cc mechanism to bundle up the current state of execution into a "continuation;" to be resumed immediately, later (perhaps from another signal handler), never, or more than once.

A snapshot of the executing ML system may be written to a file; executing that file will resume execution just at the point where the snapshot was taken. It is also possible to remove the compiler from this snapshot, to build more compact standalone applications.

Our reliance on operating system signals for garbage collection, our direct connection to system calls, our snapshot-building utility, and other useful features of the runtime system have turned out to be quite operating-system dependent. This makes it hard to port the runtime system from one machine (and operating system) to another. Perhaps as different versions of Unix become more standardized (e.g. with System V/R4) these problems will largely disappear.

10 Performance

We had several goals for Standard ML of New Jersey:

- A complete and robust implementation of Standard ML.

	Poly/ML 1.91 Compile Time	Run Time	SML/NJ 0.69 Compile Time	Run Time
Life	10	128	13	27
Lex	41	95	66	20
Yacc	abort	—	531	10
Knuth-B	19	116	30	25
Simple	44	461	124	60
VLIW	abort	—	839	45

Figure 1: Comparison of Poly/ML and SML/NJ

This table shows compile time and run time in seconds of elapsed time for each benchmark on a SparcStation 2 with 64 megabytes of memory. SML/NJ was run with the optimization settings normally used for compiling the compiler itself, and with all the input in one file to enable cross-module optimization (which makes things about 9% faster). Note that the callee-save representation is not yet implemented for the Sparc and might save an additional 7% runtime. On two of the benchmarks (as shown), the Poly/ML compiler aborted after several minutes; we believe this is caused by complicated pattern-matches tripping over an exponential-time algorithm in the Poly/ML front end.

- A compiler written in Standard ML itself, to serve as a test of ML for programming-in-the-large.

- A reasonably efficient compiler with no "bottlenecks."

- Very fast compiled code, competitive with "conventional" programming languages.

- A testbed for new ideas.

We believe we have achieved these goals. While our compiler has a few minor bugs (as does any large software system), they don't substantially detract from the usability of the system. We have found that ML is an excellent language for writing real programs. Our compiler's front end is quite carefully designed to be fast, but the back end needs (and is receiving) further work to make it compile faster. The quality of our compiled code is extremely good, as figures 1 and 2 show.

We tested Poly/ML[24] and SML/NJ on six real programs[4], whose average size was about 2000 nonblank noncomment lines of source. Figure 1 shows the results on a SparcStation 2 (the only

	Sun 3/280 16 Mbytes Run	G.C.	DEC 5000/200 16 Mbytes Run	G.C.
CAML V2-6.1	14.5	14.8	6.2	6.2
CAML Light 0.2	28.3		6.5	
SML/NJ 0.65	9.6	0.3	1.7	0.1
SML/NJ 0.65 x	8.5	0.3	1.4	0.1
LeLisp 15.23	4.1		1.4	
SunOS 3.5, cc -O	4.35			
gcc 1.37.1, gcc -O	4.22			
Ultrix 4.0, cc -O2			0.90	

Figure 2: Comparison of several different compilers

Xavier Leroy translated Gerard Huet's Knuth-Bendix program into several different languages, and ran them on two different machines[21]. This table shows non-GC run time and GC time in seconds for each version of the program. Since the program uses higher-order functions, Leroy had to do manual lambda-lifting to write the program in Lisp and C, and in some places had to use explicit closures (structures containing function-pointers).

CAML is a different version of the ML language (i.e. not Standard ML) developed at INRIA[11]; CAML V2-6.1 is a native-code compiler that shares the LeLisp runtime system, and CAML Light[20] is a compiler with a byte-code interpreter written in C. SML/NJ x refers to Standard ML of New Jersey with all modules placed in "super-module" to allow cross-module optimization.

modern platform on which they both run). Indeed, Poly/ML compiles about 43% faster (when it doesn't blow up); but SML/NJ programs run *five times faster* than Poly/ML programs, on the average (geometric mean). SML/NJ reportedly uses about 1.5 times as much heap space for execution[10]; and on a 68020-based platform (like a Sun-3), SML/NJ may not do relatively as well (since we don't generate really good code for that machine). So on obsolete machines with tiny memories, Poly/ML may do almost as well as SML/NJ.

Figure 2 compares implementations of several programming languages on a Knuth-Bendix benchmark. Standard ML of New Jersey does quite well, especially on the RISC machine (the DECstation 5000 has a MIPS processor).

11 Continuations

One of the more significant language innovations in Standard ML of New Jersey is typed first-class continuations[14]. It turned out to be possible to add a major new capability to the language merely by introducing a new primitive type constructor and two new primitive functions. The signature for first-class continuations is:

```
type 'a cont
val callcc : ('a cont -> 'a) -> 'a
val throw : 'a cont -> 'a -> 'b
```

The type int cont is the type of a continuation that is expecting an integer value. The callcc function is similar to call-with-current-continuation or call/cc in Scheme — it is the primitive that captures continuation values. The function throw coerces a continuation into a function that can be applied to invoke the continuation. Since the invocation of a continuation does not return like a normal function call, the return type of throw k is a generic type variable that will unify with any type.

The runtime implementation of first-class continuations was also quite easy and very efficient, because of the use of continuation passing style in the code generation, and the representation of continuations as objects in the heap. Bundling up the current continuation into a closure is just like what is done on the call to an escaping function, and throwing a value to a continuation is like a function call. So continuations are as cheap as ordinary function calls.

Continuations are not necessarily a good tool for routine programming since they lend themselves to tricky and contorted control constructs. However, continuations have an important "behind the scenes" role to play in implementing useful tools and abstractions. They are used in the implementation of the interactive ML system to construct a barrier between user computation and the ML system. This makes it possible to export an executable image of a user function without including the ML compiler. Another application of continuations is Andrew Tolmach's replay debugger[35], where they are used to save control states. This is the basis of the *time travel* capabilities of the debugger.

It is well known that continuations are useful for implementing coroutines and for simulating parallel threads of control[37]. Using continuations in conjunction with the signal handling mechanisms implemented by John Reppy[27] (themselves expressed in terms of continuations), one can build light-weight process libraries with preemptive process scheduling entirely within Standard ML of New Jersey. Two major concurrency systems have been implemented at this point: Concurrent ML by John Reppy[28] is based on CCS/CSP-style primitives (synchronous communication on typed channels) but introduces the novel idea of *first-class events*. ML Threads is a system designed by Eric Cooper and Greg Morrisett[12] that provides mutual exclusion primitives for synchronization. A version of ML Threads runs on shared-memory multiprocessors, where threads can be scheduled to run in parallel on separate physical processors. Both Concurrent ML and ML Threads are implemented as ordinary ML modules, requiring no enhancements of the language itself—except that ML Threads required modification of the runtime system to support multiprocessing.

12 Related projects

A number of very useful enhancements of the Standard ML of New Jersey system are being carried out by other groups or individuals. One such project is the SML-to-C translator done by David Tarditi, Anurag Acharya, and Peter Lee at Carnegie Mellon[31]. This provides a very portable basis for running ML programs on a variety of hardware for which we do not yet have native code generators, with very respectable performance.

Mads Tofte and Nick Rothwell implemented the first version of separate compilation for Standard ML of New Jersey. Recently Gene Rollins at Carnegie Mellon has developed a more sophisticated and efficient system called SourceGroups for managing separate compilation. SourceGroups builds on the primitive mechanisms provided by Tofte and Rothwell but gains efficiency by doing a global analysis of dependencies among a set of modules and minimizing redundancy when loading or recompiling the modules.

John Reppy and Emden Gansner have developed a library for interacting with the X window system. This system is based on Concurrent ML and provides a much higher-level of abstraction for writing graphical interfaces than the conventional conventional C-based libraries.

13 Future Plans

The development of Standard ML of New Jersey and its environment is proceeding at an accelerating pace. John Reppy is implementing a new multi-generation, multi-arena garbage collector that should significantly improve space efficiency. Work is in progress to improve code generation and significantly speed up the back end. Exploratory work is being done on new features like type *dynamic*, extensible datatypes, and higher-order functors.

14 Acknowledgments

Many people have worked on Standard ML of New Jersey. We would like to thank **John H. Reppy** for many improvements and rewrites of the runtime system, for designing and implementing the signal-handling mechanism[27], improving the call/cc mechanism, designing the current mechanism for calls to C functions, implementing a sophisticated new garbage collector, generally making the runtime system more robust, and implementing the SPARC code generator; and for designing and implementing the Concurrent ML system[28] and its X-windows interface[29].

Thanks to **Trevor Jim** for helping to design the CPS representation[5]; and for implementing the match compiler and the original closure-converter, the original library of floating point functions, and the original assembly-language implementation of external primitive functions.

Thanks to **Bruce F. Duba** for improvements to the match compiler, the CPS constant-folding phase, the in-line expansion phase, the spill phase, and numerous other parts of the compiler; and for his part in the design of the call-with-current-continuation mechanism[14].

Thanks to **James W. O'Toole** who implemented the NS32032 code generator, and **Norman Ramsey** who implemented the MIPS code generator.

We thank **Andrew P. Tolmach** for the SML/NJ debugger[35], and for the new pure-functional style of static environments; and **Adam T. Dingle** for the debugger's Emacs interface.

Thanks to **James S. Mattson** for the first version of the ML lexical analyzer generator; and to **David R. Tarditi** for making the lexer-generator production-quality[8], for implementing a really first-class parser generator[32], for helping to implement the type-reconstruction algorithm used by the debugger[35], and for the the the ML-to-C translator he implemented with **Anurag Acharya** and **Peter Lee**[31].

We appreciate **Lal George's** teaching the code generator about floating point registers and making floating-point performance respectable; and his fixing of several difficult bugs not of his own creation. Thanks to **Zhong Shao** for the common-subexpression eliminator, as well as the callee-save convention that uses multiple-register continuations for faster procedure calls[9].

We thank **Nick Rothwell** and **Mads Tofte** for the initial implementation of the separate compilation mechanism, and **Gene Rollins** for his recent improvements.

Finally we thank our user community that sends us bug reports, keeps us honest, and actually finds useful things to do with Standard ML.

References

[1] Andrew W. Appel. Runtime tags aren't necessary. *Lisp and Symbolic Computation*, 2:153–162, 1989.

[2] Andrew W. Appel. Simple generational garbage collection and fast allocation. *Software—Practice and Experience*, 19(2):171–183, 1989.

[3] Andrew W. Appel. A runtime system. *Lisp and Symbolic Computation*, 3(343–380), 1990.

[4] Andrew W. Appel. *Compiling with Continuations*. Cambridge University Press, 1992.

[5] Andrew W. Appel and Trevor Jim. Continuation-passing, closure-passing style. In *Sixteenth ACM Symp. on Principles of Programming Languages*, pages 293–302, 1989.

[6] Andrew W. Appel and Trevor T. Y. Jim. Optimizing closure environment representations. Technical Report 168, Dept. of Computer Science, Princeton University, 1988.

[7] Andrew W. Appel and David B. MacQueen. A Standard ML compiler. In Gilles Kahn, editor, *Functional Programming Languages and Computer Architecture (LNCS 274)*, pages 301–324. Springer–Verlag, 1987.

[8] Andrew W. Appel, James S. Mattson, and David R. Tarditi. A lexical analyzer generator for Standard ML. distributed with Standard ML of New Jersey, December 1989.

[9] Andrew W. Appel and Zhong Shao. Callee-save registers in continuation-passing style. Technical Report CS-TR-326-91, Princeton Univ. Dept. of Computer Science, Princeton, NJ, June 1991.

[10] David Berry. SML resources. sent to the SML mailing list by db@lfcs.ed.ac.uk, May 1991.

[11] CAML: The reference manual (version 2.3). Projet Formel, INRIA-ENS, June 1987.

[12] Eric C. Cooper and J. Gregory Morrisett. Adding threads to Standard ML. Technical Report CMU-CS-90-186, School of Computer Science, Carnegie Mellon University, December 1990.

[13] N. G. deBruijn. Lambda calculus notation with nameless dummies, a tool for automatic formula manipulation. *Indag. Math.*, 34:381–392, 1972.

[14] Bruce Duba, Robert Harper, and David MacQueen. Typing first-class continuations in ML. In *Eighteenth Annual ACM Symp. on Principles of Prog. Languages*, pages 163–173, Jan 1991.

[15] Carl A. Gunter, Elsa L. Gunter, and David B. Mac-Queen. An abstract interpretation for ML equality kinds. In *Theoretical Aspects of Computer Software*. Springer, September 1991.

[16] S. C. Johnson. Yacc – yet another compiler compiler. Technical Report CSTR-32, AT&T Bell Laboratories, Murray Hill, NJ, 1975.

[17] James William O'Toole Jr. Type abstraction rules for references: A comparison of four which have achieved notoriety. Technical Report 380, MIT Lab. for Computer Science, 1990.

[18] David Kranz. *ORBIT: An optimizing compiler for Scheme*. PhD thesis, Yale University, 1987.

[19] P. J. Landin. The mechanical evaluation of expressions. *Computer J.*, 6(4):308–320, 1964.

[20] Xavier Leroy. The ZINC experiment: an economical implementation of the ML language. Technical Report No. 117, INRIA, February 1990.

[21] Xavier Leroy. INRIA, personal communication, 1991.

[22] Xavier Leroy and Pierre Weis. Polymorphic type inference and assignment. In *Eighteenth Annual ACM Symp. on Principles of Prog. Languages*, Jan 1991.

[23] David B. MacQueen. The implementation of Standard ML modules. In *ACM Conf. on Lisp and Functional Programming*, pages 212–223, 1988.

[24] David C. J. Matthews. Papers on Poly/ML. Technical Report T.R. No. 161, Computer Laboratory, University of Cambridge, February 1989.

[25] Robin Milner and Mads Tofte. *Commentary on Standard ML*. MIT Press, Cambridge, Massachusetts, 1991.

[26] Robin Milner, Mads Tofte, and Robert Harper. *The Definition of Standard ML*. MIT Press, Cambridge, Mass., 1989.

[27] John H. Reppy. Asynchronous signals in Standard ML. Technical Report TR 90-1144, Cornell University, Dept. of Computer Science, Ithaca, NY, 1990.

[28] John H. Reppy. Concurrent programming with events. Technical report, Cornell University, Dept. of Computer Science, Ithaca, NY, 1990.

[29] John H. Reppy and Emden R. Gansner. The eXene library manual. Cornell Univ. Dept. of Computer Science, March 1991.

[30] Guy L. Steele. Rabbit: a compiler for Scheme. Technical Report AI-TR-474, MIT, 1978.

[31] David R. Tarditi, Anurag Acharya, and Peter Lee. No assembly required: Compiling Standard ML to C. Technical Report CMU-CS-90-187, Carnegie Mellon Univ., November 1990.

[32] David R. Tarditi and Andrew W. Appel. ML-Yacc, version 2.0. distributed with Standard ML of New Jersey, April 1990.

[33] Mads Tofte. *Operational Semantics and Polymorphic Type Inference*. PhD thesis, Edinburgh University, 1988. CST-52-88.

[34] Mads Tofte. Type inference for polymorphic references. *Information and Computation*, 89:1–34, November 1990.

[35] Andrew P. Tolmach and Andrew W. Appel. Debugging Standard ML without reverse engineering. In *Proc. 1990 ACM Conf. on Lisp and Functional Programming*, pages 1–12, June 1990.

[36] Philip Wadler and Stephen Blott. How to make ad-hoc polymorphism less *ad hoc*. In *Sixteenth Annual ACM Symp. on Principles of Prog. Languages*, pages 60–76, Jan 1989.

[37] Mitchell Wand. Continuation-based multiprocessing. In *Conf. Record of the 1980 Lisp Conf.*, pages 19–28, August 1980.

[38] Andrew K. Wright and Matthias Felleisen. A syntactic approach to type soundness. Technical Report COMP TR91-160, Rice University, April 1991.

Adding equations to NU-Prolog

Lee Naish

(lee@cs.mu.OZ.AU)

Department of Computer Science
University of Melbourne
Parkville, 3052, Australia

Abstract

This paper describes an extension to NU-Prolog which allows evaluable functions to be defined using equations. We consider it to be the most pragmatic way of combining functional and relational programming. The implementation consists of several hundred lines of Prolog code and the underlying Prolog implementation was not modified at all. However, the system is reasonably efficient and supports coroutining, optional lazy evaluation, higher order functions and parallel execution. Efficiency is gained in several ways. First, we use some new implementation techniques. Second, we exploit some of the unique features of NU-Prolog, though these features are not essential to the implementation. Third, the language is designed so that we can take advantage of implicit mode and determinism information. Although we have not concentrated on the semantics of the language, we believe that our language design decisions and implementation techniques will be useful in the next generation of combined functional and relational languages.

1 Introduction

This is a brief report describing an extension of the NU-Prolog system which allows evaluable functions to be defined using equations. The work grew out of frustration concerning many published proposals for combining functional languages and relational languages such as Prolog. Many of these proposals seemed unduly complex, and simply transforming the function definitions into Prolog seemed a much better alternative. This lead to an implementation on top of NU-Prolog several years ago. It has gradually been extended over the years, incorporating some novel implementation techniques, and now supports higher order functions, lazy evaluation and parallel execution. The whole system consists of several hundred lines of Prolog code, and the underlying Prolog system was not modified at all. The system has been give some publicity on the computer networks and is available via anonymous ftp.

However, it was considered that a more formal description of the system was long overdue. This report describes the facilities provided, the implementation techniques used, some insight into issues such as parallelism, and some general conclusions about design of combined functional and relational languages.

2 Syntax

The syntax of NUE-Prolog is identical to NU-Prolog, but some clauses are treated specially. Clauses whose heads are of the form LHS = EXPR or LHS == EXPR are treated as equations. A group of such equations with the same atom on the left hand side are considered an evaluable function definition. The following code defines the functions concat/2, sum_tree/1 and + /2.

```
        % list concatenation (append)
concat([], A) = A.
concat(A.B, C) = A.concat(B, C).

        % sum integers in a tree
sum_tree(nil) = 0.
sum_tree(t(L, N, R)) = sum_tree(L) + N + sum_tree(R).

        % addition (== forces inline expansion)
A + B == C :- C is quote(A + B). % quote prevents function evaluation
```

There are several restrictions on function definitions (for reasons which will become clear). Function definitions using == can only have a single equation. Function definitions must also be *uniform* [Jon87]. This implies that the left hand sides of all equations must be mutually non-unifiable, which makes the order of equations irrelevant to the declarative semantics. The left hand sides of equations must not contain evaluable function symbols or repeated variables as arguments. All variables appearing on the right hand side must appear on the left hand side or in the body of the clause. Error messages are displayed if these conditions are violated.

Functional code can access Prolog code, such as builtin predicates, by having equations with non-empty clause bodies, as in the definition of + above. Prolog code called in this way should be deterministic and should always succeed (currently this is left up to the programmer to ensure). The system provides a builtin non-strict if-then-else function (Goal ? TrueExpr : FalseExpr). The condition is a Prolog goal, which we feel is more natural than a function call. This provides a interface from functions to Prolog code which can succeed or fail. Because definitions must be uniform, if-then-else must sometimes be used instead of multiple conditional equations. Functions can also be called from NU-Prolog clauses and can be declared in one file but defined elsewhere. The following examples illustrate these points.

```
        % minimum defined elsewhere
?- function minimum/2.

        % maximum of two numbers
maximum(A, B) = (A >= B ? A : B).

        % sum of elements in tree T1 is greater than T2
gt_tree(T1, T2) :- sum_tree(T1) > sum_tree(T2).
```

A simple user interface is provided on top of NU-Prolog. This code is written in Prolog using the predicate eval/2 provided by the system. It enables users to type an expression and have the result of evaluating the expressions printed:

```
?= concat([1, maximum(1,2), 1 + 2], [4]).
[1, 2, 3, 4]
```

3   Semantics and implementation

The informal meaning of equations is, we hope, quite clear. However, a formal definition of the semantics is also desirable. A theoretical framework in which functions and predicates are distinct may well be preferable, especially when considering higher order functions and lazy evaluation. However, for the moment we use a simpler approach which is also very close to the implementation. We consider functions to be a special case of predicates and programs defining or using evaluable functions to be a shorthand for pure Prolog programs.

The transformation from functional code to predicate logic is known as the *flattening* transformation. Each definition of a function with N arguments is considered shorthand for a predicate definition with N+1 arguments. Each call to an evaluable function is replaced with a new variable and an extra goal is added to compute the function and bind the new variable to the result. We use a slightly modified version of this transformation for functions defined by ==. Rather than adding call to the predicate, the body of the predicate definition is unfolded inline. This can be used for defining constants in Prolog code. For example, the definition radix == 10000 allows the constant radix to be used instead of 10000 throughout a Prolog program, without incurring any runtime overhead. The definitions above result in the following NU-Prolog code, which has well defined declarative semantics. NU-Prolog's if-then-else construct is implemented soundly; it generally suspends until the condition is ground.

```
concat([], A1, A1).
concat([A1|A2], A3, [A1|A]) :- concat(A2, A3, A).

sum_tree(nil, 0).
sum_tree(t(A1, A2, A3), A) :-
        sum_tree(A1, B), C is B + A2, % +(B, A2, C) unfolded
        sum_tree(A3, D), A is C + D.

+(A1, A2, A) :- A is A1 + A2.

maximum(A1, A2, A) :- (if A1 >= A2 then A = A1 else A = A2).

gt_tree(A, B) :- sum_tree(A, C), sum_tree(B, D), C > D.
```

The transformation is currently done by a preprocessor, though is could easily be incorporated into the NU-Prolog compiler. The preprocessor reads in a file, determines the evaluable functions (either defined or declared), then performs the flattening transformation on each clause and goal in the file and outputs the result. Extra goals are also output to define what evaluable functions there are. After preprocessing and compilation the files are loaded and these goals are executed, so that at runtime eval and other predicates can distinguish between evaluable functions and normal Prolog terms (constructors).

4 Coroutining and indexing

The computation rule of NU-Prolog supports coroutining. The default execution order is left to right, but *when declarations* can be used to force suspension of calls which are insufficiently instantiated. Along with the clauses derived for each function definition, the preprocessor outputs a set of when declarations. For each clause derived from an equation, a when declaration is generated. The head as the same input arguments as the clause head. This forces a call to suspend until its input arguments are subsumed by those in one of the clause heads. Effectively, function calls suspend until they are subsumed by the left hand side of an equations.

Suspending calls with insufficiently instantiated inputs reduces the flexibility of the Prolog code in some ways. Functions cannot be run "backwards", which would be possible in some cases otherwise. However, reversible and nondeterministic procedures can be defined using the normal Prolog syntax, so this is not a great loss. Furthermore, there are many advantages in reserving functional syntax for code which is only used to compute functions. First, we believe it enhances readability of programs. It is helpful to know that a definition is used in a particular mode and is a function, and functional syntax is not a particularly natural syntax for defining more general relations.

Second, suspension can actually make code more flexible rather than less flexible. It allows functions to be used easily with coroutining code. This gives the programmer more freedom in how the logic is expressed, and can greatly reduce the search space in some cases. The reason why control information can easily be generated for functions is that the definitions implicitly state the mode of use. Similarly, functions are implicitly deterministic and this can also be used as control information. In contrast, reasonable control information for predicates is much more difficult to determine, though some heuristics can be used [Nai86].

Third, the fact that a procedure will be used in a particular mode and is completely deterministic can be used to improve the implementation in a variety of ways. It is much easier to determine this information from the syntax than to analyse programs [DW86]. An example of the use of mode information is that the NU-Prolog compiler uses when declarations to guide the choice of clause indexing.

There are two important differences between indexing in functional languages and Prolog. First, the input-output mode of calls is not normally known in advance in Prolog, so full indexing is more expensive than in functional languages (and wasteful for arguments which are always output). Second, because Prolog must support nondeterminism in the form of backtracking, imperfect indexing can have more drastic consequences. If a call only matches with a single clause but indexing is insufficient for this to be determined, a *choice point* is created [War83]. Choice point creation takes time and space and, more importantly, can prevent tail recursion optimization and garbage collection. The equivalent of a functional computation which requires constant space may require non-constant space in Prolog.

Most Prolog implementations use the compromise of indexing on the top level functor of the first argument of each procedure. Procedures which need more complex indexing to avoid choice points can be recoded to use the indexing which is available. A less desirable alternative is to use cut to remove choice points instead of avoiding choice point creation in the first place. If appropiate when declarations are present, the NU-Prolog compiler will index on multiple arguments and on multiple subterms within an argument. With uniform definitions, efficient indexing can be generated so no choice point is ever created. This allows functions to be coded in a natural way without losing the benefits of indexing. The following example illustrates the generation of when declarations and the indexing which is achieved.

```
        % add two lists of numbers, pairwise
add_lists([], _) = [].
add_lists(_._, []) = [].
add_lists(A.As, B.Bs) = (A+B).add_lists(As, Bs).
```

```
% transformed code
?- add_lists([], A, B) when ever.
?- add_lists([A|B], [], C) when ever.
?- add_lists([A|B], [C|D], E) when ever.
add_lists([], A1, []).
add_lists([A1|A2], [], []).
add_lists([A1|A2], [A3|A4], [A|B]) :- A is A1 + A3, add_lists(A2, A4, B).
```

Note that the first argument in the second equation is `_._`, rather than simply `_`. This makes the definition uniform, allowing more effective indexing. The compiled Prolog code initially checks the first argument of the call. If it is a variable, the call suspends. If it is [] the first clause is used. If it is a non-empty list then the second argument is checked. If the second argument is a variable then the call suspends; otherwise the second or third clauses are chosen. This is likely to be significantly more efficient than a straightforward coding in Prolog and can be used with coroutines. These advantages come partly from the NU-Prolog implementation, but also from the implicit mode information in the function definition.

Functional code can be run if non-uniform definitions are allowed or a Prolog system which does not support such flexible indexing is used. However, there will be a substantially decrease in efficiency of programs for which the indexing is insufficient to eliminate choice points. Any benchmarking of functional programs transformed into Prolog, for example [CvER90], should therefore take indexing into consideration. Programmers should be made aware of how the code is transformed and what indexing is done by the system so efficient code can be written. For example, with the transformation we use and the standard first argument indexing, all function definitions should have distinct functors in the first argument of the head of each equation. The function above can be recoded using an auxillary function to achieve this.

5 Parallelism

Due to their declarative semantics, both relational and functional languages are promising candidates for parallel execution. The three main forms of parallelism which can be exploited in relational programs are *or-parallelism*, *independent and-parallelism* and *stream and-parallelism* [Gre87]. Or-parallelism exploits parallelism in nondeterministic programs and independent and-parallelism exploits parallelism in conjunctions of atoms which do not share any variables. Both these forms of parallelism have been exploited in conventional Prolog.

Stream and-parallelism exploits parallelism in conjunctions of atoms which may share variables. However, every binding to a shared variable must be deterministic; it cannot be undone at a later stage. This can be achieved by procedure calls suspending until they are sufficiently instantiated and by using pruning operators like Prolog's cut. At first, stream and-parallelism was only exploited by specialized "committed choice" logic programming languages such as Concurrent Prolog and Parlog [Gre87]. These languages require the programmer to supply information such as input/output modes to control when calls suspend.

More recently, the NU-Prolog implementation has been modified so that deterministic code can be run in parallel [Nai88]. Special declarations were introduced in PNU-Prolog so that programmers can provide determinism and mode information with their predicates. The code is transformed and when declarations are added so that the resulting code is guaranteed to be completely deterministic.

The code produced by transforming equations has exactly the same properties as the PNU-

Prolog code and hence can be run in parallel. There is no reason why the output of the preprocessor could not be Parlog code. Again, this is made possible by the implicit mode and determinism information and the decision to restrict the mode of use of evaluable functions. If equations have clause bodies it is vital that the Prolog code is completely deterministic. The bodies of equations could be restricted to only call deterministic builtin predicates or PNU-Prolog procedures which are declared to be deterministic. If such procedures cannot be compiled into deterministic code, an error message is given.

The sum_tree function given earlier is a good example of how parallelism can be exploited. A sequential, innermost, left to right functional execution corresponds to a left to right Prolog execution of the flattened definitions. In a parallel functional language, we would expect the two recursive calls to proceed in parallel and when the results are returned the final addition would be done. The same effect can be achieved in Prolog by exploiting independent and-parallelism in the flattened code.

This method does not exploit all potential parallelism however. In an expression such as sum_tree(build_tree(...)), the tree is built completely before being summed. Using *dataflow parallelism* in a functional language allows the tree to be incrementally constructed and summed at the same time. This is equivalent to using stream and-parallelism in the flattened code. In the Prolog context, a partially constructed tree is simply a term containing variables. If the summing process proceeds faster than the building process, sum_tree will be called with a variable in the first argument. The when declarations will then force the process to suspend until the tree building process catches up and instantiated the variable.

6 Higher order functions

One of the attractive features of functional programming languages is the support of higher order functions. This allows high level descriptions of algorithms to be coded very concisely. Our implementation supports higher order facilities by using Prolog call to implement *apply*.

A call to apply(F, A) first checks to see if F with an additional argument is an evaluable function. If it is not, a term is constructed by adding the additional argument to the function without further evaluation; otherwise the corresponding predicate (with two additional arguments) is called to evaluate the function. For example, if + /2 is defined as before, apply(+, 1) simply returns the value +(1), whereas apply(+(1), 2) calls +(1, 2, X) and returns the value 3. The way apply treats non-evaluable functions allows us to think in terms of *curried* functions to a limited extent.

We can think of + as a function mapping a number to a function mapping a number to a number. Applying + /0 to 1 returns a function mapping a number to its successor. Whenever the result of apply is a function, a term which represents that function is returned, rather than any evaluation taking place. An advantage of currying which is not currently supported by our system is more concise definitions of functions. For example, plus can be defined to be the same function as + by the equation plus = +, rather than plus(A, B) = A + B. We are considering extending the system to allow such definitions with additional declarations which specify how many additional arguments are needed. If type declarations were supported, this would provide enough information.

The definition of apply is currently written mostly in Prolog, rather than the functional syntax. An alternative approach, due to Warren [War82] [CvER90], is to generate a definition of apply when transforming the function definitions. For each function with N arguments, N clause for apply are generated. An example using + is given below. This method results in a faster system, though it requires more space and is less convenient when the function definitions are spread across several files.

```
apply(+, A, +(A)).
apply(+(A), B, C) :- +(A, B, C).
```

Apply can be used as the basis for more complex higher order functions, as the example below illustrates. Such definitions are expanded in the normal way; the additional complexity occurs at runtime when apply is called. The use of curried functions incurs the most overhead, since a new structure representing a function must be returned. We believe the most promising way of reducing the overheads is partial evaluation of programs (Warren's method is very similar to partial evaluation of just the apply predicate). It seems simplest to do this at the functional level, though the Prolog level would be more general.

```
map(F, []) = [].
map(F, A.B) = apply(F, A).map(F, B).
```

7 Lazy evaluation

Another important feature of many functional programming languages is lazy evaluation. This allows manipulation of infinite objects and definition of a somewhat simpler semantics for programs. Although coroutining can reduce the search space of Prolog for certain kinds of programs in a similar way to lazy evaluation, it is less powerful than lazy evaluation for deterministic (including functional) programs. Coroutining can only speed up deterministic computations by early detection of failure. For successful deterministic computations, coroutining has no positive effect. In contrast, lazy evaluation can turn an infinite (deterministic) computation into a finite one.

Our system does support optional lazy evaluation however. It is done by using a more complex transformation into Prolog. When declarations are used for the purposes mentioned previously, but the code can still be run on a conventional Prolog system with a left to right computation rule. Lazy evaluation can be invoked my declaring one or more functions to be lazy, as follows.

```
        % generates (lazily) an infinite list of ones
?- lazy ones/0.
ones = 1.ones.
```

7.1 Implementation of lazy evaluation

If any function is declared lazy then all functions are transformed in a more complex way, and the way functional code and Prolog code interacts is more complex. These additional overheads are the reason why we have made lazy evaluation optional. Without lazy evaluation the code is as efficient as straightforward Prolog code, and often more efficient. This is not the case when lazy evaluation is used. With sophisticated dataflow and strictness analysis, it may well be possible to make all functions lazy with very little overhead. However, this was considered beyond the scope of our relatively simple preprocessor.

When predicates derived from lazy functions are called initially, they simply return a *closure* containing enough information to compute the value if required. If the Prolog call is p(A,B,C) then the output variable, C, will be bound to the closure $lazy($lazy$p(A, B, C1), C1). The first argument of the closure is a goal which will compute the answer, and the second argument is a variable appearing in the goal which will be bound to the answer. For each lazy function an additional predicate is generated in the same way as non-lazy functions. Typically, this partly instantiates the answer (the top-most functor, for example) then recursively calls lazy functions to compute the rest of the answer. Thus a partial answer which may contain closures is returned. The ones function above is translated as follows:

```
ones($lazy($lazy$ones(A), A)). % just returns a closure

?- $lazy$ones(A) when ever.
$lazy$ones([1|A]) :- % produces head of list
        ones(A). % returns a closure for tail of list
```

Returning lazy closures can cause two potential problems. First, the closures can be passed as arguments of functions and not match any of the left hand sides, resulting in failure. Second, they can be passed to Prolog code, including builtin predicates, which may cause failure or error messages. To solve the first problem, equations with structures in the arguments of the left hand side generate additional clauses to evaluate lazy closures. To solve the second problem, wherever the result of a function is passed to Prolog code, full evaluation is forced (by the evalLazy predicate, which is written in Prolog). The add_lists function given previously is translated as follows (we discuss the algorithm in more detail later):

```
?- add_lists($lazy(B, A), C, D) when ever.
?- add_lists([], A, B) when ever.
?- add_lists([C|D], $lazy(B, A), E) when ever.
?- add_lists([A|B], [], C) when ever.
?- add_lists([A|B], [C|D], E) when ever.
add_lists($lazy(B, A), C, D) :- call(B), add_lists(A, C, D).
add_lists([], A1, []).
add_lists([C|D], $lazy(B, A), E) :- call(B), add_lists([C|D], A, E).
add_lists([A1|A2], [], []).
add_lists([A1|A2], [A3|A4], [A|B]) :-
        evalLazy(A1, C), % fully evaluate A1
        evalLazy(A3, D), % and A3
        A is C + D, % before calling Prolog builtin
        add_lists(A2, A4, B).
```

Consider what occurs in the evaluation of an expression such as add_lists([1,2], ones). The first step is to transform the expression into a Prolog goal: ones(A), add_lists([1,2], A, B). Ones initially binds A to $lazy($lazy$ones(A1), A1). The call to add_lists matches with the third clause. This is a special clause which attempts to evaluate the closure then recursively calls add_lists with the result. The closure is evaluated by calling $lazy$ones(A1), which

binds A1 to [1 | $lazy($lazy$ones(A2), A2)]. This term is passed to the recursive call to
add_lists, which now matches with the last clause. This step is repeated, evaluating the second
element of the list of ones, then the next recursive call to add_lists matches with the second
clause. The closure in the second argument is ignored and the computation terminates with the
answer B = [2,3].

Our implementation of lazy functions is very similar to some other lazy functional language
implementations. The main difference is that a Prolog term is used to represent the closure
containing essentially a reference to some code (the $lazy$ procedure name), a set of argument
registers (the arguments to the procedure call) and an address for the returned value (the output
variable).

The interface between lazy functional code Prolog is straightforward. Any variable which
occurs on the left hand side of an equation and also in the body of a clause or the condition
of an if-then-else, must be replaced by two variables and be processed by evalLazy. This
does introduce extra overheads, especially for relatively fast operations such as arithmetic. It
could be avoided in many cases by analysing the dataflow within the program or having a more
sophisticated exception handling mechanism for Prolog builtins.

The clause head matching is very efficient, even in the presence of additional clauses for
handling closures. In many cases (add_lists, for example), exactly the same WAM instructions
are executed as in the non-lazy case, due to the clause indexing. The algorithm for generating the
extra clauses is closely related to the indexing algorithm. With the when declarations produced
by the preprocessor, the indexing algorithm constructs a tree which includes nodes for each non-
variable subterm in the clause heads. An extra case is added to each of these nodes, to deal with
$lazy/2. From this expanded tree, the clause heads for the additional clause can be extracted.
Our current implementation is actually simpler than this and makes some assumptions about
argument ordering.

7.2 Avoiding repeated evaluation

Most lazy functional language implementations avoid repeated evaluation of duplicated variables.
When using a Prolog system with a left to right computation rule this can also be achieved by
a minor change to the extra clauses we introduce to evaluate the lazy closures. Rather than
calling the goal immediately, the output variable can be checked. If it is already instantiated,
calling the goal is unnecessary (it has already been executed). In a coroutining Prolog system,
we must be more careful because functions can be called and delay before instantiating the
output variable, and output arguments can be instantiated by Prolog code before the function
is called. However, with a slightly more complex representation for closures, the problem can
be avoided. We can include an extra argument to $lazy which is a flag to indicate whether the
goal has been called yet:

```
p(...$lazy(G, V, F)...)  :- (var(F) -> F = 1, call(G) ; true), p(...V...).
```

7.3 Further optimization

The most significant gains to be made in the implementation of lazy evaluation probably come
from program analysis (for example, strictness), which lead to certain overheads being removed
completely from most code. Much of this high level analysis is independent of the low level
details of the implementation. Also, any faster Prolog can be used to speed up our system, since

we can compile to standard Prolog. Recent efforts have resulted in Prolog systems which are faster than C for small recursive programs.

There are also some optimization issues specific to our implementation on top of Prolog. The most important consideration is the indexing produced by the compiler. Whether lazy evaluation is used or not, full indexing of input arguments is very desirable. Second, call/1 is used for lazy code and call/3 is used for higher order functions; they should be implemented efficiently. Standard versions of call must deal with cut and other system constructs, increasing complexity. A specialised version which only has to deal with certain user-defined predicates can generally be implemented more efficiently. This could be done at the system level or the preprocessor could define such a predicate in a similar manner to Warren's implementation of apply. Third, if repeated evaluation is to be avoided, the conditional (var(V) -> ...) must also be implemented efficiently, without creating a choice point. All these points are useful for Prolog code also. If the Prolog implementation is to be modified for executing lazy functional code it may be desirable to have a separate tag, and perhaps representation, for the $lazy functor (as is currently done with cons). This could make indexing a little faster in functions which used other complex terms, and could save a little space. This is also suggested in the K-LEAF system, discussed later. A special construct for conditionally evaluating a closure could also be provided.

8 Related work

A great deal of work has been done on combining functional and relational languages. Due to lack of space we are only able to breifly discuss those proposals most similar to our own. In [Nai90] we give some more discussion and references, but a full paper is really needed to give justice to this topic.

The most common approach to combining functional and relational languages is to add features of relational languages (nondeterminism, multi-mode relations, logical variables) to a functional language. Functional syntax is used to define more general predicates. Most of these languages are based on narrowing [Red85] [GM85] [JD86] [BCM88] though other methods are also used [DFP85] [Fri85]. An alternative is to include some of the features of functional programs (lazy evaluation, higher order functions) into a relational language, or develop programming techniques to support these styles of programming [Nar86] [She90]. Both these approaches forego the many benefits of knowing what things are functions. A third approach, the one we take, is to support predicates and functions and restrict functional syntax to functions.

The most similar language to that we have proposed is Funlog [SY85]. However, Funlog is implemented by an interpreter written in Prolog, so it is much less efficient. When comparing high level implementation details, the most similar proposal to ours is LOG(F) [Nar88]. LOG(F) is implemented by translation into standard Prolog. However, the way lazy evaluation is implemented introduces one more level of level of functor nesting in the heads of clauses than our scheme. The standard method of indexing cannot be used to distinguish between different cases in a function definition, so choice points cannot be avoided. For simple deterministic code LOG(F) is five to ten times slower than Prolog, and uses much more space. In contrast, our system avoids choice points even with the standard indexing method if functions are coded carefully. Similar code can be obtained by partially evaluating the code produced by the LOG(F) translator. We have also exploited the better indexing of NU-Prolog and made lazy evaluation optional, allowing the convenience of functions with, if anything, increased performance.

From a low level implementation viewpoint, the most similar system to ours is K-LEAF [BCM88]. K-LEAF uses flattening plus other optimizations and is compiled into K-WAM code.

The K-WAM has special support for *prodvars* (closures) and the instructions that involve dereferencing variables (including indexing instructions) are modified. The closures are actually represented as Prolog terms, similar to those in our implementation. The main differences between the K-LEAF and NUE-Prolog implementations are that the K-LEAF compiler relies on a specialized abstract machine and includes additional optimizations, whereas NUE-Prolog is translated into Prolog and the NU-Prolog compiler does better indexing. K-LEAF also supports parallelism [BCM+90]. However, stream and-parallelism is not exploited because K-LEAF functions can be used nondeterministically. An advantage of K-LEAF is its well developed semantics.

9 Conclusions

Both relational and functional styles of programming have many merits. We believe NUE-Prolog provides a useful combination of the two, providing full NU-Prolog plus a functional language which has higher order features and (optional) lazy evaluation. We have used a pragmatic, implementation guided approach to combining the relational and functional paradigms. Our implementation takes advantage of implicit information in function definitions to provide control information, efficient indexing and parallel execution. Our method of implementing lazy evaluation entirely within standard Prolog is also new. We take advantage of the unique features of NU-Prolog, but these features are not necessary for our implementation techniques. We can transform functions into standard Prolog code and thus take advantage of faster Prolog implementations. Further work is needed on programming environment support for evaluable functions and on the semantics of NUE-Prolog. Alternatively, the NUE-Prolog implementation techniques could be applied to languages for which well defined semantics has already been established.

References

[BCM88]   P.G. Bosco, C. Cecchi, and C. Moiso. Exploiting the full power of logic plus functional programming. In Kenneth A. Bowen and Robert A. Kowalski, editors, *Proceedings of the Fifth International Conference/Symposium on Logic Programming*, pages 3–17, Seattle, Washington, August 1988.

[BCM+90] P.G. Bosco, C. Cecchi, C. Moiso, M. Porta, and G. Sofi. Logic and functional programming on distributed memory architectures. In *Proceedings of the Seventh International Conference on Logic Programming*, Jerusalem, Israel, June 1990.

[CvER90] M.H.M. Cheng, M.H. van Emden, and B.E. Richards. On warren's method for functional programming in logic. In *Proceedings of the Seventh International Conference on Logic Programming*, Jerusalem, Israel, June 1990.

[DFP85]   J. Darlington, A.J. Field, and H. Pull. The unification of functional and logic languages. In Doug DeGroot and Gary Lindstrom, editors, *Logic programming: relations, functions, and equations*, pages 37–70. Prentice-Hall, 1985.

[DW86]   Saumya K. Debray and David S. Warren. Detection and optimisation of functional computations in prolog. In Ehud Shapiro, editor, *Proceedings of the Third International Conference on Logic Programming*, pages 490–504, London, England, July 1986. published as Lecture Notes in Computer Science 225 by Springer-Verlag.

[Fri85]   Laurent Fribourg. SLOG: a logic programming language interpreter based on clausal superposition and rewriting. In *Proceedings of the Second IEEE Symposium on Logic Programming*, pages 172–184, Boston, Massachusetts, July 1985.

[GM85]   Joseph A. Goguen and Jose Meseguer. EQLOG: equality, types, and generic modules for logic programming. In Doug DeGroot and Gary Lindstrom, editors, *Logic programming: relations, functions, and equations*, pages 295–363. Prentice-Hall, 1985.

[Gre87]   Steve Gregory. *Parallel logic programming in parlog*. Addison-Wesley, Wokingham, England, 1987.

[JD86]   Alan Josephson and Nachum Dershowitz. An implementation of narrowing the RITE way. In *Proceedings of the Third IEEE Symposium on Logic Programming*, pages 187–197, Salt Lake City, Utah, September 1986.

[Jon87]   S. Peyton Jones. *The implementation of functional programming languages*. Prentice Hall International series in computer science. Prentice Hall, London, 1987.

[Nai86]   Lee Naish. *Negation and control in Prolog*. Number 238 in Lecture Notes in Computer Science. Springer-Verlag, New York, 1986.

[Nai88]   Lee Naish. Parallelizing NU-Prolog. In Kenneth A. Bowen and Robert A. Kowalski, editors, *Proceedings of the Fifth International Conference/Symposium on Logic Programming*, pages 1546–1564, Seattle, Washington, August 1988.

[Nai90]   Lee Naish. Adding equations to NU-prolog. Technical Report 91/2, Department of Computer Science, University of Melbourne, Melbourne, Australia, 1990.

[Nar86]   Sanjai Narain. A technique for doing lazy evaluation in logic. *Journal of Logic Programming*, 3(3):259–276, October 1986.

[Nar88]   Sanjai Narain. *LOG(F): An optimal combination of logic programming, rewriting and lazy evaluation*. Ph.d. thesis, Dept. of computer science, UCLA, Los Angeles CA, 1988.

[Red85]   Uday S. Reddy. Narrowing as the operational semantics of functional languages. In *Proceedings of the Second IEEE Symposium on Logic Programming*, pages 138–151 Boston, Massachusetts, July 1985.

[She90]   Yeh-Heng Sheng. HIFUNLOG: logic programming with higher-order relational functions. In *Proceedings of the Seventh International Conference on Logic Programming*, Jerusalem, Israel, June 1990.

[SY85]   P.A. Subrahmanyam and Jia-Huai You. FUNLOG: a computational model integrating logic programming and functional programming. In Doug DeGroot and Gary Lindstrom, editors, *Logic programming: relations, functions, and equations*, pages 157–198. Prentice-Hall, 1985.

[War82]   David H.D. Warren. Higher-order extensions to prolog: are they needed? In J.E Hayes, Donald Michie, and Y-H. Pao, editors, *Machine Intelligence 10*, pages 441–454. Ellis Horwood Ltd., Chicester, England, 1982.

[War83]   David H.D. Warren. An abstract Prolog instruction set. Tecnical Note 309, SRI International, Menlo Park, California, October 1983.

Extraction of Functional from Logic Program

Susumu Yamasaki

Department of Information Technology,
Okayama University, Okayama, Japan

Abstract

This paper shows a method of extracting a functional from a logic program, by means of a dataflow dealing with sequences from the set of idempotent substitutions. The dataflow is expressed as a functional involving fair merge functions in order to represent the atom set union over a sequence domain, as well as functions to act on unifiers, to reflect the unit resolution deductions virtually. The functional completely and soundly denotes the atom generation in terms of idempotent substitutions without using atom forms. Its least fixpoint is interpreted as denoting the whole atom generation in terms of manipulations on idempotent substitutions.

1.Introduction

This paper deals with a method to extract a functional from a logic program. The motivation comes from the problem of how to transform a logic program into a functional program, say, an FP program and of how to construct a semantics for the integrated, logic and functional programs.

Concerning a functional involved in the logic program, there is a work in Yamasaki (1990), in which the dataflow program is constructed from a logic program. The dataflow defines a functional over a sequence domain based on the Herbrand base. Expanding the idea described there and with refined techniques of manipulations on the substitution, this paper takes the interpretation that the logic program is regarded as dealing with idempotent substitutions instead of resolution deductions. The approach takes an interpretation of the extesionality of a predicate as a sequence from an idempotent substitution set and regards a logic program as denoting a functional over sequence variables for idempotent substitution sequences, different from the functionality in Debray and Warren (1989).

Each definite clause of a logic program is interpreted as able to generate atoms from already existing atoms by means of (unit) resolution deductions, that is, bottom-up inferences. By representing an atom set with a substitution set attached to a predicate, we might regard the definite clause as a translator of input substitutions (reflecting already existing atoms) to output substitutions (reflecting generated atoms). A function to gather existing atoms per each predicate is necessary in order to form the existing atom set (with the same predicate symbol). Through such a function, the translated output will be transformed and sent to an input of another definite clause. The above behaviour will be represented as a functional of functions as sequence variables for each logic program. Finally the functional will be shown to denote the unit resolution deduction for a logic program. At the same time, the functional is a formal expression for the fixpoint semantics of the original program. The semantic domain of the functional is defined as the set of all finite and infinite sequences from the set of idempotent substitutions with a symbol representing time delay or hiaton in Park (1983). From operational point of view, time delay is thought of as occurring when there is no output from a definite clause. It will be demonstrated that any idempotent substitution occurs in the history of some sequence variable iff the original logic program generates a corresponding atom represented with a substitution as well as some renaming of variables. The expression is a more refined and exact form than the program equation for the answer substitution sequences caused by successful SLD resolution refuatations for a logic program, which is defined by Baudinet (1988). Also the constructed functional for a logic program differs from the map for the deterministic fixpoint semantics by Fitting (1985a), which reflects answer sequences but has not been explicitly constructed, in that the functional here is an explicitly defined recursion equation set to denote answer idempotent substitution sequences. A continuous functional might be associated with a given logic program such that the least fixpoint of the functional denotes the data (that is, an idempotent substitutions set) by which all the deducible atoms are virtually represented and real forms of atoms are dismissed, although it contains time delay. The fixpoint is interpreted as a semantics of the original logic program, denoting dataflow computing mechanism for it. The semantics is unique from the point of its sequence domain consisting of finite and infinite idempotent substitutions with a symbol for time delay.

2.Preliminaries

In this paper, a logic program means a set of definite clauses. A definite clause takes the form such as $A \leftarrow B_1 \ldots B_n$ ($n \geq 0$), where $A, B_1, \ldots,$ and B_n are atoms. An atom is an expression of the form $P(t_1, \ldots, t_m)$, where P is a predicate symbol and t_i are terms. A term is recursively defined as: (i) a variable is a term, and (ii) $f(t_1, \ldots, t_k)$ ($k \geq 0$) is a term if f is a k-place symbol and t_j are terms.

In this section we have technical terms and fundamentals concerning substitutions for unifications in the resolution deductions of logic programs.

Let $Term$ be a set of terms and Var a set of variables. Note $Var \subset Term$. A substitution is a function from Var to $Term$. For a substitution θ, Var_θ denotes the set $\{ x \in Var \mid \theta(x) \neq x \}$, that is, the domain of θ. If the domain of θ is $\{ x_1, \ldots, x_n \}$ (finite), then θ is denoted by $\{x_1 \mid \theta(x_1), \ldots, x_n \mid \theta(x_n)\}$. In this paper, it is assumed that the domain of a substitution is finite. The substitution θ is especially denoted by ε if Var_θ is empty. That is, $\varepsilon(x) = x$ for any $x \in Var$.

The effect of a substitution for the term or atom, and the composition of substitutions are defined by the following formalism.

Let $Exp = Term \cup Atom$, where $Atom$ is the set of atoms.

For substitutions θ and $E \in Exp$, $E\theta$ is recursively defined:

$$E\theta = \begin{cases} \theta(x) & \text{if } E = x \text{ for } x \in Var, \\ f(t_1\theta, \ldots, t_n\theta) & \text{if } E = f(t_1, \ldots, t_n) \in Term, \\ P(t_1\theta, \ldots, t_n\theta) & \text{if } E = P(t_1, \ldots, t_m) \in Atom. \end{cases}$$

For substitutions θ, ϕ, the composition of θ and ϕ, denoted by $\phi\theta$, is defined: $\phi\theta(x) = \theta(x)\phi$ for $x \in Var$. Also we see that $(E\theta)\phi = E(\phi\theta)$ for $E \in Exp$ and substitutions θ, ϕ.

For the treatment with the substitution as unification of terms, a substitution θ is assumed to satisfy the condition that $Var_\theta \cap \{ y \mid y$ occurs in some $\theta(z)$ for $z \in Var_\theta \}$ is empty. The condition guarantees that the terms substituted for the variables do not involve any corresponding variables on which a given substitution operates. That is, the condition means for θ, $\theta \circ \theta = \theta$ (the idempotence of the substitution θ).

Definition 2.1. *Sub* means the set of all idempotent substitutions. A substitution ρ is said a permutation (a renaming of variables) if it is a bijection of *Var* into *Var*.

As Eder (1985) defined, we have:

Definition 2.2. We say that σ is more general than θ for substitutions σ, θ if there is a substitution ρ such that $\theta = \rho\sigma$. By $\sigma \preceq \theta$ it is meant that σ is more general than θ.

Note that '$\preceq$' is not a partial order.

Definition 2.3. A relation $\sim$ on *Sub* is defined: $\theta \sim \phi$ iff $\phi \preceq \theta$ and $\theta \preceq \phi$.

It is seen that $\sim$ is an equivalence relation. Note that if $\theta \sim \phi$ then there exist permutations (renamings of variables) ρ and σ such that $\rho\theta = \phi$ and $\sigma\phi = \theta$ (See Eder, 1985).

To restrict the domain of the substitution to some appropriate set, the following definition is exploited.

For a substitution σ and a set of atoms $\{A_1, \ldots, A_m\}$ ($m \geq 1$), a restriction of σ with respect to $\{A_1, \ldots, A_m\}$, that is, $[\sigma]_{\{A_1, \ldots, A_m\}}: Var \rightarrow Term$ is defined as follows.

$$[\sigma]_{\{A_1, \ldots, A_m\}}(x) = \begin{cases} \sigma(x) & \text{if } x \text{ occurs in either } A_1, \ldots, \text{ or } A_m, \\ x & \text{otherwise,} \end{cases}$$

for $x \in Var$.

Note that $[\varepsilon]_{\{A_1, \ldots, A_m\}} = \varepsilon$.

Lemma 2.4. $[\sigma]_{\{A_1, \ldots, A_m\}} \preceq \sigma$. $[\varphi]_{\{A_1, \ldots, A_m\}} \sim [\theta]_{\{A_1, \ldots, A_m\}}$ if $\varphi \sim \theta$.

The following is an extension of the restriction of a substitution.

Definition 2.5. Assume that $\Theta \subset Sub$, and $\{A_1, \ldots, A_m\} \subset Atom$ ($m \geq 1$). Then we define

$$[\Theta]_{\{A_1,\ldots,A_m\}} = \begin{cases} \{[\theta]_{\{A_1,\ldots,A_m\}} \mid \theta \in \Theta\} & \text{if } \Theta \text{ is nonempty,} \\ \text{empty} & \text{if } \Theta \text{ is empty.} \end{cases}$$

3. Translations of Substitution Sequences Based on Deductions

In this section we have a sketch on the deductions of atoms, that is, the computations for a given logic program, in terms of manipulations on Sub introduced in the previous section. Then we shall see how the translations of substitutions are performed by the deductions. In the next section, the translation will be expressed as a functional.

Assume that a set of clauses $\{C \leftarrow D_1 \ldots D_m, D_1'\varphi_1 \leftarrow, \ldots, D_m'\varphi_m \leftarrow\}$ is given, where $\varphi_1, \ldots, \varphi_m \in Sub$. Then an atom $C\theta \leftarrow$ may be derivable from the set, and θ can be calculated by means of $\varphi_1, \ldots, \varphi_m$ with the unifications between D_i and D_i' $(1 \leq i \leq m)$. We will see how θ can be decided. On the assumption that $C\theta \leftarrow$ is deducible from the above set, $C\theta' \leftarrow$ is derivable from the set $\{ C \leftarrow D_1 \ldots D_m, D_1'\varphi_1' \leftarrow, \ldots, D_m'\varphi_m' \leftarrow\}$ if $\varphi_i \sim \varphi_i'$ for $1 \leq i \leq m$ and $\theta \sim \theta'$, although a formal representation concerning it is postponed. This leads to our aim that the equivalence relation $\sim$ on the set Sub might be taken advantage of, in order to represent the deduced atoms.

Definition 3.1. Let A be a nonempty subset of $Atom$. We define $unif(A) = \{\theta \mid \forall A_1, A_2 \in A: A_1\theta = A_2\theta\}$. Also we define $mgu(A) = \{\sigma \in Sub \mid \sigma \text{ is in } unif(A) \text{ and } \sigma \text{ is most general in it }\}$. When A is empty, we regard $mgu(A)$ as $\{ \epsilon \}$, respectively.

Note that each substitution in $mgu(A)$ is a most general unifier of A in the usual sense. It is well known that $mgu(A)/\sim$ is a singleton for a nonempty $mgu(A)$. Also for the lemma to state there exists an idempotent unifier which is most general for a unifiable set, see Eder (1985).

Unit resolution is an inference to derive $C\theta \leftarrow D_1 \ldots D_{i-1}\theta \; D_{i+1}\theta \ldots D_m\theta$ from two clauses $C \leftarrow D_1 \ldots D_m$ and $E \leftarrow$, where $\theta \in mgu(\{D_i, E\})$. A unit resolution deduction from a set S of clauses is a sequence of definite clauses $G_1, \ldots, G_n$, where each G_i is either in S or infered by unit resolution from some G_j and G_k $(j, k \leq i - 1)$.

The unit deduction is interpreted as atom generations from a logic program, and as computing mechanism. We see in the following that the unit resolution deduction might be realized by some manipulations on Sub.

Definition 3.2. $consis: Sub \times Sub \to 2^{Sub}$ and $comb: Sub \times Sub \to 2^{Sub}$ are defined:

$$consis(\theta_1, \theta_2) = \{\theta \in Sub \mid \exists \sigma_1, \sigma_2(\text{substitutions}) : \sigma_1\theta_1 = \sigma_2\theta_2 = \theta\},$$
$$comb(\theta_1, \theta_2) = \{\rho \in Sub \mid \rho \text{ is in } consis(\theta_1, \theta_2) \text{ and is most general in it}\}.$$

We regard $consis(\;) = Sub$ and $comb(\;) = \{ \epsilon \}$ for the empty set in $Sub \times Sub$. $comb(\theta_1, \theta_2)$ is denoted by $\theta_1 + \theta_2$.

As easily seen, $(\theta_1 + \theta_2) + \theta_3 = \theta_1 + (\theta_2 + \theta_3)$. From now on, $(\theta_1 + \theta_2) + \theta_3$ is abbreviated by $\theta_1 + \theta_2 + \theta_3$.

Example 3.3. Let $\theta_1 = \{x \mid f(y)\}$ and $\theta_2 = \{y \mid g(z)\}$. Then $\{x \mid f(g(z)), y \mid g(z)\} \in \theta_1 + \theta_2$.

As is the case for $mgu(A)$, $\theta_1 + \ldots + \theta_n$ is either empty or consisting of substitutions which belong to an equivalence class in $Sub/\sim$. That is, we have:

Lemma 3.4. $(\theta_1 + \ldots + \theta_n)/\sim$ consists of one equivalence class for given $\theta_1, \ldots, \theta_n$.

It is easy to see the following lemma from the definition, which states the *comb* function causes the same effect for the equivalent substitutions in the sense '$\sim$' as the initially considered substitutions.

Lemma 3.5. $\theta_1 + \ldots + \theta_n = \theta_1' + \ldots + \theta_n'$ if $\theta_i \sim \theta_i'$, $1 \le i \le n$.

We can see that the *comb* function is closely related with $\uparrow$ (least upper bound) of two equivalent idempotent substitution classes under the partial oder $\preceq$, which Palamidessi (1990) defined.

The method to get an idempotent substitution in $\theta_1 + \ldots + \theta_n$, when $\theta_1, \ldots,$ and θ_n are given, see properties in Eder (1985) and Palamidessi (1990).

For nonempty Θ_1, $\Theta_2 \subset Sub$, we define $\Theta_1 + \Theta_2 = \bigcup_{\theta_1 \in \Theta_1, \theta_2 \in \Theta_2} \theta_1 + \theta_2$. If either Θ_1 of Θ_2 is empty, we regard $\Theta_1 + \Theta_2$ as empty. Then $+$ is extended to act on $2^{Sub} \times 2^{Sub}$. Furthermore, we define $\Theta_1 + \Theta_2 + \Theta_3 = (\Theta_1 + \Theta_2) + \Theta_3$.

Now for the treatment of an atom set by means of a set of idempotent substitutions, to each predicate symbol, we assign a standard form of the atom involving the predicate symbol.

Definition 3.6. Let $PRED(L)$ be the set of all predicate symbols occurring in a logic program L. For each predicate symbol $P_i \in PRED(L)$, a tuple of variables $\bar{x}_i$ is put to P_i such that

    (i) $\bar{x}_i$ contains no variable in L,
    (ii) $P_i(\bar{x}_i)$ is an atom,
    (iii) $\bar{x}_i$ shares no common variable with $\bar{x}_j$ if $i \ne j$.

$P_i(\bar{x}_i)$ is refered to as a standard atom (with P_i). For an atom D, $Stand(D)$ means the standard atom with the predicate symbol involved in D.

We examine the unit resolution deduction by means of the *comb* function.

Lemma 3.7. Let C and D be atoms without common variables. Also let $\psi = [\psi]_{\{D\}} \in Sub$, $\theta \in mgu(\{C, D\})$. Then $mgu(\{C, D\psi\}) = [\theta + \psi]_{\{C, D\psi\}}$.

Proof. Assume that $\alpha \in mgu(\{C, D\psi\})$. Since $C\theta = D\theta$ and $C\alpha = (D\psi)\alpha$, $\alpha\psi$ is a unifier of C and $D\psi$, and there exists β such that $\alpha\psi = \beta\theta$. As α is most general and $[\alpha\psi]_{\{C,D\psi\}} = \alpha$, it is conceded that $\alpha \in [\theta + \psi]_{\{C,D\psi\}}$. It follows that $mgu(\{C, D\psi\})$ $\subset [\theta + \psi]_{\{C,D\psi\}}$. Now assume $\alpha \in [\theta + \psi]_{\{C,D\psi\}}$. Then $\alpha = [\beta\theta]_{\{C,D\psi\}} = [\gamma\psi]_{\{C,D\psi\}}$ for some substitutions β, γ such that $\beta\theta = \gamma\psi = \theta + \psi$. Noting the domain of α and γ, we conclude that $\alpha = [\gamma]_{\{C,D\psi\}}$, $\gamma = \beta\theta_1$, and $\theta = \theta_2\theta_1$ for some adequate θ_1, θ_2 such that $C\theta_1$ $= (D\theta_1)\theta_2$ and $\beta\theta_2 = \beta\psi$. Therefore $C([\beta\theta_1]_{\{C,D\psi\}}) = (C\theta_1)\beta = ((D\theta_1)\theta_2)\beta = (D\psi)\gamma = (D\psi)[\gamma]_{\{C,D\psi\}} = (D\psi)([\beta\theta_1]_{\{C,D\psi\}})$. That is, $\alpha = [\gamma]_{\{C,D\psi\}}$ is a unifier of C and $D\psi$. α should be most general, as $\gamma\psi \in \theta + \psi$ and $\alpha = [\gamma]_{\{C,D\psi\}}$. This completes the proof.

Example 3.8. Let $C = P(x)$ and $D = P(f(y))$ be atoms. Also let $\psi = \{y \mid g(z)\}$. Assume that $\varphi = \{x \mid f(g(z)), y \mid g(z)\}$. As in Example 3.3, $[\varphi]_{\{C,D\psi\}} = \{x \mid f(g(z))\} \in [\theta + \psi]_{\{C,D\psi\}}$, where $\theta = \{x \mid f(y)\} \in mgu(\{C, D\})$. At the same time, it is easy to see that $[\varphi]_{\{C,D\psi\}} \in mgu(\{C, D\psi\})$.

Making use of Lemma 3.7, we have the theorem, which expresses the unit resolution deduction only in terms of substitutions, which are interpreted as attached to atoms.

Theorem 3.9. Assume a definite clause $C \leftarrow D_1 \ldots D_m$ in which $Stand(D_i) = P_i(\bar{x}_i)$ for $1 \leq i \leq m$. Also let $\psi_i = [\psi_i]_{\{P_i(\bar{x}_i)\}} \in Sub$ for $1 \leq i \leq m$. Then $C' \leftarrow$ is derivable by unit resolution deductions from $\bigcup_{1 \leq i \leq m} \{P_i(\bar{x}_i)\psi_i \leftarrow\} \cup \{C \leftarrow D_1 \ldots D_m\}$ iff $C' = C\theta$ for $\theta \in [[\delta_1 + \psi_1]_{\{D_1\}} + \ldots + [\delta_m + \psi_m]_{\{D_m\}}]_{\{C\}}$ and $\delta_i \in mgu(\{D_i, P_i(\bar{x}_i)\})$, $1 \leq i \leq m$.

Proof. We have a proof by induction on m. In case $m = 0$, the theorem holds, since $C\varepsilon \leftarrow$ is derivable.

Induction Step: Assume the theorem holds for $m \leq k - 1$. Let $m = k$ and $\varphi \in [[\delta_1 + \psi_1]_{\{D_1\}} + \ldots + [\delta_k + \psi_k]_{\{D_k\}}]_{\{C\}}$. Then $\varphi \in [\varphi_k + [\delta_k + \psi_k]_{\{D_k\}}]_{\{C\}}$ for $\varphi_k \in [[\delta_1 + \psi_1]_{\{D_1\}} + \ldots + [\delta_{k-1} + \psi_{k-1}]_{\{D_{k-1}\}}]_{\{C,D_k\}}$. For φ_k, by the induction hypothesis, $C\varphi_k \leftarrow D_k\varphi_k$ is derivable. Applying Lemma 3.7 twice, we can conclude that $(C\varphi_k)\theta \leftarrow$ is derivable from $\{P_k(\bar{x}_k)\psi_k \leftarrow, C\varphi_k \leftarrow D_k \varphi_k\}$ for $\theta \in [\varphi_k + [\delta_k + \psi_k]_{\{D_k,P_k(\bar{x}_k)\psi_k\}}]_{\{D_k\varphi_k\}}$. Since $(C\varphi_k)\theta = C[\varphi_k + [\delta_k + \psi_k]_{\{D_k\}}]_{\{C\}}$, it is conceded $(C\varphi_k)\theta = C\varphi$. That is, $C\varphi \leftarrow$ is derivable.

On the other hand, assume that $C' \leftarrow$ is derivable from $\{P_k(\bar{x}_k) \psi_k \leftarrow, C\varphi_k \leftarrow D_k\varphi_k\}$, where $\varphi_k \in [[\delta_1 + \psi_1]_{\{D_1\}} + [\delta_{k-1} + \psi_{k-1}]_{\{D_{k-1}\}}]_{\{C,D_k\}}$. Then $C\varphi \leftarrow = C' \leftarrow$ is derivable, where $\varphi = [\theta\varphi_k]_{\{C\}}$ such that $\theta \in mgu(\{D_k\varphi_k, P_k(\bar{x}_k)\psi_k\})$. By Lemma 3.7, we observe that $\theta \in [\varphi_k + [\delta_k + \psi_k]_{\{D_k,P_k(\bar{x}_k)\psi_k\}}]_{\{D\varphi_k,P_k(\bar{x}_k)\psi_k\}}$. Because $(C\varphi_k)\theta = C[\varphi_k + [\delta_k + \psi_k]_{\{D_k,P_k(\bar{x}_k)\psi_k\}}]_{\{C\}}$, it follows $\varphi = [\theta\varphi_k]_{\{C\}} = [\varphi_k + [\delta_k + \psi_k]_{\{D_k\}}]_{\{C\}} = [[\delta_1 + \psi_1]_{\{D_1\}} + \ldots + [\delta_k + \psi_k]_{\{D_k\}}]_{\{C\}}$. This completes the proof.

Theorem 3.9 states that through a definite clause $C \leftarrow D_1 \ldots D_m$ a translation of $\{\psi_1, \ldots, \psi_k\}$ to φ is performed such that $C\varphi \leftarrow$ is derivable on the basis of $\{P_1(\bar{x}_1)\psi_1 \leftarrow, \ldots, P_m(\bar{x}_m)\psi_m \leftarrow\}$. The next theorem is necessary to form $P(\bar{x})\psi$, which is equivalent to the atom $C\theta$ when the standard atom of C is $P(\bar{x})$.

Theorem 3.10. Assume that $C\theta \leftarrow$ is derivable, where $\theta = [\theta]_{\{C\}}$ for some $\theta \in Sub$. Also assume that $Stand(C) = P(\bar{x})$. Then $P(\bar{x})\psi = C\theta$ iff $\psi \in [\gamma + \theta]_{\{P(\bar{x})\}}$, where $\gamma \in mgu(\{C, P(\bar{x})\})$.

Proof. For θ, $mgu(\{C\theta, P(\bar{x})\}) = [\gamma + \theta]_{\{C\theta, P(\bar{x})\}}$ by Lemma 3.7, where $\gamma \in mgu(\{C, P(\bar{x})\})$. Thus $\psi \in [\gamma + \theta]_{\{P(\bar{x})\}}$ iff $C\theta = P(\bar{x})\psi$. This leads to the conclusion that the theorem holds.

Example 3.11. Assume a set of clauses $\{App(nil, y, y) \leftarrow, App(u.x', y', u.z') \leftarrow App(x', y', z')\}$, where (i) x', y', z', u, y are variables, and (ii) nil is a constant, and $u.x'$ and $u.z'$ are in infix notation with a function symbol '.', which is intended to be the 'cons' function. Note that the clause set means an 'append' function of lists.

Firstly $App(nil, y, y) \varepsilon \leftarrow$ is interpreted as being deducible from the set. Let $Stand(App(nil, y, y)) = App(p, q, r)$ (p, q, r: variables). For the deducible clause, $App(p, q, r)\psi \leftarrow$ is regarded as existing if $\psi = \{p \mid nil, q \mid y, r \mid y\}$. Next we observe that $App(u.x', y', u.z')\theta \leftarrow$ is derivable from $\{App(p, q, r)\psi \leftarrow, App(u.x', y', u.z') \leftarrow App(x', y', z')\}$, if $\theta = \{x' \mid nil, y' \mid y, z' \mid y\} \in [\sigma + \psi]_{\{App(u.x',y',u.z')\}}$ for $\sigma = \{p \mid x', q \mid y', r \mid z'\} \in mgu(\{App(p, q, r), App(x', y', z')\})$.

4. Extraction of Functional from Logic Program

In this section we have a functional to express the deductions for a given logic program.

As shown in Theorem 3.9, the clause $C \leftarrow D_1 \ldots D_m$ can be interpreted as an operation emitting $[[\delta_1 + \psi_1]_{\{D_1\}} + \ldots + [\delta_m + \psi_m]_{\{D_m\}}]_{\{C\}}$ for $\{\psi_1, \ldots, \psi_m\}$ (regarded as an input set) with a set $\{\delta_1, \ldots, \delta_m\}$, where $\delta_i \in mgu(\{D_i, Stand(D_i)\})$ $1 \leq i \leq m$, and $Stand(D_i)\psi_i$ $(1 \leq i \leq m)$ is assumed to be deducible. The operation for a clause can be extended to a translator of input sequences by providing an output datum for each tuple of input data. Each datum emitted by such an operation is transformable into another by means of the relationship between the head of the definite clause and its standard atom. This transformation is based on Theorem 3.10. For the transformation, there may be more than one head of definite clauses with the same predicate symbol and thus the same standard atom. Thus transformed data should be gathered as one sequence. The device of merge in Park (1983) is most adequate for such a treatment. Fairness in the merge is necessary for any datum to be transfered to other operations as an input.

To deal with the sequences, we have a base domain $Dom = Sub \cup \{\tau\}$. τ is a special symbol not in Sub, to denote a time delay or pause occurring in a sequence. It will be later exploited to manage the merge as continuous.

Dom^∞ denotes the set of partial functions from the set of natural numbers (denoted by ω) to Dom such that if $u \in Dom^\infty$ and $u(p)$ $(p \in \omega)$ is defined then $u(q)$ is always defined for $q \leq p$. That is, Dom^∞ is regarded as the set of all finite and infinite sequences from Dom. $\perp$ is the function such that $\perp(p)$ is undefined for any $p \in \omega$. Note $\perp$ is thought of as the empty sequence.

Now assume a logic program

$$L = \{Cl_1, \ldots, Cl_k\},$$

where Cl_i is $C_i \leftarrow D_{i,1} \ldots D_{i,n_i}$ $(n_i \geq 0)$ for atoms C_i and $D_{i,j}$ $(1 \leq j \leq n_j)$.

Also let

$$PRED(L) = \{P_1, \ldots, P_h\},$$

where $\{P_1(\bar{x}_1), \ldots, P_h(\bar{x}_h)\}$ be a set of standard atoms, with $Stand(D_{i,j}) = P_{i_j}(\bar{x}_{i_j})$ $(1 \leq i_j \leq h)$ for $1 \leq i \leq k$ and $1 \leq j \leq n_i$.

To equip sequence variables for the representations of outputs emitted by the operations for definite clauses, we assign $\{u_1, \ldots, u_k\}$ to $\{Cl_1, \ldots, Cl_k\}$ such that each u_i denotes a sequence in Dom^∞. Also we prepare for a set of sequence variables $\{v_1, \ldots, v_h\}$ to express the sequences whose data are connected to standard atoms, where each v_j denotes a sequence in Dom^∞.

To identify an input data tuple with a natural number, we adopt a bijection $I_m: \omega \to \omega^m$, and define a projection $J_{m,i}: \omega^m \to \omega$ by $J_{m,i}(p_1, \ldots, p_m) = p_i$. The composition $J_{m,i} \circ I_m$ is denoted by $I_{m,i}$, that is, $I_{m,i}(p) = J_{m,i}(I_m(p))$.

Then, based on Theorem 3.9, we define, for $1 \leq i \leq k$,

$$(4.1) \quad \begin{cases} u_i(p) = & \rho_i(p) \circ \varphi_i(p) \text{ for some } \varphi_i(p) \text{ such that} \\ & \varphi_i(p) \in [[\delta_{i,1} + v_{i_1}(I_{n_i,1}(p))]_{\{D_{i,1}\}} + \ldots + [\delta_{i,n_i} + v_{i_{n_i}}(I_{n_i,n_i}(p))]_{\{D_{i,n_i}\}}]_{\{C_i\}} \\ & \text{if } [\ldots]_{\{C_i\}} \text{ is defined and not empty,} \\ u_i(p) = & \tau \text{ otherwise,} \end{cases}$$

where $\rho_i(p)$ $(1 \leq i \leq k)$ is a permutation following the remark below, and $\delta_{i,j} \in mgu(\{D_{i,j}, Stand(D_{i,j})\})$ for $1 \leq j \leq n_i$.

Remark By means of Lemmas 2.4, 3.4 and 3.5, $[[\delta_{i,1} + v_{i_1}(I_{n_i,1}(p))]_{\{D_{i,1}\}} + \ldots + [\delta_{i,n_i} + v_{i_{n_i}}(I_{n_i,n_i}(p))]_{\{D_{i,n_i}\}}]_{\{C_i\}}$ consists of mutually equivalent idempotent substitutions. To avoid recursive occurrences of variables, we take just one substitution $\rho_i(p) \circ \varphi_i(p)$ from such an equivalene class, in which some term (involving no variable in $\cup_{1 \leq i \leq k}\{P(\bar{x}_i) \leftarrow\} \cup L$) is assigned to every variable in C_i and to each $p \in \omega$. It is not so difficult to see that it might be possible to choose such a substitution by exploiting a permutation $\rho_i(p)$.

Next we prepare for a sequence variable which denotes a history of equivalent substitution sets attached to a standard atom. For $1 \leq j \leq h$, we define $Pred(j) = \{i \mid C_i \leftarrow D_{i,1} \ldots D_{i,n_i} \in L \text{ and } P_j(\bar{x}_j) = Stand(C_i)\}$. The cardinal number of $Pred(j)$ is denoted by $NO(j)$. To gather the sequences (emitted by the nodes for clauses), each of which is defined as in (4.1), we need a function interleaving input sequences. Let $fairmerge^p: (Dom^\infty)^p \to Dom^\infty$ denote a 'fair merge' function of p input sequences to an output sequence obtained by interleaving inputs without neglecting any part of any input. (For fair merge, see Park, 1983.) Applying Theorem 3.10 to get sequences relevant to standard atoms, we define

$$(4.2) \quad v_j = fairmerge^{NO(j)}(w_{j_1}, \ldots, w_{j_{NO(j)}}),$$

for $1 \leq j \leq h$, where

(i) $Pred(j) = \{j_1, \ldots, j_{NO(j)}\}$,
(ii) w_{j_i} denotes a sequence in Dom^∞, and is defined by

$$\begin{cases} w_{j_i}(q) = & \sigma_{j_i}(q) \circ \varphi_{j_i}(q) \text{ for an adequate permutation } \sigma_{j_i}(q) \text{ and} \\ & \varphi_{j_i}(q) \in [\gamma_{j_i} + u_{j_i}(q)]_{\{P_{j_i}(\bar{x}_{j_i})\}} \\ & \text{if } [\ldots]_{\{P_{j_i}(\bar{x}_{j_i}))\}} \text{is defined and not empty,} \\ w_{j_i}(q) = & \tau \text{ otherwise,} \end{cases}$$

for $\gamma_{j_i} \in mgu(\{C_{j_i}, P_{j_i}(\bar{x}_{j_i})\})$.

Remark $\sigma_{j_i}(q)$ is a choice such that the range of $\sigma_{j_i}(q) \circ \varphi_{j_i}(q)$ has no common variables with Cl_k such that $Stand(D_{k,l}) = P_{j_i}(\bar{x}_{j_i})$ $(1 \leq l \leq n_k)$.

Note the above remark is in accordance with further deductions by using $P_{j_i}(\bar{x}_{j_i})(\sigma_{j_i}(q)\circ \varphi_{j_i}(q))$ with $Cl_k \equiv C_k \leftarrow D_{k,1} \ldots D_{k,n_k}$. We might see such a permutation $\sigma_{j_i}(q)$ can be obtained.

(4.1) as well as (4.2) is regarded as defining a dataflow for a logic program L.

Example 4.1. Taking the same clause set as in Example 3.11, an example for (4.1) and (4.2) is illustrated.

Two sequence variables u_1 and u_2 are necessary in accordance with the first clause $App(nil, y, y) \leftarrow$ and the second clause $App(u.x', y', u.z') \leftarrow App(x', y', z')$, respectively. A sequence variable v is used for the standard atom $App(p, q, r)$.

Since the body of the first clause is empty,

$$u_1(p) = \varepsilon \text{ for } p \in \omega,$$

by (4.1). Because the body of the second clause consists of just one atom,

$$\begin{cases} u_2(p) = & p_2(p) \circ \varphi_2(p) \text{ for an adequate permutation } \rho_2(p) \text{ and} \\ & \varphi_2(p) \in [\delta + v(I_{1,1}(p))]_{\{App(u.x',y',u.z')\}} \\ & \text{if } [\ldots]_{\{App(u.x',y',u.z')\}} \text{ is defined and not empty,} \\ u_2(p) = & \tau \text{ otherwise,} \end{cases}$$

where $\delta \in mgu(\{App(p, q, r), App(x', y', z')\})$. Note $I_{1,1}(p) = p$.

Next we have:

$$v = fairmerge^2(w_1, w_2),$$

where

(i) $$\begin{cases} w_1(q) = & \sigma_1(q) \circ \psi_1(q) \text{ for an adequate permutation } \sigma_1(q) \text{ and} \\ & \psi_1(q) \in [\gamma_1 + u_1(q)]_{\{App(nil,y,y)\}} \\ & \text{if } [\ldots]_{\{App(nil,y,y)\}} \text{ is defined and not empty,} \\ w_1(q) = & \tau \text{ otherwise,} \end{cases}$$

for γ_1 is in an equivalence class containing $\{p \mid nil, q \mid y, r \mid y\}$,

(ii) $$\begin{cases} w_2(q) = & \sigma_2(q) \circ \psi_2(q) \text{ for an adequate permutation } \sigma_2(q) \text{ and} \\ & \psi_2(q) \in [\gamma_2 + u_2(q)]_{\{App(u.x',y',u.z')\}} \\ & \text{if } [\ldots]_{\{App(u.x',y',u.z')\}} \text{ is defined and not empty,} \\ w_2(q) = & \tau \text{ otherwise,} \end{cases}$$

for γ_2 is in an equivalence class containing $\{p \mid u.x', q \mid y', r \mid u.z'\}$.

For example, by Theorem 3.10, when $\leftarrow App(p_1, p_2, p_3)$ is a goal, $App(p_1, p_2, p_3) \alpha$ is the set of deducible atoms, where $\alpha \in [\gamma + \beta]_{\{App(p,q,r)\}}$ for $\gamma \in mgu(\{App(p, q, r), App(p_1, p_2, p_3)\})$ and β is a datum regarded as attached to the standard atom $App(p, q, r)$.

(4.1) and (4.2) are regarded as a functional over functions (sequence variables) constructed from a given logic program. We will see that (4.1) and (4.2) are complete and sound in representing the substitutions with which the atoms are deduced from a given logic program.

First we have a semantics for (4.1) and (4.2), which is given by introducing a partial order on Dom^∞ and by the fixpoint approach.

Definition 4.2. A partial order $\ll$ on Dom is defined by: $\tau \ll \theta$ and $\theta \ll \theta$ for any $\theta \in Sub$. A partial order $\sqsubseteq$ on Dom^∞ is defined by: $u \sqsubseteq v$ for $u, v \in Dom^\infty$ iff $u(p) \ll v(p)$ for any $p \in \omega$ such that $u(p)$ is defined.

The partial order $\sqsubseteq$ is easily extended to act on $(Dom^\infty)^m$:

$$(u_1, \ldots, u_m) \sqsubseteq (v_1, \ldots, v_m) \text{ iff } u_i \sqsubseteq v_i \text{ for } 1 \leq i \leq m.$$

The least upper bound of $T \subset (Dom^\infty)^m$ is denoted by $\sqcup T$. Note that $\sqsubseteq$ is sequentially complete in the sense that any $\{w^0 \sqsubseteq w^1 \sqsubseteq \ldots\}$ has a least upper bound.

(4.1) shows that the finite part of u_i depends on only finite parts of inputs. Also (4.2) consists of mappings (from finite parts of inputs into finite parts of outputs) and $fairmerge$ functions. Since u_i, $1 \leq i \leq k$ and v_j, $1 \leq j \leq h$, which are defined in (4.1) and (4.2), are infinite by means of τ, the $fairmerge$ is continuous. Thus we have:

Lemma 4.3. Assume (4.1) and (4.2) for a given logic program. Let f_L: $(Dom^\infty)^{k+h} \rightarrow (Dom^\infty)^{k+h}$ be a function such that $(u_1, \ldots, u_k, v_1, \ldots, v_h) = f_L(u_1, \ldots, u_k, v_1, \ldots, v_h)$ is defined by (4.1) and (4.2). f_L is continuous and there is a least fixpoint of f_L.

Now we define the least fixpoint of f_L in Lemma 4.3 as

$$(4.3) \quad (u_1^f, \ldots, u_k^f, v_1^f, \ldots, v_h^f).$$

(4.3) is a semantics for (4.1) (with the remark) and (4.2). (4.1) and (4.2) are based on Theorems 3.9 and 3.10, respectively, which are representations of unit resolution deductions and transformations of substitutions over Sub. Therefore (4.1) and (4.2) are sound in generating atoms, which are represented virtually as data in Sub. Thus the following theorem holds.

Theorem 4.4. Assume the least fixpoint of f_L as denoted by (4.3). Then $P_j(\bar{x}_j)\psi_j \leftarrow$ is deducible from L if $\psi_j = v_j^f(p)$ for any $p \in \omega$ as long as $v_j^f(p) \neq \tau$. Also $C_i\theta_i \leftarrow$ is deducible from L if $\theta_i \in u_i^f(q)$ for any $q \in \omega$ as long as $u_i^f(q) \neq \tau$.

The completeness of (4.3) in denoting all the atoms (deducible from a given logic program L) with some permutation is guarnteed by the completeness of unit resolution deductions in atom generations, by the assurance of $fairmerge$ functions to transfer all the parts of infinite inputs, and by the representation of (4.3) for finite unit resolution deductions. Note the reason why we need some permutation to represent each deducible atom arises from the fact we have just one adequate idempotent substitution in (4.1). Therefore we have:

Theorem 4.5. Assume that $C_i\theta_i \leftarrow$ is deducible from a logic program L for $\theta_i = [\theta_i]_{\{C_i\}} \in$ Sub. Then $\theta_i \sim u_i^f(p)$ for some $p \in \omega$. Also, when $P_j(\bar{x}_j)\psi_j \leftarrow$ is assumed to be deducible from L for $\psi_j = [\psi_j]_{\{P_j(\bar{x}_j)\}} \in Sub$, $\psi_j \sim v_j^f(q)$ for some $q \in \omega$.

To remove τ, that is, time delay from a sequence, the following function $delete$: $Dom^\infty \rightarrow Sub^\infty$ is useful.

$$delete(u)(p) = if \ \ u(0) = \tau \ \ then \ delete(next(u))(p)$$
$$else \ if \ p = 0 \ then \ u(0)$$
$$else \ delete(next(u))(p-1)$$

for $u \in Dom^\infty$ and $p \in \omega$.

The $next$ function is similar to the function in Lucid (See Ashcroft and Wadge, 1976)

$$next(g)(p) = g(p+1)$$

for $g \in \omega^\omega$ and $p \in \omega$.

Finally $(delete(u_1^f), \ldots, delete(u_k^f), delete(v_1^f), \ldots, delete(v_h^f))$ is interpreted as a semantics of a given logic program over Sub^∞.

5. Concluding Remarks

We have a method to extract a functional from a logic program by forming a recursion equation set over a sequence domain, which is on the basis of the set Sub of idempotent substitutions.

The recursion equation set reflects the unit resolution deductions from the original logic program, and is regarded as involving the operations to express deductions through clauses, and the operations to transform and to merge sequences. We adopt hiaton in Park (1983) as time delay, to represent the timing at which no atom exists. Finally the constructed recursion equation set is regarded as an application of Park's dataflow to a realization of computing for a logic program. The recursion equation set is a more constructive form than the program equation by Baudinet (1988) in that the present recursion equation set is explicitly expressed by means of the $comb$ function. The application has characteristics that the dataflow manipulates only the data in Sub.

The dataflow defines a continuous functional from a direct product of sequence doamins into itself. The sequence domain is the set of all finite and infinite sequences, which are taken from Sub with the hiaton. It completely and soundly denotes the data (in Sub) with which atom sets can be virtually interpreted as being generated from the originally given logic program. It is thus concluded that the least fixpoint of the functional is a semantics of the logic program, which is much concerned with its finite computation with time delay but is different from the semantics by Apt and van Emden (1982), and by Fitting (1985b). Because it describes a concurrent computing mechanism evoked and executed by substitution manipulations and suggests a semantics for a functional program transformed from the dataflow. The semantics is defined over a sequence domain (based on the idempotent substitution set), different from the domain by Fitting (1985a), by expanding the result in Yamasaki (1990). From operational point of view, we note the datum (in Sub) is directly related to the answer substitution applied when a goal statement is enquired to a logic program. The present dataflow is expected to be straightly transformable to an FP program as well as a concurrent system in which a process of transforming substitution data is in accordance with each definite clause (that is, a procedure or an assertion), and a process of transforming and merging substitution data is assigned to each predicate name.

Acknowledgement

The author is grateful to Prof.David M.R.Park and Dr.Stephen G.Matthews for comments on dataflow during the author's visits to University of Warwick. This work was partially supported by the Royal Society of London. Also this work is supported by the Japanese Ministry of Education, Science and Culture.

References

1. Apt,K.R. and van Emden,M.H. (1982), Contributions to the theory of logic programming, J.ACM 29, 841–864.

2. Ashcroft,E.A. and Wadge,W.W. (1976), Lucid–A formal system for writing and proving programs, SIAM J. Comput. 5, 336–354.

3. Baudinet,M. (1988), Proving termination properties of PROLOG programs: A semantic approach, Res. Report STAN–CS–88–1202, Computer Science Dept., Stanford University.

4. Debray,S.K. and Warren,D.S. (1989), Functional computations in logic programs, ACM Trans. on Programming Languages and Systems 11, 3, 451–181.

5. Eder,E. (1985), Properties of substitutions and unifiers, J. Symbolic Computation 1, 31–46.

6. Fitting,M. (1985a), A deterministic Prolog fixpoint semantics, J. Logic Programming 2, 111–118.

7. Fitting,M. (1985b), A Kripke-Kleene semantics for logic programs, J. Logic Programming 2, 295–312.

8. Kahn,G. (1974), The semantics of a simple language for parallel programming, Proc. IFIP 74, 471–475.

9. Palamidessi,C. (1990), Algebraic properties of idempotent substitutions, Lecture Notes in Computer Science 443, 386–399.

10. Park,D. (1983), The "fairness" problem and nondeterministic computing networks, in "Foundations of Computer Science IV" (de Bakker and van Leeuwen, eds.), mathematisch Centrum, Amsterdam, 133–161.

11. Yamasaki.S. (1990), Recursion equation sets computing logic programs, Theoretical Computer Science 76, 309–322.

THE MAS SPECIFICATION COMPONENT

HEINZ KREDEL

UNIVERSITY OF PASSAU, GERMANY

Abstract: MAS is an experimental computer algebra system combining imperative programming facilities with algebraic specification capabilities for design and study of algebraic algorithms. MAS views mathematics in the sense of universal algebra and model theory and is in some parts influenced by category theory. We give an overview of system design and the current state of the MAS project. The main topic of this article is the informal semantics of the MAS specification component and examples.

1. Design and State. Computer algebra systems are concerned with the exact manipulation of expressions (like differentiation or integration) or the exact computation in algebraic structures (like factorization, Gröbner bases or roots of equations). Exact means without round off errors.

Starting point for the development of MAS was the requirement for a computer algebra system with an up-to-date language and design which makes the existing ALDES / SAC-2 algorithm libraries available. At this time there have been about 650 algorithms in ALDES / SAC-2 and in addition I had 450 algorithms developed on top of ALDES / SAC-2. The tension of reusing existing software in an interactive environment with specification capabilities contributes most to the evolution of MAS.

The resulting view of the software has many similarities with the model theoretical view of algebra. The abstract specification capabilities are realized in a way that an interpretation in an example structure (a model) can be denoted. This means that is is not only possible to compute in term models modulo some congruence relation, but it is moreover possible to exploit a fast interpretation in some optimized (or just existing) piece of software.

Before we turn to the discussion of these language concepts in detail we give a short summary of the design, the system components and current state of MAS.

1.1. Overview. MAS (Modula-2 Algebra System) is an experimental computer algebra system combining imperative programming facilities with algebraic specification capabilities for design and study of algebraic algorithms. The new things are: MAS brings together specifications of abstract data types, imperative programs, controlled polymorphism, compiled libraries and term rewriting. The **goal** of the MAS system is to provide:

1. an *interactive* computer algebra system
2. comprehensive algorithm *libraries*, including the ALDES/SAC-2 system [Collins 82],
3. a familiar program *development* system with an efficient compiler,
4. an algebraic *specification* component for data structure and algorithm design
5. algorithm *documentation* open to the users.

Key **attributes** of the MAS system are:

1. portability (it is portable to a computer during a student exercise 'Praktikum'), machine dependencies isolated in a small kernel,
2. extendability (it is possible to add and interface to external algorithm libraries), open system architecture,
3. transparent low level facilities: storage management (garbage collection is provided without user cooperation), stable error handling (no system break down on misspelled

expressions and runtime exceptions), input / output with streams (no changes are required to existing libraries to redirect I/O).

4. effectivity (critical parts can be compiled and still be accessed interactively)

5. expressiveness (possibility to specify abstract algebraic concepts like rings or fields)

The goals and attributes have been achieved by the following main **design concepts:**

MAS replaces the ALDES language [Loos 76] and the FORTRAN implementation system of SAC-2 by the Modula-2 language [Wirth 85]. Modula-2 is well suited for the development of large program libraries; the language is powerful enough to implement all parts of a computer algebra system and the Modula-2 compilers have easy to use program development environments.

To provide an interactive calculation system, a LISP interpreter is implemented in Modula-2 with full access to the library modules. For better usability a Modula-2 like imperative (interaction) language was defined, including a type system and function overloading capabilities. To increase expressiveness, high-level specification language constructs have been included together with conditional term rewriting capabilities. They resemble facilities known from algebraic specification languages like ASL [Wirsing 86].

Theoretical design issues are:

MAS views mathematics in the sense of universal algebra and model theory and is in some parts influenced by category theory. The MAS language and its interpreter has no *knowledge of mathematics* and mathematical objects; however it is capable to describe (specify) and implement mathematical objects and to use libraries of implemented mathematical methods. Further the imperative programming, the conditional rewriting and function overloading concepts are separated in a clean way. There exists a denotational semantics description of the MAS language in [Kredel 91].

More precisely, a *specification* construct defines a formal language L from sorts and operations Σ. Included are the capability to join specifications and to rename sorts and operations during import of specifications. This allows both the specification of abstract objects (rings, fields), concrete objects (integers, rational numbers) and concrete objects in terms of abstract objects (integers as a model of rings). Specifications can be parameterized in the sense of λ abstraction.

The *semantics* of a specification is defined by a class of Σ-structures. It can be described either by implementations, axioms or models. The *implementation* part describes (imperative) procedures and data representations. The semantics for procedures is fixpoint semantics of the respective λ-terms.

The *axioms* part describes conditional rewrite rules which define a reduction relation on the term algebra generated by the sorts and operations of the specification. The semantics is therefore the class of models of the term algebra modulo the (congruence) relation. Currently there are no facilities to solve conditional equations.

The *model* part describes the association between a specification and (several) algebraic structures (models). Models may be given by native implementations in Modula-2 libraries, imperative functions or term rewrite rules and can be associated to suitable formal languages defined by Σ. The semantics is the interpretation of the Σ functions in the associated model. The desired model for interpretation is selected according to the types of the actual function parameters.

At this point one sees a limitation of universal algebra: the functions in the models must all have the same arity as the Σ functions. This prohibits for example the usage of finite fields (of some characteristic p) as models for fields, since the functions for finite fields need an additional parameter for the characteristic. This limitation is not present in category theory, so more precisely the model part describes a functor between two categories (here between a Σ_1 and a Σ_2 structure). Additional information required for parameters can be supplied from the 'descriptors' of the actual parameters.

Since functional terms entered into the system may involve subterms in several semantics, their *evaluation* order is defined as follows: If there is a model in which the function has an interpretation and a condition on the parameters is fulfilled, then the interpretation of the function in this model is applied to the interpretation (values) of the arguments. If there is an imperative procedure, then the procedure body is evaluated in the procedure context. If the unification with the left hand side of a rewrite rule is possible and the associated condition evaluates to true, then the right hand side of the rewrite rule is evaluated. Otherwise the functional term is left unchanged.

Related systems:

In contrast to other computer algebra systems (like Scratchpad II [Jenks 85]), the MAS concept provides a clean separation of computer science and mathematical concepts. The programmer / user has complete control over all operator overloadings. MAS specifications provide multiple inheritance.

In contrast to functional programming languages (like SML [Appel 88]) which implement typed lambda calculus the types of operations are not deduced from the program text but must be explicitly defined in the specification of an operation, in a variable declaration or in a typed string expression.

Compared to pure specification systems, MAS provides the possibility that operations in algebraic structures can be implemented by fast (imperative and/or library) procedures and not only by rewrite rules in term models.

A weak point in the current MAS design is that the language is only interpreted. This is actually not a handicap in execution speed since compiled libraries can be used, but in a too weak semantic analysis of the specifications. This means that certain errors in the specifications are only detected during actual evaluation of an expression.

1.2. System components. The MAS system components are identified in figure 1. Active components (programs) are enclosed in square boxes and passive components (data) are enclosed in oval boxes. Arrows indicate flow of data and lines between boxes show that the boxes are related in some way.

As mentioned in the overview section MAS itself is a Modula–2 program. Thus the MAS program can be recompiled and linked together with other symbolic and numerical libraries by a suitable Modula–2 compiler. This is shown as an arrow from the compiler box on the right to the enclosing MAS box on the left.

On the top line the editor box both acts on the Modula–2 source code (on the right) and the MAS input data (on the left). The input is processed by the following internal components:

1. The parser for the MAS language (Parse box): character strings in concrete syntax are transformed into abstract syntax trees. Static syntax check together with variable scope analysis is performed.
2. The specification processor (Specification box) with an attached data base of declarations (Declarations box): declarations are extracted from the parse tree and stored in the declaration base, information is retrieved during interpretation. The declarations reflect the Modula–2 source code and the library structure.

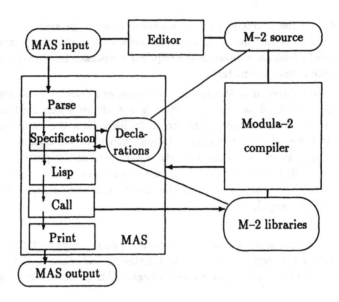

FIG. 1. *System Components*

3. The LISP interpreter (LISP box): according to the type or the function name of an S-expression innermost (that is eager) evaluation is performed.
4. The interface to the compiled library procedures (Call box): if external functions are encountered, then compiled procedures are restored and called with the appropriate parameters.
5. Finally the results are displayed by the (pretty) printing part (Print box).

1.3. Achievements and Current State. The steps towards the MAS system have been reported in [Kredel 88] (from ALDES / SAC–2 to Modula-2, storage management); in [Kredel 90] (on the implementation of a LISP interpreter with access to compiled Modula–2 procedures, the interaction language) and in [Kredel 91] (specification component, denotational semantics). Further there is a parser for the ALDES language and the MAS interpreter is now able to evaluate ALDES statements (although with low performance). Versions of the MAS system are running on Atari ST (TDI and SPC Modula–2 compilers), IBM PC/AT (M2SDS and Topspeed Modula–2 compilers) and Commodore Amiga (M2AMIGA compiler).

The ALDES / SAC–2 libraries have been implemented including the Polynomial Factorization System and the Real Root Isolation System. From the DIP system the Buchberger Algorithm System and the Ideal Decomposition and Ideal Real Root System have been implemented. Gröbner Bases are also available for non-commutative polynomial rings of solvable type. The combination of the MAS programs with numerical Modula–2 libraries has been tested. The mathematical libraries have been enlarged by a package for linear algebra.

As this overview shows, the MAS system is a comprehensive system with many facets. We turn now to a discussion of the programming language constructs.

2. Syntax. To precisely define the syntax we first specify the syntactic domains and then give the EBNF definition of the language. Note that we use the terms 'function' and 'procedure' interchangeably throughout the rest of the text.

2.1. Syntax Diagram. The syntax definition is given in extended BNF notation. That means name denotes non-terminal symbols, {} denotes (possibly empty) sequences, () denotes required entities, | denotes case selection and [] denotes optional cases. Terminal symbols are enclosed in double quotes and productions are denoted by =. The syntax diagrams are listed in table 1.

```
program    = topblock "."
topblock   = { ( unitspec | var | proc | expose ) ";" }
             statement
block      = { ( var | proc ) ";" } statement
unitspec   = { spec | implement | model | axioms }
spec       = "SPECIFICATION" header ";"
             { ( sort | import | sig ) ";" } "END" ident
implement  = "IMPLEMENTATION" header ";"
             { ( sort | import | var | proc ) ";" }
             statement "END" ident
model      = "MODEL" header ";"
             { ( sort | import | map ) ";" } "END" ident
axioms     = "AXIOMS" header ";"
             { ( sort | import | rule ) ";" }  "END" ident
sort       = "SORT" identlist
import     = "IMPORT" header [ renamings ]
sig        = "SIGNATURE" ident [ "(" [identlist] ")" ]
                                [ ":" "(" [identlist] ")" ]
var        = "VAR" identlist  ":" typeexpr
proc       = "PROCEDURE" ident ["("[identlist ]")"]
                          [":" ident]";" block ident
map        = "MAP" header "->" header [ "WHEN" header ]
rule       = "RULE" expression "=>" expression
                    [ "WHEN" condition ]
typeexpr   = header [ string ]
header     = ident [ "(" [identlist] ")" ]
renamings  = "[" { ident "/" ident ";" } "]"
expose     = "EXPOSE" ident [ "(" [actualparms] ")" ]
```

TABLE 1
Syntactic Clauses

Observe that a program is a (possibly empty) sequence of declarations, followed by a (possibly empty) statement followed by a period. A statement can be an assignment, a procedure call or a IF–, WHILE–, REPEAT–, BEGIN–statement or EXPOSE–statement. Declarations are VAR and PROCEDURE; Unit specifications are SPECIFICATION, IMPLEMENTATION, MODEL, AXIOMS. The building blocks of the unit specifications are IMPORT, SORT, SIGNATURE, MAP and RULE. The syntax of statements and expressions is fairly Modula–2 like and is not presented here.

2.2. Context conditions. In general the context conditions are assumed to be checked by the semantic functions. The only context conditions which are checked during parsing are that numbers n are in the range $-\beta < n < \beta = 2^{29}$ and the lexical scope of local variables. Other context conditions as for example the number of parameters in procedure calls, undefined procedures are not checked at this time.

3. Units. A collection of denotations which define the same class of algebraic structures is called a **unit**. A unit consists of at most one SPECIFICATION construct which defines the formal language from Σ. Optionally several constructs may accompany a specification which define the semantics of the class of Σ–structures: IMPLEMENTATION, MODEL and AXIOMS.

The pair SPECIFICATION, IMPLEMENTATION is similar to the Modula-2 pair DEFINITION MODULE, IMPLEMENTATION MODULE. The pair SPECIFICATION, AXIOMS is similar to constructs from algebraic specification languages. With the pair SPECIFICATION, MODEL native models can be associated to a language defined by Σ.

The unit declarations only collect definitions. To make the sorts and functions visible and to evaluate initialization statements, a unit must be exposed with the EXPOSE construct.

We turn now to a more detailed discussion of these constructs.

3.1. Specifications. The specification part defines a formal language from sorts and operations Σ. The basic constituents and features of a SPECIFICATION are as follows:

1. The 'header = ident (identlist)' part defines the name ident of the unit. The identlist defines the formal parameters of the specification. The semantics is λ–abstraction, that means that in the environment the actual parameters are substituted for the formal parameters.
2. The 'SORT identlist' declaration defines the identifiers as sort names.
3. The 'IMPORT header [renamings]' declaration includes an already defined specification named by the identifier in header into the actual specification. The actual specification is therefore extended by the imported specification. During import it is possible to rename functions and sorts in the imported specification. Since several specifications can be imported it is possible to join specifications.
4. The 'SIGNATURE ident (identlist): ident' declaration defines function (operator) names together with the *sorts* of the formal parameters of the function.
5. After exposition of a unit the specifications are visible in the actual environment.

These constructs allow both the specification of abstract objects (rings, fields), concrete objects (integers, rational numbers) and concrete objects in terms of abstract objects (integers as a model of rings). The specification of a concrete item like the rational numbers could be as follows:

```
SPECIFICATION RATIONAL;        (*Rational numbers specification. *)
(*1*) SORT RAT, INT, atom;
(*2*) SIGNATURE RNWRITE (RAT)     ;
      SIGNATURE RNDRD   (RAT)     : RAT;
(*3*) SIGNATURE RNone  ()         : RAT;
      SIGNATURE RNzero ()         : RAT;
(*4*) SIGNATURE RNPROD (RAT,RAT) : RAT;
      SIGNATURE RNSUM  (RAT,RAT) : RAT;
      SIGNATURE RNDIF  (RAT,RAT) : RAT;
      SIGNATURE RNNEG  (RAT)     : RAT;
      SIGNATURE RNINV  (RAT)     : RAT;
      SIGNATURE RNQ    (RAT,RAT) : RAT;
(*5*) SIGNATURE RNINT  (INT)     : RAT;
      SIGNATURE RNprec (atom)    ;
(*9*) END RATIONAL.
```

The most general unit is an object which specifies the communication (input / output) operations of objects. READ, WRITE are for handling values DECREAD, DECWRITE are for handling descriptors.

```
SPECIFICATION OBJECT;          (*Object specification. *)
(*1*) SORT obj;
(*2*) SIGNATURE READ      (obj) : obj;
      SIGNATURE WRITE     (obj) ;
(*3*) SIGNATURE DECREAD   (obj) : obj;
      SIGNATURE DECWRITE  (obj) ;
(*4*) SIGNATURE DEFAULT   (obj) : obj;
      SIGNATURE COERCE    (obj) : obj;
(*9*) END OBJECT.
```

Abstract specifications can be built from smaller pieces. For example (commutative) fields can be defined in terms of two Abelian groups, which are themselves built from Abelian monoids (which extend objects). The renamings are used to write one Abelian group 'multiplicatively', like PROD for SUM.

```
SPECIFICATION AMONO;           (*Abelian monoid specification. *)
(*1*) IMPORT OBJECT[ amono/obj ];
(*2*) SIGNATURE ZERO (amono)        : amono;
(*3*) SIGNATURE SUM  (amono,amono) : amono;
(*9*) END AMONO.


SPECIFICATION AGROUP;          (*Abelian group specification. *)
(*1*) IMPORT AMONO[ ag/amono ];
(*2*) SIGNATURE DIF  (ag,ag) : ag;
      SIGNATURE NEG  (ag)    : ag;
(*9*) END AGROUP.


SPECIFICATION FIELD;
(*Field specification joining two abelian groups. *)
(*1*) IMPORT AGROUP[ field/ag ];
      IMPORT AGROUP[ field/ag, ONE/ZERO, PROD/SUM,
                     REZIP/NEG, Q/DIF ];
(*9*) END FIELD.
```

Using the field specification, one could derive an alternative definition of the rational numbers. Note that some unique functions for rational numbers must be specified separately.

```
SPECIFICATION RATIONAL;
(*Rational numbers specification using the abstract field specification. *)
(*1*) SORT INT, atom;
(*2*) IMPORT FIELD[ RAT/field,
                    RNDRD/READ, RNWRITE/WRITE,
                    RNone/ONE, RNzero/ZERO,
                    RNSUM/SUM, RNNEG/NEG, RNDIF/DIF,
                    RNPROD/PROD, RNQ/RECIP, RNQ/Q ];
(*3*) SIGNATURE RNINT  (INT): RAT;
      SIGNATURE RNprec (atom);
(*9*) END RATIONAL.
```

3.2. Implementations. The *implementation* part describes (imperative) procedures and data representations. The basic constituents and features of an IMPLEMENTATION are as follows:

1. Procedures and sorts are invisible outside the implementation, except those defined in the specification.

2. The SORT declaration defines additional identifiers as local sort names.

3. The IMPORT declaration makes the sorts and operations of a specification visible inside the implementation. Its semantics correspond to the EXPOSE statement.

4. The 'VAR identlist : typeexpr' declaration defines names for local variables and associates type information with them.

5. The 'PROCEDURE ident formals; block ident' declaration defines an imperative (or functional) implementation of a procedure. block is a sequence of declarations followed

by a statement.

6. An implementation defines a closed environment for the contained variable and procedure declarations (so called closures).

7. The Modula–2 library functions exist a priori and can be accessed without further implementation definitions.

8. A statement can be a BEGIN–statement and is executed during the exposition of the unit.

The implementations can be used to define concrete procedures or abstract procedures. The imperative language constructs (like assignments and loops) are fairly standard and are not discussed here.

In case of the rational number unit just some gaps left by the library functions need to be filled.

```
IMPLEMENTATION RATIONAL;
        VAR s: atom;
(*1*) PROCEDURE RNone();
      BEGIN RETURN(RNINT(1)) END RNone;
(*2*) PROCEDURE RNzero();
      BEGIN RETURN(RNINT(0)) END RNzero;
(*3*) PROCEDURE RNWRITE(a);
      BEGIN IF s < 0 THEN RNWRIT(a) ELSE RNDWR(a,s) END;
            END RNWRITE;
(*4*) PROCEDURE RNprec(a);
      BEGIN s:=a END RNprec;
(*8*) BEGIN
            s:=-1;
(*9*) END RATIONAL.
```

Here procedures RNone, RNzero, RNprec and RNWRITE are defined in terms of the library functions RNINT, RNWRIT and RNDWR. The other functions RNSUM etc. defined in the specification are library functions and need no further implementations.

Abstract functions are those which use function names of abstract specifications to implement something. For example in a ring one could have an abstract exponentiation function EXP.

```
IMPLEMENTATION RING;
(*1*) PROCEDURE EXP(X,n);
      VAR   x: ring; VAR   i: atom;
      BEGIN
(*1*) IF n <= 0 THEN x:=ONE(X); RETURN(x) END;
(*3*) i:=n; x:=X;
          WHILE i > 1 DO i:=i-1;
                  x:=PROD(x,X) END;
          RETURN(x)
(*9*) END EXP;
(*9*) END RING.
```

Here ONE and PROD denote (abstract) functions from the ring. The operators <=, > and - are used on atoms (integers k in the range $-\beta < k < \beta = 2^{29}$).

3.3. **Models.** The *model* part describes the association between a specification and (several) algebraic structures (models). The basic constituents and features of a MODEL are as follows:

1. The SORT declaration defines additional identifiers as local sort names.

2. The IMPORT declaration makes the sorts and operations of a specification visible inside the implementation. Its semantics correspond to the EXPOSE statement.

3. The 'MAP header -> header [WHEN header]' declaration defines the interpretation of an operation on the left hand side of '->' by a function of a model on the right hand side. In the parameter list of a function on the left hand side the sort names of

a model are specified. In the parameter list of a function of a model the two selectors VAL and DESC can appear. The i-th VAL selects the value of the i-th abstract function parameter. The i-th DESC selects the descriptor of the i-th abstract function parameter.

4. Operations in models can be compiled functions, user defined imperative functions or term rewrite rules.

5. Conditional interpretation can be expressed by a WHEN clause following the real function. In the condition the VAL and DESC selectors can be used.

Observe that the model interpretation can also be viewed as function overloading. The abstract functions are sometimes also called generic functions or polymorphic functions.

```
MODEL FIELD;        (*Rational numbers are a model for fields. *)
    (*1*) IMPORT RATIONAL;
    (*2*) MAP READ(RAT)       -> RNDRD();
          MAP WRITE(RAT)      -> RNWRITE(VAL);
    (*3*) MAP ONE(RAT)        -> RNone();
          MAP ZERO(RAT)       -> RNzero();
    (*4*) MAP PROD(RAT,RAT)   -> RNPROD(VAL,VAL);
          MAP SUM(RAT,RAT)    -> RNSUM(VAL,VAL);
          MAP DIF(RAT,RAT)    -> RNDIF(VAL,VAL);
          MAP NEG(RAT)        -> RNNEG(VAL);
          MAP Q(RAT,RAT)      -> RNQ(VAL,VAL);
          MAP REZIP(RAT)      -> RNINV(VAL);
    (*9*) END FIELD.
```

This reads as follows: the product function PROD is interpreted in the model of rational numbers (two rational numbers as parameters RAT) as the concrete function RNPROD (from the abstract parameters the values are to be taken according to the VAL selector). The other functions on the left hand side are defined in the rational number specification and are partly library functions or implemented in the rational number implementation part.

For finite fields as models of fields an example using descriptors and conditional interpretation is as follows.

```
MODEL FIELD;          (*Modular integers are a model for fields. *)
    ...
    (*2*) MAP READ(MI)    -> MIREAD(DESC);
          MAP DECREAD(MI) -> IREAD();
    ...
    (*4*) MAP PROD(MI,MI) -> MIPROD(DESC,VAL,VAL) WHEN EQ(DESC,DESC);
    ...
    (*9*) END FIELD.
```

MI denotes the modular integer $Z/(p)$ sort. MIPROD denotes the modular integer product where the first parameter is the modulus p selected by DESC. The WHEN clause specifies that only elements from the same finite field are to be multiplied (that is their descriptors must be equal (EQ)). Descriptors can be specified in the type expression string in VAR declarations provided a DECREAD function has been defined.

3.4. Axioms. The *axioms* part describes conditional rewrite rules. The basic constituents and features of an AXIOMS clause are as follows:

1. The SORT declaration defines additional identifiers as local sort names.

2. The IMPORT declaration makes the sorts and operations of a specification visible inside the implementation. Its semantics correspond to the EXPOSE statement.

3. The 'RULE expr => expr [WHEN cond]' declaration defines a rewrite rule. If the left hand side of a rule can be unified with the expression under consideration, then the variables on the right hand side are substituted according to the unification. Then the right hand side replaces the actual expression.

4. Conditional rewriting can be expressed by a WHEN clause following the right hand side of the rewrite rule. The condition is evaluated with the variables substituted according to the actual unification of the left hand side.

5. Variables need not be declared and are assumed to be universally quantified and unbound.

6. Currently there are no facilities to solve conditional equations since there is no backtracking of unsuccessful rewritings.

7. There are also no provisions to check if the rewrite system is confluent or Noetherian.

For example the Peano structure can be specified as follows:

```
SPECIFICATION PEANO;
(*Peano structure specification. *)
(*1*) SORT nat, bool;
(*2*) SIGNATURE null  ()        : nat;
      SIGNATURE one   ()        : nat;
   (* SIGNATURE succ  (nat)     : nat; *)
      SIGNATURE add   (nat,nat) : nat;
      SIGNATURE prod  (nat,nat) : nat;
(*3*) SIGNATURE equal (nat,nat) : bool;
(*9*) END PEANO.
```

Observe that the succ (successor) constructor need not be defined in the specification, since no rule has succ on the first functional term in the right hand side expression. The Peano axioms can then be coded as rewrite rules as follows:

```
AXIOMS PEANO;
(*Axioms for Peano system. *)
      IMPORT PROPLOG;
      RULE one()                  => succ(null());
(*1*) RULE equal(X,X)             => TRUE();
      RULE equal(succ(X),null())  => FALSE();
      RULE equal(null(),succ(X))  => FALSE();
(*2*) RULE equal(succ(X),succ(Y)) => equal(X,Y);
(*3*) RULE add(X,null())          => X;
      RULE add(null(),X)          => X;
(*4*) RULE add(X,succ(Y))         => succ(add(X,Y));
(*5*) RULE prod(X,null())         => null();
      RULE prod(null(),X)         => null();
(*6*) RULE prod(X,succ(Y))        => add(prod(X,Y),X);
(*9*) END PEANO.
```

PROPLOG denotes a propositional logic specification not listed here. There the constant functions TRUE() and FALSE() are defined. During unification the variables X and Y are bound in the left hand side and then substituted in the right hand side.

3.5. Exposition of units. Once this specification apparatus has been set up one wants to see how it works and what benefits are obtained. The language constructs discussed so far modify only the declaration data base. To access the defined functions they must first be exposed in the top level environment with the 'EXPOSE ident (actualparms)' statement. ident is the name of the unit which is to be exposed. The parameters can be any valid MAS expressions which are meaningful inside the unit.

The exposition of some of the example units is as follows:

```
EXPOSE RATIONAL.
EXPOSE PEANO.
EXPOSE FIELD.
```

From then on the functions like PROD can be used in expressions or top level statements and procedures.

4. Expression Evaluation. We now turn to the evaluation of arbitrary expressions. Expressions are transformed to functional terms by the parser. The evaluation of functional terms is defined as follows:

1. If there is a model in which the function has an interpretation and a condition on the parameters is fulfilled, then the interpretation of the function in this model is applied to the interpretation (values) of the arguments.
2. If there is an imperative procedure, then the procedure body is evaluated in the procedure context.
3. If the unification with the left hand side of a rewrite rule is possible and the associated condition evaluates to true, then the right hand side of the rewrite rule is evaluated.
4. Otherwise the functional term is left unchanged.

To understand the following example we must mention a feature of the parser: For convenience of the users, the MAS parser can be instructed to generate generic function names for the arithmetic operators. However some care is needed since then also a specification of the atoms structure is required to access the built-in primitive arithmetic.

This feature is set by the pragma switch GENPARSE. Thus PRAGMA(GENPARSE) activates / inactivates the generic code generation of the parser. The operators correspond to the following functions: + to SUM, - to DIF or NEG, * to PROD, / to Q and ^ to EXP.

```
VAR r, s: RAT.            ANS: RAT().
r:="2222222222.777777777777777".
ANS: "2222222222777777777777777/1000000000000000".
s:=r/r.                   ANS: "1".
s:=r^0 + s - "1": RAT.    ANS: "1".
```

The first line declares the variables r and s to be of type RAT, that is to be rational numbers. The second line is a so-called generic assignment. Depending on the type of r the character string on the right hand side is read (or converted to internal form). Recall that the interpretation of READ(RAT) was defined as RNDRD() which reads a rational number in decimal representation.

Internally an object with type, value and descriptor information is created. This information is then used by the generic write function WRITE(RAT) for displaying the result in the next line.

The fourth line shows the computation of r/r. According to the type information of r the corresponding generic function Q(RAT,RAT) is determined. Then RNQ(VAL,VAL) is computed where the values of the data objects are substituted. Finally the information on the output parameters of RNQ namely RAT is used to create a new typed object. This object is then bound to the variable s and finally it is displayed.

The last line shows the computation of a more complex expression r^0 + s - "1": RAT. The term "1": RAT denotes a constant from the rational numbers, namely 1. The contents of the character string are read by the generic function READ(RAT) and a new typed object is created Note further that r^0 is computed by an abstract function (namely EXP) of the abstract RIN(implementation. Then the computation proceeds as expected.

The declaration of elements from a finite field could be as follows: VAR m: MI "7"; m:="9" ANS: "2". That means m is an element of $Z/(7)$, during assignment 9 is reduced mod 7 to 2.

5. Conclusions. We have have given an overview of the MAS system design and its his tory. Further we presented the MAS specification component in an informal way. The MA system combines three programming language concepts in a clean way: imperative program ming, term rewriting and explicit models. The overall glue of these concepts is provided by th specification capabilities.

Compared to pure specification systems, MAS provides the possibility that operations in algebraic structures can be implemented by fast (imperative and/or library) procedures and not only by rewrite rules in term models. Compared to computer algebra systems, MAS supports concepts from specification languages, which are useful for the construction of computer algebra software.

A weak point in the current MAS design is that the language is only interpreted and has therefore a too weak (too late) semantic analysis of the specifications. This means that certain errors in the specifications are only detected during actual evaluation of an expression and not immediately during the declaration of the specification.

In the way of programming methodology MAS provides not only implementation by refinement of (abstract) specifications, it also supports the opposite way starting with a model (a rapid prototype) and then abstracting a general specification and an abstract Σ-structure.

The MAS development is also a good exercise in software reusability. Although it is in some respect easier to start a system from scratch, it is also obvious that comprehensive, well-tested and 'error free' software is hard to obtain. Once the storage management problem had been resolved it was very fruitful to reuse the ALDES / SAC-2 system, which had already a sophisticated module structure.

A strong point of the MAS language and the MAS system is that it runs on small computers and therefore makes specification capabilities available to a wide range of users. Due to the Modula-2 concept and modern program development systems it is further easy to recompile and taylor the MAS system.

Not included was a discussion of the specification of relevant algebraic structures. The development of specifications is currently in work. As a first impression it seems to be at least as expressive as the Scratchpad II language (restricted to specifications). For a treatment of this topic, including an interesting discussion on the relation between constructive and non-constructive mathematics, see [Davenport 90].

Acknowledgements: Many thanks to all who made contributions or influenced this project: R. Loos, B. Buchberger, W. Kynast, M. Pesch, B. Deyle. Especially I which to thank M. Wirsing, T. Becker, M. Gengenbach and V. Weispfenning for various discussions on the design and presentation of the system. Further I thank the referees for helpful comments and suggestions.

REFERENCES

[Appel 88] A. W. Appel, R. Milner, R. W. Harper, D. B. MacQueen, *Standard ML Reference Manual (preliminary draft)*, University of Edinburgh, LFCS Report, 1988.

[Collins 82] G.E. Collins, R. Loos, *ALDES/SAC-2 now available*, SIGSAM Bulletin 1982, and several reports distributed with the ALDES/SAC-2 system.

[Davenport 90] J. H. Davenport, B. M. Trager, *Scratchpad's View of Algebra I: Basic Commutative Algebra*, Proc. DISCO 90 Capri, LNCS 429, pp 40-54, Springer, 1990.

[Jenks 85] R. D. Jenks et al., *Scratchpad II Programming Language Manual*, Computer Algebra Group, IBM, Yorktown Heights, NY, 1985.

[Kredel 88] H. Kredel, *From SAC-2 to Modula-2*, Proc. ISSAC'88 Rome, LNCS 358, pp 447-455, Springer, 1989.

[Kredel 90] H. Kredel, *MAS Modula-2 Algebra System*, Proc. DISCO 90 Capri, LNCS 429, pp 270-271, Springer, 1990.

[Kredel 91] H. Kredel, *Semantics of the MAS Language*, University of Passau 1991.

[Loos 76] R. G. K. Loos. *The Algorithm Description Language ALDES (Report)*, SIGSAM Bulletin 14/1, pp 15-39, 1976.

[Wirsing 86] M. Wirsing, *Structured Algebraic Specifications: A Kernel Language*, Theoretical Computer Science 42, pp 123-249, Elsevier Science Publishers B.V. (North-Holland) (1986).

[Wirth 85] N. Wirth, *Programming in Modula-2*, Springer, Berlin, Heidelberg, New York, 1985.

Domesticating Imperative Constructs So That They Can Live in a Functional World

T.C. Nicholas Graham

GMD Karlsruhe
Vincenz-Prießnitz-Str. 1
D-7500 Karlsruhe 1
Federal Republic of Germany
E-Mail: graham@karlsruhe.gmd.dbp.de

Gerd Kock

University of Karlsruhe
Postfach 6980
D-7500 Karlsruhe 1
Federal Republic of Germany
E-Mail: kock@ira.uka.de

Abstract

Many problems in program analysis and transformation are simplified in the context of pure functional languages. The desire for efficiency and notational flexibility often leads, however, production versions of functional languages to include ad-hoc imperative extensions that defeat program analysis. This paper shows how a hierarchy of restricted imperative constructs can be introduced into a functional language. These constructs preserve a local impurity property where the detrimental effects of non-functional constructs are limited to the functions in which they appear. The properties of the language are investigated in the context of a formal semantics.

1   Introduction

Purely functional languages have been used as the basis of considerable theoretical work in programming languages. For example, work in program transformation[2][13][17], program complexity analysis[18], visual programming languages[3], and the exploitation of implicit parallelism in programs[1][7][11] is greatly simplified if the language being used is purely functional. However, when actually trying to write programs, it is often highly inconvenient to be limited to a purely functional language. In fact, most commercial implementations of functional languages provide non-functional constructs, which are in practice widely used.

This paper shows how non-functional constructs can be safely introduced into a functional language. The underlying philosophy is that normally, most of a program can be written in a purely functional style; however, for reasons of efficiency and notational clarity, parts of a program may often be better coded in an imperative style. For these limited parts of the program, non-functional constructs can be used, provided they do not destroy the functional nature of the remainder of the program: i.e., the impure properties of the constructs should be localized to the parts of the program where they are used. The restricted imperative constructs presented in this paper syntactically guarantee this localization of impurity, continuing to provide a great deal of the expressiveness of the non-functional style at minimal cost to the analysability of the program.

The constructs defined in this paper have been implemented using the TXL program transformation system [5].

```
procedure sortInPlace (inout L)
    procedure swap (exists L, P1, P2)
        [L@P1, L@P2] := [L@P2, L@P1];
    end swap;

    var ChangesMade := true;
    while ChangesMade
        var P := 1;
        ChangesMade := false;
        while lessThan (P, length(L))
            if greaterThan (L@(P-1), L@P) then
                swap(P, P-1);
                ChangesMade := true;
            endif;
            P := P + 1;
        endwhile;
    endwhile;
end sortInPlace;
```

```
function split (L)
    % Splits a list in half
    ...
end split;

function mergeSort (L)
    if lessEqual (length (L), 1000) then
        sortInPlace (L);
    else
        var [L1, L2] := split (L);
        L := merge (mergeSort (L1), mergeSort (L2));
    endif;
    L
end mergeSort;

function merge (L1, L2)
    % Standard functional merge: returns L1
    % merged with L2 in ascending order.
    ...
end merge;
```

Figure 1: *A merge sort program intended for parallel execution. Despite using the procedure sortInPlace, the function mergeSort is still referentially transparent, and therefore amenable to parallel execution.*

2 Overview of the Functional/Imperative Language

The critical defining aspect of functional languages is *referential transparency*, that is, the guarantee that whenever a function f is applied to a particular argument X, the result will be the same. In a non-functional language, this need not be the case, since f may refer to some global state which may be altered between applications (perhaps even by f itself). Referential transparency is the property which guarantees the declarative nature of functional programs, and is the property upon which many algorithms to analyse functional programs depend.

In spite of the mathematical elegance of pure-functional languages, most commercial implementations of functional languages in fact permit non-functional constructs. This is for two main reasons: efficiency and notational clarity. In order to achieve referential transparency, functions may not observe or modify any non-local state. This means non-local variables and assignments cannot be permitted. These constructs, however, allow the creation of efficient programs, and application programmers therefore demand and use them.

The aim of this paper is to present a notation where the (relatively small) speed-critical parts of a program can be coded imperatively, while provably not disturbing the declarative properties of the remainder of the program.

2.1 An Example: Parallel Merge Sort

Figure 1 shows an example of a program written based on the constructs defined in this paper. The program is a merge sort intended for parallel execution. In order to allow easy partitioning of the program over multiple processors, the program is written functionally. The main *mergeSort*

Language	Facility	Example Languages	Cost	Proposed Construct
t_0	Pure Functional Facilities	Pure Lisp, Hope, FP, Miranda		
t_1	Local Variables and Assignments		Sequential Reduction of Function Bodies	Variable Declaration, Assignment Statement
t_2	Read-only Imports of Non-local State	Functional Subset of Euclid/Turing	Lose Referential Transparency	Existential Parameters
t_3	Var Imports of Non-local State	Nial, Lisp with 'set' function, C	Lose Side-Effect Freedom	Procedures, Writable Existential Parameters
t_4	Reference Parameters	Ada, Pascal	Lose Expression-Based Language	*inout* Parameters

Figure 2: *Hierarchy of Non-Functional Constructs. Each Level in the hierarchy moves further from the pure-functional ideal. Constructs are suggested to implement each facility, and the cost of each construct is given.*

function is written using the traditional algorithm, where both halves of the list are recursively sorted, and then merged. Because of the Church-Rosser property of functional languages, the two recursive invocations of *mergeSort* can be executed in parallel.

Beyond a certain point (here arbitrarily chosen to be 1,000 elements), however, it is cheaper to use an efficient sequential sort algorithm on one processor than to continue the parallel merge sort. In *mergeSort*, therefore, small lists are sorted using the *sortInPlace* procedure, an imperative bubble sort. *sortInPlace* uses imperative features such as direct array updates (the notation $A@I$ means the list A indexed at position I), *while* loops, and reference parameters.

In an ad-hoc integration of functional and imperative features, sophisticated static analysis would be required to prove that *mergeSort* is in fact still referentially transparent, allowing its continued use in a parallel context. As will be described in the next section, the imperative constructs are defined in such a way as to syntactically *guarantee* the referential transparency of any function. This allows efficient and simple implementation of programs, as well as giving the programmer intuitive rules as to what effects various constructs will have on referential transparency.

The language presented in this paper is a subset of the language that has been implemented, in order to simplify its formal definition and reasoning about its properties. This example contains a few constructs *not* defined in the paper, particularly array indexing (e.g. $L@P1$), the while loop, intermixed declarations and statements, and syntactic short-cuts such as multi-assignments and initializing assignments.

$$
\begin{aligned}
program &::= fBody \\
fBody &::= fpDeclarations\ expression \\[6pt]
fpDeclarations &::= \{fpDeclaration;\} \\
fpDeclaration &::= \textbf{function}\ id\ formalParmList\ fBody\ \textbf{end}\ id \\
formalParmList &::= \text{``("}[formalParm\{\text{``,"}formalParm\}]\text{")"} \\
formalParm &:= id \\[6pt]
expression &::= constant \\
expression &::= id \\
expression &::= functionApplication \\
expression &::= \textbf{if}\ expression\ \textbf{then}\ expression\ [\textbf{else}\ expression]\ \textbf{endif} \\[6pt]
functionApplication &::= id\ actualParmList \\
actualParmList &::= \text{``("}\ [actualParm\ \{\text{``,"}\ actualParm\}]\ \text{")"} \\
actualParm &::= expression
\end{aligned}
$$

Figure 3: *The Grammar of the t_0 Language. The grammar is structured to support incremental modification leading up to the t_4 language.*

2.2 A Hierarchy of Non-Functional Constructs

There is a middle-ground between a purely functional language and a language permitting general global variables and assignments. Figure 2 shows a hierarchy of non-functional constructs. Each level in this hierarchy takes us further from the pure-functional ideal, but affords greater notational flexibility. At each level, a construct is proposed to be added into functional languages to achieve the desired facility. These constructs provide a restricted aspect of the facility named at each level. The restrictions *localize* the effect of imperative constructs, so that if a pure function f uses an imperative function g, then f will still be referentially transparent. This localization property we call *local impurity*, defined precisely in section 4.2.

As a basis for the introduction of these constructs, a severely restricted functional language is used (figure 3). The language allows function definition and prefix application. Functions take tuples as arguments. The language is untyped. This language was chosen only for its simplicity and readability; the given constructs should be adaptable to any functional language. Figure 2 shows a hierarchy of non-functional extensions to this base language; we shall refer to this base language as t_0, and the successive extended languages as t_1 through t_4. Each level has an explicit cost associated with its use; for example, t_2 functions are not referentially transparent, and t_3 functions have side effects.

3 The Combined Language

The following sections informally introduce the new constructs proposed in figure 2.

$$fBody \quad ::= \quad varDeclarations\ fpDeclarations\ statements\ expression$$

$$varDeclarations \quad ::= \quad \{varDeclaration\ ;\}$$
$$varDeclaration \quad ::= \quad \textbf{var}\ id$$

$$statements \quad ::= \quad \{statement\ ;\}$$
$$statement \quad ::= \quad id\ \text{“}:=\text{”}\ expression$$
$$statement \quad ::= \quad \textbf{if}\ expression\ \textbf{then}\ statements\ [\textbf{else}\ statements]\ \textbf{endif}$$

Figure 4: *New Constructs Introduced In the t_1 Language*

$$formalParm \quad ::= \quad \textbf{exists}\ id$$

Figure 5: *New Constructs Introduced In the t_2 Language*

3.1 t_1: Local Variables and Assignments

The t_1 language introduces the concept of a statement, an imperative construct operating on variables local to the body of a function (figure 4). Variable declarations and assignments are introduced using the standard syntax.

A variable is visible only in the scope in which it is declared; i.e., functions may not import variables. Given this restriction, referential transparency is maintained, since a function may neither reference nor modify its surrounding scope.

The cost of introducing variables and statements is that the Church-Rosser property is lost within function bodies: in general, a list of statements must be executed sequentially.

3.2 t_2: Read-Only Imports

Consider we wish to maintain a variable *TokenStream* with a list of tokens, and to have a function *currentToken* which delivers the first element of this list. The traditional solution to this problem would be to import the variable *TokenStream* into the function:

```
var TokenStream;
function currentToken ()
    first (TokenStream)
end currentToken;
...
currentToken()
```

This solution fails to provide the local impurity property. For example, we can construct a function that tests if the current token matches its parameter:

```
function match (MatchToken)
    equal (currentToken(), MatchToken)
end match
```

match is a t_0 function, but through its indirect dependency on *TokenStream* is not referentially transparent.

Existential parameters [6] are a slightly weaker construct that provide local impurity (figure 5). We shall rewrite our example as:

```
var TokenStream;
function currentToken (exists TokenStream)
    first (TokenStream)
end currentToken;
...
currentToken()
```

Here *TokenStream* is an existential parameter, meaning that rather than being explicitly named at the call site, a variable named *TokenStream* will be assumed to be visible at the call site, and will be automatically inserted into the actual parameter tuple. The *match* function is then also obliged to name the existential parameter, since in order to apply *currentToken*, there must be a *TokenList* variable in the current scope:

```
function match (exists TokenList, MatchToken)
    equal (currentToken(), MatchToken)
end match
```

Within a t_2 function, existential parameters are *read-only*, i.e., it is syntactically illegal to assign to them.

The local impurity property is obtained by the fact that if a function f uses a function g where $g \in t_2$, then existential parameters in g can refer only to data local to f, therefore not affecting the referential transparency of f itself.

Existential parameters do not provide the same power as general imports of variables, since the scope of the existential parameter is the scope of application of the function, not the scope of definition. In practice, the notational flexibility of existential parameters comes from groups of functions that share a variable by each naming it as an existential parameter.

At a first glance, existential parameters appear to use dynamic scoping; the techniques differ in that the call-site scoping of existential parameters is resolved statically (this is made explicit in the formal definition in section 4.1). To the functional programmer, call-site scoping is intuitively preferable to the declaration-site scoping of imperative languages, since it in effect provides a syntactic short form for passing state parameters explicitly, as one is typically obliged to do in functional notations.

3.3 t_3: Var Imports

The next level in our hierarchy of non-functional features is the ability of functions to reference and modify non-local data (figure 6). This permits the definition of functions that not only are not referentially transparent, but also have side effects. The traditional mechanism for achieving this feature is to permit functions to import non-local data structures *var*.

The existential parameters introduced in the last section can be extended to provide a facility that is slightly weaker than var imports. Syntactically, what we permit is the assignment to

$$
\begin{aligned}
fpDeclaration &\quad ::= \quad \text{procedure } id \, formalParmList \, pBody \text{ end } id \\
pBody &\quad ::= \quad var Declarations \, fpDeclarations \, statements
\end{aligned}
$$

$$
\begin{aligned}
statement &\quad ::= \quad procedureCall \\
procedureCall &\quad ::= \quad id \, actualParmList
\end{aligned}
$$

Figure 6: *New Constructs Introduced In the t_3 Language*

$$
\begin{aligned}
formalParm &\quad ::= \quad \text{inout } id \\
actualParm &\quad ::= \quad \text{\& } id
\end{aligned}
$$

Figure 7: *New Constructs Introduced In the t_4 Language*

existential parameters within a function. In order to syntactically differentiate which functions are permitted to have side-effects, we choose to call such functions *procedures*, and to restrict the procedure application to locations where assignments are permitted. An example procedure is:

```
procedure push (exists Stack, Val)
    Stack := [Val, Stack];
end push;

var Stack;
...
push (10);
push (12);
...
```

Here the *Stack* parameter is instantiated from the scope of application, and consequently the assignment to *Stack* in the procedure modifies the *Stack* variable visible at the call site.

It is syntactically forbidden to pass a read-only identifier as a writable existential parameter; i.e., a t_2 function with existential parameter q cannot pass on q as a writable t_3 existential parameter.

This solution satisfies the local impurity property, since the procedure can only modify data local to the calling function.

3.4 t_4: Reference Parameters

Up to now, the only way a procedure may return a value is to modify one of its existential parameters. For many applications, this is an unacceptable loss of abstraction. The solution to this is to introduce reference parameters, i.e. parameters for which the user of the procedure is expected to supply a variable rather than a value, and where the procedure is permitted to modify that variable (figure 7). The cost of introducing this feature is that a distinction is introduced between references and values, a distinction that does not exist in pure functional languages. We introduce reference parameters using the Ada notion of inout parameters. These type of parameter can be modified within the body of the procedure. It is illegal to pass a read-only identifier as an inout parameter. The following procedure increments its parameter:

```
procedure p (inout X)
    X := plus (X, 1);
end p;

var Y;
Y := 10;
p (&Y);
Y
```

Through the introduction of reference parameters, it becomes possible to introduce aliases. As in Euclid, these aliases are simply banned[12]. This restriction can be efficiently checked[4].

4   Formal Properties of the Language

The table in figure 2 attributes properties to the various constructs in the language: most important of these attributes is that t_0 and t_1 functions are referentially transparent (even if their bodies contain calls to arbitrary procedures), and that t_0, t_1 and t_2 functions are side-effect free.

In order to reason about such qualities of the language, a side-effect semantics for the language is defined. Under this semantics, constructs are mapped to a *view set*, a pair of sets expressing which non-locally declared variables might be viewed or modified by the construct. In terms of this semantics, a precise definition can be given to the *local impurity* property, and the constructs proposed in this paper can be shown to posess it.

In section 4.3, the terms *referentially transparent* and *side effect-free* are precisely defined. As corollaries to the local impurity property, the properties of figure 2 can then be shown.

4.1   Semantic Domains and Semantic Functions

Because of space limitations, we give only an outline of the side-effect semantics. We present the semantic domains and the signatures of those semantic functions that we need to discuss the *local impurity* property. The full semantic specification, especially the omitted definitions of semantic functions, can be found in [9].

The semantics of each expression and statement is defined as a *viewSet*, a pair of sets of locations. The first of these sets is used to represent the non-locally declared variables that the construct potentially *views*, and second the non-locally declared variables the construct potentially *sets*, i.e. modifies. A location is simply a unique integer assigned to each identifier in the program to avoid naming conflicts:

$$viewSet \stackrel{\text{def}}{=} \mathcal{P}(location) \times \mathcal{P}(location)$$
$$location \stackrel{\text{def}}{=} \mathbb{N}_\perp$$

Environments are defined to collect declarations. An environment is represented by a tuple of three functions, the first defined over function and procedure identifiers, and the second and third over non-local and local variables respectively:

$$env \;\overset{\text{def}}{=}\; env_f \times env_g \times env_l$$

$$env_v \;\overset{\text{def}}{=}\; ident \rightarrow location$$
$$env_g \;\overset{\text{def}}{=}\; env_v$$
$$env_l \;\overset{\text{def}}{=}\; env_v$$
$$env_f \;\overset{\text{def}}{=}\; ident \rightarrow fpVal$$

$$ident \;\overset{\text{def}}{=}\; id_\perp$$

The value of a function is a mapping from the recursively defined domain $fpDom$ to a view set. The $fpDom$ domain contains mappings from arbitrarily long sequences of identifiers and a variable environment to a view set. The sequences of identifiers will be required in the definition of procedures with reference parameters.

$$fpVal \;\overset{\text{def}}{=}\; fpDom \rightarrow ViewSet$$
$$fpDom \;\overset{\text{def}}{=}\; env_v + ident \times fpDom$$

There are two main semantic functions, $\mathcal{D}[\cdot]$ for defining declarations, and $\mathcal{E}[\cdot]$ for defining expressions and statements. $\mathcal{D}[\cdot]$ updates an environment to include a new declaration. $\mathcal{E}[\cdot]$ gives the set of non-local variables potentially referenced or modified by the evalution of an expression or statement. For our purposes, we need only the expression form of $\mathcal{E}[\cdot]$:

$$\mathcal{E}[\cdot] \;:\; expression \rightarrow env \rightarrow viewSet$$

To define the local impurity property, we require the semantic function $\mathcal{LI}[\cdot]$ (which also serves as a helper function in the definition of $\mathcal{E}[\cdot]$). $\mathcal{LI}[\cdot]$ is used to define what variables non-local to the call site are potentially viewed or modified when applying a function f (or procedure p).

$$\mathcal{LI}[\cdot] \;:\; functionApplication \rightarrow env \rightarrow viewSet$$
$$\mathcal{LI}[\cdot] \;:\; procedureCall \rightarrow env \rightarrow viewSet$$

4.2 Local Impurity

Through the side-effect semantics of section 4.1, the local impurity property can be defined. This property states intuitively that if a function $f \in t_i$, then f suffers from the effects of t_i only, even if f applies some function $g \in t_j$ for $j > i$. For example, a t_1 function is always referentially transparent, even if it contains a procedure call. Local impurity is formulated by specifying that a function or procedure can only possibly view or modify those non-local variables that are specified in its interface.

In the following definitions, allow F to stand for the function declaration:

$$F = \textbf{function } f\,(\textbf{exists } q_1, \ldots, \textbf{exists } q_m, p_1, \ldots, p_n)\, b\ \textbf{end } f$$

and P for the procedure declaration:

$$P = \textbf{procedure } p\,(\textbf{inout } r_1, \ldots, \textbf{inout } r_k, \textbf{exists } q_1, \ldots, \textbf{exists } q_m, p_1, \ldots, p_n)\, b\ \textbf{end } p$$

The *interface* of a function or procedure is the view set that is obtained from looking only at the formal parameter list. A function interface contains all the existential parameters named in the functions parameter list; a procedure interface also contains the procedure's inout parameters. We define the function

$$\mathcal{I}[\,\cdot\,] : fpDeclaration \rightarrow fpVal$$

to indicate what a function/procedure states in its interface.

Definition 4.1 (Interface) *Given function declaration F (of f), and procedure declaration P (of p), the interface function is defined as:*

$$\mathcal{I}[F] \stackrel{\text{def}}{=} \lambda\mu_v.\langle\{\mu_v q_1,\ldots,\mu_v q_m\},\emptyset\rangle$$
$$\mathcal{I}[P] \stackrel{\text{def}}{=} \lambda\langle r_1',\ldots,r_k',\mu_v\rangle.\langle V,S\rangle$$
$$\text{where } V = \{\mu_v r_1',\ldots,\mu_v r_k',\mu_v q_1,\ldots,\mu_v q_m\}$$
$$\text{and } S = \{\mu_v r_1',\ldots,\mu_v r_k'\}$$

A function is locally impure if it only views/sets the variables specified in its interface:

Definition 4.2 (Local Impurity) *Given function declaration F (of f), and procedure declaration P (of p), then f is locally impure iff for all possible applications $f(a_1,\ldots,a_n)$ of f under environment $\langle\mu_f,\mu_g,\mu_l\rangle$,*

$$\mathcal{LI}[f(a_1,\ldots,a_n)]\langle\mu_f,\mu_g,\mu_l\rangle \subseteq \mathcal{I}[F]\mu_v \text{ where } \mu_v = \mu_g \uplus \mu_l$$

and p is locally impure iff for all possible applications $p(\&s_1,\ldots,\&s_m,a_1,\ldots a_n)$ of p under environment $\langle\mu_f,\mu_g,\mu_l\rangle$,

$$\mathcal{LI}[p(\&s_1,\ldots,\&s_m,a_1,\ldots a_n)]\langle\mu_f,\mu_g,\mu_l\rangle \subseteq \mathcal{I}[P]\langle s_1,\ldots,s_k,\mu_v\rangle \text{ where } \mu_v = \mu_g \uplus \mu_l$$

In this, $\mu_g \uplus \mu_l$ stands for the disjoint union of functions μ_g and μ_l.

Theorem 4.1 (Local Impurity) *Given a well-defined program in the languages t_0 through t_4, all function and procedure declarations within that program are locally impure.*

Proof: By structural induction over the t_4 language. $\square$

4.3 Hierarchy Revisited

The hierarchy of figure 2 indicates that functions defined in the t_0 and t_1 languages are referentially transparent, and that t_2 functions are side-effect free. These properties are direct consequences of the local impurity property.

Referential transparency means that a function neither views nor sets non-local data:

Corollary 4.1 (Referential Transparency) *Given a well-defined program in the languages t_0 through t_4, All t_0 and t_1 functions of that program are referentially transparent. I.e., if f is such a function, then for all applications $f(a_1,\ldots,a_n)$ under environment μ, we have*

$$\mathcal{LI}[f(a_1,\ldots,a_n)]\mu = \langle\emptyset,\emptyset\rangle$$

Side-effect freedom means that a function modifies no non-local data:

Corollary 4.2 (Side-Effect Freedom) *Given a well-defined program in the languages t_0 through t_4, all t_0, t_1 and t_2 functions in the program are side-effect free. I.e., for all such applications $f(a_1, \ldots, a_n)$ under environment μ, and for some set of locations V, we have*

$$\mathcal{E}[f(a_1, \ldots, a_n)]\mu = \langle V, \emptyset \rangle$$

Corollaries 4.1 and 4.2 both follow directly from theorem 4.1 and definition 4.1.

5 Related Work

There have been many ad-hoc approaches to the integration of imperative features into functional languages. For example, the various dialects of Lisp include a variety of features ranging from assignment (*set*), loops (*prog*) and in-place update (*replaca* and *replacd*). Nial[10] and ML[14] both include variables and general assignment; Nial and Hope+C[16] include call facilities to arbitrary routines (with arbitrary side-effects) written in C.

Gifford and Lucassen[8] have identified a hierarchy of imperative constructs similar to the one presented in this paper. The constructs do not have, however, the restrictions given here, and therefore do not necessarily have a local impurity property.

Odersky[15] has investigated an abstract interpretation based technique for determining if a program written in a wide spectrum language is confluent. Odersky's approach has the advantage that more control is available over when updates are performed in-place as opposed to through copying, whereas in the method presented here, procedures must be used to gain such control. The cost of this control is that Odersky's language does not provide simple syntactic rules defining the legality of a program, as are provided by the locally impure constructs presented in this paper.

6 Conclusion

This paper has presented a series of constructs designed to introduce imperative features into functional programs in as non-destructive a manner as possible. The constructs are introduced in a hierarchy, where each level takes us further away from the pure functional ideal. A compiler may choose to optimize these imperative forms using traditional implementation methods.

This work can be extended in both the theoretical and practical directions. An optimizing implementation would be interesting to compare the execution time of programs coded with imperative versus purely functional constructs. The side-effect semantics should be extended to include variable references with index expressions (i.e. the '@' operation from figure 1). This would allow a copying/sharing analysis that would be necessary in an efficient implementation.

Acknowledgements

This work was partially supported by ESPRIT Basic Research Action 3147, the Phoenix Project. Many thanks to Uwe Aßmann, Jim Cordy, Roland Dietrich, Birgit Heinz and Hendrik Lock for careful reading and criticism of various drafts of the paper. Thanks are also due to Jim Cordy for his aid with the TXL language, and to the Nial group at Queen's University, Canada, for their support with the Nial interpreter used to implement the t_0 language.

References

[1] G.L. Burn. Overview of a parallel reduction machine: Project II. In *Proceedings of PARLE 89*, pages 385–396, 1989.

[2] R.M. Burstall and J. Darlington. A transformation system for developing recursive programs. *Journal of the Association of Computing Machinery*, 24(1):44–67, January 1977.

[3] Luca Cardelli. Two-dimensional syntax for functional languages. In *Proceedings of Integrated Interactive Computing Systems*, pages 107–119, 1983.

[4] James R. Cordy. Compile-time detection of aliasing in Euclid programs. *Software Practice and Experience*, 14(8):755–768, August 1984.

[5] James R. Cordy, Charles Halpern, and Eric Promislow. TXL : A rapid prototyping system for programming language dialects. In *IEEE International Conference on Computer Languages*, October 1988.

[6] Gordon V. Cormack and Andrew K. Wright. Type-dependent parameter inference. In *Proceedings of the 1990 SIGPLAN Conference on Design and Implementation of Programming Languages*, 1990.

[7] John Darlington, Peter Harrison, Hessam Khoshnevisan, Lee McLoughlin, Nigel Perry, Helen Pull, Mike Reeve, Keith Sephton, Lyndon While, and Sue Wright. A functional programming environment supporting execution, partial execution and transformation. In *Proceedings of PARLE 89*, pages 286–305, 1989.

[8] David K. Gifford and John M. Lucassen. Integrating functional and imperative programming. In *Proceedings of the 1986 SIGPLAN Conference on Design and Implementation of Programming Languages*, pages 28–38, 1986.

[9] T.C. Nicholas Graham and Gerd Kock. Domesticating imperative constructs for a functional world. Technical report, GMD, 1991.

[10] Michael A. Jenkins, Janice I. Glasgow, and Carl McCrosky. Programming styles in Nial. *IEEE Software*, January 1986.

[11] Simon L. Peyton Jones, Chris Clack, and Jon Salkild. High-performance parallel graph reduction. In *Proceedings of PARLE 89*, pages 193–206, 1989.

[12] B.W. Lampson, J.J. Horning, R.L. London, J.G. Mitchell, and G.L. Popek. Report on the programming language Euclid. *SIGPLAN Notices*, 12(2), 1977.

[13] Carl McCrosky. The elimination of intermediate containers in the evaluation of first-class array expressions. In *IEEE International Conference on Computing Languages*, pages 135–142, 1988.

[14] Robin Milner. The standard ml core language. *Polymorphism*, 2(2):1–28, October 1985.

[15] Martin Odersky. How to make destructive updates less destructive. In *ACM Principles of Programming Languages*, 1990.

[16] Nigel Perry. I/O and inter-language calling for functional languages. In *Proceedings of the Ninth International Conference of the Chilean Computer Science Society and Fifteenth Latin American Conference on Informatics*, July 1989.

[17] Simon Thompson. Functional programming: Executable specifications and program transformation. *Fifth International Workshop on Software Specification and Design*, pages 287–290, May 1989.

[18] Paul Zimmermann and Wolf Zimmermann. The automatic complexity analysis of divide-and-conquer algorithms. Technical Report 1149, INRIA, December 1989.

Logic–Based Specification of Visibility Rules

Arnd Poetzsch–Heffter
Institut für Informatik der TU München
Arcisstrasse 21, D–8000 München 2
poetzsch@lan.informatik.tu-muenchen.de

Abstract

The paper describes a new, declarative method for the formal specification of visibility rules. In contrast to common methods that are based on the specification of a symboltable (or environment), of appropriate update operations and of passing rules, the presented method is related to visibility descriptions in language reports. It consists of three steps: First, the program entities are specified, i.e. the candidates for the meaning of an identifier; the next step defines the ranges where the program entities are valid or hidden; finally, visibility is expressed in terms of these ranges. To formally define the semantics of these visibility specifications, a modeltheoretic view of abstract syntaxtrees is sketched. Using this framework, we give a fixpoint semantics for such specifications.

1 Introduction

Formal specifications play an increasingly important role in the design and definition of programming languages (PL's). This has two main reasons: The PL's become more and more complex, and the requirements concerning portability and standardization increase. A programming language definition usually consists of four more or less seperated parts:

- the lexical syntax (regular expressions)

- the context–free syntax (context–free grammars)

- the context–dependent syntax (attribute grammars or functional specifications)

- the semantics (e.g. denotational semantics)

By context–dependent syntax, we mean the visibility and type rules and similar contextual constraints (cf.[Wat84]). This paper concentrates on specifications of visibility rules. It proposes a new specification technique that allows considerable shorter and better to read formal specifications than e.g. attribute grammars.

1.1 State of the Art

There are two problems with specification techniques for context–dependent syntax:

- How can we get efficient implementations from specifications?

- How can we decrease the size of these often voluminous specifications?

A lot of work has been done to solve the first problem, especially in the field of compiler generation ([Jon80]). Most of the developed methods make use of the same concept: They gather declaration information by constructing an appropriate data structure (symboltable, environment,..) and pass this data structure to the applications of program entities in the syntaxtree. Thus, the visibility definition consists of three parts: The specification of the often very complex data structure, the specification of the update operations, and the passing rules (cf. [Ua82]). Unfortenately, such specifications are hard to read, are a bad basis to prove language properties, and give a rare support for error handling and treatment of program fragments. Recent approaches tried to meet these requirements by using predicate logic or specially designed specification languages. Let us briefly sketch the so far proposed methods:

- S. Reiss (see [Rei83]) presented a specification language for the definition of very elaborate symboltable modules. These modules are developed for visibility rules like those in Ada. The symboltable modules provide functions that have to be used to define the visibility rules in an attribute grammar like framework.

- The work of G. Snelting (see [SH86]) concentrates on type rules and similar constraints and their checking in incremental structure editors. The visibility rules are defined in a specially designed specification language based on a fixed scope and visibility concept like that in PASCAL.

- J. Uhl (see [Uhl86]) and M. Odersky (see [Ode89]) provide a general framework for the specification of context–dependent syntax. As in this paper, they regard a syntax tree of the programming language being specified as a first–order logical structure and define the contextual constraints via first–order formulae. In contrast to our approach, they do not provide special means to define visibility rules. This is less comfortable, leads to less structured specifications, and does not allow special implementation techniques for the identification process, that are indispensable to generate sufficiently efficient context analysers for realistic validation purposes.

1.2 A New Approach to Visibility Specification

This paper presents a new method for the specification of visibility rules. The method is part of a general framework for the specification of context–dependent syntax. It provides a formal logical basis and is related to visibility descriptions in language reports. Visibility specifictions have to answer the following question:

What identifiers are visible at a program point and what is their meaning there?

A specification according to the proposed method, answers this question in three steps:

i) The first step specifies what the meaning of an identifier can be; this is what we call a *program entity*, i.e. a variable, procedure, function, type, label, selector, etc..

ii) The second step specifies the program constructs influencing the visibility, specifies the ranges of this influence, and specifies, how these constructs influence the visibility: A construct can make valid identifier–entity–bindings and/or it can hide such bindings in the specified range.

iii) Finally, visibility is defined in terms of the ranges for validity and hiding.

The main advantage of the method compared to attribute grammar or functional specifications is that symboltable mechanisms — the crucial aspect of those techniques — can be avoided. The method is more flexible than the fixed visibility models of Reiss [Rei83] and Snelting [SH86]. On the other hand, it is sufficiently restricted to get much better implementations than [Ode89].

1.3 Paper Overview

The paper is organized as follows: Section 2 presents the specification method in more detail, describes the underlying model, and explains the visibility clauses, i.e. the construct used for the visibility specification. Section 3 provides the formal semantics for the visibility clauses by mapping them into first-order predicate definitions. As these definitions are recursive, a fixpoint-semantics is given. Section 4 contains a sketch of the whole framework for syntax specification.

2   Specification Method

This section presents the 3-step specification method sketched in the introduction. The aim of such specifications is to define a predicate *is_visible* for given abstract syntax trees (AST). The predicate takes two arguments: A so-called binding consisting of an identifier string ID and a program entity PE (see above), and as second argument a program point PP. It yields true iff the program entity PE is visible under ID at PP. To illustrate this, let us consider the following PASCAL-fragment:

```
(1)        procedure P;
(2)            type T1  =  record ... end;
(3)                T2  =  T1;

(4)            procedure P;
(5)                type T1  =  T2;
(6)            begin  ...  end;

(7)        begin  ...  end;
```

We are interested in the effects of the type declarations: Line (2) introduces a record type under the identifier T1. In line (3), this type additionally gets the name T2. What is the effect of the pathological[1] type renaming in line (5) ? It hides the binding between T1 and the record type of line (2) from the beginning of line (5) up to the end of line (6). And it makes just the same binding valid from the end of line (5) to the end of line (6) (cf. [ANS83]). Before we show how these informal statements can be described more precisely, we have to say some words about the representation of programs.

As we need the program structure for the visibility specification, we use abstract syntax trees to represent programs. To model the relation between the tree nodes, we provide functions like *father, firstson*, etc. and selection by names of nonterminals or terminals.

---

[1] We use a somewhat pathological program to demonstrate some problems with visibility specifications by a tiny example.

Let us for instance consider the AST of the above PASCAL–program (figure 1). If R denotes the root of the tree, $R.DfId.string$ denotes the name of the outermost procedure, i.e. "P". The dotted notation is used throughout this paper as a convenient equivalent to unary function application. So an equivalent notation for $R.DfId.string$ would be $string(DfId(R))$. To express the grammatical properties, we consider the nonterminals as types and provide type predicates of the form $<nonterminal\_name>$ [-] (cf. section 3).

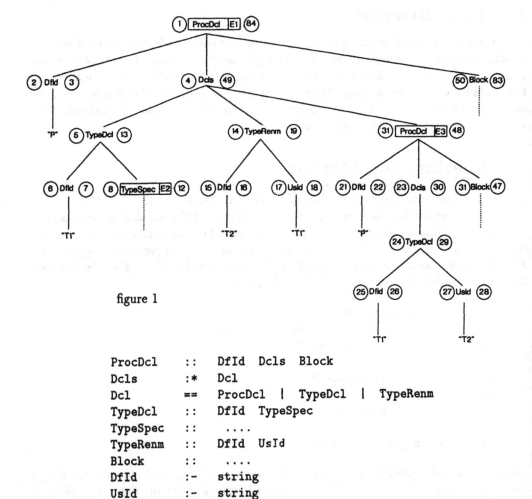

figure 1

```
ProcDcl     ::    DfId   Dcls   Block
Dcls        :*    Dcl
Dcl         ==    ProcDcl  |  TypeDcl  |  TypeRenm
TypeDcl     ::    DfId   TypeSpec
TypeSpec    ::    ....
TypeRenm    ::    DfId   UsId
Block       ::    ....
DfId        :-    string
UsId        :-    string
```

figure 2

2.1 Specification of Program Entities

To keep our example specification small, we consider PASCAL–programs consisting only of parameterless procedure and type declarations as in the example above. The corresponding grammar is given in figure 2. In this PASCAL–subset, we have two kinds of program entities, namely procedures and types. Given an AST, we can represent the program entities by tree nodes. For procedures, we use the corresponding $ProcDecl$-nodes, for types the *TypeSpec*–nodes; i.e. a program entity is a procedure declaration node or type specification node. The entities of the example are surrounded by a box in figure1. In our framework, we express this by the class production

```
ProgEntity  ==    ProcDcl  |  TypeSpec
```

2.2   Specification of Visibility Ranges

Certain constructs in a programming language influence the visibility. These are typically the declarations making the declared entities valid and hiding others. But there are other such constructs as well: e.g. the with– and use–clauses in Ada, renaming constructs, selections, etc. The influence usually ranges over a certain part of the program called a "range". To model ranges, we introduce program points. For each node N of an AST, there are two program points, which we denote by "$before(N)$" and "$after(N)$". The program points are linearly ordered as shown in figure 1 by the numbering. A range is specified by its starting and end point.

As the method should be strong enough to define overloading and renaming, the following situations can occur at a given program point: Several entities are visible under the same identifier; an entity is visible under several identifiers; an entity is not visible. That's why we describe the visibility by bindings between identifiers and program entities. Thereby, the basic idea of the specification method can be put as follows: Define the ranges where bindings are valid (called v-ranges) and where they are hidden (called h-ranges); then combine these ranges to get the points where an binding is visible (s. next section). The distinction between v-ranges and h-ranges is necessary, because in most languages they are different and independent, i.e. it is usually not possible to derive h-ranges from corresponding v-ranges. Let us e.g. consider the visibility in PASCAL. A declared program entity is valid (under its identifier) from the end of its declaration to the end of the directly enclosing block (except for procedures and pointer types), but hides bindings with the same identifier in the entire block (cf. [ANS83]). This has the following consequence for our program fragment: If there was a type visible under the name T2 outside the outermost procedure, it would not be allowed to reference this type in the type specification of line (2), because it is hidden there by the type renaming in line (3).

To specify visibility, we use a special language construct, called "visibility clause". A visibility clause consists of three parts: The first part specifies the program construct influencing the visibility (starting with keyword **vis**); the second part describes the bindings that become valid or are hidden (starting with keywords **valid** or **hidden**); and the third part specifies the range (indicated by the keywords **from** and **to**). E.g. the first visibility clause in figure 3 can be read as follows: A procedure declaration makes valid the binding between the procedure identifier and the procedure itself in the specified range; this range starts after the DfId–node and ends after the procedure node, if the procedure node is the root of the AST, otherwise it ends after the block of the enclosing procedure. The visibility clause for type renamings makes valid the binding between the lefthandside identifier and the type that is visible under the identifier on the righthandside of the equation in the specified range. As the hiding rules are the same for all declarations, we can specifiy them by one visibility clause as shown by the last one of figure 3: A declaration (except the outermost procedure) hides all bindings with the same identifier as the declared identifier in the range extending from before the corresponding declaration list to after the corresponding block.

```
vis     ProcDcl PD {
valid   Binding( PD.DfId.string, PD )
from    after( PD.DfId )
to      if is_root[PD]    then   after(PD)
                          else   after(PD.father.father.Block) fi }


vis     TypeDcl TD {
valid   Binding( TD.DfId.string, TD.TypeSpec )
from    after( TD )          to   after(TD.father.father.Block)  }


vis     TypeRenm TR {
valid   Binding( TR.DfId.string, TS ){
            TypeSpec[TS]
          ∧ is_visible[ Binding(TR.DfId.string,TS), before(TR.UsId) ] }
from    after( TR )          to   after(TR.father.father.Block)  }


vis     Dcl D : ¬ is_root[D] {
hidden  Binding( D.DfId.string, PE )
from    before(D.father)   to   after(D.father.father.Block)  }
```

figure 3

That is all we need to specifiy visibility. Especially, we do not need any symboltable data structure with update routines. The semantics for visibility clauses is given in section 3. Up to now, it should be noticed that the predicate *is_visible* is used in the visibility clauses, although it will be defined by them; i.e. we have a recursive definition.

2.3   Specification of Visibility

Finally, we define visibility in terms of the specified ranges: A binding *BD* is said to be visible at a program point *PP*, if

- there is a range $\langle SP, EP \rangle$ containing *PP*, where *BD* is valid, and

- there is no range containing *PP* that is part of $\langle SP, EP \rangle$, where *BD* is hidden.

This is a more precise formulation of language report statements like "an entity is visible at a given place, if it is valid and not hidden". It covers as well cases in which a binding is made visible in a range where it is hidden. This is illustrated by our example: The type renaming in line (5) hides all bindings with identifier T1 that are defined outside the procedure P in line (4), but not the binding that is made valid by the declaration itself.

Even though the presented specification method is rather simple, it is very powerful: It can capture the different scope rules of block structured languages as well as named scopes and mutual recursive declarations. For example, if we want to allow mutual recursive definitions of functions in a declaration sequence or in a letrec–expression of a functional language, we only have to make valid the bindings for the functions from before the declaration sequence or the letrec–expression respectively. The main advantage of the method is that it leads to smaller and more natural specifications. In an experience with a PASCAL–subset, the specification with our method was about five times smaller than the corresponding attribute grammar, mainly because we need not specify symboltable data structure, update routines, and passing rules, and because we can handle many productions by one visibility clause as shown by the hiding clause in figure 3.

3 Semantics

The given tiny example specification consists of three parts: the abstract syntax, the class production for ProgEntity, and the visibility clauses. The following subsections define the semantics of these parts.

3.1 Syntax Trees as First–Order Structures

A *signature* of a first–order structure consists of two families of finite sets of symbols, the predicate symbols $(PRED_s)_{s \in \mathbb{N}}$ and the fuction symbols $(FUNC_s)_{s \in \mathbb{N}}$. A *first–order structure* S with signature Σ is given by a set U called the universe of S and two families of mappings $(\varphi_s)_{s \in \mathbb{N}}$ and $(\pi_s)_{s \in \mathbb{N}}$,

$$\varphi_s : FUNC_s \to \mathcal{F}(U^s, U) \quad \text{and} \quad \pi_s : PRED_s \to \mathcal{P}(U^s) ,$$

where $\mathcal{F}(U^s, U)$ denotes the functions from U^s to U and $\mathcal{P}(U^s)$ denotes the power–set of U^s. For more details about first–order structures see [End72], p. 79.

The semantics of a grammar as shown in figure 2 is given by a class of first–order structures. The signature of these structures consists of

- a fixed part containing the function symbols *father(_)*, *firstson(_)*, *rightbrother(_)*, *root()*, *after(_)*, *before(_)*, and the predicate symbols *is_root[_]*, *Node[_]*, *Point[_]*, [_≤_];

- a grammar–dependent part with the predicate symbols for the nonterminals, like *ProcDcl[_]*, *Dcl[_]* in our example, and the function symbols to denote son–selection via nonterminal or terminal name, like *DfId(_)*, *TypeSpec(_)*, *string(_)*.

The class of first–order structures for a grammar can be regarded as a representation of the set of abstract syntax trees. Each abstract syntax tree is modelled by one structure. The universe of such a structure is the union of the tree nodes, the terminal values, and the program points, where we have two program points for each tree node, as indicated in figure 1. Additionally, the universe contains an extra element called *undef* to handle partial functions. The interpretation of predicate and function symbols is defined as follows:

- *father(_)*, *firstson(_)*, *rightbrother(_)*, *root()* are interpreted according to the structure of the syntax tree; in cases where their evident meaning is not defined, they yield *undef*;

- if the argument is a node, *after(_)* and *before(_)* yield the program points after and before the node; otherwise they yield *undef*;

- whether an element of the universe is the root, a node, or a program point is expressed by the predicates *is_root*[_], *Node*[_], *Point*[_]; the order on the program points is modelled by the predicate [_≤_];

- whether a node is marked by a certain nonterminal, is expressed by the corresponding predicate; e.g. the predicate *ProcDcl*[_] is exactly true for all ProcDcl–nodes, and the predicate *Dcl*[_] is true, iff a node is a ProcDcl–node, a TypeDcl–node, or a TypeRenm–node;

- the interpretation of the selection functions, like *DfId*(_), *TypeSpec*(_), *string*(_), is as follows: They are only defined for nodes that have exactly one son of the corresponding terminal or nonterminal type; if they are defined, they yield this son, otherwise *undef.*

Thus, the abstract syntax trees with program points are represented by a class of first order structures with the same signature. We call these structures *program models*. The class production for ProgEntity enriches each program model by the predicate *ProgEntity*[_] defined by

$$\forall N : ProgEntity[N] \leftrightarrow ProcDcl[N] \lor TypeSpec[N]$$

Finally, a binding is a pair that has a string as first and a program entity as second component. For bindings, we provide the constructor *Binding*(_,_) and the predicate *is_binding*[_] that test whether an element is a binding. For details, how such classes of enriched program models can be formally defined and implemented see [PH91].

3.2   Semantics for Visibility Clauses

As already mentioned, the visibility clauses enrich each program model by the visibility predicate *is_visible*[_,_]. Their semantics will be given by transforming them into a recursive definition for this predicate. This is done in four steps:

1. Define an auxiliary predicate *is_valid* that corresponds to the visibility clauses with the keyword valid, the so–called *v–clauses*.

2. Define an auxiliary predicate *is_hidden* that corresponds to the visibility clauses with the keyword hidden, the so–called *h–clauses*.

3. Express the predicate *is_visible* in terms of *is_valid* and *is_hidden*.

4. Expand the definition of *is_visible* by the auxiliary definitions.

We demonstrate this transformation by our example. For the three v–clauses, we get the following definition:

$is\_valid[\,BD,\ SP,\ EP\,]$ $\Leftrightarrow_{def}$

$\quad is\_binding[BD] \wedge Point[SP] \wedge Point[EP]$

$\wedge\ (\ (\exists PD:\ ProcDcl[PD]$

$\qquad \wedge\ BD\ =\ Binding(PD.DfId.string, PD)$

$\qquad \wedge\ SP\ =\ after(PD.DfId)$

$\qquad \wedge\ EP\ =\ \textbf{if}\ is\_root[PD]\ \textbf{then}\ after(PD)$

$\qquad\qquad\qquad\qquad\qquad\qquad \textbf{else}\ after(PD.father.father.Block)\ \textbf{fi}$

$\quad)$

$\vee\ (\exists TD:\ TypeDcl[TD]$

$\qquad \wedge\ BD\ =\ Binding(TD.DfId.string, TD.TypeSpec)$

$\qquad \wedge\ SP\ =\ after(TD)\ \wedge\ EP\ =\ after(TD.father.father.Block)$

$\quad)$

$\vee\ (\exists TR:\ TypeRenm[TR]$

$\qquad \wedge\ (\exists TS:\ BD\ =\ Binding(TR.DfId.string, TS)$

$\qquad\qquad \wedge\ TypeSpec[TS]$

$\qquad\qquad \wedge\ is\_visible[\,Binding(TR.DfId.string, TS),\ before(TR.UsId)\,]\,])$

$\qquad \wedge\ SP\ =\ after(TR)\ \wedge\ EP\ =\ after(TR.father.father.Block)$

$\quad)$

$)$

In just the same way, we get the definition for the auxiliary predicate is_hidden :

$is\_hidden[\,BD,\ SP,\ EP\,]$ $\Leftrightarrow_{def}$

$\quad is\_binding[BD] \wedge Point[SP] \wedge Point[EP]$

$\wedge\ \exists D:\ Dcl[D] \wedge \neg\, is\_root[D]$

$\quad \wedge\ (\exists PE:\ BD\ =\ Binding(D.DfId.string, PE)\,)$

$\quad \wedge\ SP\ =\ before(D.father)$

$\quad \wedge\ EP\ =\ after(D.father.father.Block)$

Then, the predicate $is\_visible$ is defined in terms of is_valid and is_hidden following exactly the informal description in section 2.3:

$is\_visible[\,BD,\ PP\,]$ $\Leftrightarrow_{def}$

$\quad is\_binding[BD] \wedge Point[PP]$

$\wedge\ (\exists\, SP,\ EP:\ Point[SP] \wedge Point[EP]$

$\quad \wedge\ SP\ \le\ PP \wedge PP\ \le\ EP$

$\quad \wedge\ is\_valid[\,BD,\ SP,\ EP\,]$

$\quad \wedge\ (\not\exists SPH,\ EPH:\ Point[SPH] \wedge Point[EPH]$

$\qquad\qquad \wedge\ (\,(SP < SPH \wedge EPH \le EP) \vee (SPH \le PP \wedge PP < EPH)\,)$

$\qquad\qquad \wedge\ is\_hidden[\,BD,\ SPH,\ EPH\,]$

$\quad)$

$)$

Finally, we expand the occurrences of *is_valid* and *is_hidden* in the above equivalence. The result is a recursive definition of *is_visible*.

We will give a fixpoint-semantics for such definitions. We call a predicate occurrence in a formula *positive* (*negative*), if there is an even (odd) number of negations on the path from the predicate occurrence to the root in the abstract syntax tree of the formula. If we claim that all occurrences of *is_visible* in the v-clauses are positive and those in the h-clauses are negative, then all occurrences of *is_visible* in the defining equivalence are positive. This restriction is fulfilled by nearly all visibility rules of existing programming languages; a detailed discussion of this aspect and a semantics for visibility clauses violating this restriction can be found in [PH91]. With this restriction, we get the following fixpoint-definition.

Let *PROG* be a program model with universe U and let us denote the righthandside of the defining equivalence for *is_visible* by $\alpha[BD, PP]$ (the visibility clauses must guarantee that $\alpha[BD, PP]$ has no free variables except BD and PP). Considering 2-ary predicates as subsets of U^2, we define a mapping

$$\tau \ : \ \mathcal{P}(U^2) \rightarrow \mathcal{P}(U^2)$$

as follows: Let $Q \subseteq U^2$ and $PROG_Q$ be the enrichment of *PROG* by the predicate *is_visible* such that the interpretation of *is_visible* is given by Q. Then:

$$\tau(Q) \ =_{def} \ \{\, (v, w) \in U^2 \mid \alpha[v, w] \text{ is valid in } PROG_Q \,\} \ .$$

It is not hard to show that the positivity of $\alpha[BD, PP]$ with respect to *is_visible* implies the monotonicity of τ. As $(\mathcal{P}(U^2), \subseteq)$ is a complete lattice, the Knaster-Tarski theorem [Tar55] ensures that τ has a least fixpoint. As this holds for every program model, we can define the semantics of *is_visible* by the least fixpoint of the corresponding τ. (For a more formal treatment of fixpoint definitions in first-order logic and further references to related problems see e.g. [GS86].)

4 Application

As already pointed out, the visibility clauses are only part of a comprehensive method for syntax specification. In this section, we sketch the rest of this method and shortly discuss implementation aspects of visibility clauses.

Comprehensive Specification Framework A specification consists of five parts:

- the specification of the abstract syntax defining the program models;

- the specification of the concrete syntax defining the relation between program texts and program models;

- the visibility specification defining the meaning of used identifier occurrences;

- the type rules;

- further contextual constraints.

A visibility specification itself consists of three parts. The specification of the program entities as shown in section 2.1, the visibility clauses, and the specification of a visibility function *meaning* taking a UsId–node as argument and yielding the corresponding program entity; for languages without overloading, we would have specifications like

```
function  meaning ( UID: UsId ) ProgEntity:
  that  ProgEntity PE: is_visible[ Binding(UID.string, PE), before(UID) ]
```

Implementation Aspects In the discussion of Odersky's approach in section 1.1, we critized specification methods that do not even have implementations for realistic validation purposes. What is the advantage of the presented approach in this respect? The visibility clauses are a specialized specification construct for the definition of visibility rules. They are sufficiently powerful to precisely describe the visibility of common programming languages in a very natural way. On the other hand, the restrictions compared with arbitrary specification in first–order logic (as in [Uhl86] and [Ode89]) permit specialized and therefore more efficient implementations.

The visibility clauses give strong hints how to implement the corresponding part of a context checker:

- generate a matching mechanism that finds the constructs influencing the visibility according to the part after the keyword **vis**;

- provide a general and global datatype that manage the visibility information: For each identifier–entity–binding, we have to know, where it is valid and where it is hidden;

- a global function is then able to extract all program entities that are visible at a given program point.

Even if such generated symboltable mechanisms will probably be less efficient than hand-coded ones, they are certainly much better than general implementations of pure first–order logic. And there is another advantage of the proposed approach. After the identification process is correctly finished, we have a simple and formal representation of the identified abstract syntax tree: The function *meaning* is sufficient to get the declaration information; it can be simply implemented by pointers from the applications to the corresponding declarations.

5   Conclusions

A new, declarative method to formally specify visibility rules of programming languages was developed. This method is related to visibility descriptions in language reports and does not use complex data structures to pass information through the abstract syntax tree. We presented a logic–based fixpoint semantics for such specifications. As a side–effect, the paper reveals the intrinsic recursion hidden in visibility rules.

The presented method is only part of a project for the specification of context-dependent syntax. We view the context–dependent analysis as a partial mapping from abstract syntax trees to syntax DAG's containing arcs from used program entities to their declaration. We described this mapping only for programming languages where the visibility rules are independent of the typing rules. For languages in which overloading resolution depends on user–defined types, we will get mutual recursive definitions for the predicates expressing the visibility and typing. Of course, this is no problem for the presented approach, because we only have to generalize the mapping τ, so that it can handle several predicates.

References

[ANS83] ANSI. *Pascal Computer Programming Language*, ansi/ieee 770 x3.97–1983 edition, 1983.

[End72] H. B. Enderton. *A Mathematical Introduction to Logic*. Academic Press, 1972.

[GS86] Y. Gurevich and S. Shelah. Fixed–point extensions of first–order logic. *Annals of pure and applied logic*, 32, 1986.

[Jon80] N. D. Jones, editor. *Semantics–Directed Compiler Generation*, volume 94 of *Lecture Notes in Computer Science*. Springer Verlag, 1980.

[Ode89] M. Odersky. *A New Approach to Formal Language Definition and its Application to Oberon*. PhD thesis, Swiss Federal Institute of Technology (ETH) Zürich, 1989. Diss. ETH No. 8938.

[PH91] A. Poetzsch-Heffter. *Context-dependent Syntax of programming languages: A New Specification Method and its Application*. PhD thesis, Technische Universität München, 1991. (to appear in german).

[Rei83] S. Reiss. Generation of compiler symbol processing mechanisms from specifications. *ACM Transactions on Programming Languages and Systems*, 5(2), 1983.

[SH86] G. Snelting and W. Henhapl. Unification in many–sorted algebras as a device for incremental semantic analysis. *Conference Record of the Thirteenth ACM Symposium on Principles of Programming Languages*, 1986.

[Tar55] A. Tarski. A lattice–theoretical fixpoint theorem and its application. *Pacific Journal of Mathematics*, 5, 1955.

[Ua82] J. Uhl and andere. An Attribute Grammar for the Semantic Analysis of Ada. *Lecture Notes in Computer Science 139*, 1982.

[Uhl86] J. Uhl. *Spezifikation von Programmiersprachen und Übersetzern*, volume 161 of *GMD-Bericht*. R. Oldenbourg Verlag, 1986.

[Wat84] D. A. Watt. Contextual constraints. In B. Lorho, editor, *Methods and Tools for Compiler Construction*, pages 45–80. Cambridge University Press, 1984.

Optimal Instruction Scheduling using Constraint Logic Programming

M. Anton Ertl

DMS Decision Management Systems Ges.m.b.H.
Wallnerstraße 2, A-1014 Wien
ertl@vip.at

Andreas Krall

Institut für Computersprachen
Technische Universität Wien
Argentinierstraße 8, A-1040 Wien
andi@mips.complang.tuwien.ac.at

Abstract

Instruction scheduling is essential for the efficient operation of today's and to-morrow's processors. It can be stated easily and declaratively as a logic program. Consistency techniques embedded in logic programming enable the efficient solution of this problem.

This paper describes an instruction scheduling program for the Motorola 88100 RISC processor, which minimizes the number of pipeline stalls. The scheduler is written in the constraint logic programming language ARISTO and uses a declarative model of the processor to generate an optimal schedule. The model uses lists of domain variables to represent the pipeline stages and describes the dependencies between instructions by constraints in order to ensure correct scheduling. Although optimal instruction scheduling is NP-complete, the scheduler can be applied to real programs because of the speed gained through consistency techniques.

1 Introduction

Current RISC processors achieve their high performance by exploiting parallelism through pipelining and multiple execution units. As a consequence, the results of previous instructions are sometimes not available when the next instruction is executed. If the next instruction needs the result, it has to wait. The problem of arranging the instructions in a way that reduces the number of wait cycles is known as instruction scheduling or instruction reordering. Microcode compaction is a related problem.

Instruction scheduling can have a drastic impact on performance: On the Motorola 88100 one floating point multiplication can be started at every cycle, but the result is only available after six cycles. Even a simple formulation of optimal instruction scheduling is an NP-complete search problem [HG83]. Scheduling is further complicated by the interactions between the execution units. E.g., on the Motorola 88100 only one result at a time can be written back to the register file. Since up to three execution units may want to write a result, the scheduler must also consider the priority scheme implemented in the hardware. Scheduling is even more important for the superscalar and VLIW processors now being developed which can execute multiple instructions per cycle.

The existing algorithms make use of an explicit dependency graph. The scheduler determines the path length, heuristically selects one of the instructions having no predecessor, appends it to the instruction sequence, and removes it from the graph. The usual heuristic procedure chooses the instruction with the longest path length. Hu [Hu61] developed an early algorithm for a similar problem. Hennessy and Gross [HG83] present an algorithm with $O(n^4)$ worst-case complexity for a simple instruction scheduling problem: The results of the instructions are available after a fixed amount of time. Gibbons and Muchnick [GM86] describe an algorithm with $O(n^2)$ worst-case and observed linear complexity, that produces slightly worse schedules. [GH88] and [BEH91] integrate instruction scheduling and register allocation. These algorithms work on basic blocks, whereas Fisher [Fis81] introduces trace scheduling for global microcode compaction. The same technique is used for VLIW machines [CNO+88]. Another technique to achieve better scheduling by transcending basic block boundaries is software pipelining [Lam88], which can also be combined with loop unrolling [LKB91]. A short overview of the field is given in [Kas90, chapter 8.5].

The use of consistency techniques combined with tree searching for solving combinatorial problems has been an Artificial Intelligence research topic for a long time [Wal72, Nud83]. Problems are represented as networks of constraints on variables. The domains of the variables are represented explicitly. Constraints are used actively to remove values from the domains which cannot appear in a solution. In this way the search tree is pruned *a priori*.

Van Hentenryck [VH89, VHD87] integrated consistency techniques in logic programming. He added a new data type, the domain variable. It behaves like an ordinary logic variable except that it can be instantiated only with values from its domain. Constraints are used to reduce these domains and thus the search tree.

For example, given the variables X with the domain $\{1, 2, \ldots 6\}$ and Y with the domain $\{4, 5, \ldots 9\}$, the constraint X #> Y (Declaratively #> means the same as >) immediately reduces the domains to $\{5, 6\}$ and $\{4, 5\}$ respectively. The constraint keeps watching the variables and becomes active again if the domains are further reduced.

Domain variables and constraints combine nicely with the search capabilities inherent in logic programming languages and form an efficient approach to solving combinatorial problems. The resulting language has been used to efficiently solve a wide range of toy and real-world problems, among them scheduling problems [VH89, DSVH90, Ert90]. Programs for these purposes look like generate&test-programs with the tests coming first.

The experiences gained with this method suggest its application for instruction scheduling. Furthermore the use of constraint logic programming eases the integration of other code generation and optimization techniques, e.g. [Gan89].

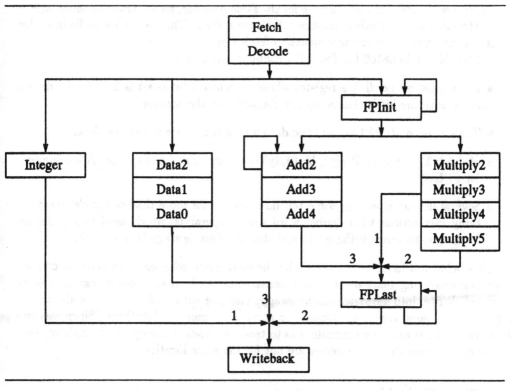

Figure 1: Pipeline stages in the Motorola 88100. Numbers indicate priorities.

2 The Problem

A modern microprocessor consists of several resources, which can be used by only one instruction at a time. These resources include pipeline stages, buses and register file ports. The goal of an instruction scheduler is to achieve high throughput by maximizing resource utilization.

As an example, take a look at the Motorola 88100 processor [Mot90] (see figure 1): It has

- a two-stage instruction fetch and decode unit,

- a one-stage integer execution unit,

- a three-stage data unit for accessing data memory,

- a five-stage floating-point add pipeline and a six-stage floating-point multiply unit, which share the initial and final stages,

- three 32-bit buses, and

- one write port to the register file.

Scoreboarding (an 'in use'-bit for every register) is used to ensure that instructions that access a register are not executed before the register is up to date.

When several instructions compete for the same resource, an arbitration scheme selects one instruction for processing and the others are stalled. This can cause stalls in earlier stages of the pipeline up to the instruction fetch stage.

On the Motorola 88100 the following conflicts can occur:

- An instruction needing a register whose scoreboard bit is set is held in the decode stage until the bit is cleared by a writeback into the register.

- The integer unit, FPLast and the data unit compete for writeback slots.

- Multiply3 (integer multiply), Multiply5 (floating point multiply) and Add4 compete for FPLast.

- Several instructions need some pipeline stage(s) for more than one cycle, most notably instructions with double word source operands, which need two cycles for fetching data through the buses and stall the decode stage for one cycle.

This behaviour must be considered by the instruction scheduler. To preserve correctness, the reordering must obey several constraints: Only reads from a register can be swapped. Writes into a register can be swapped neither with reads nor with writes to this register. The same reasoning applies for accesses to memory locations. Since memory accesses can be aliases, the scheduler has to treat the whole memory like a single register, unless it can prove that the accesses are not for the same location.

3   The Solution

We have written an instruction scheduler for the Motorola 88100 in ARISTO. ARISTO is an industry-level constraint logic programming language, that employs consistency techniques for the solution of combinatorial problems [Ert90]. The scheduler is implemented as a logic program that models the execution structure of the processor.

The scheduling program consists of three parts. The first part reads the assembly language source and splits it into basic blocks[1]. Moreover, simple peephole optimization is performed. The second part works on basic blocks. It collects the constraints (test phase) and searches for an optimal solution (generate phase). This part is displayed in figure 2; The subgoals are explained in the subsequent sections. The final part uses the resulting ordering information and outputs the reordered instructions as assembly language source.

The basic data structure used by the scheduler is the domain variable. For every instruction i in the basic block there is a domain variable D_i representing its decode cycle. These variables are later instantiated by the generate part. In the same way, for every instruction and every pipeline stage it uses, there is a variable representing the cycle during which it resides in the stage. For some stages, where an instruction can stall, there are variable pairs Start .. End. The scheduler works by assigning cycles to the domain variables and thus to the instructions.

---

[1]It is assumed that jump destinations either are defined labels or follow subroutine calls.

```
schedule_block(Instructions, Decode):-
    clear_scoreboard(Scoreboard),
    collect(Instructions, Scoreboard, Decode, FPU, Data, Writeback),
    global_constraints(Decode, FPU, Data, Writeback),
    minimize(Decode, FPU, Data, Writeback).
```

Figure 2: The basic block scheduling predicate takes a list of instructions and returns a list of optimally ordered decode cycles for these instructions.

```
collect(['fadd.sss'(Rd, Rs1, Rs2)|Instructions],
        Scoreboard,
        [D|Decode],
        fpu([S1..E1|FP1], [S2..E2|Add2], [S3..E3|Add3], [S4..E4|Add4],
            Mul2, Mul3, Mul4, Mul5, [S5..E5|FPLast]),
        Data, [W|WriteBack]):-
    read(Rs1, D, Scoreboard),
    read(Rs2, D, Scoreboard),
    write(Rd, W, D, Scoreboard, NewScoreboard),
    S1#=D+1, S2#=E1+1, S3#=E2+1,              %pipeline structure
    S4#=E3+1, S5#=E4+1, W#=E5+1,
    collect(Instructions, NewScoreboard, Decode,
            fpu(FP1, Add2, Add3, Add4, Mul2, Mul3, Mul4, Mul5, FPLast),
            Data, WriteBack).
```

Figure 3: The instruction description for a single-precision floating-point addition

3.1 Collecting the constraints

The constraints stated in the second part model the execution structure of the processor and the dependencies between the instructions.[2]

The constraint collector sequentially processes the instructions of the basic block, generates inequality constraints to enforce correct ordering and collects the domain variables for each pipeline stage into a list. These lists are then used in global constraints like "Only one instruction can be in this stage at a time".

The predicate collect/6 (see figure 3) takes a list of instructions and the current state of the scoreboard and outputs the list of decode cycle variables for these instructions, a structure fpu(FP1,... , FPLast) containing lists of domain variables for every FPU pipeline stage, a similar structure for the data unit (Data), and the list of variables for the writeback stage (WriteBack). In this program the scoreboard is not represented by a bit for every register, but a pair of variables for every register r. One variable (LastWrite$_r$) represents the time when the scoreboard bit for the register was cleared by the completion of the last write. The other variable (NextCheckWrite$_r$) stands for the time when the scoreboard bit will be set by the next write. clear_scoreboard/1

[2]We assume 0 wait cycles for memory accesses.

initializes the scoreboard structure.

The `collect/6` predicates consists of instruction descriptions like the one in figure 3. The `read/3` predicate generates the constraints that force the instruction to be executed after the last write to the register and before the next write to the register, e.g.

LastWrite, #=< D+1, D+1 #=< NextCheckWrite,

where D is the decode cycle of the instruction. Similarly, `write/5` generates

LastWrite, #=< D+1, W #=< NewNextCheckWrite,

and unifies D+1 and W with the corresponding variables in `Scoreboard` and `NewScoreboard`. These constraints are an implicit representation of the dependency graph used by existing algorithms.

`collect/6` also produces pipeline structure constraints that describe the relations of the stages (and the variables) within one instruction. E.g. `S2#=E1+1` means that the instruction enters the Add2 stage one cycle after it leaves FP1.

The other constraints are produced by the predicate `global_constraints/5`.

Only one instruction may be in a stage at a time. This fact is represented by an `alldifferent/1` constraint for each of the variable lists, that are collected for the pipeline stages. `alldifferent(L)` ensures that the variables in L get different values. For lists of start/end pairs an `alldisjoint/1` constraint is used. It ensures that two start/end pairs do not cover the same range.

Branches may only appear as last or second-to-last (delayed branch) instruction. This is enforced by a constraint of the form D_{branch} #>= D_i -1 for every instruction i in the basic block. This (and the `alldifferent/1` constraint on the decode cycles) ensures that only one instruction can be executed after the branch. Nondelayed branches are given a two cycle cost to make delayed branches preferable.

3.2   Searching for an optimal solution

`minimize/4` searches for a solution that respects the constraints and minimizes the basic block execution time.

A solution can be found by instantiating all domain variables with values from their respective domains. Instantiating a variable may cause the execution of constraints, and therefore, failure. On backtracking, another value from the domain has to be chosen. This procedure is called labeling. For a more detailed discussion of the basic generator see [VH89]. The heuristic we used for choosing the next variable to be instantiated is: Choose the one with the smallest domain and the largest number of constraints. No specific heuristic is used to determine what value of the domain is chosen first.

In order to compute the optimal solution, the time when the last instruction leaves the decode stage is minimized. This is done by restricting all decode cycles to be less than the variable `MaxCycle` by the constraint D_i #=<`MaxCycle`. First, `MaxCycle` is instantiated with the number of decode cycles of the basic block, a lower bound. Then, a labeling is tried. If it fails, successively higher values are given to `MaxCycle`. The first solution found is optimal.

```
for (i=1; i<=n; i++) {
  dy[iy] += da*dx[ix];
  ix += incx;
  iy += incy;
}
```

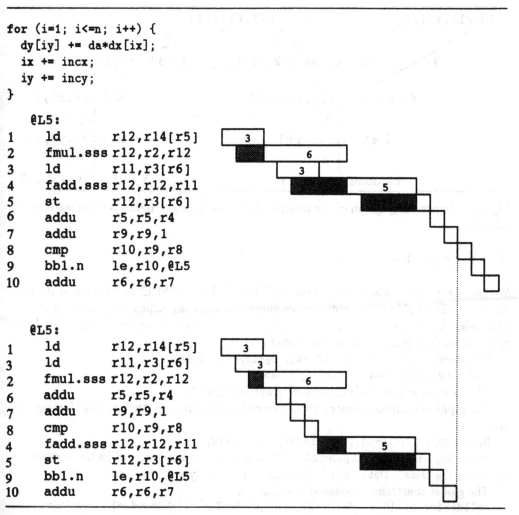

```
   @L5:
1     ld        r12,r14[r5]
2     fmul.sss  r12,r2,r12
3     ld        r11,r3[r6]
4     fadd.sss  r12,r12,r11
5     st        r12,r3[r6]
6     addu      r5,r5,r4
7     addu      r9,r9,1
8     cmp       r10,r9,r8
9     bb1.n     le,r10,@L5
10    addu      r6,r6,r7
```

```
   @L5:
1     ld        r12,r14[r5]
3     ld        r11,r3[r6]
2     fmul.sss  r12,r2,r12
6     addu      r5,r5,r4
7     addu      r9,r9,1
8     cmp       r10,r9,r8
4     fadd.sss  r12,r12,r11
5     st        r12,r3[r6]
9     bb1.n     le,r10,@L5
10    addu      r6,r6,r7
```

Figure 4: An ANSI C version of a Linpack loop, GNU C 1.37 output before and after scheduling (bb1.n is a delayed branch, .sss means single precision, the numbers in the boxes indicate instruction latencies)

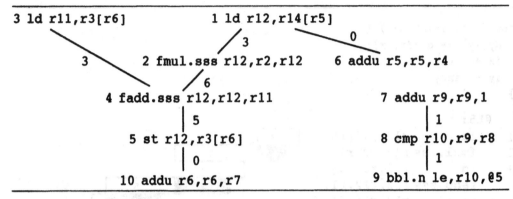

Figure 5: Dependency graph of the program in figure 4; Edge lengths > 0 are instruction latencies.

4 Example

We translated the Linpack loop given in [Mot90][3] into ANSI C and compiled it[4] (see figure 4). The collect/6 predicate produces the following dependency constraints for this code. D_i represents the decode stage for instruction i, W_i represents the writeback cycle (redundant constraints are not shown).

$D_2+1\#>=W_1$, $D_4+1\#>=W_2$, $D_4+1\#>=W_3$, $D_5+1\#>=W_4$, $D_6\#>D_1$,
$D_8+1\#>=W_7$, $D_9+1\#>=W_8$, $D_{10}\#>D_3$, $D_{10}\#>D_5$

This corresponds to the dependency graph in figure 5.

The pipeline structure constraints produced by collect/6 for the first load instruction are:

$D_1+1\#=S2_1$, $E2_1+1\#=S1_1$, $E1_1+1\#=S0_1$, $E0_1+1\#=W_1$

where $S2_1..E2_1$, $S1_1..E1_1$ and $S0_1..E0_1$ represent the timespans when the instruction resides in the Data2, Data1 and Data0 stage respectively.

The global constraints produced look like this:

```
alldifferent([D_1,D_2,D_3,D_4,D_5,D_6,D_7,D_8,D_9,D_10]),   %decode+1
alldifferent([W_1,W_2,W_3,W_4,W_5,W_6,W_7,W_8,W_10]),       %writeback
alldisjoint([S1_2..E1_2,S1_4..E1_4]),                      %FPInit
...                                                         %other stages
```

Finally, the $D_9\#>=D_i-1$ constraints for branch placement are produced.

The scheduler restricts the decode cycle of the last instruction by $D_i\#=<MaxCycle$ constraints.

At this point the constraints have reduced the domains of some variables: they have removed all values < 4 from the domain of D_2 (fmul),... and all values < 16 from D_{10} (addu r6,r6,r7) and therefore all values < 16 from MaxCycle, too.

The scheduler then instantiates MaxCycle to 16. By reducing their domains to a single value the constraints instantiate D_5 (st) and D_9 (bbl) to 15. Since this is incompatible with alldifferent/1, this attempt fails.

The next attempt instantiates MaxCycle to 17. The labeling then selects D_9 (bbl)

[3]The assembly language code given there is scheduled incorrectly.
[4]GNU C 1.37 (gcc -ansi -O)

Program	lines C	lines Assembly	measured speedup	scheduling time
example	19	84	1.17	1.6s
fft	101	288	1.17	7.9s
dhrystone	779	835	1.03	13.7s
WAM	2073	3481	1.06	69.1s
VAM	2647	4436	1.05	91.3s

Table 1: The test programs

for instantiation and tries to instantiate it with 15. This fails, because the st and addu r6,r6,r7 would have to share cycle 16. Therefore D_9 is instantiated to 16; This causes the variables of the instructions on the critical path to be instantiated. Then the remaining instructions are scheduled without backtracking by labeling their variables.

Note that no integer instruction is scheduled for the third cycle. It would cause a collision with the writeback of the first ld and thereby would delay the execution of fadd. Conventional schedulers like the one in the Harris C compiler do not consider this.

5   Results

The instruction scheduler is written declaratively and shares all advantages of logic programming, among them ease and flexibility of programming and short development time, because constraint propagation and tree search programming are abstracted away from the programmer.

The schedules produced are optimal in the sense that there is no basic block that: contains the given instructions, respects the dependencies, and executes in shorter time, when all registers are initially available and there are no memory waits. Hennessy and Gross [HG83] report that their scheduling algorithm removes 85% of the removable stalls on the simpler MIPS processor.

Our main goal was to create a working example, so we did not try to make the program retargetable. However, developing the machine dependent parts for a machine like the Motorola 88100 takes about one person-day, so the retargetability of the scheduler is about as good as that of specialized tools.

We used the scheduler on a few programs: the Linpack loop of section 4, a fast fourier transformation routine, Dhrystone 1.1, and two Prolog abstract machine emulators (WAM and VAM [KN90]) running naïve reverse. All of these programs were compiled with gcc-1.37 -O, scheduled and run on a Data General AViiON 5000 (20 MHz 88100) with 16 MB RAM under DG/UX 4.32. Table 1 gives some information on these programs. The scheduling time includes I/O and instrumentation (mainly computing the old basic block length). We gathered some statistical data on these programs. Figure 6 shows the achieved speedup.

Since most basic blocks produced by GNU C are very short, many of them cannot be improved (81%). For the rest, speedups of up to 1.75 were achieved. The overall static speedup (the ratio of the cumulated old and new basic block durations) is 1.04. A higher

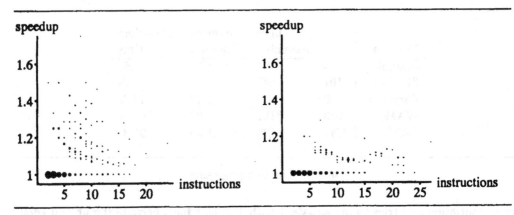

Figure 6: Speedups through optimal scheduling of code produced by GNU and Harris C compilers vs. basic block length. The area of the dots is proportional to the number of basic blocks.

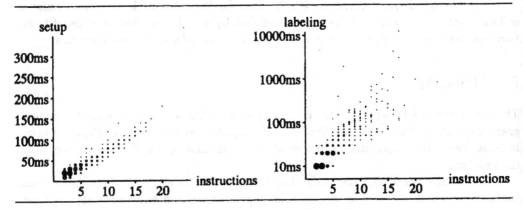

Figure 7: Timing behaviour of the scheduler

proportion of floating-point code would result in a higher speedup.

The Harris compiler includes an instruction scheduler and generally generates longer basic blocks. 8.5% of the basic blocks it produced can be improved by scheduling, the overall static speedup is 1.02. The main cause for suboptimal scheduling in the Harris compiler is writeback collisions.

The running time of the scheduler is acceptable, but should be improved. The current, untuned version takes about twice as long as compilation with gcc. Figure 7 shows the timing behaviour of the scheduler on the GNU C output. Constraint setup time and the number of variables and constraints are linear with the number of instructions, with floating-point code taking about twice as long as integer code. Labeling takes a short time for most basic blocks, but in some cases the NP-completeness of the problem results in exponential behaviour. This happens mainly in longer blocks with few dependencies. In seven cases, the scheduler ran into timeout before finding a solution. In such cases the basic block was divided and the pieces were scheduled. The figures 6 and 7 show statistics on the resulting basic blocks.

6 Further Work

Although long basic blocks were rare in the example programs, the scheduler should handle them in a better way, since it is counterproductive to divide basic blocks produced by an unrolling compiler. This could be achieved by using more restrictive constraints, which can be produced by a better analysis and by using a more problem-specific labeling procedure. Van Hentenryck has solved scheduling problems with 300 tasks [VH89].

Currently the scheduler works on just one basic block at a time. This local view can lead to avoidable pipeline stalls at basic block boundaries. Therefore the scheduler should also consider the adjacent blocks.

Our experience with compiler generated assembly language code has shown that the consideration of scheduling at earlier stages of the compilation process is essential for fast code. In general an optimizing compiler tries to minimize the usage of registers in a basic block. This increases the dependencies between instructions and prevents a good schedule. The declarative nature of the scheduler should make the integration of other parts of a compiler back end, e.g. register allocation, easy.

Acknowledgements

We wish to acknowledge the efforts of several others who contributed to the work in this paper. The anonymous referees, Manfred Brockhaus, Andreas Dieberger, Andreas Falkner, Thomas Graf, Max Hansch, Ulrich Neumerkel and Wolfgang Slany commented on earlier drafts; Paul Beusterien of Harris Corp. compiled our benchmarks with the Harris compiler; Martin Laubach supplied the fast fourier transformation routine.

References

[BEH91]   David G. Bradlee, Susan J. Eggers, and Robert R. Henry. Integrating register allocation and instruction scheduling for RISCs. In *Architectural Support for Programming Languages and Operating Systems*, pages 122–131, 1991.

[CNO+88] Robert P. Colwell, Robert P. Nix, John J. O'Donnel, David B. Papworth, and Paul K. Rodman. A VLIW architecture for a trace scheduling compiler. *IEEE Transactions on Computers*, 37(8):318–328, August 1988.

[DSVH90] Mehmet Dincbas, Helmut Simonis, and Pascal Van Hentenryck. Solving large combinatorial problems in logic programming. *The Journal of Logic Programming*, (8):75–93, 1990.

[Ert90]   M. Anton Ertl. Coroutining und Constraints in der Logik-Programmierung. Master's thesis, Technische Universität Wien, 1990.

[Fis81]   Joseph A. Fischer. Trace scheduling: A technique for global microcode compaction. *IEEE Transactions on Computers*, 30(7):478–490, July 1981.

[Gan89]   Mahadevan Ganapathi. Prolog based retargetable code generation. *Computer Languages*, 14(3):193–204, 1989.

[GH88]   J. R. Goodman and W.-C. Hsu. Code scheduling and register allocation in large basic blocks. In *International Conference on Supercomputing*, 1988.

[GM86]   Phillip B. Gibbons and Steve S. Muchnick. Efficient instruction scheduling for a pipelined architecture. In *Proceedings of the SIGPLAN '86 Symposium on Compiler Construction*, pages 11–16, 1986.

[HG83]   John Hennessy and Thomas Gross. Postpass code optimization of pipeline constraints. *ACM Transactions on Programming Languages and Systems*, 5(3):422–448, July 1983.

[Hu61]   T. C. Hu. Paralell sequencing and assembly line problems. *Operations Research*, 9(6):841–848, 1961.

[Kas90]   Uwe Kastens. *Übersetzerbau*. R. Oldenbourg Verlag, München, 1990.

[KN90]   Andreas Krall and Ulrich Neumerkel. The Vienna Abstract Machine. In P. Deransart and J. Małuzyński, editors, *Programming Language Implementation and Logic Programming (PLILP'90)*, pages 121–136. Springer LNCS 456, 1990.

[Lam88]   Monica Lam. Software pipelining: An effective scheduling technique for VLIW machines. In *Proceedings of the SIGPLAN '88 Conference on Programming Language Design and Implementation*, pages 318–328, 1988.

[LKB91]   Roland L. Lee, Alex Y. Kwok, and Fayé A. Briggs. The floating-point performance of a superscalar SPARC processor. In *Architectural Support for Programming Languages and Operating Systems*, pages 28–37, 1991.

[Mot90]   Motorola, Inc. *MC88100 RISC Microprocessor User's Manual*, second edition edition, 1990.

[Nud83]   Bernard Nudel. Consistent labeling problems and their algorithms: Expected complexities and theory-based heuristics. *Artificial Intelligence*, 21:135–178, 1983.

[VH89]   Pascal Van Hentenryck. *Constraint Satisfaction in Logic Programming*. Logic Programming Series. The MIT Press, Cambridge, Massachusetts, 1989.

[VHD87]   Pascal Van Hentenryck and Mehmet Dincbas. Forward checking in logic programming. In *Logic Programming: Proceedings of the Fourth International Conference*, pages 229–256, 1987.

[Wal72]   D. Waltz. Generating semantic descriptions from drawings of scenes with shadows. Technical report AI271, MIT, 1972.

AN ARCHITECTURAL MODEL FOR OR-PARALLELLISM ON DISTRIBUTED MEMORY SYSTEMS

F. Baiardi , D.M. Di Bella

Dipartimento di Informatica, Università di Pisa
Corso Italia 40, 56100 Pisa - Italy
email : baiardi@dipisa.di.unipi

Abstract

A model for OR-parallel execution of logic programs on highly parallel, distributed memory architectures is proposed. The model aims to reduce the overhead due to a parallel execution by using a parallel decomposition of a WAM into three units devoted to, respectively, memory management, unification, and subtree scheduling. A further unit may be introduced to handle message routing in the case of partial interconnection networks.

The proposed model can be applied independently of the method to handle the multiple bindings for a variable.

After discussing the parallel decomposition of a WAM, the implementation of the model is considered. We show that the implementation mainly consists of the mapping of the units onto the processing elements of the target architecture. Some performance figures of a prototype implementation on a Transputer based system are presented and discussed.

1. INTRODUCTION

Most OR-parallel Prolog implementations recently proposed [1, 11, 24] are based on an abstract shared memory architecture involving a number of WAMs (Warren Abstract Machines [20]) which concurrently visit distinct subtrees of the OR-search tree. Communication and synchronization among WAMs are usually implemented through shared data, i.e. a global stack and some structures for load balancing. These systems also need a method to manage multiple

Supported by "Prog. Finalizzato Informatica e Calcolo Parallelo",CNR, sottoprogetto 3.

bindings in the global environment. Two broad classes can be distinguished: *dangling import methods* [2, 4, 16, 21, 24] and *binding import* methods [1, 9, 11, 21, 22, 23]. In the former class, the overhead for OR parallelism is paid when a variable is accessed, in the latter one, when a processor performs a task switch [7]. The resulting abstract machines map naturally onto shared memory multiprocessors but the bandwidth of the shared memory and that of the processor-memory interconnection structure limit the number of processors and, hence, the maximal degree of parallelism. Since, in the case of OR parallelism, the cooperation among the WAMs mainly requires read accesses to the shared data, the percentage of read-write or write-write conflicts is very low, zero in some cases. Based on this, systems of both classes have been improved through cache memories [8, 15]. This strongly reduces the number of accesses to the shared memory and, in general, the communication overhead. However, through the adoption of cache memories, an increase of less than one order of magnitude in the number of WAMs may be achieved [14].

On the other hand, a distributed memory system offers a larger processor-memory bandwidth [14, 19] and hence it may support a larger degree of parallelism. Among the few OR-parallel models proposed for distributed memory systems one of the most interesting is the Oregon Model on Closed Environments [5], where locality of references is achieved through a *closure* operations on the local environment and a copy of the closed environment at each task switch. In such a way, the overhead for OR parallelism is transferred to task creation.

This work proposes a new architectural model PWIM (Parallel WAM with Intelligent Memories) for a shared memory, OR-parallel interpreter. The goal of the model is to support the development of a parallel interpreter for a highly parallel, distributed memory, MIMD system including a set of processing elements, PEs (CPU+local memory), connected by a partial, point-to-point network. In PWIM, an OR-parallel interpreter is seen as the interconnection of *parallel logical nodes*. A logical node implements a WAM and includes *three parallel units*:
- an unification processor (UP);
- an intelligent memory unit (IMU);
- a scheduling unit (SU).

Such a decomposition of a WAM supports a parallel execution of some activities that usually reduce the performance of an OR-parallel Prolog system. As a matter of fact, the UP visits a subtree of the OR-tree by executing a cycle of goal unifications and it finds any information in the IMU of its logical node. The IMU manages the data structures of a WAM concurrently with the operations of the UP. In this way, the most critical operations for an OR-parallel interpreter, i.e. task creation and deletion, task switching, variable access, have, as far as the UP is concerned, a cost independent of the size of the OR-tree and of the number of processors visiting it [7]. The IMUs globally emulate a shared memory through a highly scalable, distributed memory. Theoretical results show that this approach may achieve good scalability for a fairly wide number of PEs [12, 13, 18]. The SUs cooperate to determine the subtree to be visited by each node. The introduction of a further unit supports the definition of more effective scheduling policies and avoids that the operations of an UP are suspended to supply a task to another node.

In terms of PWIM, the implementation of an OR-parallel interpreter on a highly parallel system mainly consists of a mapping of the logical nodes, and the parallel units within each node, onto the PEs. The optimal mapping depends on the bandwidth of the interconnection network, the amount of memory in each PE, as well as on the processor performance.

Sect. 2 of this paper describes the main features of PWIM. The mapping of an instance of the model onto a physical system is discussed in sect. 3. In particular, a mapping strategy based on excess parallelism is considered [19]. The last section describes a prototype implementation of PWIM on an INMOS ITEM. Performance results are presented.

2. AN ARCHITECTURAL MODEL FOR OR-PARALLELISM

This section introduces the architectural model PWIM. The goals of PWIM are:
a) to support the design of an OR-parallel interpreter able to fully exploit the large degree of parallelism of a distributed memory system;
b) to reduce the influence of the overhead due to non costant-time operations.
According to PWIM, an OR-parallel interpreter includes a set of logical nodes, each emulating a WAM, interconnected by a partial point-to-point network. A logical node is decomposed into three parallel units. After considering the parallel decomposition of a WAM and the advantages it offers, we discuss a modular implementation of the interconnection network.

2.1 Logical Nodes and Intelligent Memories

Most of the data structures of a WAM has been introduced to store the bindings produced by unification. In the same way, the most crucial operations in an OR-parallel implementation are primarily operations on bindings [3, 7, 17]. Variable access (scan of hash windows), task switch (initialization of binding arrays), and also task creation (closure operation) are memory operations on bindings. Because of these operations, in an OR-parallel execution, a WAM has to delay the beginning or suspend the visit of a subtree. Independently of the method to handle multiple bindings, the time required by a memory operation can be reduced by decomposing a

WAM into two parallel units:

i)   an abstract intelligent memory, IMU, managing data structures of a WAM;

ii)   an abstract processor that has a copy of the whole program.

After a subtree switch, the IMU can cooperate with other IMUs to build and manage a local copy of the environment, autonomously from the abstract processor. The abstract processor can start the visit of the subtree immediately. The visit is suspended if and only if the processor requires data that the IMU has not copied yet. Since some of the functionalities previously implemented by the WAM have been assigned to the IMU, the abstract processor is mostly devoted to unification, hence in the following it will be referred to as *unification processor* UP.

Since the cooperation among the IMUs does not involve the UPs, the parallel implementation of a WAM also simplifies some operations of the WAM itself. Moreover, functionalities such as complex term management and garbage collection, that in traditional architectures may force to suspend the visit of the subtree, can be executed by the IMU autonomously and concurrently with the operations of the UP.

2.2 Communication Locality and Scheduling

Even if the introduction of the IMUs supports the parallel execution of the visit of a subtree and of the operations for bindings management, subtree assignment remains a critical issue. The complexity of the assignment is increased because of the partial interconnection structure among nodes. The design of a strategy for subtree scheduling should take into account that:

- several logical nodes should be able to execute a subtree switch in parallel; this implies the definition of distributed scheduling strategies;
- the interconnection structure should not be overloaded; the load on the interconnection structure is proportional to both the *distance* between the logical nodes exchanging data, i.e. the number of links to be crossed, and the amount of data to be exchanged;
- a node should not become idle soon after a subtree switch or after assigning subtrees to other nodes. This is related to "fake" and "useless" OR parallelism. Fake OR parallelism occurs in deterministic computations where no actual parallelism arises, e.g. because of the termination clauses of recursive procedures or of non logical constructs such as cut, and it depends upon the considered program and query [10]. Useless OR parallelism is due to clauses leading to very short computations and it is a dynamic phenomenon, usually appearing in the final steps of a computation.

It is well known that strategies trying to achieve an optimal trade off among all these constraints quickly become sources of large overheads themselves. This problem may be tackled by introducing a *scheduling unit, SU*, into a logical node. The SU is devoted to subtree scheduling and it determines the next subtrees to be visited by examining the choicepoints created in each node, in parallel with the current visit. In the model, the creation and management of choicepoints may be assigned to the SU. In order to gather scheduling information each SU is directly connected to the interconnection structure and also to both the UP and the IMU of its logical node, see Fig. 1.

2.3 The Modular Interconnection Structure

To be able to connect a large number of nodes, the interconnection network has to be defined according to a modular approach. Hence, we introduce into each logical node a communication unit, CU, supporting the communications with other nodes. The communications may be related either to bindings management or to subtree scheduling. Each CU is directly connected to the IMU and the SU of its logical node and to the CUs of other nodes. It routes messages to/from the IMU and SU of the node it belongs to, as well as among logical nodes that are not directly connected. The CU allows a logical node to route messages concurrently with the resolution of a (sub)query. Neither the topology of the interconnection network nor the message routing strategy are defined by the architectural model.

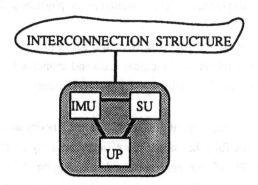

Fig. 1.   The Parallel Decomposition of a WAM: UP is the Unification Processor, IMU is the Memory Module and SU is the Scheduling Unit.

3. IMPLEMENTING THE PWIM MODEL

The implementation of a system based on PWIM onto a distributed memory system includes two main steps:
- the *configuration* of an instance of the model;
- the *mapping* of logical nodes, and units within each node, onto the PEs.

The configuration step, related to the query to be solved, determines the largest degree of parallelism that is exploited and, hence, the number of nodes and their interconnection [6].

The mapping step is more crucial and evaluates the trade off between parallelism and resources utilization. Before briefly reviewing some solutions to the mapping problem, we notice that the mapping of the CUs is strongly related to the capabilities of the physical interconnection network. If the network supports message routing among PEs that are not directly connected, then the CUs can be mapped onto the network itself. Otherwise, the CUs have to be mapped onto the PEs.

A *simple mapping* is an one-to-one mapping that assigns distinct units of a logical node to distinct PEs. If the link bandwidth is large, this strategy can maximize the speed-up with respect to the logical nodes but, since the resources utilization is neglected, the efficiency of the whole system can be very low.

If the overhead due to the dispatching of several units on a single PE is very low, a many-to-one mapping can improve resources utilization and, hence, the efficiency. Since both the amount of memory used and that of data to be transmitted increase, the optimal number of units to be mapped onto a single PE is strongly related to the physical architecture.

A *folding mapping* assigns distinct units of the same logical node to one PE. This solution constraints the parallelism into a logical node and cannot achieve a larger speed-up than a simple mapping, at least if the units are assigned to PEs connected as shown in Fig. 1.

In an *excess parallelism mapping*, units of distinct logical nodes are mapped onto a single PE. Even if the load is not fully balanced, the efficiency surely increases because of the overlapping, into a single PE, of communication and computation. A PE is idle if and only if all the units mapped onto it are waiting to send or receive a message. Moreover, if units with the same functionalities are mapped onto the same PE, some optimizations are possible. As an example, the heap management or the scheduling strategy can be simplified if information is shared among the SUs or the IMUs mapped onto the same PE. As a counterpart, the mapping onto the same PE of units with distinct functionalities can support a better resources utilization without reducing the parallelism.

4. THE OPTIMA PROTOTYPE

This section describes the OPTIMA (OR-Parallel Transputer-based Intelligent Memories

Architecture) prototype, a first implementation based on PWIM. The parallel system used is an INMOS ITEM with 40 PEs, each including an INMOS T414 transputer and 256 kbytes of memory. The PEs are connected according to a folded torus topology as shown in Fig. 2.

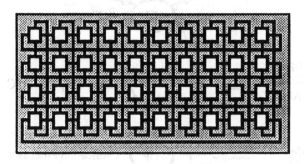

Fig. 2. The Machine Configuration. Each Box is a T414 plus 256 kbyte of Local Memory.

A mapping strategy based on pure excess parallelism has been adopted in all the experiments. A set of logical nodes has been mapped onto four processing elements, i.e. a column in Fig. 2. The CUs are mapped onto the first element of a column, the IMUs on the second one, the UPs on the third and the SUs on the last one, see Fig. 3. Configurations with a different number of logical nodes, and hence different degrees of excess parallelism, have been considered.

4.1 Implementation of the Units

To enhance modularity, each unit of a logical node has been implemented through a set of Occam processes. A binding import strategy has been adopted to manage multiple bindings [23]. This choice is completely arbitrary; any other method can be implemented. The binding array is managed by the IMU together with the extended trail, the binding environment and the heap of complex terms. The trail and the binding array are managed by a process, BaM, which has also to build the environment after a subtree switch. BaM determines the portion of the binding array that has to be copied and supplies the values for the bindings as soon as they are available. Another process, the heap manager, HpM, implements a heterogeneous heap that collects the frames of the shared environment and the complex terms. Two other functionalities have been assigned to HpM: complex terms unification and garbage collection. A further process, SolC, collects the solutions produced during the visit of a subtree.

The process implementing the UP only manages the WAM structures recording the code of the whole program. The UP executes a cycle of unifications and has no visibility of OR parallelism. In the current version, UP is an unoptimized unifier able to interpret all the Prolog built-in predicates but those that update the program.

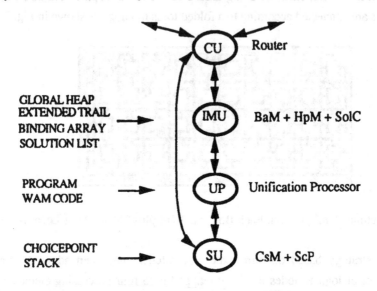

Fig. 3. Implementation and Mapping of the Parallel Units onto the Physical System.

In order to guarantee that an SU owns all the data required to schedule subtrees, it is split into two processes: CsM and ScP. CsM manages a stack of choicepoints and ScP implements the scheduling strategies. In some strategies, ScP manages a private data structure recording the global load of the system.

4.2 Scheduling Strategies

Three scheduling strategies have been tested in the OPTIMA prototype. Each one determines a subtree assignment according to some heuristic strategies. We recall that a task corresponds to a subtree of the OR-tree and the subtrees to be scheduled are recorded as entries in the choicepoint stack of each logical node. The following heuristics are used:

i) the deeper subtree in the choicepoint stack is assigned to an idle node; this increases the probability that a larger subtree is assigned;

ii) to reduce useless parallelism, a lower bound L is established to the number of choicepoints in the choicepoint stack of each logical node. L can be dynamically updated;

iii) an SU searches a task if its choicepoint stack has less than L elements (idle node);

iv) an SU can assign a task to another SU if its choicepoint stack contains more than L elements (busy node). This attempt can be always successful or *abortions* of assignments can be allowed.

No strategy exploits a measure of the distance between current subtree and the next one [10, 11, 16, 22, 24]. Experiments have shown that this measure is not required because of the parallelism between bindings management and OR-tree visit. The schedulers are distinguished according to the methods to collect information on the system load.

In the *structure-independent strategy* , point-to-point messages between any pair of SUs are used to transmit load information.

In the *ring strategy* , a token circulating in a ring is used.

Both these strategies minimize the distance, measured with metrics dependent upon the topology of interconnection structure, between a busy node and an idle one.

The third strategy, the *random* one, does not take into account system load and it assigns randomly a subtree to a node. If the target node is idle then its SU accepts the task, else the SU assigns the task to another, randomly chosen, node.

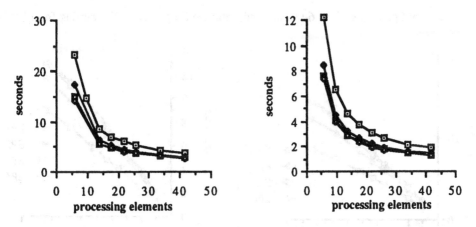

Fig. 4. Execution Times for the 6-Queens and Database Programs; EP ranges from 1 to 4.

4.3 Experimental Results

The evaluations shown in this section may be considered as a worst-case evaluation of the model. First of all, because of the simple interconnection structure among PEs, no message routing is supplied by the network and hence, some PEs have to be devoted to such task. Furthermore, the units of a logical node cannot be mapped to a subset of fully connected PEs. We have chosen to directly connect the UP to both the SU and the IMU, while messages exchanged between the SU and the IMU are routed through the CU of the same node. A PE executing a CU is connected only to other two PEs executing CUs. Moreover, the use of serial links definitely introduces a serious penalty on the actual performance. This does not affect the scalability of the architecture, but certainly makes a difference in its absolute performance.

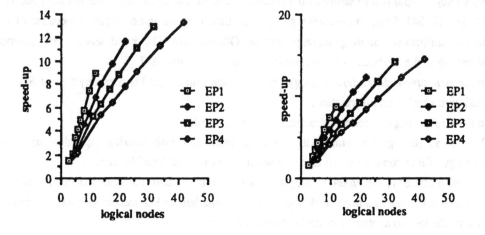

Fig. 5. Logical Speed ups of the 6-Queens and Database Programs for EP varying from 1 to 4.

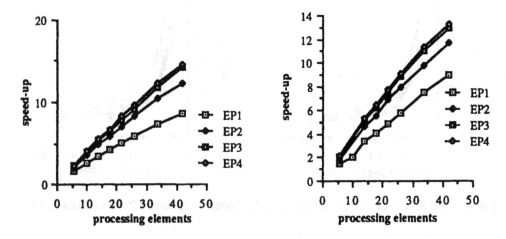

Fig. 6. Speed ups with respect to the PEs for the 6-Queens and Database Programs.

The test problems we have chosen are the 6-queens problem and a query on a database reporting the surfaces and the populations of 25 countries, to find all the pair of countries with a population density different of at most 5%. For both the programs, the time to compute all the solutions is considered. The values shown have been averaged over at least 10 runs.

In all the experiments, the topology independent strategy is the best one, but the performance of the other ones is very similar. We noted that no improvement could be achieved with an excess parallelism degree (EP) larger than 4.

Fig. 4 shows the execution times for the 6-queens and database programs with values of EP ranging from 1 to 4. The corresponding speed ups with respect to the number of logical nodes are shown in Fig.5. In Fig. 6 we plot the speed ups with respect to the number of PEs.

4.4 Future Work

Even if, in the experiments, all the curves show a linear behaviour, the efficiency of the current implementation is not satisfactory. A better mapping strategy is needed to improve the utilization of links, processors and memory without reducing the parallelism among units. Furthermore, the absence of scheduling strategies whose overhead does not depend upon the number of PEs prevents, at this time, the development of a proper highly scalable OR-parallel interpreter. The design of a scheduling strategy for a system with a very large number of PEs has not been considered yet.

As a further step, we plan to include restricted AND-parallelism in the future versions of the model and of the prototype [6, 24].

References

[1] K.A.M. Ali, R. Karlsson, *The MUSE OR-Parallel Prolog Model and its Performance*, Proc. of the 1990 NACLP, pp. 757-776, 1990.

[2] P. Borgwardt, *Parallel Prolog Using Stack Segments on Shared Memory Multiprocessors*, Proc. of the 1984 Int. Symposium on Logic Programming, IEEE, Atlantic City, pp. 2-11, 1984.

[3] J. Chassin, U.C. Baron, W. Rapp, M. Ratcliff, *Performance Analysis of a Parallel Prolog: a correlated approach*, PARLE '89, vol. II, pp. 151-164, Eindhoven, 1989.

[4] A. Ciepielewsky, S. Haridi, *A Formal Model for OR-parallel Execution of Logic Programs*, Proc. of IFIP 83, Mason ed., North Holland, 1983.

[5] J.S. Conery, *Binding Environments for Parallel Logic Programs in Non-Shared Memory Multiprocessors*, Proc. of the Int. Conference on Parallel Processing, IEEE, pp. 457-467, 1987.

[6] B.S. Fagin, A.M. Despain, *The Performance of Parallel Prolog Programs*, IEEE Transaction on Computers, vol. 39, No. 12, pp. 1434-1445, 1990.

[7] G. Goupta, B. Jayaraman, *On Criteria for Or-Parallel Execution Models of Logic Programs*, Proc. of the 1990 NACLP, pp. 737-756, 1990.

[8] S. Haridi, E. Hagersten, *The Cache Coherence Protocol of the Data Diffusion Machine*,

PARLE '89, vol. I, pp. 1-18, Eindhoven, 1989.

[9] B. Hausman, A. Ciepielewsky, S. Haridi, *OR-parallel Prolog Made Efficient on Shared Memory Multiprocessors*, Proc. of the Int. Conference on Parallel Processing, pp. 69-79, 1987.

[10] B. Hausman, *Pruning and Scheduling Speculative Work in OR-Parallel Prolog*, PARLE '89, vol. II, pp. 133-150, Eindhoven, 1989.

[11] Lusk et al., *The Aurora OR-Parallel Prolog System*, New Generation Computing, 7, 243-271, 1990.

[12] S. Peyton Jones, C. Clack, J. Salkild, M. Hardie, *GRIP: a High Performance Architecture for Parallel Graph Reduction*, Functional Programming Languages and Computer Architecture, LNCS 274, September 1987.

[13] A.G. Ranade, *How to Emulate Shared Memory*, Proc. of 28th IEEE Symposium on Fundations of Computer Science, pp. 185-194, 1987.

[14] C.L. Seitz, *Concurrent VLSI Architectures*, IEEE Trans. on Computers, C33, 12, 1984.

[15] E. Tick, *Memory Performance of Prolog Architectures*, Kluwer Academic, 1987.

[16] P. Tinker, G. Lindstrom, *A Performance Oriented Design for OR-parallel Logic Programming*, Proc. of the Int. Conference on Logic Programming, Melbourne, 1987.

[17] H. Touati, A. Despain, *An Empirical Study of the Warren Abstract Machine*, Proc. of the Int. Conference on Parallel Processing, IEEE, 1987.

[18] E. Upfal, *Efficient Schemes for Parallel Communication*, Journal of ACM, vol. 31, No. 3, pp. 507-517, 1984.

[19] L.G. Valiant, *General Purpose Parallel Architecture*, Technical Report TR-07-89, Harward University, Cambridge, 1989.

[20] D.H.D. Warren, *An Abstract Prolog Instruction Set*, Technical Report 309, SRI International, AI Center, 1983.

[21] D.H.D. Warren, *OR-parallel Execution Models of Prolog*, Proc. of the Int. Joint Conference on Theory and Practice of Software Development, Pisa, pp. 243-259, 1987.

[22] D.H.D. Warren, *The SRI Model of OR-parallel Execution of Prolog. Abstract Design and Implementation Issues*, Proc. of the Int. Conference on Parallel Processing, pp. 92-102, 1987.

[23] D.S. Warren, *Efficient Memory Management for Flexible Control Strategies*, Proc. of the Int. Conference on Parallel Processing, IEEE, pp. 198-202, 1984.

[24] H. Westphal, P. Robert, J. Chassin, J.Syre, *The PEPSys Model Combining Backtracking, AND- and OR-parallelism*, Proc. of the Int. Conference on Parallel Processing, IEEE, pp. 436-448, 1987.

FRATS: A parallel reduction strategy for shared memory

K.G. Langendoen   W.G. Vree

University of Amsterdam,

Kruislaan 403, 1098 SJ Amsterdam, The Netherlands

e-mail: koen@fwi.uva.nl

Abstract

FRATS is a strategy for parallel execution of functional languages on shared memory multiprocessors. It provides fork-join parallelism through the explicit usage of an annotation to (recursively) spark a set of parallel tasks. These tasks are executed by ordinary sequential graph reducers which share the program graph. FRATS avoids the consistency problem of graph reducers updating shared nodes by a special evaluation order: Before sparking a set of tasks, all (sub) redexes in those tasks are reduced to normal forms. Then the tasks can proceed in parallel without any synchronisation (e.g., locks) because tasks only share normalised graph nodes. The eager evaluation of shared redexes, however, does not preserve full laziness which might result in superfluous or, worse, infinite computation. The paper presents in detail program transformations to enforce termination and avoid superfluous computation. Analysis of a benchmark of parallel applications shows that these transformations are necessary and effective with negligible costs. Sometimes they even increase performance.

1 Introduction

Functional languages have been suggested as ideal candidates for programming parallel machines because of their referential transparency. Despite considerable effort, however, the performance of functional language implementations can not compare with their imperative competitors. The disappointing performance is mainly caused by the manipulation of computation graphs which are needed to support two other important properties of most functional languages: higher order functions and laziness. Laziness is the major source of inefficiency because it introduces delayed computations which will be stored in, and fetched from a heap during execution. It is impossible to use a stack because it is unknown at compile time when a delayed computation will be evaluated. References to delayed computations (i.e. pointers to graph nodes) can be duplicated and stored in the heap as part of other delayed computations. This results in a graph of delayed computations, and forces the usage of destructive updates in the graph to share the evaluation of a delayed computation with all references to it. As a side effect of these updates, the graph reduction method to implement functional languages is itself not referentially transparent. Therefore it is no surprise that shared delayed computations complicate parallel graph reduction. The difficulties, however, are quite different for shared memory and local memory architectures, as will be elaborated on in the following two paragraphs:

A number of different designs have been proposed to implement functional languages on shared memory multiprocessors, see for example [Goldberg88, Augustsson89, George89], but the basic execution strategy is the same: all processors perform (sequential) graph reduction on different parts of one single program graph which resides in shared memory. The destructive updates of delayed computations with their result have to be atomic to provide all processors with a consistent view of the graph. Furthermore a graph reducer has to mark a delayed computation when evaluating it to prevent others from evaluating the same computation. The common approach is to extend a graph node with a lock bit to enforce mutual exclusion and a special tag to mark a "computation in progress". To prevent processors from busy waiting on a result, the nodes include a pointer to a waiting list. This enables a processor to suspend the computation, add it to the waiting list, and continue with other work. The locking overhead can be kept small, for example by using test-and-set instructions as demonstrated in [Augustsson89]. It is, however, encountered for every access of a delayed computation even though only a small percentage of the nodes is actually shared [Hartel88]. The lock operations also have a bad effect on cache performance because each node has to be fetched as exclusive data.

Graph reduction on a local memory architecture usually starts by loading the initial program graph into one memory. The associated processor reduces the graph and whenever it detects a computation suitable for parallel evaluation it sends the computation to another processor, which starts a graph reducer that detects some more parallel computations, etc. Sending a computation to a processor essentially copies a piece of graph from one memory to another. A potential danger of shared computations is that a computation is duplicated and evaluated at several processors because of the work distribution. A common solution, see for example [Loogen89, Kingdon89], is to keep the shared computations at home and send global pointers instead; whenever a graph reducer needs the value of such a global pointer it sends a request message to the processor holding the referenced data. The request is not handled directly, but added to the local task-pool to avoid consistency problems in the graph. The performance of this class of distributed graph reduction critically depends on locality in the application: [Loogen89] reports severe performance degradation, even for small programs, when parallel tasks share some frequently accessed expressions. To avoid such problems the design of APERM ([Hertzberger89]) has taken a rather different approach: it copies the tasks in their entirety. Before copying the complete graph to another processor, however, all shared computations contained in it are evaluated to basic values. This clearly prevents multiple evaluations of shared computations and minimises process synchronisation, but at the expense of sacrificing some laziness.

In summary shared redexes complicate parallel graph reduction because, essentially, each shared node represents a synchronisation point for the parallel tasks. In case of shared memory designs this results in lock bits to synchronise the parallel processes, whereas local memory designs use a task pool. To achieve a set of independent tasks, FRATS avoids implicit synchronisation through graph nodes altogether by adopting APERM's idea of eagerly evaluating shared redexes beforehand. Since the APERM reduction strategy was developed for a distributed memory machine, it had to be tailored to work for shared memory as well. For example, APERM does not normalise expressions at function positions to avoid non-termination. This crude solution is tolerable in a distributed memory machine at the expense of superfluous computations, but it is unacceptable for shared memory where this could cause serious inconsistency problems. Therefore FRATS has to eagerly evaluate all shared redexes as described in section 2. As a consequence FRATS is more likely to evaluate a superfluous or non-terminating computation. Section 4 presents program transformations to enforce termination and minimise superfluous computations. Measurements of a benchmark of parallel functional programs are presented in section 5 to quantify the effectiveness of these transformations.

2   Parallel reduction strategy

The design of FRATS aims at implementing functional languages on shared-memory multiprocessors equipped with caches. The performance of these architectures critically depends on the effectiveness of the caches. Task synchronisation has a negative effect because the associated lock operations have to show through to all owners. Therefore FRATS's starting point is to generate a set of tasks which can be executed in parallel with minimal synchronisation; the graph reducers operate on disjoint parts of the shared program graph. The programmer has to use the *sandwich* annotation [Vree90] to express parallelism in an application. This provides FRATS with explicit control of all synchronisation.

$$\text{sandwich F } (\text{G } a_1 \ldots a_m) (\text{H } b_1 \ldots b_n) \ldots$$

Figure 1: Syntax of sandwich annotation.

The sandwich annotation can be put in front of any function application and results in parallel evaluation of the function arguments. To prevent synchronisation between tasks FRATS uses the following parallel reduction strategy:

1. All function bodies G, H, ... and their corresponding arguments $a_1 \ldots a_m$, $b_1 \ldots b_n$, ... are (sequentially) evaluated to normal forms.

2. A set of tasks is sparked to evaluate F's arguments (G $a_1 \ldots a_m$), (H $b_1 \ldots b_n$), ... to normal form in parallel.

3. After termination of all tasks from step 2, the function F is invoked with the computed argument values.

The tasks sparked by FRATS do not need to synchronise on graph nodes because of the eager evaluation in step 1 and explicit synchronisation (i.e. waiting) in step 3. The evaluation in step 1 ensures that the tasks sparked in step 2 do not share any redex. Hence, these tasks cannot modify any part of the graph accessible by others. Even though the tasks modify F's arguments in the graph, no consistency problem arises because F is suspended during their execution. Henceforth, the preprocessing in step 1 is referred to as "squeezing". Note that FRATS differs from APERM in step 1 where it also reduces the function bodies G, H, ... to normal forms, and that it suffices to squeeze all but one task to reduce all shared redexes.

Programs executed under FRATS show typical fork-join parallelism since tasks that execute a sandwich have to wait for the termination of all subtasks. This results in a tree of sleeping interior tasks and independently executing leaf tasks. Since divide-and-conquer applications show similar behaviour, it is rather easy to annotate them with the sandwich. The following example lists the SASL[Turner79]-code of a quicksort function.

```
qsort ()            = ()
qsort (key:lst)     = append (qsort left) (key : qsort right)
                      WHERE  left : right = split key lst

split key ()        = () : ()
split key (n:lst) = n<key  ->  (n:left) : right
                      left : (n:right)
                      WHERE  left : right = split key lst
```

Unfortunately, direct insertion of the sandwich annotation in the definition of qsort does not result in the parallel evaluation of the two recursive calls.

```
qsort (key:lst) = sandwich append (qsort left) (cons key (qsort right))
```

First the FRATS evaluation strategy squeezes all shared redexes out of append's arguments. This results in the evaluation of qsort, left, cons, key, and (qsort right). Hence one recursive call is evaluated before sparking any task. A modification of qsort that lifts the (qsort right) call one level results in a correct parallel version. Now the argument squeeze just evaluates qsort, left, and right, which results in the evaluation of the shared split key lst computation. After the execution of both recursive calls the function join will glue the results together.

```
qsort ()            = ()
qsort (key:lst) = sandwich (join key) (qsort left) (qsort right)
                      WHERE  left : right = split key lst

join key left right = append left (key : right)
```

This last definition is not an efficient parallel program because it will generate a lot of tiny tasks. Further transformation is required to derive a version that guarantees some minimal grainsize. We will not present the final grainsize controlled version, but rather refer to [Vree90] for details.

3   Benefits

The FRATS reduction strategy has some benefits for parallel graph reduction on shared memory multiprocessors compared to the more traditional designs. Firstly, the "squeeze" removes all shared updatable nodes from the computation graph, which allows parallel tasks to freely access graph nodes. The traditional designs have to resort to locks (and waiting lists) to enforce mutual exclusive node access in order to guarantee graph consistency. Besides the overhead of additional instructions, locking has a negative impact on the overall performance when the multiprocessor is equipped with caches: each lock operation has to show through to all other caches (e.g., write through to memory). Note that the overhead is usually incurred in vain because few nodes are actually shared.

Secondly, we can exploit FRATS' tree structure of sleeping interior tasks and independent runnable leaf tasks by equipping each task with its own private heap:

- Each leaf task can now perform garbage collection locally because it is impossible for leaf tasks to refer to a graph node in another leaf's heap; since leaf tasks do not share

If infinite is used by a sandwich task then the redex 'rec_fun value' will be evaluated by FRATS. An important observation is that FRATS reduces the arguments of a
task separately. Therefore evaluation of the offending redex can be prevented by placing
the function and value part in different task arguments. Of course the new task has to
restore the original redex by applying the function to its argument value during execution.
This solution requires a modification of the program code, but it suffices to lift the value
part of the redex to "sandwich level" only. The needed program transformation is similar
to lambda lifting:

value-lifting: Let fun be a function definition that contains an infinite computation
denoted by 'rec_fun value'. Take out value as an extra parameter of fun and
replace all occurrences of fun with 'fun value'. Repeat lifting until the value
appears as a task argument inside a sandwich annotation.

In the following example the function bin computes a binomial coefficient and uses the
sandwich annotation to compute three factorials in parallel. The definition of factorial is
based on the equation $fac\ n = 1 * 2 * \ldots * n$; it takes a list of the first n natural numbers
and uses the higher order function product to multiply them.

```
from n    = n : from (n+1)
product  = foldr times 1
naturals = from 1
fac n    = product (take n naturals)
bin n p  = sandwich form (fac n) (fac p) (fac (n-p))
             WHERE  form fn fp fn_p = fn / (fp * fn_p)
```

Direct execution under FRATS would result in a non-terminating squeeze of fac since
it uses the infinite list of natural numbers (naturals). The infinite list is generated by
the from 1 redex, and according to the value-lifting transformation the value 1 is lifted
through naturals and fac inside the sandwich annotation:

```
naturals v = from v
fac v n    = product (take n (naturals v))
bin n p    = sandwich form (fac 1 n) (fac 1 p) (fac 1 (n-p))
             WHERE  form fn fp fn_p = fn / (fp * fn_p)
```

The transformed program can be safely executed since the infinite list (from 1) is
now generated as part of the factorial computation. As a consequence (part of) the list
naturals will be computed three times, once for each factorial computation. In essence
the value-lifting transformation provides each independent task with its own set of infinite
datastructures, hence sharing between tasks is impossible. In principle the performance
loss could be severe, but the analysis in section 5 shows that it is negligible for the
benchmark programs.

```
ones = 1 : ones
```

It is not always necessary to perform value-lifting on recursive data: repeating patterns
like ones above are compiled to cycles in the program graph. These cycles won't be
unrolled into infinite lists because FRATS records which nodes have already been visited
during a squeeze. This also prevents multiple scans of shared data between tasks. After
the squeeze the graph cycle does not contain any redexes and can be safely shared between
parallel tasks.

redexes, it is impossible that a task updates a shared graph node with a pointer into its private heap. Tasks do, of course, refer to nodes in their ancestors heaps, but since these are sleeping no problems occur. The advantage of local garbage collection is that tasks will never be interrupted during execution. In case of global garbage collection all processors have to participate in a joined effort to reclaim unused graph nodes, which requires strict synchronisation of all processors.

- There is no need for keeping the cached graph nodes consistent between all caches in the multiprocessor since leaf tasks do not share data in their private heaps. Only when a task sparks a set of child tasks, its graph nodes have to be exported to the "universe". This could be accomplished by a flush to memory, or in case of the MC88200 cache controller by only setting the "global bit" in the page tables. Both schemes have the performance advantage that ordinary graph accesses in the private heaps do not affect neighbouring caches in the multiprocessor (i.e. no cache cycles lost because of snooping). In the traditional shared memory designs, on the contrary, each access of a graph node results in synchronisation of all caches.

4 Transformations

The reduction of a sandwich annotated function application as described in section 2 puts some extra demands on the program compared to ordinary lazy reduction. The functions F and G, H, ... from the sandwich definition in Figure 1 have to be "extremely" strict in all their arguments. It is not enough to demand strictness, in the usual sense of needing a head normal form, since FRATS will completely evaluate those arguments to normal forms to squeeze out all shared redexes. If the sandwich annotation is used with a non-strict function then evaluation under FRATS can result in evaluation of unneeded expressions. In the worst case FRATS fails to terminate because it enters an infinite computation. This section presents program transformations to overcome these pitfalls of the FRATS reduction strategy.

In functional programs datastructures like lists are often used as glue between modules that result from functional decomposition. Two modules connected in a producer-consumer relation can communicate via an infinite datastructure because of lazy evaluation semantics. Such an infinite producer causes problems when it is present in a task argument without the consumer as in the following example:

```
sandwich join (consumer infinite_producer) ....
```

FRATS starts to completely evaluate the infinite datastructure to squeeze out all shared redexes, but will fail to terminate since it runs out of memory. Fortunately a mechanical transformation suffices to change a non-terminating program under FRATS into a terminating one. Without loss of generality we may assume that infinite computations/datastructures are defined by the application of a recursive function to one single data value.

```
infinite    = rec_fun value
rec_fun par = ... rec_fun ...
```

4.1   Curried functions

The usage of curried functions complicates the recognition of infinite datastructures in the program source because they can generate such expressions at runtime.

```
range a b = take (b-a+1) (from a)
trouble p = ... sandwich foo (bar ... (range p) ...) ...
```

Figure 2: Producer-consumer version of range.

As an example the function range of Figure 2 returns the list a : ... : b by taking a prefix of the infinite list a : a+1 : ... as generated by (from a). The definition of range can be seen as two processes connected through a list: a producer part (from a) and a consumer part (take (b-a+1)). Evaluation of the term trouble 13 results in FRATS evaluating the curried function range 13 to normal form. In a fully lazy implementation this leads to the instantiation of the producer (from 13) because it only depends on the first parameter (a) of range. The consumer part, of course, can not proceed without the second parameter (b). Hence, FRATS will continue to completely evaluate the infinite list 13 : 14 : ....

Curried functions themselves can unfold into infinite datastructures as will be shown by the following example. The producer and consumer of range in Figure 2 have been merged into one definition:

```
range a b = a>b  ->  ()
            a : range (a+1) b
```

Although it looks as if range does not contain an infinite producer, the term range 13 still represents an infinite datastructure. Note that the subexpression range (a+1) does not depend on parameter b and therefore can be evaluated as soon as parameter a is present. This is made explicit by performing fully lazy lambda-lifting [Hughes82] on the range definition, which results in:

```
range a           = range0 a (range (a+1))
range0 a next b = a>b  ->  ()
                  a : next b
```

Squeezing all redexes out of the expression range 13 results in an infinite chain of curried functions (range0):

```
range 13 = range0 13 (range 14)
         = range0 13 (range0 14 (range 15))
         = range0 13 (range0 14 (range0 15 (range 16)))
         = ...
```

As with static infinite datastructures we can use the value-lifting transformation to enforce termination by breaking the dynamically generated redex. With the range examples we could lift p out of the redex range p in Figure 2. For curried functions, however, a simpler transformation is possible.

4.1.1 Order Changing

A fundamental observation about the range examples is that the consumer part of the definition (e.g., take (a-b+1)) could not be initiated because it lacked a parameter while the producer of the infinite datastructure did have enough arguments to be evaluated. A simple reversal of the parameters suffices to make the producer dependent on the lacking parameter of the consumer:

```
range a b     = rev_range b a
rev_range b a = take (b-a+1) (from a)
```

Now it is impossible for the term range 13 to generate the redex (from 13) because it needs the parameter b to "call" function rev_range. This is due to the underlying semantics of functional languages. Again sharing is lost, but this transformation does not suffer the performance loss of dragging an extra parameter around. To minimise loss of sharing we should not modify the general definition of range but just the calls that cause non-termination of FRATS's squeeze phase. This can easily be accomplished by inserting the higher order function delay at those places in the program source:

```
delay f a b   = converse f b a
converse f b a = f a b
```

The function delay will only call f when all arguments are present. Hence, the usage of delay with the range examples will prevent the squeeze from evaluating range p:

```
trouble p = ... sandwich foo (bar ... (delay range p) ...) ...
```

4.1.2 Cycle naming

As with ordinary datastructures, the squeeze of a curried function can result in an infinite chain of one repeated curried function: a partial application of the same function and arguments. For example, the evaluation of the higher order function map applied to one argument.

```
map f ()     = ()
map f (h:t) = f h : map f t
```

Again we will perform fully lazy lambda lifting for clarity:

```
map f                = map0 f (map f)

map0 f next ()     = ()
map0 f next (h:t) = f h : next t
```

The squeeze of the expression map sqrt will result in an infinite repeating chain:

```
map sqrt = map0 sqrt (map sqrt)
         = map0 sqrt (map0 sqrt (map sqrt))
         = map0 sqrt (map0 sqrt (map0 sqrt (map sqrt)))
         = ...
```

When this chain is represented as a cycle in the graph, FRATS's squeeze does terminate and no program transformation is necessary, just as with the ones example. In general, however, compilers do not generate code to create a cycle, but code to build a fresh node with the same curried application. This requires a program transformation to stop FRATS from endlessly building new partial applications. Fortunately, a little help from the programmer suffices to get the desired cycle in the graph: explicitly naming the cycle through a local function definition suffices. The following definition of map forces our SASL compiler to generate combinator code to construct a cycle at runtime. An extra advantage is that the local function mf has one parameter less than the original map, which results in fewer reduction steps.

```
map f = mf   WHERE  mf ()    = ()
                    mf (h:t) = f h : mf t
```

The class of cyclic unfolding functions is relatively large because in functional programs functions often carry some global state, which rarely changes, around in parameters.

4.2   Transformation methodology

Whenever the FRATS reduction strategy causes problems, either non-termination or superfluous computation, the programmer has to apply one of the transformations described before. Value-lifting is the most general transformation and can always be applied, but it is also the most drastic one because usually a lot of function definitions have to be changed to lift the "value" to sandwich-level. The other two transformations operate on a single function definition, but can only be applied in a limited number of cases. In general the programmer should proceed in the following way:

1   Locate the offending redex (say R).
2   Determine if R is a curried function (say fun val)
   yes   If fun makes a direct call to itself with an unchanged parameter (val) then perform Cycle-naming else perform Order-changing on fun's definition.
   no   Apply Value-lifting on R.

The difficult part is finding the redex R in the first step, the rest can be done automatically by some software tool.

5   Performance consequences

In general FRATS's eager reduction strategy, to squeeze out shared redexes, will result in superfluous computation when a task is not strict in all its arguments, the worst case being that FRATS tries to evaluate an infinite datastructure. This requires a modification of the program to delay the computation by applying a transformation (value-lifting, order-changing, or cycle-naming) from section 4. The same transformations can be applied to prevent superfluous computations. If several tasks share a computation that needs to be transformed, sharing will be lost since the transformation causes each task to compute a private version during execution. To quantify the performance consequences of FRATS, superfluous computation and loss of sharing, we have analysed a benchmark of parallel functional programs. The benchmark is derived from the APERM project and consists of six programs already annotated with the sandwich.

To measure the performance effects of FRATS our SASL implementation was extended

qsort	Quicksort on a list of 1024 elements (sin 1, ..., sin 1024).
fft	Fast fourier transform on 512 points, matrixes are represented as lists of lists.
wang	Wang's algorithm for solving a tri-diagonal system of linear equations. The matrix is divided into fixed blocks and two passes (elimination and fill-in) are performed on the blocks in parallel.
wave	A mathematical model of the tides in the North Sea. Consists of a sequence of iterations that updates matrix parts in parallel.
sched	A program to find the optimal schedule of a set of jobs on a number of processors. Implemented as a parallel tree search algorithm.
range	A program to answer a set of queries on a database that is divided in separate pieces. Each lookup is performed in parallel [Hartel89].

Figure 3: Benchmark programs.

to include a SANDWICH combinator. The sandwich annotation is directly translated into a SANDWICH call which has access to the internal interpreter state. On execution the SANDWICH combinator reduces the task arguments to normal form, simulates the parallel task execution by sequential evaluation, and registers task specific properties like number of reduction steps, size of task graph, etc. The exact behaviour of the SANDWICH combinator has been parameterized to investigate the performance effects of various reduction strategies (e.g., APERM, FRATS).

program	total reduction steps			
	lazy	APERM	FRATS	transf+FRATS
qsort	558387	558429	558408	426320
fft	437197	441275	∞	423674
wang	121273	121524	∞	121166
wave	236362	238637	236603	231630
sched	191934	194773	207455	198706
range	9871499	10640802	∞	7789470

Table 1: Benchmark results

The first run of the benchmark was performed without any squeezing of arguments to measure the pure run length of the programs. The results are listed in the column labeled "lazy" of Table 1. These values will be used as a reference to derive the amount of superfluous computation encountered by the other reduction strategies. The next column labeled "APERM" contains the results of using the APERM reduction strategy. Comparison with the first column shows that only the **range** program incurs non-negligible superfluous computation (8%). The FRATS reduction strategy is more strict than APERM: it also reduces the function bodies G, H, ... in the sandwich definition of Figure 1. This shows in the third column in table 1 where three applications fail to terminate under FRATS. The **sched** program takes considerably more reduction steps, whereas the other two need somewhat fewer steps than under APERM. This last decrease is caused by a small optimization in FRATS that only squeezes $n - 1$ sandwich arguments, whereas APERM processes all arguments.

Next the cycle-naming transformation was applied to each benchmark program. As a result all transformed programs do terminate under FRATS. **Sched** and **range** also needed a value-lifting respectively an order-changing transformation to limit superfluous computations. Especially **range** was sensitive because the queries did not cover the whole database, which would be completely evaluated by FRATS without any transformation. The applied transformations are listed in Table 2, and it shows that only a small number was needed.

program	transformations
qsort	cycle-naming (2 ×)
fft	cycle-naming (2 ×)
wang	cycle-naming (2 ×)
wave	cycle-naming (2 ×)
sched	order-changing (1 ×)
range	cycle-naming (5 ×) + value-lifting (1 ×)

Table 2: Applied transformations

program	speed-up		
	lazy	APERM	FRATS
qsort	3.58	3.51	4.81
fft	4.55	3.83	4.09
wang	4.46	4.07	4.14
wave	1.85	1.82	1.84
sched	12.14	9.76	11.23
range	3.91	1.90	1.85

Table 3: Parallel performance compared to column "lazy" in table 1

The final performance of the transformed programs is listed in the last column of Table 1 labeled "transf+FRATS". A remarkable observation is that all programs except **sched** require fewer reduction steps than the original version. This is due to the cycle-naming transformation which uses a local function with one parameter less. The compilation of SASL (bracket abstraction, see [Turner79]) is very sensitive to the number of parameters: worst case code size is exponentially proportional to the number of parameters. It is also the cause of the decreased performance of sched: order-changing was applied to a function with six parameters whose last one needed to be swapped with the first. This parameter shuffle is completely responsible for the incurred overhead.

Our SASL interpreter generates a task description file. It contains enough information to compute the runtime on an ideal parallel machine (∞ processors, no task set-up time, etc). Table 3 contains the computed parallel speed-ups for the benchmark under various reduction strategies. It shows that FRATS performs slightly better than APERM in most cases and approaches the ideal values in the "lazy"-column. In case of **qsort** FRATS outperforms the original program because of the cycle-naming transformation. The disappointing performance of the **range** application is a simple loss of parallelism caused by the normalisation of the database before the parallel queries. Further research is needed to improve FRATS performance in this case.

6 Conclusions and future work

The FRATS reduction strategy aims at the implementation of functional languages on a shared memory multiprocessor. Compared to other shared-memory designs, FRATS's tasks explicitly synchronise at process level in stead of implicitly through locks on graph nodes. This is accomplished by squeezing all shared redexes out before sparking parallel child tasks and blocking parent tasks until termination of all its offspring. This results in fork-join parallelism with independently executing leaf tasks.

The danger of squeezing the shared redexes out of child tasks is that FRATS evaluates computations not strictly needed by its children. This results in superfluous work or, worse, in non-termination. The transformations presented in section 4 to enforce termination were demonstrated to be both successful and necessary: three out of six benchmark programs did not terminate when executed under FRATS, whereas all transformed programs ran to completion without any significant overhead. The cycle-naming transformation actually improved performance by generating better code: up to 20% decrease in executed reduction steps.

The explicit synchronisation of the FRATS reduction strategy obviates the need for implicit low-level synchronisations, like locking and cache coherence mechanisms, when tasks access nodes in the computation graph. A detailed performance model and/or a real implementation is needed to assess these low-level performance advantages of FRATS.

7 Acknowledgements

We would like to thank Henk Muller, Rutger Hofman, and Pieter Hartel for their comments and advice on a draft version of this paper.

References

[Augustsson89] L. Augustsson and T. Johnsson, *"Parallel Graph Reduction with the $< \nu,G>$-machine"*, Proc. Functional Programming Languages and Computer Architecture 1989, pp 202-213.

[George89] L. George, *"An Abstract Machine for Parallel Graph Reduction"*, Proc. Functional Programming Languages and Computer Architecture 1989, pp 214-229.

[Goldberg88] B. Goldberg, *"Buckwheat: Graph Reduction on a Shared Memory Multiprocessor"*, Proc. ACM Conf. on LISP and Functional Progr. 1988, pp 40-51.

[Hartel88] P.H. Hartel and A.H. Veen, *"Statistics on graph reduction of SASL programs"*, Software practice and experience, Vol 18, no 3, pp 239-253, 1988.

[Hartel89] P.H. Hartel, M.H.M. Smid, L. Torenvliet, and W.G. Vree, *"A parallel functional implementation of range queries"*, Computing Science in the Netherlands, pp 173-189, 1989, eds. P.M.G Apers, D. Bosman, and J. van Leeuwen.

[Hertzberger89] L.O.H. Hertzberger and W.G. Vree. *"A Coarse Grain Parallel Architecture for Functional Languages"*, PARLE '89; Parallel Architectures and Languages Europe, Vol I, pp 269-285, Springer-Verlag LNCS 365, 1989.

[Hughes82] R.J.M. Hughes *"Super combinators - A new implementation method for applicative languages"*, ACM Symp. on Lisp and functional programming, pp 1-10, 1982.

[Loogen89] R. Loogen, H. Kuchen, K. Indermark, and W. Damm, *"Distributed Implementation of Programmed Graph Reduction"*, PARLE '89; Parallel Architectures and Languages Europe, Vol I, pp 136-157, Springer-Verlag LNCS 365, 1989.

[Kingdon89] H. Kingdon, D.R. Lester, and G.L. Burn, *"A Highly Distributed Graph Reducer for a Transputer Network"*, Tech. Rep. 123, CEG Hirst Research Centre, 1989.

[Turner79] D.A. Turner, *"A new implementation technique for applicative languages"*, Software practice and experience, Vol 9, no 1, pp 31-49, 1979.

[Vree90] W.G. Vree, *"Implementation of Parallel Graph Reduction by Explicit Annotation and Program Transformation"*, Mathematical Foundations of Computer Science 1990, LNCS 452, pp 135-151.

Narrowing as an Incremental Constraint Satisfaction Algorithm

María Alpuente [†]   Moreno Falaschi [‡]

1 Introduction

In the last few years several approaches to the integration of logic and equational programming have been developed [3,4,8,9,11,14,20]. One relevant approach [11] defines equational logic programs as logic programs which are augmented by Horn equational theories. These programs admit least model and fixpoint semantics. Interpreted function symbols can appear as arguments of relations and existentially quantified variables can occur as arguments of functions. Function definition and evaluation are thus embedded in a logical framework.

To properly cope with the equational theory, the conventional SLD-resolution mechanism based on a (syntactical) unification algorithm of logic programs has to be modified. The operational semantics of an equational logic language is based on some special form of equational resolution (such as SLDE resolution [7,9,11]), where SLD-resolution is usually kept as the only inference rule but the syntactical unification algorithm is replaced by equational (semantic) unification ($\mathcal{E}$-unification [23]). It is well known that the set of $\mathcal{E}$-unifiers of a pair of terms is only semidecidable, so that the process of semantic unification may run forever even if the equational theory is unconditional and canonical. Moreover, there is in general no single most general $\mathcal{E}$-unifier. Infinitely many incomparable mgu's over $\mathcal{E}$ of a pair of terms may exist. Three approaches are relevant to the problem of computing the set of $\mathcal{E}$-unifiers of two terms, namely flat SLD-resolution [2], complete sets of transformations [12] and paramodulation [5] or some special form of it, such as superposition [4] or narrowing [2,13,14,21]. In [11], a lazy resolution rule is defined in order to overcome the problem of nontermination of the $\mathcal{E}$-unification procedure.

Recently, the logic programming paradigm has been generalized to the framework of Constraint Logic Programming (CLP) [15,16]. CLP is a generic logic programming language scheme which extends pure logic programming to include constraints. Each instance CLP($\mathcal{X}$) of the scheme is a programming language that arises by specifying a structure $\mathcal{X}$ of computation. The structure defines the underlying domain of discourse and the operations and relations on this domain, thus giving a semantic interpretation to it. The implementations of one instance of the CLP scheme can differ for the choice of the specific constraint solvers. The existence of a canonical domain of computation, least and greatest model semantics, the existence of a least and greatest fixpoint semantics and the soundness and completeness results for successful derivations are inherited by any extension which can be formalized as an instance of the scheme. Suitable models which correspond to different observable behaviours have been developed for the CLP scheme [6]. These results apply to each instance of the scheme.

[†]Departamento de Sistemas Informáticos y Computación, Universidad Politécnica de Valencia, Camino de Vera s/n, Apdo. 22012, 46020 Valencia, Spain.
[‡]Dipartimento di Informatica, Università di Pisa, Corso Italia 40, 56125 Pisa, Italy.

In this paper we are concerned with an instance of the CLP scheme specialized in solving equations with respect to a Horn equational theory $\mathcal{E}$ [11]. The intended structure is just given by the finest partition induced by $\mathcal{E}$ on the Herbrand Universe $\mathcal{H}$ over a finite one sorted alphabet Σ. = is the only predicate symbol for constraints (equational constraints) and is interpreted as semantic equality over the domain. In the following, we will refer to such a structure as $\mathcal{H}/\mathcal{E}$.

The advantages of this approach are that, since the language is an instance of the scheme, all the above mentioned fundamental properties are automatically inherited by it. Besides, if an efficient algorithm to solve the constraints is developed it can easily be embedded into a general CLP interpreter and can cooperate with other constraint solvers.

Let us notice that for the language to be formally based on the semantics of the scheme, the structures to be considered in CLP must be *solution compact* as defined in [15,16]. Informally, solution compactness means that every element of the domain can be described by a constraint and that the language of constraints must be precise enough to distinguish any object which does not satisfy a given constraint from those ones which do. As pointed in [16], any structure which has no limit elements is trivially *solution compact*. This includes, in particular, the structure $\mathcal{H}/\mathcal{E}$.

A narrowing algorithm or some other $\mathcal{E}$-*unification* procedure can be considered the kernel of the constraint solver which semidecides the solvability of the constraints in the structure $\mathcal{H}/\mathcal{E}$. Solvability has to be tested but the equations do not need to be completely solved at every computation step.

This work first deals with an abstract description of an *incremental constraint solver* for equational logic programming which relies on a narrowing calculus. Our constraint solver not only checks the solvability but also simplifies the constraints. Then we describe a calculus for a narrowing procedure which is heuristically guided from the discarded substitutions, while looking for solutions to new constraints. We discuss the following issues:

How to verify the solvability of constraints in the structure $\mathcal{H}/\mathcal{E}$ by using some sound and complete semantic unification procedure, such as narrowing. How to simplify constraints in a computation sequence. How to achieve incrementality in the computation process. How to profit from finitely failed derivations as a heuristic for optimizing the algorithms to achieve an intelligent narrowing.

The paper is organized as follows. In Section 2 we briefly recall the basic concepts of the CLP framework, conditional rewrite systems and universal unification. In Section 3 we define CLP($\mathcal{H}/\mathcal{E}$) logic programs and an incremental constraint solver as a kernel of an operational semantics for them. Section 4 is devoted to the heuristic narrowing calculus. Finally Section 5 concludes.

All the proofs of the lemmata and theorems presented in this paper can be found in [1].

2 Preliminaries

In this paper we refer to a language which is an instance of CLP, as defined in [15]. We assume the reader to be familiar with logic programming [19], constraint logic programming [6,15,16], equations and conditional rewrite systems [18] and universal unification [23]. We first recall the basic concepts of the CLP framework. By Σ, Π and V (possibly subscripted) we denote denumerable collections of function symbols, predicate symbols and variable symbols with their signatures. We assume that each sort is non-empty. $\tau(\Sigma \cup V)$ and $\tau(\Sigma)$ denote the sets of terms and ground terms built on Σ and V, respectively. $\tau(\Sigma)$ is usually called the Herbrand Universe ($\mathcal{H}$) over Σ. A (Π, Σ)-atom is an element $p(t_1, \ldots, t_n)$ where $p \in \Pi$ is n-ary and $t_i \in \tau(\Sigma \cup V)$, $i = 1, \ldots, n$. A (Π, Σ)-constraint is a possibly empty or infinite set of (Π, Σ)-atoms. Intuitively, a constraint is a conjunction of (Π, Σ)-atoms. The empty constraint will be denoted by *true*. The symbol $\tilde{\ }$ will denote a finite sequence of symbols. A structure $\Re(\Pi, \Sigma)$ over Π and Σ can be defined as in [15,16].

Definition 2.1 (CLP programs) *[15] Let $\Pi = \Pi_C \cup \Pi_B$ and $\Pi_C \cap \Pi_B = \emptyset$. A (Π, Σ)-program is a set of clauses of the form $H \leftarrow c\square.$ or $H \leftarrow c\square B_1, \ldots, B_n.$ where c is a possibly empty finite (Π_C, Σ)-constraint, H (the head) and $B_1, \ldots, B_n$ (the body), $n \geq 0$, are (Π_B, Σ)-atoms. A goal is a program clause with no head.*

(Π, Σ)-programs, (Π_B, Σ)-atoms and (Π_C, Σ)-constraints will often be called programs, atoms and constraints. The equational representation of a substitution $\theta = \{x_1/t_1, \ldots, x_n/t_n\}$ is the set of equations, $\hat{\theta} = \{x_1 = t_1, \ldots, x_n = t_n\}$. The empty substitution is denoted by ϵ. For any substitution θ and set of variables V, we define θ restricted to V, denoted by $\theta_{|V}$, to be $\{(x/t) \mid (x/t) \in \theta \wedge x \in V\}$. $Var(e)$ is the set of distinct variables occurring in the expression e.

The notions of application, composition and relative generality are defined in the usual way [19], e.g., we say that a substitution θ is more general than γ, denoted by $\theta \leq \gamma$, if there exists a (Π, Σ)-substitution δ such that $\gamma = \theta\delta$. Let c be a set of equations. We denote by $mgu(c)$ the set of most general unifiers of c [19].

In this paper we follow the operational approach to semantics which is based on transition systems following the standard terminology as in Plotkin's approach [22].

Definition 2.2 CLP($\mathcal{X}$) configurations

Let P be a constraint logic program and $G = \leftarrow c\square A_1, \ldots, A_n.$ be a goal over $\mathcal{X} = \Re(\Pi, \Sigma)$. We define a CLP($\mathcal{X}$) configuration C as a pair $C = \langle \leftarrow s[c] \diamond A_1, \ldots, A_n \rangle$ where $s[c]$ denotes a constraint solver state, whose structure is left unspecified as it depends on the specific constraint solver, but which includes at least the constraint c. When the constraint c is clear from the context $s[c]$ will simply be denoted by s.

Definition 2.3 CLP($\mathcal{X}$) transition relation $\rightarrow_{CLP}$

The rule describing a $(P, \Re)$-computation step from a configuration $\langle \leftarrow s_i \diamond A_1, \ldots, A_n \rangle$ where s_i includes c_i is given by

$$\frac{s_i \overset{\tilde{c}}{\hookrightarrow} s_{i+1}}{\langle \leftarrow s_i \diamond A_1, \ldots, A_n \rangle \rightarrow_{CLP} \langle \leftarrow s_{i+1} \diamond \tilde{B}_1, \ldots, \tilde{B}_n \rangle}$$

if there exist n variants of clauses in P, $H_j \leftarrow c'_j \square \tilde{B}_j., j = 1, \ldots, n$, with no variables in common with $\leftarrow c_i \square A_1, \ldots, A_n.$ and with each other and $\tilde{c} = \{c'_1, c'_2, \ldots, c'_n, A_1 = H_1, \ldots, A_n = H_n\}$. The condition $s_i \overset{\tilde{c}}{\hookrightarrow} s_{i+1}$ means that some constraint solver can make a move verifying the $\Re$-solvability of the constraint $c_i \cup \tilde{c}$ and returning the new constraint solver state s_{i+1} which includes (a possibly simplified version of) this constraint, c_{i+1}.

Definition 2.4 CLP($\mathcal{X}$) initial configuration (C_0)

Let $G_0 = \leftarrow c_0 \square \tilde{B}.$ be a goal and s_0 be the 'empty' CLP constraint solver state. If c_0 is not an empty constraint and $s_0 \overset{c_0}{\hookrightarrow} s$ then $C_0 = \langle \leftarrow s \diamond \tilde{B} \rangle$ is the CLP($\mathcal{X}$) initial configuration. If c_0 is empty then $C_0 = \langle \leftarrow s_0 \diamond \tilde{B} \rangle$.

Definition 2.5 CLP($\mathcal{X}$) terminal configurations (C)

A terminal configuration C has the form $C = \langle \leftarrow s[c] \diamond \quad \rangle$, where c represents the answer constraint.

Let us notice that when the constraint solver is designed for simply testing the $\Re(\Pi, \Sigma)$-*Solvability* of the constraint $c_i \cup \tilde{c}$, the above defined transition relation becomes the standard $(P, \Re)$-derivation step as defined in [16]:

$$\frac{\Re(\Pi, \Sigma) \models \exists (c_i \cup \tilde{c})}{\langle \leftarrow c_i \square A_1, \ldots, A_n. \rangle \rightarrow_{CLP} \langle \leftarrow c_i \cup \tilde{c} \ \square \tilde{B}_1, \ldots, \tilde{B}_n. \rangle}$$

In this case, the concepts of $(\Gamma, \Re)$-derivation and successful and finitely failed $(P, \Re)$-derivations can be defined in the usual way [15]. The constraint in the last goal of a successful derivation is the answer constraint of the derivation. Let us next briefly recall some basic notions and results about equations, conditional rewrite systems and universal unification. Full definitions can be found in [18,23].

Definition 2.6 *A Σ-equation $s = t$ is a pair of terms s and t of the same sort $(s, t \in \tau(\Sigma \cup V))$. A Horn equational Σ-theory $\mathcal{E}$ consists of a finite set of equational Horn clauses of the form $l = r \leftarrow e_1, e_2, \ldots, e_n$, $n \geq 0$, where the Σ-equation $l = r$ in the head is implicitly oriented from left to right and the literals e_i in the body are ordinary non-oriented Σ-equations. Σ-equations and Σ-theories will often be called equations and theories, respectively.*

A Horn equational theory $\mathcal{E}$ can be viewed as a term rewriting system $\mathcal{R}$ where the rules are the heads and the conditions are the respective bodies. If all clauses in $\mathcal{E}$ have an empty body then $\mathcal{E}$ and $\mathcal{R}$ are said to be unconditional, otherwise they are said to be conditional. The equational theory $\mathcal{E}$ is said to be canonical if the binary one-step rewriting relation $\rightarrow_{\mathcal{R}}$ defined by $\mathcal{R}$ is noetherian and confluent. For syntactical characterizations of confluent conditional theories refer to [20]. Each Horn equational theory $\mathcal{E}$ generates a smallest congruence relation $=_{\mathcal{E}}$ called $\mathcal{E}$-equality on the set of terms $\tau(\Sigma \cup V)$. We will denote by $\mathcal{H}/\mathcal{E}$ the finest partition $\tau(\Sigma)/=_{\mathcal{E}}$ induced by $\mathcal{E}$ over the set of ground terms $\tau(\Sigma)$.

Definition 2.7 *Given two terms s and t, we say that they are $\mathcal{E}$-unifiable (or $\mathcal{E}$-equal) iff $\exists$ a substitution σ such that $s\sigma$ and $t\sigma$ are in the congruence $=_{\mathcal{E}}$, i.e. such that $s\sigma =_{\mathcal{E}} t\sigma$ or, equivalently[1], $\mathcal{E} \models s\sigma = t\sigma$. The substitution σ is called an $\mathcal{E}$-unifier of s and t.*

Roughly speaking S is a complete set of unifiers for e if any other unifier can be obtained by instance from some element of S (see [23] for a formal definition). Complete sets of minimal $\mathcal{E}$-unifiers do not always exists [23]. $\mathcal{E}$-unification can be viewed as the process of solving an equation within the theory $\mathcal{E}$. Since $\mathcal{E}$-unification is only semidecidable, an $\mathcal{E}$-unification algorithm can be viewed as a semidecision procedure for testing the solvability of equational constraints over the quotient $\mathcal{H}/\mathcal{E}$. Each instance of an $\mathcal{E}$-unifier represents a solution over this structure (an $\mathcal{H}/\mathcal{E}$-solution).

A $\mathcal{E}$-unification procedure is complete if it generates a complete set of $\mathcal{E}$-unifiers for all input equations. A number of $\mathcal{E}$-unification procedures have been developed in order to deal with conditional equational theories [3,4,8,12,14,18]. For instance, conditional narrowing has been shown to be complete for conditional theories satisfying different restrictions [3,11,18].

Definition 2.8 *Let $\mathcal{E}$ be a Horn equational theory. A function symbol $f \in \Sigma$ is called irreducible iff there is no clause $(l = r \leftarrow e_1, e_2, \ldots, e_n) \in \mathcal{E}$ such that f occurs as the outermost function symbol in l or l is a variable; otherwise it is a defined function symbol.*

---

[1]We assume that interpretations obey the axioms of equality, i.e. the reflexivity, symmetry, transitivity and the substitutivity for functions and predicates. Hence, satisfiability and logical consequence are defined w.r.t. these axioms.

The use of irreducible function symbols is justified from the several optimization techniques defined in [14]. This affects the size of the search tree of the $\mathcal{E}$-unifiers for a set of equations. In the best case an infinite search tree can be reduced to a finite one. In theories where the above distinction is made, the signature Σ is partitioned as $\Sigma = C \uplus F$, where C is the set of irreducible function symbols and F is the set of definite function ones. The members of C are also called *constructors*. Terms are viewed as labelled trees in the usual way [13]. We will use, among others, the following standard notions: Occurrences are represented by sequences, possibly empty, of naturals. The empty sequence is represented by λ. t/u is the subterm at the occurrence u of t. $t[u \leftarrow r]$ is the term t with the subterm at the occurrence u replaced with r.

3 The language CLP($\mathcal{H}/\mathcal{E}$) and its Operational Semantics

In this section we define the language CLP($\mathcal{H}/\mathcal{E}$), which is an instance of CLP that caters for solving constraints in the structure $\mathcal{H}/\mathcal{E}$, defined as the finest partition induced by a canonical Horn equational theory without extra variables $\mathcal{E}$ on the Herbrand Universe $\mathcal{H}$ over a finite one sorted alphabet Σ. $=$ is the only predicate symbol for constraints and is interpreted as semantic equality over the domain. This structure is *solution compact* since has no limit element [16]. It is known that the equational theory $\mathcal{E}$ and the structure $\mathcal{H}/\mathcal{E}$ correspond in the sense defined in [16], i.e. $\mathcal{H}/\mathcal{E} \models \mathcal{E}$, i.e. $\mathcal{H}/\mathcal{E}$ models $\mathcal{E}$ and $\mathcal{H}/\mathcal{E} \models \tilde{\exists}c$ implies $\mathcal{E} \models \tilde{\exists}c$ for all constraint c.

We assume that (Π_C, Σ)-constraints are sets of Σ-equations to be solved in $\mathcal{E}$ by a suitable conditional narrowing algorithm which will be described in the following.

Definition 3.1 CLP($\mathcal{H}/\mathcal{E}$) programs
Let $\Pi_C = \{=\}$, $\Pi = \Pi_C \cup \Pi_B$ and $\Pi_C \cap \Pi_B = \emptyset$. We define a CLP($\mathcal{H}/\mathcal{E}$) (Π, Σ)-program as a (Π, Σ)-program augmented by a canonical Horn equational Σ-theory without extra variables $\mathcal{E}$.

3.1 CLP($\mathcal{H}/\mathcal{E}$) Incremental Constraint Solver

The standard way to integrate narrowing and resolution [9] is using narrowing to generate the solutions which are then tested by the logic program. This method does not fit in the CLP scheme and requires some complex interactions between backtracking of the narrowing and resolution algorithms. Instead, narrowing can be used as a procedure to test the consistency of the new constraint, looking for a solution, since the constraint is proved consistent once a solution has been found. Thus we consider how to reuse incrementally the work done in the previous step to keep smaller the cost of solving the new set. An approach that does not try to exploit the information which can be gathered from the previous steps to guide the search would be impractical [10].

In the CLP($\mathcal{H}/\mathcal{E}$) framework it seems very natural to reuse the solution θ to the constraint c found in the previous step to check the solvability of the new constraint $c \cup \tilde{c}$.

In [10] the matter of incremental constraint satisfaction is also considered. [10] shows that an approach based on backtracking is inadequate since it can only use passively the new constraints to test if they are satisfied by the newly generated tentative solutions. In the context of CLP it seems to be more appropriate using actively the new constraints to guide the search towards a solution. Accordingly, [10] presents a scheme based on reexecution and pruning of the search tree of resolution. We have a similar problem, yet related to the search tree of the constraint solving algorithm. In this section we are concerned with an incremental algorithm for testing the solvability and an appropriate simplified representation for constraints. In the following we describe a constraint solver for CLP($\mathcal{H}/\mathcal{E}$) programs by means of a transition system labeled on (Π_C, Σ)-constraints and specially tailored to be incremental.

Definition 3.2 *ICS-representation of a solvable constraint*

Let c be a solvable constraint. Let us suppose that c can be written in the form $c_1 \cup c_2$ (either of which can be empty) where c_1 has a unique mgu θ over $\mathcal{E}$. Then $r_c = (\hat{\theta}, c_2)$ is the ICS-representation of the simplified constraint $\hat{\theta} \cup c_2$. The ICS-representation of an empty constraint is $(\emptyset, \emptyset)$, which will be abbreviated as $()$.

Let us notice that if $r_c = (\hat{\theta}, \emptyset)$ then the constraint c has a unique mgu θ over $\mathcal{E}$.

Definition 3.3 *ICS-states*

We define an Incremental Constraint Solver (ICS) state s as a triple $s = \langle \theta, f, r_c \rangle$, where r_c is the ICS-representation of a (Π_C, Σ)-constraint c, θ represents a $\mathcal{E}$-unifier of c and f is a set of (Π, Σ)-substitutions.

Roughly speaking, for the Incremental Constraint Solver which will be described next, r_c represents the accumulated simplified constraint and f represents the set of substitutions which have already been unsuccessfully tried by the constraint solver and that are useful for a heuristic search of other solutions.

Definition 3.4 $\langle \epsilon, \emptyset, () \rangle$ *is the empty ICS-state. Any ICS-state can be a terminal ICS-state.*

Definition 3.5 *An ICS label $\tilde{c}$ is a (Π_C, Σ)-constraint. It represents the new constraint $\tilde{c}$ to be added to the accumulated simplified constraint c in order to test the solvability of the new set $c \cup \tilde{c}$.*

In the following we describe the *ICS* transition system. Roughly speaking, let $s = \langle \theta, f, r_c \rangle$ be an *ICS*-state. In order to achieve incrementality, the costs of testing the solvability of the new set of constraints $c \cup \tilde{c}$ should be as close as possible to the cost of solving the new constraint $\tilde{c}$ plus the cost of combining the new solution with the previous one. With this aim, since the substitution θ represents an $\mathcal{E}$-unifier of the accumulated solvable constraint c, we search for an $\mathcal{E}$-unifier to the new set $c \cup \tilde{c}$ looking for the solutions to $\tilde{c}\theta$. If $\tilde{c}\theta$ has no solution, we must start from scratch, looking for the solutions to the whole constraint $c \cup \tilde{c}$ and add θ to the set of unsuccessfully tried substitutions. We use a *Heuristic Conditional Narrowing* procedure $\mathcal{N}$ which will be described later for the search of a solution θ' of this constraint and combine adequately θ and θ' to obtain a new accumulated solution and a new accumulated simplified constraint.

Definition 3.6 *ICS* transition relation[2] $\xrightarrow{\tilde{c}}_{ICS}$
Single Solution Rules

$$(1) \quad \frac{\mathcal{N}(\tilde{c}\theta, f) \text{ has a unique solution } \theta'}{\langle \theta, f, (\hat{\theta}, \emptyset) \rangle \xrightarrow{\tilde{c}}_{ICS} \langle \theta\theta', f, (\widehat{\theta\theta'}, \emptyset) \rangle}$$

$$(2) \quad \frac{\mathcal{N}(\tilde{c}\theta, f) \text{ has a unique solution } \theta' \wedge \mathcal{N}((c_2 \cup \tilde{c})\theta_1, f) \text{ has a unique solution } \theta''}{\langle \theta, f, (\hat{\theta}_1, c_2) \rangle \xrightarrow{\tilde{c}}_{ICS} \langle \widehat{\theta_1\theta''}, f, (\widehat{\theta_1\theta''}, \emptyset) \rangle}$$

Multiple Solution Rules

$$(3) \quad \frac{\mathcal{N}(\tilde{c}\theta, f) \text{ has a unique solution } \theta' \wedge \mathcal{N}((c_2 \cup \tilde{c})\theta_1, f) \text{ has a first solution}}{\langle \theta, f, (\hat{\theta}_1, c_2) \rangle \xrightarrow{\tilde{c}}_{ICS} \langle \theta\theta', f, (\hat{\theta}_1, c_2 \cup \tilde{c}) \rangle}$$

---

[2]For the sake of simplicity, in this set of rules c_2 denotes a non empty constraint

(4)
$$\frac{\mathcal{N}(\tilde{c}\theta, f) \text{ has a first solution } \theta'}{\langle\ \theta, f,\ (\hat{\theta}_1, c')\ \rangle \xrightarrow{\tilde{c}}_{ICS} \langle\ \theta\theta', f,\ (\hat{\theta}_1, c' \cup \tilde{c})\ \rangle}$$

Start from Scratch Rule

(5)
$$\frac{\mathcal{N}(\tilde{c}\theta, f) \text{ has no solution } \wedge\ \langle\ \epsilon, f \cup \theta_{|Var(G_0)},\ ()\ \rangle \xrightarrow{(c_2 \cup \tilde{c})\theta_1}_{ICS} s}{\langle\ \theta, f,\ (\hat{\theta}_1, c_2)\ \rangle \xrightarrow{\tilde{c}}_{ICS} s}$$

Let us notice that each transition of this system depends on the termination of the $\mathcal{N}$arrowing procedure $\mathcal{N}$. We assume that if $\mathcal{N}$arrowing procedure $\mathcal{N}(c, f)$ terminates returning the output *no solution* then the constraint c has no $\mathcal{E}$-unifier, if the output *unique solution* θ then θ is the unique mgu over $\mathcal{E}$ of c and if the output is *first solution* θ then θ is an $\mathcal{E}$-unifier to c but, possibly, it is not the unique mgu over $\mathcal{E}$ of c. In the following we will define a transition system for $CLP(\mathcal{H}/\mathcal{E})$ which depends on the ICS solver. The transition relation for $CLP(\mathcal{H}/\mathcal{E})$ will depend on one single ICS transition step. Therefore, we do not rely on a termination proof for the ICS system.

The following theorem establishes the correctness of the Incremental Constraint Solver.

Theorem 3.7 (Correctness of ICS)

If an ICS transition $\langle\ \theta, f,\ r_c\ \rangle \xrightarrow{\tilde{c}}_{ICS} \langle \theta', f',\ r'_c\ \rangle$ ican be proved then the constraint $c \cup \tilde{c}$ is solvable in $\mathcal{H}/\mathcal{E}$. r'_c represents the simplified form of this constraint and θ' represents an $\mathcal{E}$-unifier to it. f' represents the set of substitutions unsuccessfully tried.

Obviously the incremental constraint solver as defined above is, in general, not complete because narrowing may loop forever in the evaluation of the condition of a rule not only when the constraint is not $\mathcal{E}$-*unifiable* but also when it has a unique mgu over $\mathcal{E}$ that has already been found and it tries to obtain another solution.

3.2 CLP($\mathcal{H}/\mathcal{E}$) Operational Model

The following definitions instantiate definitions 2.3, 2.4 and 2.5 to the case of CLP($\mathcal{H}/\mathcal{E}$) programs.

Definition 3.8 CLP($\mathcal{H}/\mathcal{E}$) transition relation $\rightarrow_{CLP(\mathcal{H}/\mathcal{E})}$

Let P be a $CLP(\mathcal{H}/\mathcal{E})$ (Π, Σ)-program. The $CLP(\mathcal{H}/\mathcal{E})$ Transition Relation is given by the following rule describing a computation step from a configuration $(\leftarrow s_i \diamond A_1, \ldots, A_n)$ where $s_i = \langle\ \theta_i,\ f_i,\ r_{c_i}\ \rangle$:

$$\frac{s_i \xrightarrow{\tilde{c}}_{ICS} s_{i+1}}{(\leftarrow s_i \diamond A_1, \ldots, A_n) \rightarrow_{CLP(\mathcal{H}/\mathcal{E})} (\leftarrow s_{i+1} \diamond \tilde{B}_1, \ldots, \tilde{B}_n)}$$

if there exist n variants of clauses in P, $H_j \leftarrow c'_j \Box \tilde{B}_j, j = 1, \ldots, n$, with no variables in common with $\leftarrow c_i \Box A_1, \ldots, A_n$. and with each other and $\tilde{c} = \{c'_1, c'_2, \ldots, c'_n, A_1 = H_1, \ldots, A_n = H_n\}$.

Definition 3.9 CLP($\mathcal{H}/\mathcal{E}$) initial configuration ($\mathcal{C}_0$)

Let $G_0 = \leftarrow c_0 \Box \tilde{B}$. be a goal. If c_0 is not an empty constraint and the empty ICS-state $s_0 \xrightarrow{c}_{ICS} s$ then $\mathcal{C}_0 = (\leftarrow s \diamond \tilde{B})$ is the $CLP(\mathcal{H}/\mathcal{E})$ initial configuration. If c_0 is empty then $\mathcal{C}_0 = (\leftarrow s_0 \diamond \tilde{B})$.

Definition 3.10 CLP($\mathcal{H}/\mathcal{E}$) terminal configurations ($\mathcal{C}$)

A terminal configuration C has the form $C = \langle \leftarrow s \diamond \rangle$, where $s = \langle\ \theta, f,\ (\hat{\theta}_1, c_2)\ \rangle$ is an ICS-state and $c = (\hat{\theta}_1 \cup c_2)_{|Var(G_0) \cup Var(c_2)}$ is the answer constraint.

As in the generic case, when the constraint solver is designed for simply testing the solvability of constraints, the above defined transition relation $\rightarrow_{CLP(\mathcal{H}/\mathcal{E})}$ becomes the standard definition of derivation of goals. We give an example that illustrates the definitions above.

Example 1 *Let us consider the following CLP($\mathcal{H}/\mathcal{E}$) program[3] $P \cup \mathcal{E}$:*
0/1 Knapsack problem

$P = \{$ *(p1)* $knapsack(M, L, W) \leftarrow addweight(M) = W \;\square\; sublist(M, L).$
      (p2) $sublist([\,], Z).$
      (p3) $sublist([X \mid Y], [X \mid Z]) \leftarrow \;\square\; sublist(Y, Z).$
      (p4) $sublist(Y, [X \mid Z]) \leftarrow \;\square\; sublist(Y, Z). \}$
$\mathcal{E} = \{$ *(e1)* $0 + y = y.$ *(e2)* $s(x) + y = s(x + y).$
      (e3) $weight(a) = 1.$ *(e4)* $weight(b) = 1.$
      (e5) $weight(c) = 2.$ *(e6)* $addweight([\,]) = 0.$
      (e7) $addweight([X \mid Y]) = weight(X) + addweight(Y). \}$

% *knapsack(M, L, W) states that items in the sublist M of a given list L can be packed into a knapsack such that the knapsack weights exactly W %*

Let us consider the initial goal $G_0 = \leftarrow \;\square\; knapsack(K, [a, b], 2)$. Since the constraint in G_0 is empty, the initial CLP($\mathcal{H}/\mathcal{E}$) configuration is
    $C_0 = \langle \leftarrow \langle \epsilon, \emptyset, (\,) \rangle \;\Diamond\; knapsack(K, [a, b], 2) \rangle$
considering (p1), the new constraint $\tilde{c}$ is $\tilde{c} = \{addweight(M) = W, K = M, L = [a, b], W = 2\}$
since ($\tilde{c}\epsilon$) has a first solution $\theta = \{M/[a, a], K/[a, a], L/[a, b], W/2\}$, applying (4),
    $C_0 \rightarrow_{CLP(\mathcal{H}/\mathcal{E})} C_1 = \langle \leftarrow \langle \theta, \emptyset, (\emptyset, \tilde{c}) \rangle \;\Diamond\; sublist(M, L) \rangle$
considering (p3), the new constraint $\tilde{c}'$ is $\tilde{c}' = \{M = [X \mid Y], L = [X \mid Z]\}$
since ($\tilde{c}'\theta$) has a unique mgu $\theta' = \{X/a, Y/[a], Z/[b]\}$ and $\tilde{c} \cup \tilde{c}'$ has a first solution
$\{M/[a, a], K/[a, a], L/[a, b], W/2, X/a, Y/[a], Z/[b]\}$, *applying (3),*
    $C_1 \rightarrow_{CLP(\mathcal{H}/\mathcal{E})} C_2 = \langle \leftarrow \langle \theta\theta', \emptyset, (\emptyset, \tilde{c} \cup \tilde{c}') \rangle \;\Diamond\; sublist(Y, Z) \rangle$
considering again (p3), the new constraint $\tilde{c}''$ is $\tilde{c}'' = \{Y = [X' \mid Y'], Z = [X' \mid Z']\}$
since ($\tilde{c}''\theta\theta'$) has no solution and $\tilde{c} \cup \tilde{c}' \cup \tilde{c}''$ has a unique mgu
$\theta'' = \{M/[a, b], K/[a, b], L/[a, b], W/2, X/a, Y/[b], Z/[b], X'/b, Y'/[\,], Z'/[\,]\}$, *applying (5),*
    $C_2 \rightarrow_{CLP(\mathcal{H}/\mathcal{E})} C_3 = \langle \leftarrow \langle \theta'', \theta\theta', (\tilde{\theta}'', \emptyset) \rangle \;\Diamond\; sublist(Y', Z') \rangle$
considering (p2), the new constraint $\tilde{c}'''$ is $\tilde{c}''' = \{Y' = [\,], Z' = Z''\}$
since ($\tilde{c}'''\theta''$) has a unique mgu $\theta''' = \{Z''/[\,]\}$, applying (1),
    $C_3 \rightarrow_{CLP(\mathcal{H}/\mathcal{E})} C_4 = \langle \leftarrow \langle \theta''\theta''', \theta\theta', (\theta''\theta''', \emptyset) \rangle \;\Diamond\; \rangle$
finally, since a final ICS-configuration has been reached, the answer constraint associated to this transition sequence is $(\theta''\theta''' \cup \emptyset)_{[Var(G_0) \cup Var(\emptyset)]} = (K = [a, b])$

As shown in the example, the accumulated solvable constraint is simplified when rules (1), (2) and (5) are applied and the solution found in the previous step is always reused to try to prove that the new accumulated constraint is solvable. Thus the test of the solvability results less expensive than a solvability test for the whole (non simplified) constraint.

4   Heuristic Conditional Narrowing

In this section, we present a *H*euristic *C*onditional *N*arrowing Calculus (*HCNC*), an adaptation of the Conditional Narrowing Algorithm CNA presented in [14] which was shown complete for noetherian and confluent rewrite systems without extra variables. We extend it by actively using heuristic information to quickly converge towards a solution.

---

[3]For convenience we often abbreviate $s^n(0)$ to n.

Let us briefly recall the Conditional Narrowing Algorithm CNA presented in [14]. An expression of the form $c \, with \, \tau$ where c is a set of equations and τ is a substitution is called a (sub-)goal. The substitution τ is assumed to be restricted to $Var(c)$ and the variables in the equational clauses are standardized apart, i.e. the set of variables in c and in the clauses are disjoint. To solve c, the algorithm starts with the subgoal $c \, with \, \epsilon$ and tries to derive subgoals (recording applied substitutions in the with-parts) until a terminal goal of the form $\emptyset \, with \, \theta$ is reached. Each substitution θ in a terminal goal is an $\mathcal{E}$-$unifier$ of c. In abuse of notation, it will often be called solution. Let $O'(c)$ be the set of nonvariable occurrences of a constraint c. We define

$$narred(c, u, k, \sigma) \Leftrightarrow (\ u \in O'(c) \ \wedge \ (l_k = r_k \leftarrow \tilde{e}_k) \in \mathcal{E} \wedge \ \sigma = mgu(c/u, l_k) \ )$$

The calculus CNA is defined by the following two rules:

CNA unification rule:

$$\frac{c \text{ syntactically unifies with } mgu \ \sigma}{c \, with \, \tau \quad \rightarrow_{CNA} \quad \emptyset \, with \, \tau\sigma}$$

CNA narrowing rule:

$$\frac{narred(c, u, k, \sigma) \text{ holds}}{c \, with \, \tau \quad \rightarrow_{CNA} \quad (\tilde{e}_k, c[u \leftarrow r_k])\sigma \, with \, \tau\sigma}$$

This calculus defines an algorithm, since all the CNA-derivations (sequences of states) from a given $\mathcal{N}$-state can be easily enumerated. These derivations can be represented by a (possibly infinite) finitely branching tree. The nodes of this tree have to be visited following a complete search strategy (e.g. breadth first) while looking for unifiable nodes.

In the following, we present the $\mathcal{H}$euristic $\mathcal{C}$onditional $\mathcal{N}$arrowing $\mathcal{C}$alculus ($\mathcal{HCNC}$).

Definition 4.1 $\mathcal{N}$-state

An $\mathcal{N}$-state is a triple $\langle n, \theta, L \rangle$, where n is a positive integer $\in \{0, 1, 2\}$, θ is a (Π, Σ)-substitution and L is a (possibly empty) list $[g_i]_{i=1}^n$ of subgoals $g_i = c_i \, with \, \tau_i$, where c_i is a set of (Π_c, Σ)-constraints (a conjunction of equations) and τ_i is a (Π, Σ)-substitution. List constructors are denoted by $[\ ]$ and $\bullet$. For abuse of notation, we assume $\bullet$ to be homomorphically extended to concatenation of lists.

Roughly speaking, the first component of an $\mathcal{N}$-state represents the current number of solutions. The second component represents the solution to be returned and the third component represents the list of subgoals yet to be narrowed. The nodes of the narrowing search tree are stored in the list. This list is treated as a queue to emulate a breadth first search strategy of the search tree.

To solve the conjunction of equations c using the heuristic information f, the algorithm starts with the initial $\mathcal{N}$-state $\mathcal{N}_0 = \langle 0, \epsilon, [\ c \, with \, \epsilon \ ] \rangle$ and tries to derive subgoals (recording applied substitutions in the with-parts) until a terminal $\mathcal{N}$-state $= \langle n, \theta, L \rangle \not\rightarrow_{\mathcal{HCNC}}$ is reached. Terminal $\mathcal{N}$-states are characterized by $L = [\ ]$. The set f of substitutions unsuccessfully tried by the constraint solver are used for a heuristic search of the solutions. If ρ is a substitution which belongs to f, then, by the compactness theorem it can not solve the accumulated constraint c because there is a subset of c which can not be solved. Moreover, no substitution ρ' such that, restricted to the variables of ρ, is an instance of ρ can solve c either.

Let us define the relation $\preceq$ over substitutions: $\rho \preceq \rho' \Leftrightarrow (\exists \alpha)$ such that $\rho'_{|Var(\rho)} = \rho\alpha$. Let $\langle n, \theta, L \rangle$ be a $\mathcal{N}$-state and $c_i \, with \, \tau_i$ be a subgoal in L. Let $S = \{(u, k, \sigma) \mid narred(c_i, u, k, \sigma)\}$. If $\langle u', k', \sigma' \rangle \in S$ and $(\exists \rho \in f)$ such that $\rho \preceq \tau_i\sigma'$ then the corresponding subtree of the search tree which has as root the subgoal $(\tilde{e}_{k'}, c_i[u' \leftarrow r_{k'}])\sigma' \, with \, \tau_i\sigma'$ does not contain any solution. Since it is useless to explore this subtree, we will 'prune the search tree' by removing the representation of its root $(\langle u', k', \sigma' \rangle)$ from S.

Definition 4.2 *Heuristic Conditional Narrowing Calculus ($HCNC$)*
The $HCNC$ calculus is defined by means of a stratified transition system[4]:
Heuristic Branching Rules

(1)
$$\frac{\{\langle u, k, \sigma\rangle \mid narred(c, u, k, \sigma) \} = \emptyset}{c \; with \; \tau \rightarrow_{Branch} [\,]}$$

(2)
$$\frac{s\_to\_l(\{\langle u, k, \sigma\rangle \mid narred(c, u, k, \sigma) \wedge \{\rho \in f \mid \rho \preceq \tau\sigma\} = \emptyset\}) = [\langle u_i, k_i, \sigma_i\rangle]_{i=1}^{n}}{c \; with \; \tau \rightarrow_{Branch} [\; (\tilde{e}_{k_i}, c[u_i \leftarrow r_{k_i}])\sigma_i \; with \; \tau\sigma_i \;]_{i=1}^{n}}$$

Success Rules

(3)
$$\frac{c \; with \; \tau \rightarrow_{Branch} L' \; \wedge \; c \; \text{syntactically unifies with } mgu \; \sigma}{\langle 0, \epsilon, \; c \; with \; \tau \bullet L\rangle \rightarrow_{HCNC} \langle 1, \tau\sigma, \; L \bullet L'\rangle}$$

(4)
$$\frac{c \; with \; \tau \rightarrow_{Branch} L' \; \wedge \; c \; \text{syntactically unifies with } mgu \; \sigma \; \wedge \; (\tau\sigma) \not\preceq \theta \; \wedge \; \theta \not\preceq (\tau\sigma)}{\langle 1, \theta, \; c \; with \; \tau \bullet L\rangle \rightarrow_{HCNC} \langle 2, \theta, \; [\,]\rangle}$$

Narrowing Rules

(5)
$$\frac{c \; with \; \tau \rightarrow_{Branch} L' \; \wedge \; c \; \text{syntactically unifies with } mgu \; \sigma \; \wedge \; \theta \preceq (\tau\sigma)}{\langle 1, \theta, \; c \; with \; \tau \bullet L\rangle \rightarrow_{HCNC} \langle 1, \theta, \; L \bullet L'\rangle}$$

(6)
$$\frac{c \; with \; \tau \rightarrow_{Branch} L' \; \wedge \; c \; \text{syntactically unifies with } mgu \; \sigma \; \wedge \; (\tau\sigma) \preceq \theta}{\langle 1, \theta, \; c \; with \; \tau \bullet L\rangle \rightarrow_{HCNC} \langle 1, \tau\sigma, \; L \bullet L'\rangle}$$

(7)
$$\frac{c \; with \; \tau \rightarrow_{Branch} L' \; \wedge \; mgu(c) = \emptyset \; \wedge \; n \in \{0, 1\}}{\langle n, \theta, \; c \; with \; \tau \bullet L\rangle \rightarrow_{HCNC} \langle n, \theta, \; L \bullet L'\rangle}$$

where the function $s\_to\_l(S)$ returns a list composed by the elements of the set S).

Definition 4.3 Behaviour of the $HCNC$ calculus
Let N_0 be an initial N-state. We define the function:

$$N(c, f) = \begin{cases} no \; solution & if \; N_0 \rightarrow_{HCNC}^{*} \langle 0, \epsilon, [\,]\rangle \\ unique \; solution \; \theta & if \; N_0 \rightarrow_{HCNC}^{*} \langle 1, \theta, [\,]\rangle \\ first \; solution \; \theta & if \; N_0 \rightarrow_{HCNC}^{*} \langle 2, \theta, [\,]\rangle \end{cases}$$

Let us notice that if $f = \emptyset$ then the above algorithm explores (the prefix of) the search tree of CNA which would have been explored following a breadth first search strategy until the conditions characterizing a terminal N-state would have been reached. If $HCNC$ does not terminate or terminates with $L = [\,]$ then $N(c, f)$ is equivalent to CNA as it explores exactly the same search tree. When $f \neq \emptyset$ the heuristic pruning based on f only removes subtrees which do not contain any solution so that it is guaranteed that no solution is lost.

Theorem 4.4 (Correctness of the *Heuristic Conditional Narrowing*)
Let $\mathcal{E}$ be a canonical conditional Horn equational theory without extra variables. Let c be a constraint. Let f be a set of substitutions such that no instance of any of them can solve c. If $N(c,f)$ returns the output no solution then the constraint c is not solvable. If the output is unique solution θ then θ is the unique mgu over $\mathcal{E}$ of c. If the output is first solution θ then θ is a solution over $\mathcal{E}$ to c.

---

[4]The states of the transition system for Branch are subgoals or list of subgoals

Let us notice that when $\mathcal{N}(c, f)$ returns the output *first solution* θ, θ might be the unique mgu over $\mathcal{E}$ of c. In fact, the solutions are checked not to be syntactic instances of one another while only two solutions which are not $\mathcal{E}$-instances of one another should be distinguished. This means that the incremental constraint solver may simplify the constraint less than it would be possible.

The Heuristic Conditional Narrowing as defined above is a procedure to test the solvability of constraints over the structure $\mathcal{H}/\mathcal{E}$. If the constraint is solvable then the procedure finds a solution (an $\mathcal{E}$-*unifier*) to it and looks for another incomparable $\mathcal{E}$-*unifier*. If the constraint is not $\mathcal{E}$-*unifiable* or if there is not a second solution, the procedure may run forever.

Even in the case of canonical theories, the use of narrowing as a semantic unification procedure presents several drawbacks since it may not terminate when there are no (more) solutions and the generation of a independent set of $\mathcal{E}$-*unifiers* is not guaranteed. An infinite set of solutions can be derived even if the equation admits a finite complete minimal one [13]. This problem was also inherited by the various techniques developed for equational resolution, as SLDE-resolution, where a single resolution step may loop forever without producing any answer. In [13], a sufficient condition for termination is given for canonical non-conditional theories.

To overcome these difficulties, several proposals have been made. In [11] a lazy resolution rule is introduced. With this rule, the problem of the unification of a set of equations within the theory $\mathcal{E}$ must not be solved before the rule can be applied but the equations are added as a constraint to the derived goal clause without testing its solvability.

Another idea is to delay the constraint solving until all other goals are solved and then submit it to a complete $\mathcal{E}$-*unification* algorithm. This could be considered similar to the postponement of the satisfiability test of non linear constraints in CLP($\mathfrak{R}$) [17].

5 Conclusions and further research

We have presented, in a formal setup, several strategies to obtain an incremental narrower as constraint solver for canonical Horn equational theories. Our methods fits in the CLP scheme, as opposed to the standard way to integrate narrowing and resolution [9] where narrowing generates the solutions which are then tested by the logic program. Then we have presented a calculus which allows to reuse the substitutions discarded as an heuristic to look for other solutions.

We are integrating these results in a unique framework and giving declarative semantics to the resulting language [1]. In [1] we have also defined an optimized incremental constraint solver based on a technique which allows to cut the search space.

6 Acknowledgement

We gratefully acknowledge Pierpaolo Degano and Maurizio Gabbrielli for many helpful discussions.

References

[1] M. Alpuente, M. Falaschi, and G. Levi. Incremental Constraint Satisfaction for Equational Logic Programming. Technical report, Dipartimento di Informatica, Università di Pisa, 1991. in preparation.

[2] P. Bosco, E. Giovannetti, and C. Moiso. Narrowing vs. SLD-resolution. *Theoretical Computer Science*, 59:3–23, 1988.

[3] N. Dershowitz and A. Plaisted. Logic Programming cum Applicative Programming. In *Proc. First IEEE Int'l Symp. on Logic Programming*, pages 54–66. IEEE, 1984.

[4] L. Fribourg. Slog: a logic programming language interpreter based on clausal superposition and rewriting. In *Proc. Second IEEE Int'l Symp. on Logic Programming*, pages 172–185. IEEE, 1985.

[5] U. Furbach, S. Hölldobler, and J. Schreiber. Horn equality theories and paramodulation. *Journal of Automated Reasoning*, 5:309–337, 1989.

[6] M. Gabbrielli and G. Levi. Modeling answer constraints in Constraint Logic Programs. In K. Furukawa, editor, *Proc. eighth Int'l Conf. on Logic Programming*. The MIT Press, 1991. to appear.

[7] J.H. Gallier and S. Raatz. Extending SLD-resolution to equational Horn clauses using E-unification. *Journal of Logic Programming*, 6:3–43, 1989.

[8] E. Giovannetti, G. Levi, C. Moiso, and C. Palamidessi. Kernel Leaf: A Logic plus Functional Language. *Journal of Computer and System Sciences*, 42, 1991.

[9] J. A. Goguen and J. Meseguer. Eqlog: equality, types and generic modules for logic programming. In D. de Groot and G. Lindstrom, editors, *Logic Programming, Functions, Relations and Equations*, pages 295–262. Prentice Hall, Englewood Cliffs, NJ, 1986.

[10] P. Van Hentenryck. Incremental Constraint Satisfaction in logic programming. In D.H.D. Warren and P. Szeredi, editors, *Proc. Seventh Int'l Conf. on Logic Programming*, pages 189–202. The MIT Press, Cambridge, Mass., 1990.

[11] S. Hölldobler. *Foundations of Equational Logic Programming*, volume 353 of *Lecture Notes in Artificial Intelligence*. Springer-Verlag, Berlin, 1989. subseries of Lecture Notes in Computer Science.

[12] S. Hölldobler. Conditional equational theories and complete sets of transformations. *Theoretical Computer Science*, 75:85–110, 1990.

[13] J.M. Hullot. Canonical Forms and Unification. In *5th Int'l Conf. on Automated Deduction*, volume 87 of *Lecture Notes in Computer Science*, pages 318–334. Springer-Verlag, Berlin, 1980.

[14] H. Hussman. Unification in conditional-equational theories. Technical report, fakultät für mathematik und informatik, Universität Passau, 1986.

[15] J. Jaffar and J.-L. Lassez. Constraint Logic Programming. Technical report, Department of Computer Science, Monash University, 1986.

[16] J. Jaffar and J.-L. Lassez. Constraint Logic Programming. In *Proc. Fourteenth Annual ACM Symp. on Principles of Programming Languages*, pages 111–119. ACM, 1987.

[17] J. Jaffar and S. Michaylov. Methodology and Implementation of a CLP System. In J.-L. Lassez, editor, *Proc. Fourth Int'l Conf. on Logic Programming*, pages 196–218. The MIT Press, 1987.

[18] S. Kaplan. Conditional Rewrite Rules. *Theoretical Computer Science*, 33:175–193, 1984.

[19] J.W. Lloyd. *Foundations of logic programming.* Springer-Verlag, Berlin, 1987. second edition.

[20] J.J. Moreno and M. Rodriguez-Artalejo. BABEL: A Functional and Logic Programming Language based on a constructor discipline and narrowing. In I. Grabowski, P. Lescanne, and W. Wechler, editors, *Algebraic and Logic Programming*, volume 343 of *Lecture Notes in Computer Science*, pages 223–232. Springer-Verlag, Berlin, 1988.

[21] W. Nutt, P. Réty, and G. Smolka. Basic narrowing revisited. *Journal of Symbolic Computation*, 7:295–317, 1989.

[22] G. Plotkin. A structured approach to operational semantics. Technical Report DAIMI FN-19, Computer Science Department, Aarhus University, 1981.

[23] J.H. Siekmann. Universal unification. In *7th Int'l Conf. on Automated Deduction*, volume 170 of *Lecture Notes in Computer Science*, pages 1–42. Springer-Verlag, Berlin, 1984.

The Implementation of Lazy Narrowing

Manuel M. T. Chakravarty and Hendrik C. R. Lock
GMD Forschungsstelle an der Universität Karlsruhe
Vincenz Prießnitz Straße 1, D-7500 Karlsruhe 1, FRG
email: (*chak* or *lock*)@karlsruhe.gmd.de

Abstract

Lazy narrowing has been proposed as the operational model of functional logic languages. This paper presents a new abstract machine which implements lazy narrowing. The core of this machine consists of a conventional stack based architecture like the one used for imperative languages. Almost orthogonal extensions of this core implement the different concepts of functional logic languages. This simplifies the machine design to a great deal and reduces the instruction set which has been particularly designed to support the application of standard code generation techniques. By its orthogonality, it is achieved that unused features introduce only minimal overhead. As a result, when performing ground term reduction the machine enjoys the same characteristics as efficient graph reduction machines.

1 Introduction

Functional logic languages are obtained when free logical variables are allowed in functional expressions [BL86, Red86]. Several methods of combination have been proposed. For instance, logical variables may appear in the guards of rules defining functions [MR88], or inside *absolute set abstractions* [DFP86]. Independent of their embedding, the problem is that expressions containing free variables cannot be evaluated using graph reduction which is the operational model of functional languages. Narrowing as the operational model for evaluating such expressions has been independently proposed by [Red85, DFP86].

An analysis of the WAM [War83, AK90] and lazy graph reduction machines [FW87, Mei88, PS88] showed that they are based on a common core which is a conventional stack based architecture [ASU86]. This core supports deterministic first order expression evaluation. In addition, the WAM, which was designed for implementing SLD resolution, supports unification and non-determinism. On the other hand, lazy graph reduction machines which implement non-strict functional languages additionally support laziness (i.e. call-by-need) as well as higher order functions. Thus, these machines can be seen as almost orthogonal extensions of the core machine. One of the main contributions of this work is to show how they can be integrated into a single abstract machine.

This paper presents a new abstract machine, called the JUMP machine, which is designed for implementing lazy narrowing. A particular aim was to design a powerful and reduced instruction code which in connection with the conventional core allows the application of traditional code generation and standard optimization techniques. The paper further describes how to translate programs to the instruction code (*JCode*) of the machine.

The paper is organized as follows. The second section introduces the syntax of a basic functional logic language. The third section presents an introduction to narrowing and identifies three issues: *rule* and *redex non-determinism* as well as *dependent updates*. The main part of the paper is covered in section four and reflects in its argument the design principles of the machine: it first presents the conventional core based on which we discuss the representation of terms and function evaluation, and then presents the implementation of unification and non-determinism, as well as that of laziness and higher order functions. Finally, it shows how the translation of patterns influences the termination property of lazy narrowing. The fifth section discusses optimizations, in particular fast stack deallocation, deterministic computations, and memory mapping. Related work and the conclusions are contained in section six and seven, and the appendix contains the translation schemes.

2 Syntax

This section defines an almost functional notation which contains a simple functional logic language construct: the *equational goal* whose free variables are implicitly existentially quantified logical variables. This notation is used in examples and as a source for the translation. The sets of function symbols $f, g, ..$, data constructors $c, d, ..$ and variable identifiers $X, Y, ..$ are disjoint. Only variables symbols are written in capitalized letters.

$$
\begin{array}{lll}
progr & ::= & rule^* \ goal \\
rule & ::= & f \ pat^* \ '=>' \ expr \\
pat & ::= & X \mid c \mid c \ '(' \ pat, \ldots, pat \ ')' \\
term & ::= & X \mid c \mid c \ '(' \ expr, \ldots, expr \ ')' \\
expr & ::= & term \mid f \ expr^* \mid \ 'let' \ (rule \mid pat = expr)^+ \ 'in' \ expr \\
goal & ::= & '?' \ data\_term = expr
\end{array}
$$

Usual functional specifications can be mapped to this basic notation, through local definitions in let-expressions need not to be λ-lifted (this side issue will be discussed later). It is assumed that data constructors are declared and that programs are already type checked. The types of logical variables are restricted to first order types thus ensuring first order unification.

3 An Informal Introduction to Narrowing

Narrowing has been introduced as a general E-unification algorithm used to prove existentially quantified equational formulas [Fay79, Hul80] in equational logic. Later, a restricted algorithm has been independently proposed by [Red85, Red86] and by [DFP86] (although the latter does not use the term narrowing) as the operational model for evaluating functional expressions containing free variables. It can be understood as functional term reduction where pattern matching is replaced by syntactic unification performing bi-directional data flow.

Lazy narrowing has been developed as an operational model for functional logic languages with a non-strict semantics [DG89, MKLR90]. Throughout this paper, *lazy narrowing* is understood as being the mechanism which narrows outermost redexes, and narrows at inner redexes only if they are *demanded* by unification. It also preserves *sharing*. Its formal definition can be found in [MKLR90] and it is taken as the operational semantics of the language introduced in section 2.

Below, we introduce the terms *rule non-determinism*, *redex non-determinism* and *dependent updates*, each characterizing an aspect to be implemented by the abstract machine.

Given a redex in a ground expression and a functional specification, at most one of the non-overlapping rules matches the redex. Given a redex in a non-ground expression, i.e. one which contains logical variables, more than one rule may unify. Thus, unification is the source of what is called *rule non-determinism*. Non-determinism, and the facility to compute all solutions of a non-ground expression is the main requirement for an implementation of lazy narrowing, and it determines the main difference to lazy reduction. We now illustrate this by an example.

$$
\begin{array}{ll}
\text{f } 1 \ => 6. & \text{g } 6 \ => 5. \\
\text{f } 2 \ => 7. & \text{g } 7 \ => 4. \\
\text{f } 3 \ => 6.
\end{array}
$$

The outcome of an evaluation is the pair of result expression and answer substitution for free variables. According to the specification, the ground application g (f 2) has the outcome $(4, \epsilon)$, and evaluation is just lazy reduction. However, the non-ground application g (f X) has three outcomes: $(5, X \mapsto 1), (4, X \mapsto 2), (5, X \mapsto 3)$. The inner redex f X contributes to the matching of g's patterns, hence, it is narrowed first. The interesting point is that after selecting g 6 and then g 7, the third narrowing of f X makes it necessary to select g 6 again. Rule non-determinism is obtained when the application unifies with more than one rule, and hence, different solutions are found. By this, each narrowing of an inner redex may lead to reevaluation of an outer redex, known as *redex non-determinism* (coined in [MKLR90]). One of the central issues for a simple machine design is to obtain redex non-determinism through rule non-determinism.

The next example motivates a problem which arises from the interaction of logical variables, laziness, and sharing. Let g be defined as above.

```
h 6 Y  => Y.
h 7 Y  => Y.
```

Consider the goal ? 4 = h X (g X) in which X appears free. Laziness requires that the evaluation of g X is postponed and the expression is passed as a suspension to h. The first evaluation step produces the binding X↦6 which is propagated to the suspension. Then, the body of h evaluates g 6, and the failure of 4=5 causes that h is retried, this time resulting in the binding X↦7. Sharing requires updating the suspension by its result. However, the problem is that this update becomes invalid when the binding of free variables on which it depends become invalid as well. We term that a *dependent update*. The solution is to reverse the dependent update when bindings are undone.

4   The Abstract Machine

The description of the JUMP machine in this section reflects its design as a systematic extension of a conventional stack based architecture. First its components and the representation of machine objects will be presented. These form a core machine able to perform function evaluation. This core is extended by choice points which implement unification and backtracking. The last two subsections discuss the translation of patterns which is important for unification, and address the implementation of sharing.

Small examples illustrate the use of the instruction code (JCode). JCode is a block structured intermediate language abstracting from labels, memory allocation and the direct manipulation of machine registers such as first proposed in [PS88]. This is obtained through identifying values by intermediate variables and supported by a small set of powerful instructions. Hence, JCode gives scope for exploiting traditional code generation techniques such as described in [ASU86].

4.1   The Architecture and its Components

The JUMP machine is stack based and contains the following components:
- a **stack** which stores activation records and choice points,
- a **heap** which stores closures and environments,
- a **code area** and a program counter which are as usual,
- a **trail stack** which records the bindings of logical variables and updates of suspensions (during backtracking it is used to reverse these operations), and
- a set of **global registers** used to reference the current activation record, the environment, the topmost choice point, the active closure, and the count of pending arguments respectively (however, JCode abstracts from the use of these registers, and also do we throughout this paper).

As illustrated in figure 1, each function call creates its own activation record (also called *frame*) which contains all relevant information for the function to compute and to restore the previous context when it returns. This corresponds to the call sequence used for the implementation of imperative languages (pp. 404, [ASU86]) except for environments and the argument count.

User defined local variables and intermediate variables are stored in the local area. Those variables which appear free in suspended expressions (not to be confused with free logical variables) are stored in environments on the heap since suspensions tend to survive their defining context. Environments are also used to store free variables of local function definitions which makes λ-lifting unnecessary [MP90]. A detailed definition of the JUMP machine is given in [Loc91].

local area
access link
control link
caller's env
argument count $= k$
return address
arguments $a_1..a_k$

environment pointer for free variables
pointer to the previous stack frame
environment pointer

Figure 1: *The stack frame of a function*

4.2   Terms, logical variables, and suspensions

Several types of machine objects are used to encode terms. For instance, the type *logical variable* implements an unbound logical variable. A *suspension* is an object encoding an expression whose evaluation is postponed. A *partial application* represents the function resulting from instantiating some of its arguments. Further types are *data terms* and *indirection nodes*.

A uniform and consistent representation for those objects is required since logical variables may appear inside terms, suspension may be bound to them, and their evaluation is lazy. The integration is achieved by representing all objects by executable closures. This has several advantages [PS88]: First, a function call is conceptually the same as evaluating an object. Second, tag testing and untagging can be replaced by vectored return operations, thus eliminating interpretive overhead. And third, the code pointer can be used for garbage collection as well.

As usual, a closure consists of a code pointer and an environment relative to which the code is executed. Respectively, the environment contains the arguments of a term, the binding reference of a logical variable, or it defines the variables which appear free in the suspension. A closure is entered by setting a node register to it and by executing its code. Its environment is accessible via that node register.

a term or a suspension:

code pointer	ref to environment

an unbound variable or an indirection node:

code pointer	ref to a closure

The code of a term performs the instruction RET TERM c node which returns the constructor c and indicates a term. The code of an unbound logical variable performs RET VAR node indicating its unbound state. This information will be used for unification. The code of a suspension executes the suspended expression. As will be shown in part 4.3, partial applications are a special case of suspension closures which execute the suspended function as soon as enough arguments are supplied.

A logical variable which becomes bound or a suspension which becomes updated by its result are both converted into an indirection node by overwriting their code and argument field. Then, the new code simply jumps along the indirection pointer. These updates are recorded on the trail stack such that during backtracking the updated objects are reset to their previous state.

4.3   Functions and partial application

The following example illustrates function definition and application.

```
kcomb X Y => X.
? Result = kcomb 1 2.
```

The function kcomb and the goal ? Result = kcomb 1 2 are translated into:

```
FUN kcomb [a1,a2] {JUMP a1 []}
FUN main [] { x1 := create_term (1, []); x2 := create_term (2, []);
              result := CALL kcomb [x1,x2] }
```

FUN defines a function whose arguments are identified by the list [a1,a2]. When entered, the function creates an activation record on top of the stack and enters the closure identified by a1. The JUMP performs a *tail call* which is the optimized form of z:=CALL a1 []; RET z. The RET and JUMP instructions eventually deallocate an activation record only when it is not protected by choice points (see the next section). The main code calls kcomb with the argument terms 1 and 2.

Both CALL and JUMP instructions place their actual arguments, which are identified by the second parameter list, on the stack. Their first parameter denotes either a function or a variable. Conceptually, calling a function is the same as entering an object.

In a type system which guarantees first order unification *currying* can be allowed. This means that a function defined on n parameters can be applied to less than n arguments, e.g. in let X=kcomb 1 in X 2. The result of such *partial application* is a function over the remaining parameters which is represented by a closure of type *partial application (papp)*. A papp is a variant of a suspension containing the function label, the pending arguments, and the free variables of the function body. When a papp is entered its code executes an *uncurry* function which tests whether sufficient arguments are supplied. In that case, it unfolds the pending arguments on the stack and executes the function. Otherwise, a new papp is created which collects the new arguments. Each pairing of function arity and pending argument count is encoded by a separate uncurry function.

Summary

- Function evaluation is based on a conventional stack based architecture where stack frames store arguments, saved registers, and local values. A function returns a partial application in case not enough arguments have been supplied. There is no difference in entering a closure or calling a function which is the base to represent suspended function calls by closures.

4.4 Unification, choice points, and backtracking

Choice points are used to record states of computation to which evaluation can backtrack in order to try an alternative. The new perspective developed here, in contrast to the WAM [AK90], is that a choice point manages all bindings for one logical variable. And that bindings are defined by the cases of a SWITCH instruction (see below). By this means rule non-determinism is implemented.

SWITCH statements are obtained from translating patterns (see section 4.5), and in general they can be nested.

```
SWITCH y OF { TERM c_1 [x1,..,xn_1] : { (* first case *)     }
              ...
              TERM c_k [x1,..,xn_k] : { (* the k-th case *)   }
              ELSE x                : { RETRY (* backtrack *) }   }
```

When the argument of a SWITCH instruction returns a data constructor a case analysis is performed over the patterns c_1 to c_k. The variables x1,..,xn_j identify subterms. When applied to an unbound variable, unification is performed. The SWITCH defines all alternative bindings for the logical variable, according to its cases. In order to manage these bindings the SWITCH creates a choice point on the stack. A choice point contains a link to the previous one, a code label which contains the next alternative, an activation record for that code, a pointer to the trail segment which records all side effects performed during that choice, and a pointer to the logical variable it binds (see the left hand side of figure 2). A choice point protects all frames below it against deallocation since they may be reused during backtracking.

The task of the choice point is to generate and bind the terms c_1(..) to c_k(..) in sequence. Each binding updates the logical variable by an indirection node pointing to the new term. The arguments of each term are initialized by unbound logical variables which are identified by x1,...

When SWITCH instructions are nested, each inner SWITCH generates the bindings for the logical variable in an argument of a term generated by an outer SWITCH. This organization ensures that all possible binding combinations become systematically enumerated, as sketched on the right hand side of figure 2.

A RETRY instruction performs backtracking. It is integral part of the SWITCH and executed when pattern matching fails or after the last binding has been generated. In contrast to the above example, the else branch containing RETRY is usually left implicit. RETRY enters the topmost choice point, reverses all updates recorded on the trail, deallocates the whole stack down to that choice point, and executes the alternative code relatively to the parent frame. If a choice point is retried after its last

label of alternative	h's choice point of level k ↑ stack growth
parent frame	... (h's choice points)
trail segment	h's choice point of level 1
logical variable	h's activation record
previous choice point	:

Figure 2: *The choice point organization on the stack*

binding has been generated it resets the logical variable to an unbound state and executes RETRY. This activity is called a *return by failure*.

The next example shows the translation of f and g from section 3 and illustrates backtracking in between two function calls.

```
FUN f [a1] { SWITCH a1 OF {
              TERM 1 [] : {z1 := create_term(6, []); RET TERM 6 z1}
              TERM 2 [] : {z1 := create_term(7, []); RET TERM 7 z1}
              TERM 3 [] : {z1 := create_term(6, []); RET TERM 6 z1} } }

FUN g [a2] { SWITCH a2 OF {
              TERM 6 [] : {z1 := create_term(5, []); RET TERM 5 z1}
              TERM 7 [] : {z1 := create_term(4, []); RET TERM 4 z1} } }
```

In the goal ? X = g f Y appears Y free, hence, it is initialized by a logical variable:

```
FUN main [] { z1 := SUSP [] {y := create_var(); JUMP f [y]};
              x := CALL g [z1]; }
```

The discussion of suspensions defined by the SUSP instruction is postponed to section 4.6.

Since f is called with a logical variable it creates a choice point on top of the activation records for g and f (see figure 3). The SWITCH creates term 1 and binds it to the logical variable identified by a1. Then, RET returns term 6 to the caller such that g deterministically selects its first branch. We observe that f's frame needs to be protected since it is referenced by f's the choice point, and g's frame needs to be protected since f's alternative will return to it.

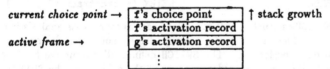

current choice point →	f's choice point ↑ stack growth
	f's activation record
active frame →	g's activation record
	:

Figure 3: *The stack situation after function f returns to g*

Now, suppose that some RETRY is executed. It backtracks into f's choice point and, accordingly, the term 2 is bound to the logical variable and this time f returns with result term 7. It is important to notice that the return address fixed by the first call is valid for all choices. Therefore, each time f is retried it returns into the same SWITCH instruction inside g such that all of g's alternatives can be retried. This becomes even more apparent when a second RETRY is issued because then f returns with 6 and the first branch in g is selected again. Although the SWITCH in g performs deterministic selection due to the returned constructor terms, and although g does not create a choice point, f's choice point protects g's activation record and makes g backtrackable. Thus, an elegant and simple solution to redex non-determinism is obtained.

Finally, when a third RETRY is issued all choices defined by the SWITCH have been tried. Accordingly, f's choice point returns by failure whilst deallocating its choice point, its activation record and all protected activation records underneath including g's. Hence, the whole expression g (f Y) fails.

Summary

- Rule heads of function definitions are translated into SWITCH instructions which perform unification columnwise (see also part 4.5). Since the rules are left-linear it is never necessary to call a general E-unification procedure. Thus, a SWITCH either performs deterministic pattern matching on a non-variable input term, or generates bindings for an unbound logical variable. In that case, a choice point manages all possible bindings and the alternative results of the function call belonging to those bindings. Backtracking enumerates that set of results.

- So far, the frame of a function is only discarded by backtracking. Section 5.1 addresses the issue how to free activation records as soon as possible.

- When a SWITCH evaluates a backtrackable expression, all alternatives return into that SWITCH which thereby becomes reevaluated for each alternative result. This gives an elegant implementation of redex non-determinism. For that reason, a choice point must protect all frames beneath because they may become reevaluated during backtracking.

4.5   Translating patterns

The lazy narrowing process misses solutions when non-terminating subexpressions are demanded too early by unification. To ensure that lazy narrowing finds all solutions, a transformation of function definitions into a *canonical form* is required. Furthermore, the transformation makes single unification steps explicit which prepares the translation to JCode and removes common subexpressions.

After giving some necessary definitions the transformation is illustrated by an example; its formal description is out of the scope of this paper, but contained in [Loc91]. A similar transformation has been first proposed in [MKLR90]. Finally, we present the translation of canonical forms into JCode, in particular into assignments and nested SWITCH instructions.

An intermediate language level is used which introduces a choice operator [] and a guard operator ->. In a -> e_1 [] b -> e_2, a and b are the guards of the branches e_1 and e_2.

Functional specifications are non-overlapping which means that their rules head do not pairwise unify. Two rule heads are *subunifiable* if there exists a position whose respective subpatterns unify. For instance, the first and third rule of h in figure 4 are subunifiable due to 1 and Y. A function definition in the intermediate language is *uniform* if all subunifiable columns are guarded by non-subunifiable ones. For instance, the second matrix in figure 4 is not uniform since the column identified by Z is subunifiable. However, the third matrix is uniform since Z=1 and Z=Y are guarded by the non-subunifiable column A2. Hence, the evaluation of Z=1 and Z=Y becomes postponed. A uniform matrix whose equations are flat is *canonical*.

The first transformation phase flattens nested patterns, thereby generating flat equational guards which represent single unification steps. In particular, Z=Y denotes assignment and Z=c(..) denotes the unification of term c(..). The second phase eliminates redundancies by coalescing overlapping equations and postpones subunifiable positions. Both steps are illustrated in figure 4.

```
h c(1) 2 => e_1.
h d(X) 2 => e_2.
h c(Y) 3 => e_3.
```

$$
h\ A1\ A2\ \Rightarrow \left\{ \begin{array}{l} \ A1=c(Z),\ Z=1,\ A2=2\ ->\ e\_1 \\ []\ A1=d(Z),\ Z=X,\ A2=2\ ->\ e\_2 \\ []\ A1=c(Z),\ Z=Y,\ A2=3\ ->\ e\_3 \end{array} \right\}
$$

$$
h\ A1\ A2\ \Rightarrow \left\{ \begin{array}{l} A1=c(Z)\ ->\ \left\{ \begin{array}{l} \ A2=2\ ->\ Z=1\ ->\ e\_1 \\ []\ A2=3\ ->\ Z=Y\ ->\ e\_3 \end{array} \right\} \\ []\ A1=d(Z)\ ->\ Z=X\ ->\ A2=2\ ->\ e\_2 \end{array} \right\}
$$

Figure 4: *Transforming a function definition into its canonical form*

Consider the expression h c(f A) 3 and suppose that ? 1 = f A has no finite solution. Narrowing the expression h c(f A) 3 relatively to the original definition is in danger to evaluate the subgoal f A = 1, whereas narrowing relatively to the canonical definition first matches 3 by A2=3 and then performs an assignment according to Z=Y. This shows how evaluation of an argument expression is demanded only if it cannot be avoided.

It is easy to prove by induction that any set of non-overlapping rule heads can be transformed into a canonical form and that non-uniform positions can be moved to the inside. In this way the termination property of the narrowing implementation is improved.

The leading column of a canonical matrix which consists of non-trivial equations is translated into a SWITCH, whereas trivial equations are translated into assignments (see also scheme $\mathcal{N}_{unify}$ in appendix A). Hence, h's code looks like:

```
SWITCH a1 OF {
    TERM c [z] : SWITCH a2 OF {
                    TERM 2 [] : SWITCH z OF { TERM 1 [] : { (* e_1 *) } }
                    TERM 3 [] : { y := z; (* e_3 *) }  }
    TERM d [z] : {  x := z;  SWITCH a2 OF { TERM 2 [] : (* e_2 *) } }   }
```

The SWITCH corresponds to the index instruction of the WAM which is often used as an optimization for unifying the first argument. The columnwise translation used in our approach not only optimizes but also improves the termination properties further, as illustrated below.

```
m s(s(o)) => s(o).          nth o   [X|XS] = X.
m s(N)    => s(m N).         nth s(X) [_|XS] = nth X XS.
```

Obviously, narrowing on ? o = min Y will search forever. Thus, when patterns are compiled rulewise from left to right into single unification instructions, as done in most WAM based approaches, e.g. [Han90, Loo91, Sch91], evaluation of ? 2 = nth (min Y) [1,2,3] would not terminate either. When compiling columnwise, however, the second case of nth can be immediately matched and the solution Y↦s(s(o)) is found (the example is due to JJ. Moreno-Navarro.)

4.6 Sharing and dependent updates

Sharing means that an expression is evaluated at most once. This can be achieved by updating a suspension by an indirection node pointing to the result of its evaluation in order to prevent another reference reevaluating the suspension. We discuss three issues: the suspension's code, dependent updates, and updating partial applications.

The instruction SUSP list code (as used in the last example) not only creates a suspension at run-time but also statically defines a parameterless function (a *thunk*). The suspension consists of the thunk's label and of an environment. The thunk executes code which encodes a suspended application in the context of an environment which is passed by the suspension. Its environment defines the free variables of code which are identified by list.

When executed, the thunk creates an activation record. After executing code, the update operation converts the suspension object into an indirection node which points to the result. Since the evaluation of the suspension may depend on bound logical variables, the suspension is recorded on the trail stack by its code pointer, environment pointer and its reference. During backtracking, this information is used to restore the old suspension object. Finally, the thunk tail calls the result object.

The next example illustrates updating a suspension by a partial application.

```
let X = f Y Z in X U V
{ x := SUSP [y,z] { JUMP f [y,z] };  JUMP x [u,v];}
```

JUMP x forces the evaluation of the suspension identified by x. The issue is that this suspension must be updated by a papp representing f Y Z before it consumes the pending arguments [u,v]. This is possible because the activation record of the thunk protects [u,v] and the evaluation of f Y Z is performed on top of that activation record without noticing the presence of [u,v]. After the update, the thunk must test whether its activation record is protected. If so, it copies all pending arguments to the top of stack, otherwise it can discard its own activation record. In both cases, the pending arguments (e.g. [u,v]) are made visible for the papp which finally is entered by a tail call.

Summary

- Strictly speaking, two types of suspensions exist: in general, the result of evaluating a suspension depends on bindings of logical variables. Thus, their update is termed **dependent**. The other type only appears during ground term reduction, in which case the update is termed **fixed**. Dependent updates are required to be reversed during backtracking.

- The technique of protecting pending arguments by an activation record is novel, simpler and more efficient than the update markers used in [PS88, MP90]. Update markers require that each access to an argument on the stack first tests whether it is protected by an update marker. Then, argument consumption stops and the marked node becomes updated by the current partial application. Afterwards, the marker is unlinked and argument consumption proceeds.

5 Optimizations

This section addresses the issues of fast stack deallocation, deterministic computations and efficient memory mapping.

5.1 The last call optimization (LCO)

During narrowing, activation records cannot be automatically discarded at a function exit since they may in general be protected. However, the following condition is sufficient to detect deallocation:

the active choice point is below the current activation record

A test on this condition is included in each RET and JUMP operation thus improving space consumption by the cost of one branch for each call and return. Otherwise, stack space is deallocated by backtracking.

Furthermore, when executing the last alternative of a function its activation record may still be protected although it is never referenced again except for its return address. If no subcall has created a choice point, then the only choice point protecting the activation record is the one belonging to that last alternative. Thus, the choice point can be savely discarded when the last alternative is entered in order to deallocate the frame earlier. At the same time, the pointer to the logical variable stored in that choice point needs to be pushed to the trail stack such that backtracking can perform the unbinding. Accordingly, a *last call* is a tail call or a return operation in the last choice.

5.2 Green Cuts for free

When a SWITCH encounters a non-variable term it perfroms a deterministic case analysis instead of creating a choice point which manages bindings. Therefore, in the case of ground term reduction each tail call is a last call. When performing deterministic computations the JUMP machine is in its behaviour closely related to a stack based graph reduction machine like the STGM [PS88]. This clearly shows that the main difference between a graph reduction and a narrowing machine is given by the treatment of non-determinism by means of choice points, which is not surprising.

5.3 Memory Mapping

A further critical issue is the space consumption of the generated code in the heap as well as on the stack. Heap space is never explicitly freed, instead a stop-and-copy garbage collector reclaims unreferenced space.

Stack space consumption mainly depends on two factors: on the frequency with which frames will be discarded which has been discussed previously, and on their size. The naive way to determine the size of f's frame is to count all local and intermediate variables which are defined in f's code. This size is an upper bound of the actual size required to execute f. This approach is easily improved when the size is determined such that the local values of all execution paths fit into that frame. The most sophisticated approach is to use a register allocation algorithm which takes the lifetimes of variables into account. With that, two different values (identified by different variables) can be mapped to the same location when their lifetimes do not overlap.

A different approach is to demand (and discard) stack space dynamically. This technique is definitely adequate for performing deterministic computations. However, matters become complicated when choice points protect frames. Then, the only way to extend a protected frame is to demand a new fraction on top of the choice point, and to link it into the current frame. Since this may happen repeatedly, new fractions are linked into previous ones, and hence frames become non-flat and access to variables becomes indirect and more expensive.

In the same sense the calculation of the size of environments influences the consumption of heap space. Furthermore, two different environment techniques exist: by *environment sharing* static scopes are directly reflected by means of nested environments, and by *environment copying* only the relevant parts are copied into a new flat environment. The first technique tends to create large and heavily shared environments which may be hard to be garbage collected. The second one produces small single-referenced environments, however, it tends to consume heap space much more quickly.

It is not yet clear which of both is the superior technique. It seems to be good advice to use a mix determined by the code generator on the base of static information about the program.

6 Related Work

Based on an equivalence between narrowing and SLD-resolution [BGM89] functional logic programs can be transformed into Horn clauses which are then compiled into code for an extended WAM (KWAM) [BCM89]. The extension is an suspension mechanism which supports lazy evaluation.

Extensions of functional graph reduction machines have been extensively investigated for implementing *Babel* [MR88], e.g. [MKLR90, Loo91]. The machine proposed in [Loo91] adds choice points to a stack based reduction machine. Its design which is quite related to the WAM embodies the idea to perform deterministic computations more efficiently by eliminating choice points at run-time.

The implementation of innermost basic narrowing in the class of general confluent term rewriting systems has been investigated by [Müc90, Han90] and [Sch91]. The contribution of [Sch91] is to extend a WAM by a simple referencing scheme which during normalization handles the propagation of basic redexes into the substitution part, whereas in [Han90] an additional occurrence stack in a WAM keeps track of basic occurrences. This problem does not occur for a lazy evaluation strategy.

All these approaches, except [Müc90], use a WAM oriented instruction code. In our opinion, this has the drawback that traditional code generation techniques do not work very well. Even worse is that a rulewise unification in WAM style terminates less often than a columnwise regime would.

7 Conclusions and Future Work

This paper presented the implementation of lazy narrowing on a new abstract machine. It has been designed as a systematic extension of a conventional stack based architecture via the principles of graph reduction machines. An important point was the design of the instruction code such that it supports the application of standard compilation techniques, in particular the memory mapping. This approach ensures an efficient implementation. Furthermore, we developed a new understanding of choice points. Instead of statically assigning a choice point for rule selection, it is dynamically created only when an unbound variable is matched, and then it controls all alternative bindings of that variable. With that, a deterministic computation will automatically avoid unnecessary choice point creation. By representing objects as executable closures, a clean integration of logical and functional machine features was achieved. Furthermore, the interaction of choice points and activation records gave an elegant and efficient solution to the problem of redex non-determinism.

Laziness lead to the issue of *dependent updates*. The question arises whether situations can be detected where *dependent updates* need not yet to be reversed when execution backtracks over the binding of a logical variable on which they do not depend. We call that *sharing across choices*, and we are thinking of a kind of static dependency analysis.

We are currently engaged in writing a compiler which translates a functional logic language using the techniques presented in this paper. Further work will investigate the implementation of narrowing for general term rewriting systems.

8 Acknowledgements

This work was funded by the CEC (BRA 3147, PHOENIX). We owe special thanks to all members of the project for the stimulating atmosphere, in particular to Erik Meijer whose work inspired us from the beginnings two years ago, to Yike Guo whose work on constraint functional programming gave us many insights, and to Jim Cordy for his advice. Furthermore, we are grateful to those who participated in the compiler project: Gabi Keller, Thomas Bantle and Volker Wollmann.

References

[AK90]   Hassan Ait-Kaci. The WAM: a real tutorial. Technical report, DEC research center, Paris, 1990.

[ASU86]   Alfred V. Aho, Ravi Sethi, and Jeffrey D. Ullman. *Compilers – Principles, Techniques and Tools*. Addison-Wesley, 1986.

[BCM89]   P.G. Bosco, C. Cecchi, and C. Moiso. An extension of WAM for K-LEAF: a WAM- based compilation of conditional narrowing. In *Poceedings of the 6th Int. Conf. on Logic Programming*, page ?, 1989.

[BGM89]   P.G. Bosco, E. Giovanetti, and C. Moiso. Narrowing vs. SLD-resolution. J. of Theoretical Comp. Sci., Vol. 59, No. 1-2 pages 3–23, 1988.

[BL86]   Marco Bellia and Giorgio Levi. The Relation Between Logic and Functional Languages: a Survey. *Journal of Logic Programming*, 3(3), Oct. 1986.

[DFP86]   J. Darlington, A.J. Field, and H. Pull. The unification of functional and logic languages. In D. DeGroot and G. Lindstrom, editors, *Logic Programming*, pages 37–70. Prentice-Hall, Englewood Cliffs, New Jersey, 1986.

[DG89]   John Darlington and Yi-ke Guo. Narrowing and unification in functional programming - an evaluation mechanism for absolute set abstraction. In *Int. Conf. on Term Rewriting*. LNCS 355, 1989.

[Fay79]   M. Fay. First-order unification in an equational theory. In *"CADE" '79*, pages 161–167, 1979.

[FW87]   Jon Fairbairn and Stuart Wray. TIM - A Simple Machine to Execute Supercombinators. In *Conference on Functional Programming Languages and Computer Architecture*, LNCS 274, 1987.

[Han90]   Michael Hanus. Compiling logic programs with equality. In *PLILP*, LNCS 348, Springer Verlag, pages 387–401, 1990.

[Hul80]   Jean-Marie Hullot. Canonical forms and unification. In *5th Conf. on Automated Deduction*. LNCS 87, 1980.

[Loc91]   Hendrik C.R. Lock. The implementation of functional logic programming languages, April 1991. manuscript of a forthcoming dissertation, Universität Karlsruhe.

[Loo91]   Rita Loogen. From reduction machines to narrowing machines. In *CCPSD, TAPSOFT*, LNCS 494, pages 438–454, 1991.

[Mei88]   Erik Meijer. The Dump Environment Mix Machine. Informatics Department, University of Nijmegen, Nov. 1988. unpublished manuscript.

[MKLR90] Juan J. Moreno-Navarro, Herbert Kuchen, Rita Loogen, and Mario Rodriguez-Artalejo. Lazy narrowing in a graph machine. In *Conf. on Algebraic and Logic Programming*, LNCS 463, 1990. also appeared as report $N°$ 90-11 at RWTH Aachen.

[MP90]   Erik Meijer and Ross Patterson. Down with λ-lifting. University of Nijmegen and Imperial College, London, Sept. 1990. unpublished manuscript.

[MR88]   Juan J. Moreno-Navarro and Mario Rodriguez-Artalejo. BABEL: A functional and logic programming language based on constructor discipline and narrowing. In *First Int. Workshop on Algebraic and Logic Programming*, number 343 in LNCS, pages 223–232, 1988.

[Müc90]   A. Mück. How to compile narrowing. In *PLILP*, LNCS 348, pages 16–39, 1990.

[PS88]   Simon L. Peyton-Jones and J. Salkild. The Spineless Tagless G-Machine. In *Workshop on Implementations of Lazy Functional Languages*, Aspenas,Sweden, Sept. 1988. appeared also in 1989 ACM Conf. on Functional Progr. Languages and Computer Architecture.

[Red85]   Uday S. Reddy. Narrowing as the operational semantics of functional languages. In *IEEE Int. Symposium on Logic Programming*, pages 138–151. IEEE, 1985.

[Red86]   Uday S. Reddy. Functional Logic Languages, Part 1. In J.H. Fasel and R.M. Keller, editors, *Poceedings of a Workshop on Graph Reduction, Santa Fee*, LNCS 279, pages 401–425, 1986.

[Sch91]   Carlo Scharnhorst. Entwurf eines übersetzenden Narrowing-Systems basierend auf der WAM. Universität Karlsruhe, April 1991. Diplomarbeit.

[War83]   D.H.D Warren. An abstract Prolog instruction set. Techn. Note 309, SRI International,Menlo Park,Calif., October 1983.

A   Translation Rules for the JUMP machine

This appendix presents a translation scheme to JCode for the grammar given in part 2. Some details have been simplified for readability. In particular, without formalizing them here, $\mathcal{F}lat_n$ and $\mathcal{Z}ip$ transform a function definition into its canonical form (as illustrated in part 4). The full translation is given in [Loc91]. Let P_f be the set of rules defining function f.

$$\text{FUN } f \; [\text{x1},..,\text{xn}] \quad \{ \; \mathcal{N}_{unify}[\mathcal{Z}ip \; \mathcal{F}lat_n[P_f]] \; [\,] \; \}_{block}$$

Let $exp = \text{Y=c\_1}(..)\text{->e\_1} \; [] \; ... \; [] \; \text{Y=c\_k}(..)\text{->e\_k}$ be a column in a canonical definition.

$$\mathcal{N}_{unify}[exp] \; Args \;\; = \;\; \text{SWITCH y OF } \{ \quad \text{TERM } c_1 \; [..] : \mathcal{N}_{unify}[e_1] \; Args$$
$$\cdots$$
$$\text{TERM } c_k \; [..] : \mathcal{N}_{unify}[e_k] \; Args \quad \}$$
$$\mathcal{N}_{unify}[\; \text{Y=X -> } e] \; Args \;\; = \;\; \{ \; \text{x} := \text{y}; \; \mathcal{N}_{unify}[e] \; Args \; \}_{block}$$
$$\mathcal{N}_{unify}[main] \; Args \;\; = \;\; C[main] \; Args \qquad\qquad \textit{if it is a main expression}$$

A main expression is translated by scheme C. Let $k \geq 0$, and id a variable or function identifier.

$$C[id] \; [y_1,..,y_k] \qquad\qquad = \;\; \{ \; \text{JUMP } [y_1,..,y_k] \; id \; \}_{block}$$
$$C[c(t_1,..,t_n)] \; [\,] \qquad\quad = \;\; \{ \; A[t_1](z_1); \; ...; \; A[t_n](z_n);$$
$$\text{x} := \text{create\_term}(c,[z_1,..,z_m]); \; \text{RET TERM } c \; \text{x} \; \}_{block}$$
$$C[id \; e_1 \; .. \; e_n] \; [y_1,..,y_k] = \;\; \{ \; A[t_1](z_1); \; ...; \; A[t_n](z_n); \; C[id \;] \; [z_1,..,z_n, \; y_1,..,y_k] \; \}_{block}$$

Non-strict argument expressions are translated by scheme A. Let $[z_1,..,z_m]$ be the set variables which appear free inside exp, but which are defined by some enclosing function definition.

$$A[c(t_1,..,t_n)](x) \;\; = \;\; \{ \; A[t_1](z_1); \; ...; \; A[t_n](z_n); \; \text{x} := \text{create\_term}(c,[z_1,..,z_m]); \; \}_{block}$$
$$A[y](x) \qquad\qquad = \;\; \{ \; \text{x} := \text{y} \; \}_{block}$$
$$A[exp](x) \qquad\quad\; = \;\; \{ \; \text{x} := \text{SUSP } [z_1,..,z_m] \; C[exp] \; [\,] \; \}_{block}$$

Local pattern matching and local function definitions can be sorted out of a let-expression. Let D_f be the set of local rules defining function f, and let $[z_1,..,z_m]$ be the set of its free variables. Note, that only function symbols are allowed to be recursive. We define:

$$C[\; \text{let } pat = exp, \; D_f \; \text{in } main] \; Args =$$
$$\{ \; A[exp](y); \; \mathcal{N}_{unify}[\mathcal{F}lat_n[\text{Y=pat-> } main]] \; Args$$
$$\text{FUN } f \; [\text{x1},..,\text{xn}] \; [z_1,..,z_m] \; \{ \; \mathcal{N}_{unify}[\mathcal{Z}ip \; \mathcal{F}lat_n[D_f] \; [\,]] \; \}_{block} \; \}_{block}$$

The Instruction Set

- $\text{x}:=\text{create\_term}(c,[p1,...,pn])$ creates a a term with arguments identified by $p1,...,pn$. $\text{x}:=\text{create\_var}()$ creates a closure which represents an unbound logical variable.
- RET TERM c x returns constructor c and the term identified by x.
- CALL id $[\text{x1},...,\text{xn}]$ calls id with the actual arguments $\text{x1},...,\text{xn}$.
- JUMP id $[\text{x1},...,\text{xn}]$ tail calls id with the actual arguments $\text{x1},...,\text{xn}$.
- SWITCH y OF $\{..\}$ performs a case analysis over the type and constructors of the object y. When y is an unbound variable, a choice point is created which manages the bindings of y.
- RETRY enters the topmost choice point and tries its alternative.
- $\text{z}:=$ SUSP $[\text{x1},..,\text{xn}]$ code defines a thunk which executes code in the environment $[\text{x1},...,\text{xn}]$, and creates a suspension at run time which then is identified by z.
- FUN f $[\text{x1},...,\text{xn}]$ $[\text{v1},...,\text{vm}]$ code is a function definition for f whose parameters are identified by $[\text{x1},...,\text{xn}]$. In the case of a local function definition $[\text{v1},...,\text{vm}]$ is non-empty and identifies the free variables of f which are defined in the enclosing definition. When f is called, that environment is passed as an implicit parameter identified by a machine register.

Semantics-Directed Generation of a Prolog Compiler*

Charles Consel   Siau Cheng Khoo

Yale University

1 Introduction

In [22], Nicholson and Foo presented the denotational semantics for the language Prolog. As in [15, 11] the aim was to formalize the core of Prolog. Beyond the theoretical interest, it has long been argued that denotational definitions can be used to derive interpreters or compilers [24, 27, 21]. This approach is attractive because it closely relates the formal specification and its implementation.

This paper describes how the denotational semantics of Prolog can be used to *interpret* and to *compile* Prolog programs. Furthermore, it is shown how a *compiler* for the language can be *automatically generated*.

Our approach can be decomposed as follows. The semantic equations of Prolog are coded in a functional programming language: a side-effect free dialect of Scheme [23]. The result can be viewed as an interpreter and consequently be executed by a Scheme processor.

Then, compilation is achieved, using partial evaluation [1], by specializing the interpreter with respect to a program. Because partial evaluation is semantic preserving [17], the target code has the same behavior as the interpretation of the original program. Moreover, since a partial evaluator is a static semantic processor [24], compile-time actions are executed, and the result solely represents dynamic operations as shown in [18]. A source program and the corresponding compiled code are displayed in Appendix A.

Although the compiled code has been found to run about six times faster than the interpreted code, the compilation phase might be slow [18]. However, our experiment is based on Schism [4, 5], a self-applicable partial evaluator for a side-effect free dialect of Scheme. As such, Schism can generate compilers; this is done by specializing the partial evaluator with respect to an interpreter [17]. As a result, the compilation phase is about twelve times faster than specialization of the interpreter.

Our approach improves on previous work [18, 12, 27] in that: (i) it enables compiler generation and consequently speeds up the compilation process, and (ii) it goes beyond the usual mapping from syntax to denotations by processing the static semantics of the language definition.

The success of a semantic-directed compilation depends on the ability to distinguish between the static semantics (the usual compile-time actions) and the dynamic semantics. In the context of partial evaluation, this distinction is achieved automatically by a preliminary phase called *binding time analysis* [17].

Mix [16] was the first partial evaluator able to generate compilers as well as a compiler generator. It partially evaluates first order recursive equations. Schism handles both higher order functions and data structures [6]; and, it is based on a *polyvariant* binding time analysis unlike the existing binding time based-partial evaluators. As a result, it extends the class of applications that can be tackled by a self-applicable partial evaluator. This is illustrated by the partial evaluation of a Prolog definition. Indeed, because Prolog is defined by a continuation semantics, the corresponding interpreter requires the partial evaluator to handle higher order functions. Because some semantic arguments in the language definition can be both compile-time and run-time, a polyvariant binding time analysis is essential (see Section 4). As a consequence, the interpreter input to the partial evaluator is almost a direct transliteration of the semantic definition.

---

*This research was supported in part by NSF and DARPA grants CCR-8809919 and N00014-88-K-0573, respectively. The second author was also supported by a National University of Singapore Overseas Graduate Scholarship. Address: Department of Computer Science, Yale University, P.O. Box 2158, New Haven, CT 06520, USA. Email: {consel,khoo}@cs.yale.edu.

The paper is organized as follows. Section 2 introduces partial evaluation, presents Schism and lists the related work. Section 3 discusses the language Prolog by first briefly presenting its denotational definition, and then by describing its representation in Schism. Section 4 investigates the partial evaluation aspects of the specification. The paper concludes with an assessment in Section 5; a method to improve the compiled programs by specializing them further is also described.

2 Partial Evaluation

Partial evaluation aims at specializing a program with respect to some of its input. The partial evaluation phase can be seen as a staging of the computations of the program: expressions that only operate on available data are executed during this phase; for the others, a residual expression is generated. This staging improves the execution time of the specialized program compared to the original program.

Using binding time analysis, these early computations can be identified independently of the actual values of the input. In essence, this phase determines when the value of a variable is available: if the value is known at partial evaluation-time, it is said to be *static*; if it is not known until run-time it is *dynamic*. This information characterizes the computations that may be performed at partial evaluation-time — *static expressions* — or at run-time — *dynamic expressions*. Semantically speaking, binding time analysis determines the static and the dynamic semantics of a program for a given description of its inputs (*i.e.*, known/unknown). This division greatly simplifies the partial evaluation process. Indeed, to process the static semantics, the partial evaluator simply follows the binding time information to reduce the static expressions of the program. This simplification of the partial evaluation process is crucial to self-application [17].

Self-application is achieved by specializing the partial evaluator with respect to an interpreter and yields a compiler. A compiler generator can be obtained by specializing the partial evaluator with respect to itself. Beyond the unusual aspects of these applications, they are of practical interest: compilation and generation of compilers are improved. Indeed, compilation using a generated compiler is about twelve times faster than compilation by specialization of an interpreter with respect to a program. A comparable speed-up is obtained for the generation of a compiler by applying the compiler generator to an interpreter rather than by specializing the partial evaluator with respect to an interpreter.

2.1 Schism

Schism [4, 5] is a *self-applicable* partial evaluator for a side-effect free dialect of Scheme. As such, it can generate a compiler out of the interpretive specification of a programming language. The source programs are written in pure Scheme: a weakly typed, applicative order implementation of lambda-calculus. Schism handles *higher order functions* as well as the *data structures* manipulated by the source programs, even when they are only *partially known*.

2.2 Related Work

Partial evaluation of Prolog was taken up in [19]. Since then, several partial evaluators of Prolog have been developed, but mostly written in Prolog (*e.g.*, [13, 14, 26]). In [14] and [13] self-application of Prolog is addressed. The former discusses the problem of self-application and proposes some solutions. The latter describes an implementation which is said to be small and minimal in functionality; the results obtained are limited. None of them address the binding time analysis of Prolog.

In [20], Komorowski presents a partial evaluator for Prolog programs. An *Abstract Prolog Machine* is defined in Lisp to describe the operational semantics of Prolog. Then, formally correct program transformations based on partial evaluation are introduced by enhancing this abstract machine. Self-application is not investigated.

I	$\in$	Ide	Identifiers
B	$\in$	Con	Constants symbols
F	$\in$	Fun	Function/Predicate symbols
G	$\in$	$Goals$	Goal lists
P	$\in$	$Pred$	Predicates and terms
A	$\in$	Arg	Argument lists
C	$\in$	$Clause$	Clause
S	$\in$	$Prog$	Sentences (or Programs)
Q	$\in$	$Input$	Queries

$$S ::= Q, C$$
$$Q ::= :-G$$
$$C ::= C_1; C_2 \mid P \mid P:-G$$
$$G ::= P, G_1$$
$$P ::= I \mid B \mid F(A)$$
$$A ::= P, A_1 \mid P$$

Figure 1: Syntactic Domains and Syntactic Rules

Kahn and Carlsson describe in [18] a partial evaluator that partially evaluates a Prolog interpreter, written in Lisp, with respect to a Prolog program, yielding an equivalent Lisp program. This program is then compiled into machine language using an existing Lisp compiler. The resulting programs are said to be efficient, however, the compilation phase appeared to be slow. They suggested that self-application could solve the problem but did not explore this issue.

Felleisen presents in [12] an implementation of Prolog in Scheme based on macro-expansion. Prolog entities are transliterated into corresponding Scheme constructs on a one-to-one basis. The efficiency of the code produced depends highly on how each Prolog entity is being transliterated. The approach does not address compile-time processing (no static reductions).

3 Specification of Prolog

This section first gives an overview of the denotational semantics of Prolog as presented in [22]. Then, its representation for Schism is described.

3.1 Denotational Semantics

The denotational definition of the language consists of three parts: the abstract syntax (Figure 1), the semantics domains (Figure 2) and the valuation functions (Figure 3).

The crucial aspect of this semantics is the control. Traditionally, it can be modeled by a semantic argument called *continuation*. As described in [24], a backtracking facility can be integrated into a language by using a *failure continuation*. The continuation representing the usual evaluation sequence is called the *success continuation*. These two continuations capture the main aspects of the control of the denotational definition. The continuation functions are displayed in Figure 2.

The meaning of a program is given by the functions defined in Figure 3-a. Functions $\mathcal{D}$ and $\mathcal{C}$ are responsible for setting up the database and declaring the clauses respectively. Function $\mathcal{G}$ processes queries and premises. Function $\mathcal{S}$ specifies the semantics of a program. Functions $\mathcal{P}$ and $\mathcal{A}$ assign new locations for identifiers in the current clause. *Unify* defines the unification process. It unifies two terms and returns a substitution list containing the variable instantiations created during the unification.

Figure 3-b gives a general idea of the call graph of the valuation functions. Note that for convenience, functions $\mathcal{P}$ and $\mathcal{A}$ appear twice in this figure. Indeed, they first manipulate goals (upper part of the diagram) and then clauses (lower part).

3.2 Interpreter

This section presents the Schism representation for the semantics described above. To ease the coding of the denotational definition, our system provides constructs that define and manipulate types; these are simplified versions of ML constructs. The construct defineType defines a product or a sum depending on

Compound Domains

$\tau \in Tv$	$=$	$Var + Con + [Fun \times Av]$	Terms
$\pi \in Av$	$=$	Tv^*	Argument lists
$\rho \in Env$	$=$	$Ide \rightarrow Var$	Identifier environments
$\theta \in Subs$	$=$	$Var \rightarrow [Tv + uninstantiated]$	Substitutions

Continuation Functions

$\psi \in Qc$	$=$	$Subs \rightarrow Res$	Substitution continuation
$\zeta \in Ec$	$=$	$Env \rightarrow Qc$	Continuation with Environment
$\upsilon \in Tc$	$=$	$Tv \rightarrow Env \rightarrow Qc$	Terms
$\omega \in Ac$	$=$	$Av \rightarrow Env \rightarrow Qc$	Argument lists
$\kappa \in Kc$	$=$	$Res \rightarrow Qc$	Continuation with failure
$\gamma \in Gc$	$=$	$Env \rightarrow Res \rightarrow Qc$	Goal list
$\delta \in Db$	$=$	$Tv \rightarrow Kc \rightarrow Kc$	Database

Figure 2: Semantic Domains

whether it contains one clause or more. The constructs let and let* create new bindings, as in Scheme, but in addition may perform destructuring operations on elements of products. The construct caseType is a conditional on the injection tag of the element of a sum and allows the destructuring of this element.

Except for a few technical details, this interpreter is a direct coding in Scheme of the valuation functions presented above. As a result, the implementation is precise and easy to reason about.

Figure 4 displays the data types corresponding to the abstract syntax and the domain of denotable values defined in the denotational semantics.

The interpretation functions, corresponding to the valuation functions, are shown in Figure 5. Notice that the filter construct specifies how calls to a given function should be treated. This issue is addressed in Section 4.2. The unification function as well as the auxiliary functions are omitted; they correspond to the functions presented by Nicholson and Foo in [22].

4 The Interpreter from a Partial Evaluation Point of View

The Prolog interpreter receives two inputs: the first is the Prolog program, called the database; the second input is the query. The functionality of the main function of the interpreter is $S : Prog \rightarrow Input \rightarrow Res$. The first input is static, and the second is dynamic.

During partial evaluation, the interpreter is specialized with respect to a Prolog program. The resulting residual program accepts an input query and yields a result. The functionality of the main function of the specialized program is $S_{Prog} : Input \rightarrow Res$.

4.1 Multiple Binding Time Signatures

To explain how the static and the dynamic computations of the Prolog interpreter are determined, we first need to introduce the notion of *binding time signature*. The binding time analysis determines a binding time signature for each function of a program. The binding time signature specifies the binding time value of each parameter of a function and the binding time value of the result. For clarity, we simplify the binding time signatures presented in this paper. The symbols used to represent a binding time signature are *St*, *Dy*, *Cl* and *Ps*; they denote respectively the binding time value static, dynamic, closures and partially static data. The binding time value of the result of a function is assumed to be dynamic.

$S : Prog \rightarrow Input \rightarrow Res$
$\quad S[C][:-Q.] = G[Q](D[C]) \, printall \, (\lambda\iota.unbound) <> (\lambda\upsilon.unused)$
$D : Clause \rightarrow Tv \rightarrow Kc \rightarrow Kc$
$\quad D[C] = fixedpoint(C[C])$
$C : Clause \rightarrow Db \rightarrow Tv \rightarrow Kc \rightarrow Res \rightarrow Subs \rightarrow Res$
$\quad C[C_1, C_2]\delta\tau\kappa\phi\theta = C[C_1]\delta\tau\kappa(C[C_2]\delta\tau\kappa\phi\theta)\theta$
$\quad C[P.] = P[P](\lambda\tau_1\rho. \, unify \, \tau\tau_1(\kappa\phi)\phi)(\lambda\iota.unbound)$
$\quad C[P:-G] = P[P](\lambda\tau_1\rho_1. \, unify \, \tau\tau_1(G[G]\delta(\lambda\rho_2.\kappa)\rho_1\phi)\phi)(\lambda\iota.unbound)$
$G : Goals \rightarrow Db \rightarrow Gc \rightarrow Env \rightarrow Res \rightarrow Qc$
$\quad G[P]\delta\gamma\rho\phi = P[P](\lambda\tau\rho_1.\delta\tau(\gamma\rho_1)\phi)\rho$
$\quad G[G_1, G_2]\delta\tau = G[G_1]\delta(G[G_2]\delta\gamma)$
$P : Pred \rightarrow Tc \rightarrow Ec$
$\quad P[I]\upsilon\rho = \rho I = unbound \rightarrow newvar(\lambda\tau.\upsilon\tau\rho[\tau/I]), \upsilon(\rho I)\rho$
$\quad P[F(A)]\upsilon = A[A]\lambda\zeta.\upsilon < F,\zeta >$
$A : Arg \rightarrow Ac \rightarrow Env \rightarrow Qc$
$\quad A[P]\omega = P[P]\lambda\tau.\omega < \tau >$
$\quad A[P, A]\omega = P[P]\lambda\tau.A[A]\lambda\zeta.\omega(\tau.\zeta)$
$Unify : Tv \rightarrow Tv \rightarrow Kc \rightarrow Res \rightarrow Subs \rightarrow Res$

(a)

(b)

Figure 3: Valuation Functions and The Call Graph

```
(defineType Atom          (defineType Clause         (defineType Term
   (Constant      value)      (Seq  Clause1 Clause2)     (Const         value)
   (Identifier    name)       (Fact            atom)     (Var        name ref)
   (Predicate function arguments))  (Rule conclusion premises))  (Pred function arguments))
```

Figure 4: Abstract Syntax and Denotable Values

Initially, the main function S calls function G with a *dynamic* input query and a *static* database. Then, to satisfy subgoals function G is called *recursively*, but with *static* goals, i.e., the premises (see function C in Figure 5). Therefore, function G and the inner functions are first called in a context where the query is dynamic, and then, in a context where it is static. [1]

This is summarized in Figure 6 where the binding time signatures of the interpretation functions are displayed. The functions have more than one binding time signatures to reflect the fact that they are called in different binding time contexts. To distinguish each context the function names are prefixed with S, D or P.

Note that partially static data arises because of terms containing both static and dynamic parts. In the Prolog interpreter, this is illustrated by the environment which binds static identifiers to dynamic variables.

Since each function has multiple binding time signatures, if a binding time analysis maps each function to only one binding time signature, it will have to "fold" these two binding time signatures into one. This is the case for a *monovariant* binding time analysis. As a result for the Prolog interpreter, almost all the computations are considered run-time and specialization performs very poorly. To avoid this situation, one can duplicate the original set of interpretation functions: one set of functions deals with the initial query (dynamic), the other manipulates the premises (static). Because each function of the resulting program is then called with only one pattern of binding time values, the binding time analysis does not do any "folding" and consequently more static computations are detected.

---

[1] Notice that, in addition to the contexts described above, function G is also called in a context where the initial environment is static; this situation occurs when function G is called by function S.

```
(define (S database queries)
  (filter SPECIALIZE (list database queries))
  (G queries (lambda (t gc fc env res subs) (D database t gc fc env res subs))
             (lambda (env s1 subs) (update-result s1 (Q:inst env subs)))
             (lambda (s1 subs) s1) (init-env) (init-result) (init-subs) ))

(define (G goals db gc fc env res subs)
  (filter SPECIALIZE (list goals db DYN fc env DYN subs))
  (cond ((null? goals) (gc env res subs))
        (else (P (car goals) (lambda (goal e subs1)
                               (db goal (lambda (env1 res1 subs2) (G (cdr goals) db gc fc e res1 subs2))
                                   fc env res subs1))
                 env subs))))

(define (D database t gc fc env res subs)
  (filter UNFOLD VOID)
  (C database (lambda (t' gc' fc' env' res' subs') (D database t' gc' fc' env' res' subs'))
     t gc fc env res subs))

(define (C' cl cls db t gc fc env res subs)
  (filter SPECIALIZE (list cl cls db t gc fc env res subs))
  (C cls db t gc fc env (C cl db t gc fc env res subs) subs))

(define (C clause db t gc fc env res subs)
  (filter UNFOLD VOID)
  (caseType clause
    ([Seq cl cls] (C' cl cls db t gc fc env res subs))
    ([Fact term]
     (P term (lambda (tv e subs1) (unify tv t (lambda (s) (gc e res s)) (lambda (s) (fc res s)) subs1))
        (init-env) subs))
    ([Rule conclusion premises]
     (P conclusion
        (lambda (tv e subs1)
          (unify tv t (lambda (s) (G premises db gc fc e res s)) (lambda (s) (fc res s)) subs1))
        (init-env) subs)) ))

(define (P t tc env subs)
  (filter (if (dyn? t) SPECIALIZE UNFOLD) (list t DYN DYN DYN))
  (caseType t
    ([Constant number] (tc (Const number) env subs))
    ([Identifier name] (Let ( [varlist (associate name env)] )
                         (if (null? varlist)
                             (lete ( [v0 (New-var subs)] [v (Var name v0)] )
                                   (tc v (cons (cons name v) env) (update-subs subs v 'uninstantiated)))
                             (tc (cdr varlist) env subs))))
    ([Predicate fn args] (A args (lambda (arglis e subs1) (tc (Pred fn arglis) e subs1)) env subs)) ))

(define (A args v env subs)
  (filter (if (dyn? args) SPECIALIZE UNFOLD) (list args DYN env DYN))
  (if (null? args) (v '() env subs)
      (P (car args) (lambda (t e1 subs1)
                      (A (cdr args) (lambda (arglis e subs2) (v (cons t arglis) e subs2)) e1 subs1))
         env subs)))
```

Figure 5: Interpretation Functions

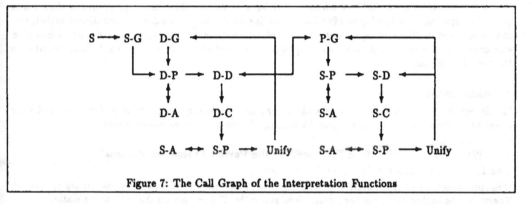

$$
\begin{array}{ll}
\text{(G goals db gc fc env res subs)} \\
\quad S-G: \quad Dy \times Cl \times Cl \times Cl \times St \times Dy \times Dy \to Dy \\
\quad D-G: \quad Dy \times Cl \times Cl \times Cl \times Dy \times Dy \times Dy \to Dy \\
\quad P-G: \quad St \times Cl \times Cl \times Cl \times Ps \times Dy \times Dy \to Dy \\
\text{(D database t gc fc env res subs)} \\
\quad D-D: \quad St \times Dy \times St \times Cl \times Cl \times Dy \times Dy \times Dy \to Dy \\
\quad S-D: \quad St \times Ps \times St \times Cl \times Cl \times Ps \times Dy \times Dy \to Dy \\
\text{(C clause db t gc fc env res subs)} \\
\quad D-C: \quad St \times Cl \times Dy \times Cl \times Cl \times Dy \times Dy \times Dy \to Dy \\
\quad S-C: \quad St \times Cl \times Ps \times Cl \times Cl \times Ps \times Dy \times Dy \to Dy
\end{array}
\qquad
\begin{array}{ll}
\text{(P t tc env subs)} \\
\quad D-P: \quad Dy \times Cl \times Dy \times Dy \to Dy \\
\quad S-P: \quad St \times Cl \times Ps \times Dy \to Dy \\
\text{(A args v env subs)} \\
\quad D-A: \quad Dy \times Cl \times Dy \times Dy \to Dy \\
\quad S-A: \quad St \times Cl \times Ps \times Dy \to Dy
\end{array}
$$

Figure 6: Binding Time Signatures for the Functions

Figure 7: The Call Graph of the Interpretation Functions

In Schism, since the binding time analysis is *polyvariant* this duplication happens automatically. A new instance of a function is created whenever it is called with a new pattern of binding time values. The finiteness of the duplication process is ensured by a function of the binding time analysis which determines when two patterns differ. A more detailed description of this polyvariant binding time analysis can be found in [7].

4.2 Termination of Partial Evaluation

Partial evaluating function calls has two common pitfalls: infinite call unfolding and infinite function specialization [17, 5]. Infinite call unfolding may occur when recursive calls to a function under dynamic control are systematically unfolded. In the Prolog interpreter this situation happens for the recursive calls to functions G and unify (see Figure 7). Therefore, we use the filter construct to instruct the partial evaluator to create specialized versions of these functions. As an example, consider the interpretation function G displayed in Figure 5. This function has a filter whose first part instructs specialization. Notice that, the filter of function C' causes the partial evaluator to specialize this function with respect to each clause of the database.

Infinite function specialization may arise when recursive calls are systematically suspended and some arguments always have a different static value. As show by Consel and Danvy in [8], this situation is caused typically by a static accumulator under dynamic control. To ensure termination of specialization, the accumulator needs to be generalized. Notice that for continuation passing style programs, the continuation usually represents an accumulator in the sense that it accumulates computations to be performed once the continuation is applied.

In the Prolog interpreter, we have detected two accumulators; they are denoted by variables gc and res, both in function G. Variable gc is the success continuation, it accumulates computations to be performed once the current goal is proved. Variable res accumulates the components of the final answer. These two variables are initially bound to static values since the main function of the interpreter, S, calls function G with a static initial value for gc and res. To ensure termination of specialization, the second part of the filter of function G prevents the value of gc and res from being propagated.

To improve the size of the compiled code (*i.e.*, the residual program), we have generalized some variables of other interpretation functions (*e.g.* functions P and A). In doing so we avoid propagating values that are not essential to compilation.

Finally, let us point out that the existing strategies for automatically annotating programs are too limited to handle the Prolog interpreter. Indeed, these strategies are either restricted to first order programs [25] or they only prevent from infinite unfolding [2]. Furthermore, if the binding time analysis is polyvariant, the annotation construct needs to be *conditional*, that is, it needs to yield different program transformations (unfolding/specialization) and generalizations depending on the available values. As mentioned earlier, this can be achieved in Schism using a filter. This annotation construct contrasts with the existing annotation strategies (*e.g.*, the strategy used by Mix [16]) where an annotation denotes an unconditional directive to the partial evaluator.

5 Assessment

In this section we discuss what has been achieved by partial evaluation, evaluate its performance, and show a way to produce better residual program by refining the input to the partial evaluator.

5.1 What Has Actually Been Processed by the Partial Evaluation Process?

The Failure Continuation has been Eliminated

Compiled programs have an interesting property: *the failure continuation has been completely eliminated*. Therefore, the backtracking has been determined statically. This is because the database is static.

Consider the Prolog program in Appendix A and its corresponding residual program. The compilation of the backtracking continuation can be illustrated by comparing the traversal of the database that the interpreter would performed (Figure 8-a) with the traversal of the database represented by the compiled program (Figure 8-b). These diagrams also include the accumulation of a result which is denoted by the symbol res. The labels $c1, \ldots, c4$ correspond to the clauses of the source program. Each clause in Figure 8-b has attached its corresponding specialized function. Figure 8 clearly shows that the intermediate backtracking has been eliminated in the compiled code.

Lookup Operations in the Environment are Compiled

Because the environment is a partially static data (static identifiers and dynamic variables), access to a given identifier/variable pair has been compiled. The resulting expression is a sequence of operations to access the variable in this pair.

Part of the Unification Process is Compiled

When the interpreter manipulates premises, part of the unification process can be performed. Indeed, in this context, the type of terms to unify is static and thus processing depending on this information can be performed. What remains then is the unification between the arguments of two predicate terms.

5.2 What are the Dynamic Operations?

Some operations cannot be performed during the partial evaluation process because some data are not available. The printing of the result is deferred to run-time (function Q:inst). Unification process (function

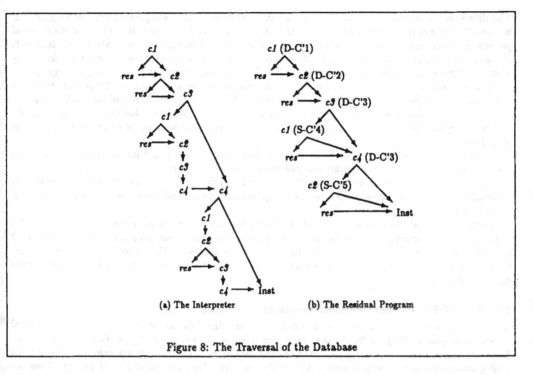

(a) The Interpreter   (b) The Residual Program

Figure 8: The Traversal of the Database

unify) is frozen when terms to be unified are dynamic. Since the variables cannot be generated until run-time, the substitution list is dynamic. Therefore, operations that manipulate the substitution list are frozen.

Table 1 summarizes the above explanations by classifying the interpretation functions as static if they are eliminated during partial evaluation and dynamic otherwise. Notice that functions lookup-env and Unify are listed in both columns; this means that they are treated differently under different context. We could have labeled these different instances with different names, as we did for the interpretation functions.

Static Functions	Dynamic Functions
S-C	D-C'
S-P, S-A	D-P, D-A
unify	unify, unifyargs
lookup-env	lookup-env
	S, S-G, D-G, P-G
	lookup-subs, update-subs
	New-var, Q:inst

Table 1. Static vs. Dynamic Functions

Prolog Program		Speed-up
Number of Facts	Number of Rules	
4	4	4
6	2	5
8	8	7

Table 2. Speed-up with Compilation

5.3 Performance of Partial Evaluation

Programs compiled by partial evaluation have been found to be about six times faster than interpretation. Note that this speed-up is more important with other languages. For Algol-like programs, Consel and Danvy reported in [9] that compiled code is more than a hundred times faster than the interpreted source program.

This difference is due to the fact that the static semantics of Prolog is not as important as other languages. It is especially difficult to see how unification could be further processed statically without introducing special purpose program transformations. As we have seen in the example of the compiled code, the unification is the major component of the dynamic semantics. It is interesting to notice that partial evaluators for Prolog written in Prolog do not deal with unification either. Indeed, unification is part of the target language.

It is generally difficult to fairly evaluate the performance improvement obtained by partial evaluation and the size of the resulting programs since they strongly depend on the specificity of the Prolog program. In particular, we notice that the speed-up with compilation is related to the number of possible unsuccessful unifications contained in the source program. Indeed, the intermediate backtracking have been removed by partial evaluation.

Some run-time results are displayed in Table 2; they are obtained with both the interpreter and the residual programs compiled into machine language using a Scheme compiler.

The size and the structure of compiled code are not surprising: the residual program represents the traversal of the database. A specialized function is generated for each unification clause. This is illustrated in Appendix A.

Further work needs to be done to extend the Prolog interpreter for a larger subset of the language. To improve the size of target code, we want to investigate combinator based-semantics [24]. This approach could capture more compactly the dynamic semantics and be more abstract with respect to its implementation. Consequently, different strategies for implementing the dynamic semantics could be explored to improve the run-time of target code.

5.4 Specializing Further Compiled Programs

So far, we consider the input query as being dynamic and the database static. Therefore, the residual programs are general: they can handle any query. One might want to obtain more specific residual programs, *i.e.*, programs dedicated to a query restricted to some predicates. This could be achieved by extending the Prolog interpreter to deal with incompletely specified queries. They are called *partial queries*. They are used by the partial evaluator to eliminate irrelevant clauses and yield a residual program dedicated to this partial query. As a result, the size of the residual program will be drastically reduced. This technique of using partial query has been investigated in [20] for source-to-source program transformations; the resulting program is called a *pruned version*. The necessary changes to the interpreter are presented in [10].

5.5 Incorporating Cut Operation

The cut operation is used to control backtracking. It can be modeled semantically by another continuation function. This requires few modifications to the Prolog semantics, as shown in [22]. The modified Prolog interpreter is described in [10].

When specializing the resulting interpreter with respect to a program including the cut operation, the cut continuation is eliminated. This is not surprising since the cut operation is part of the program it can be treated statically.

Acknowledgments

This work has benefited from David Schmidt's interest and insightful comments. Thanks are also due to Olivier Danvy and Paul Hudak for their thoughtful comments.

References

[1] D. Bjørner, A. P. Ershov, and N. D. Jones, editors. *Partial Evaluation and Mixed Computation*. North-Holland, 1988.

[2] A. Bondorf and O. Danvy. Automatic autoprojection of recursive equations with global variables and abstract data types. DIKU Research Report 90/04, University of Copenhagen, Copenhagen, Denmark, 1990. To appear in Science of Computer Programming.

[3] W. F. Clocksin and C. S. Mellish. *Programming in Prolog*. Springer-Verlag, 1981.

[4] C. Consel. New insights into partial evaluation: the Schism experiment. In H. Ganzinger, editor, *ESOP'88, 2nd European Symposium on Programming*, volume 300 of *Lecture Notes in Computer Science*, pages 236–246. Springer-Verlag, 1988.

[5] C. Consel. *Analyse de Programmes, Evaluation Partielle et Génération de Compilateurs*. PhD thesis, Université de Paris VI, Paris, France, June 1989.

[6] C. Consel. Binding time analysis for higher order untyped functional languages. In *ACM Conference on Lisp and Functional Programming*, pages 264–272, 1990.

[7] C. Consel. *The Schism Manual*. Yale University, New Haven, Connecticut, USA, 1990. Version 1.0.

[8] C. Consel and O. Danvy. For a better support of static data flow. In *FPCA'91, 5th International Conference on Functional Programming Languages and Computer Architecture*, 1991. To appear.

[9] C. Consel and O. Danvy. Static and dynamic semantics processing. In *ACM Symposium on Principles of Programming Languages*, pages 14–23, 1991.

[10] C. Consel and S. C. Khoo. Semantics-directed generation of a Prolog compiler. Research Report 781, Yale University, New Haven, Connecticut, USA, 1990. Extended version.

[11] S. K. Debray and P. Mishra. Denotational and operational semantic for prolog. *Journal of Logic Programming*, 5:61–91, 1988.

[12] M. Felleisen. Transliterating Prolog into Scheme. Technical Report 182, Indiana University, Bloomington, Indiana, 1985.

[13] H. Fujita and K. Furukawa. A self-applicable partial evaluator and its use in incremental compiler. In Y. Futamura, editor, *New Generation Computing*, volume 6 of *2,3*. OHMSHA. LTD. and Springer-Verlag, 1988.

[14] D. A. Fuller and S. Abramsky. Mixed computation of Prolog. In D. Bjørner, A. P. Ershov, and N. D. Jones, editors, *Partial Evaluation and Mixed Computation*. North-Holland, 1988.

[15] N. D. Jones and A. Mycroft. Stepwise development of operational and denotational semantics for Prolog. In *IEEE International Symposium on Logic Programming*, pages 289–298, 1984.

[16] N. D. Jones, P. Sestoft, and H. Søndergaard. An experiment in partial evaluation: the generation of a compiler generator. In J.-P. Jouannaud, editor, *Rewriting Techniques and Applications, Dijon, France*, volume 202 of *Lecture Notes in Computer Science*, pages 124–140. Springer-Verlag, 1985.

[17] N. D. Jones, P. Sestoft, and H. Søndergaard. Mix: a self-applicable partial evaluator for experiments in compiler generation. *Lisp and Symbolic Computation*, 2(1):9–50, 1989.

[18] K. M. Kahn and M. Carlsson. The compilation of Prolog programs without the use of Prolog compiler. In *International Conference on Fifth Generation Computer Systems*, pages 348–355, 1984.

[19] H. J. Komorowski. A specification of an abstract Prolog machine and its application to partial evaluation. Linkoping studies in science and technology dissertations n° 69, Linkoping University, Linkoping, Sweden, 1981.

[20] H. J. Komorowski. Partial evaluation as a means for inferencing data structures in an applicative language: A theory and implementation in the case of Prolog. In *ACM Symposium on Principles of Programming Languages*, 1982.

[21] P. Lee and U. F. Pleban. On the use of Lisp in implementing denotational semantics. In *ACM Conference on Lisp and Functional Programming*, pages 233–248, 1986.

[22] T. Nicholson and N. Foo. A denotational semantics for Prolog. *ACM Transactions on Programming Languages and Systems*, 11(4), 1989.

[23] J. Rees and W. Clinger. Revised³ report on the algorithmic language Scheme. *SIGPLAN Notices*, 21(12):37–79, 1986.

[24] D. A. Schmidt. *Denotational Semantics: a Methodology for Language Development*. Allyn and Bacon, Inc., 1986.

[25] P. Sestoft. Automatic call unfolding in a partial evaluator. In D. Bjørner, A. P. Ershov, and N. D. Jones, editors, *Partial Evaluation and Mixed Computation*. North-Holland, 1988.

[26] R. Venken. A Prolog meta-interpreter for partial evaluation and its application to source to source transformation and query-optimisation. In T. O'Shea, editor, *ECAI'84*. North-Holland, 1988.

[27] M. Wand. A semantic prototyping system. *SIGPLAN Notices, ACM Symposium on Compiler Construction*, 19(6):213–221, 1984.

A   Prolog Program and Its Residual

A Source Program :

```
c1    Male (Adam),
c2    Female (Eve),
c3    Person (x) :- Male (x),
c4    Person (x) :- Female (x).
```

The Residual Program (sugared) :

```
(define (SO queries) (state-result (D-G1 queries '() (init-state) (init-subs))))

(define (D-G1 goals env stt subs)
  (cond ((null? goals) (update-state-result stt (Q:INST env subs)))
        (else (D-P (car goals)
                   (lambda (goal2 e3 st1) (D-D2 goal2 (lambda (e2 subs1 st2) (D-G1 (list-tail goals 1) subs1 st2 e3))
                                          env st1 subs)) env stt))))

(define (D-D5 t gc env stt subs)
  (lete ([(list v0 st1) (new-var 'x stt)] [v1 (Var 'x v0)])
    (D-Unify (Pred 'person (list v1)) t (lambda (r1) (S-G1 gc (list (cons 'x v1)) st1 r1))
                                        (lambda (r1) st1) subs)))

(define (D-D4 t gc env stt subs)
  (lete ([(list v0 st1) (new-var 'x stt)] [v1 (Var 'x v0)])
    (D-Unify (Pred 'person (list v1)) t
             (lambda (r1) (D-D5 t gc env (S-G2 gc (list (cons 'x v1)) st1 r1) subs))
             (lambda (r1) (D-D5 t gc env stt subs st1)) subs)))

(define (D-D3 t gc env stt subs)
  (D-Unify (Pred 'female '((Const eve))) t (lambda (r1) (D-D4 t gc env (gc '() stt r1) subs))
                                           (lambda (r1) (D-D4 t gc env stt subs)) subs))

(define (D-D2 t gc env stt subs)
  (D-Unify (Pred 'male '((Const adam))) t (lambda (r1) (D-D3 t gc env (gc '() stt r1) subs))
                                          (lambda (r1) (D-D3 t gc env stt subs)) subs))

(define (S-G1 gc env stt subs)
  (S-D2 (Pred 'female (list (list-tail (car env) 1))) (lambda (e st2 su1) (gc env st2 su1)) env stt subs))

(define (S-G2 gc env stt subs)
  (S-D1 (Pred 'male (list (list-tail (car env) 1))) (lambda (e st2 su1) (gc env st2 su1)) env stt subs))

(define (S-D2 t gc env stt subs)
  (unifyargs '((Const eve)) (list-ref t 2)
             (lambda (r1) (list-ref (new-var 'x (list-ref (new-var 'x (gc '() stt r1)) 1)) 1))
             (lambda (r1) (list-ref (new-var 'x (list-ref (new-var 'x stt) 1)) 1)) subs))

(define (S-D1 t gc env stt subs)
  (unifyargs '((Const adam)) (list-ref t 2)
             (lambda (r1) (list-ref (new-var 'x (list-ref (new-var 'x (gc '() stt r1)) 1)) 1))
             (lambda (r1) (list-ref (new-var 'x (list-ref (new-var 'x stt) 1)) 1)) subs))
```

A framework to specify database update views for Prolog

Egon Boerger

Dip. di Informatica Universita di Pisa Cso Italia 40,I-56 100 PISA
boerger@dipisa.di.unipi.it

Bart Demoen

Dept. Computer Science Celestijnenlaan 200A B-3001 Leuven Belgium
BIM Kwikstraat 4 B-3078 Everberg Belgium
bimbart@cs.kuleuven.ac.be

Abstract

A formal specification of a set of reasonable database update views is presented. Each reasonable database update view is a superset of the minimal view and a subset of the maximal view. Both the logical view and the immediate update view are special cases. It is then argued that the only view attractive for implementing as well as for explaining to users of a Prolog system, is the minimal view.

1 Introduction

WG17, in its attempt to standardize Prolog, found it impossible to impose a specific view on database updates, like the logical view, or the immediate update view. Still, the effect of assert, retract and the execution of dynamic predicates had to be specified, preferably in such a way that at least the two afore mentioned views were allowed by the standard, and so that every implementation had to behave sensibly. At first, some attempts were made to construct a set of verbal rules with these characteristics. As was to be expected, it turned out that it was nearly impossible to make the verbal rules at the same time understandable, precise and complete. So, the alternative approach, constructing a formal specification, was attempted. We give here a completely formal but simple specification which follows closely the underlying intuitive understanding which is also described. Our specification can easily be embedded into the framework of the Prolog formalization by Gurevich's evolving algebras [Gurevich] as developed in [Boerger a,b,c], even though that description was given with the intent to describe the logical database update of view in accordance with the position of WG17 at that moment [Scowen]. We nevertheless will present our specification here from scratch, so that it is independent from any other reading, though acquaintance with [Lindholm] and [Moss] for a correct understanding of the motivation of the logical view is surely desirable.

The paper is organized as follows:

Section 2 contains a formal specification, within a framework closely related to [Boerger a,b,c].

Section 3 contains a set of verbal rules which attempt to express the same as the formal specification. Also a classification of views is presented.

Section 4 outlines the transition from the specification to an implementation: it contains only well-known implementation issues.

Section 5 addresses a problem related to the garbage collection of retracted clauses in two of the extreme views, from which the problems of a third one follow.

2 The formal specification.

In order to give a complete description of database update views, we need to describe the following objects and actions:

a clause   a choice point
asserta   assertz   retract
calling a predicate   backtracking

These will be described in the detail which is needed here, but not further.

We define in the system a global variable - or function with no arguments - which we call CLOCK, and which is initialized to 0. This CLOCK is incremented by 1 before every assert and before every retract. CLOCK can be thought of as a timing device, but 'database update counter' is closer to its meaning.

Clauses in the Prolog data base (PDB) are 5-tuples, consisting of

cl : the representation of a term as a clause
begin : integer
retracted : boolean
end : integer
a-asserted : boolean

The value of *end* will never be needed if *retracted* == false.

A clause ∈ PDB is uniquely defined by its *begin*.

In formal specification, the PDB as presented is not partitioned. In an implementation of course, partitioning the PDB - e.g. according to the main functor of the predicates - is crucial for efficiency.

The description of the actions:

> **asserta(cl)**
> > CLOCK := CLOCK + 1
> > add (cl,CLOCK,false,0,true) to PDB

> **assertz(cl)**
> > CLOCK := CLOCK + 1
> > add (cl,CLOCK,false,0,false) to PDB

> **retract((cl,b,false,0,any))**
> > CLOCK := CLOCK + 1
> > redefine the retracted field of the clause to true
> > redefine the end field of the clause to CLOCK

In this way, there is at most one clause CL ∈ PDB with *begin*(CL) = T for any $0 < T <= $ CLOCK, and none for T outside that range.

We define the function *before* : PDB x PDB → {true,false} by:

> *before*(CL1,CL2) ==

> > if (*a-asserted*(CL1) == true and *a-asserted*(CL2) == true and
> > > > > > *begin*(CL1) > *begin*(CL2)) then true

> > else
> > if (*a-asserted*(CL1) == true and *a-asserted*(CL2) == false) then true
> > else
> > if (*a-asserted*(CL1) == false and *a-asserted*(CL2) == false and
> > > > > > *begin*(CL1) < *begin*(CL2)) then true

> > else false

The function *before*, is a total order on PDB.

Any subset of PDB now has a minimal element for this order, so we can define:

minimal : P(PDB) → PDB   where P(PDB) is the power set of PDB
by
minimal(S) = the (unique) CL ∈ S such that *before*(CL,x) for all x ∈ S\{CL}

We also define *next* : PDB → PDB ∪ {none} by:

next(CL) ==
    if ({CL1 | CL1 ∈ PDB and *before*(CL,CL1) == true} == Ø) then none
    else *minimal*({CL1 | CL1 ∈ PDB and *before*(CL,CL1) == true})

Let B denote a choice point. We need two components in B, from which the following
functions are derived:
    activation_time : B → integers
    current_active : B → PDB

The value of these functions is defined later under backtracking.

calling a predicate
    if {CL | CL ∈ PDB and *retracted*(CL) == false} == Ø then goto fail
    else

        CL := *minimal*({CL | CL ∈ PDB and *retracted*(CL) == false})
        create a choice point B
        define *activation_time*(B) := CLOCK
        define *current_active*(B) := CL
        USE(CL)

backtracking
    CL := *current_active*(B)
    a := *activation_time*(B)
label:
    if *next*(CL) == none then remove choice point B and goto fail
    CL := *next*(CL)
    define *current_active*(B) := CL
    if (*retracted*(CL) == true and *end*(CL) <= a) then DO_NOT_USE
    else if (*begin*(CL) <= a)
        then if (*retracted*(CL) == false) then USE(CL)
                        else CHOICE1
        else if (*retracted*(CL) == true) then DO_NOT_USE
                        else CHOICE2

DO_NOT_USE means 'goto label'

CHOICE1 and CHOICE2 can have only the values: DO_NOT_USE or USE(CL), but their value can depend on the state of the Prolog engine, and even on how the Prolog engine got into this state: at any moment, the choice is completely free. In our opinion, any choice leads to a reasonable behavior of the Prolog system, although some choices are more coherent than others. In section 3.2 a classification of views is presented.

3 The verbal description

3.1 A set of verbal rules

The following is an attempt to capture verbally the formalization given above.

First some notions are defined:

> a clause is ALIVE from the moment it is asserted until it is retracted

> a clause is DELETED if it is retracted; note that retracting a clause does not end its presence in the Prolog data base: it merely changes one of the properties of the clause; retract ends the life of a clause

> a clause c1 COMES LATER THAN a clause c2 if
> c1 was a-assert'ed before c2 was a-assert'ed
> or
> c2 was z-assert'ed before c1 was z-assert'ed
> or
> c1 was z-assert'ed and c2 was a-assert'ed

The following holds for clauses whose head has the same predicate indicator as the called predication:

> db1. Clauses that were never alive during the activation period, shall not be used to resolve with.

> db2. Clauses that are alive during the complete activation period, must be used to resolve with.

db3. A clause which was alive at the beginning of the activation period and which was still alive when a later clause is used, must be used.

db4. If two clauses A and B are both used and B comes later than A, then A must be used before B.

db5. A clause that has been asserted after the beginning of the activation period of the predication, and retracted before it was selected, shall not be used.

db6. If rules db1 to db5 do not specify about a particular clause that it must be used or that it should not be used, then it is implementation dependent whether this clause is used or not.

Note that only db4 deals with the order in which clauses must be executed. The other rules deal with the decision whether to use a clause or not.

The beginning of the activation period is defined as follows:

b1. The beginning of the activation period is the moment at which the predicate is called.

The following restrictions on the end of the activation period hold:

e1. the execution of a cut in the body of a clause of the predication ends the activation period if it was not ended before

e2. the execution of a cut in the continuation of the execution of a body of a clause of the predication ends the activation period if it was not ended before

e3. in the absence of cuts, the activation period ends not earlier than after the last non-deleted clause which was alive at the beginning of the activation period, has been used.

Some undefined notions are clear in the formal specification. On the other hand, the notion of activation period is not explicit in the formal specification. Still, we chose to introduce it in the verbal description, because it is widely used and it is clearly representable in the formal specification which nevertheless uses only part of the notion, namely the start of the activation period (i.e. the activation time) and the current selection time.

3.2 A classification

The table below shows how systematic choices in the functions CHOICE1 and CHOICE2, lead to the identification of 4 different views: the logical and immediate view are amongst them. The other 2 have been named 'minimal' and 'maximal' because they represent the view that uses the smallest set of clauses required by the formal specification, and the view that uses the largest set of clauses allowed by the formal specification.

	MINIMAL	LOGICAL	IMMEDIATE	MAXIMAL
CHOICE1	DO_NOT_USE	USE	DO_NOT_USE	USE
CHOICE2	DO_NOT_USE	DO_NOT_USE	USE	USE

The formal specification allows these 4 views, but does not require any of these views, as the choices for CHOICE1 and CHOICE2 can be made at random.

Both the minimal and the logical view end the activation period with the usage of the last clause that was alive when the activation period started. The maximal and immediate update view both end the activation period only when *next* becomes *none* on backtracking.

3.3 Some more explanation

The essentials of the above formal specification can be verbalized as follows:

> A clause that was dead before the activation period started, should not be executed. This corresponds to db1.

> Let CL be a clause that was alive at the beginning of the activation period.
> If CL is not retracted before it comes to its turn for being selected, then it must be executed. (see db2)
> If CL was retracted before it comes to its turn for being selected, then an implementation is free to execute CL or not (CHOICE1).

> Finally, any other clause CL was asserted after the beginning of the activation period.
> If CL was z-asserted before it comes to its turn for being selected, and retracted before it comes to its turn for being selected, then CL cannot be executed. (db5)
> If CL was z-asserted before it comes to its turn for being selected, and not retracted before it comes to its turn for being selected, then an implementation is free to execute CL or not (CHOICE2).

Rule db4 about the order of clauses, is implemented by the *next* function. In particular: *a-asserted* clauses are not reached by *next*, so cannot be executed. db4 implies indeed that clauses *a-asserted* after the beginning of the activation period cannot be executed.

Note that it follows from the specification that the update of a predicate does not influence the selection for use of other predicates.

4 From specification to implementation.

If the specification were implemented literally, it would be unacceptably slow: the *next* function is the culprit. It can be readily optimized by keeping the clauses in a chain, where the *next* element in the chain, is the result of the *next* function applied to a clause.

The maximal and immediate update view do not allow for early determinacy detection, or alternatively one can say that they must keep the choicepoint longer than the other views. It follows that their choicepoint must indeed contain the current_active clause. In the minimal and logical view, the choicepoint can contain *next(current_active)* and in this way, early determinacy detection can be achieved: a proof that for the minimal and logical view to have *next(current_active)* in their choicepoint is equivalent to the formal specification above, is in a separate report [Demoen].

It is nice to see that the implementation of the logical view as described in [Lindholm], follows readily from the formal specification, or to put it in a different way: the description of the logical view in [Lindholm] is easily adapted to form a framework for describing other interesting views.

See [Boerger d] for a formalization and a proposal for uniform implementation along these lines, using an evolving algebra description of the or-structure of Prolog with dynamic code.

5 Garbage collection of retracted clauses.

There are two levels at which retracted clauses can be 'garbage collected':

1. retracted clauses can be made inaccessible, i.e. no longer in the range of *next*
2. the space occupied by retracted clauses can be made reusable

Both are of interest: the first one saves time, because if a retracted clause can be made inaccessible, the *next* function of the formal specification will never have as value this retracted clause and backtracking does not need to spend time on deciding whether to execute this clause or not.

The second is not possible without the first of course.

Simple examples show that without extra effort - like traversing the choicepoint stack on every assert or retract, or keeping a reference count in clauses (slowing down cut !) - a retracted clause cannot be made immediately inaccessible in the logical view and that also in the immediate update view, execution of a call potentially has to spend time visiting retracted clauses, even if these clauses were retracted before the activation period of that call started.

5.1 Example for the logical view

Suppose a/1 has 4 clauses

```
a(1) .
a(2) .
a(3) .
a(4) .
```

and

```
retr :- retract(a(3)) , ! .
retr .
```

the query

```
?- a(X) , retr , a(Y) .
```

Then, the second activation of a/1 must be able to go from clause 2 to clause 4, but the first activation must still go from 2 to 3, so, the links between the clauses cannot be updated when a clause is retracted. It means that in the logical view, retracted clauses can't be made inaccessible immediately.

5.2 The immediate update view

Suppose a/1 has 3 clauses:

```
a(1) :- ...
a(2) :- ...
a(3) :- ...
```

and during the execution of a(2), the definitions of a(2) and a(3) are retracted and a(4) . is z-asserted. Suppose the pointers in the chain of clauses are updated immediately when a clause is asserted and/or retracted.

Then the pointers will look like:

$$1 \to 4$$
$$2 \to 3$$

so that 4 is no longer reachable from 2, but the immediate view requires 4 to be reachable.

By never removing the last clause in the chain, this problem can be solved: a retracted clause in the chain is now represented by its number followed by a 'd'; after retracting 2 and 3, the chains look like

$$1 \to 3d$$
$$2d \to 3d$$

and when 4 is asserted, a link from 3d to 4 is created:

$$1 \to 3d \to 4$$
$$2d \to 3d \to 4$$

By repeatedly retracting the last clause and z-asserting a new clause, one can make the length of the chain of deleted clauses arbitrarily long.

The point here is that new calls to this predicate, have to traverse this chain of deleted clauses.

5.3 The most attractive view.

Both in the immediate update view and the logical view, the execution of a (new) call can suffer from an earlier assert and/or retract. Since the maximal view uses all clauses used by the immediate update view and the logical view, it suffers from the same problem. It seems that only the minimal view has no such problem: indeed, in the minimal view, a retracted clause can be made inaccessible immediately (for a proof see [Demoen]).

Moreover, the minimal view does not interfere badly with indexing and not even with assert at an arbitrary place in the PDB (like in ProLog by BIM with assert/2).

Also the minimal view is easily explained to users: "of the set of clauses which are present at the start of the activation period, the ones are used that are not retracted before being selected". And the behavior of retract as a backtracking built-in predicate, is still easy to explain and consistent with the view.

This leads us to the conclusion that the minimal view is to be preferred by implementors and users.

6 Conclusion

We have presented a framework for specifying formally different database update views for Prolog. The well-known immediate update and logical view fit into the framework. Two more extreme views are identified: the minimal and the maximal view. It is argued that the most attractive view is the minimal view.

7 Acknowledgment

We thank an anonymous referee for pointing out a slight imprecision in the preliminary version of this paper. B.D. thanks DPWB for support through project RFO/AI/02.

8 References

[Boerger a] E. Boerger "A Logical Operational Semantics of full Prolog. Part I:Selection Core and Control" in: CSL'89. 3d Workshop on Computer Science Logic, Springer LNCS 440, 36-64.

[Boerger b] E. Boerger "A Logical Operational Semantics of full Prolog. Part II: Built-in Predicates for Database Manipulations" in: MFCS'90. Mathematical Foundations of Computer Science (B.Rovan,Ed.), Springer LNCS 452, 1-14.

[Boerger c] E. Boerger "A Logical Operational Semantics of full Prolog. Part III: Built-in Predicates for Files, Terms, In-Output and Arithmetic" in: Proc. Workshop Logic from Computer Science (Ed.Y.Moschovakis), MSRI Proceedings,Springer (to appear 1991)

[Boerger d] E. Boerger, D. Rosenzweig "An analysis of Prolog database views and their uniform implementation." CSE-TR-89-91, Computer Science and Engineering Division, University of Michigan, Ann Arbor, pp.44, April 1991.

[Demoen] B. Demoen, "From a specification to an implementation of a database update view for Prolog: a correctness proof" CW-report 121, K.U.Leuven

[Scowen] in N64 "Prolog, Draft for Working Draft 4.0", R.S. Scowen; NPL Teddington, England

[Gurevich] Y. Gurevich "Logic and The Challenge of Computer Science." in: E. Boerger (Ed.): Trends in theoretical Computer Science, Computer Science Press, pp.1-57, 1988; see also by the same author "Evolving Algebras, A tutorial Introduction", in EATCS Bulletin 43, 1991, 264-284

[Lindholm] T. Lindholm, R.A. O'Keefe, "Efficient Implementation of a Defensible Semantics for Dynamic PROLOG Code" ICLP 87 21-39, Melbourne 1987

[Moss] C. Moss "Cut and Paste - Defining the impure primitives of Prolog" ICLP 86 686-694, 1986 London

Concepts for a Modular and Distributed Prolog Language

Szabolcs FERENCZI
Multilogic Computing Ltd.
1119 Budapest, Vahot u. 6.
Hungary

Abstract

The paper introduces new language concepts for building modular and distributed Prolog programs. The language concepts are synthesized from the analysis of the language concept of Distributed Processes and in parts from the concepts of Object Oriented Programming, while Prolog serves also as an implementation language. In this concept, a Prolog program is solved by a number of communicating Prolog objects. Each object can have an own initial goal. Objects communicate by remote predicate calls. A remote predicate call results in an additional Prolog thread at the target object to prove the goal. Consequently, objects execute multiple Prolog programs in an interleaved manner. A general form of Prolog data base operations is proposed for communication of concurrently evaluating Prolog programs in the internal scope of an object. Non-determinism arising from concurrency is handled by help of an adapted form of the Guarded Commands concept of Dijkstra. A program under execution consists of concurrently acting nested objects while the description of the behaviour of objects is achieved in a modular way.

Introduction

This paper introduces new language concepts for an extension of Prolog. The extension lays in the direction of modularity and concurrency, thus arriving to a distributed and modular Prolog language. The solution is also similar to the Object-Oriented Programming (OOP) paradigm.

One of the main consequences of the paper is that the concurrent extensions of object-oriented programming languages should be designed by inheriting advanced results of concurrent programming like the concept of Distributed Processes (DP) and the access rights of Concurrent Pascal, rather than introducing asynchronous messages and semaphores; tools characteristic of concurrent programming in the early '70s. For synchronization of asynchronously evolving processes, the semaphore is a general but low level engineering tool while the guarded commands construction, which is the base of DP and also proposed in this paper, is a high level and declarative language tool. Consequently the object-oriented methodology should be applied in the design phase of OOP languages too: inheriting established results rather than redesigning everything from the scratch.

In this paper a synthesis is made from literature, and the most useful parts are assembled from different sources. The Prolog language is taken as a carrier of the concepts, since it is a declarative language and proved to be a useful tool in rapid prototyping. The concept of DP and Prolog is merged in this paper. The guarded commands construction is a means in DP to handle non-determinism, so it is also inherited and merged with Prolog's non-deterministic construction. DP is massive with respect to concurrency but is very weak in textual modularity and encapsulation. To solve these, the class concept and the access rights of Concurrent Pascal are adapted.

The guide-line in assembling was security. A lesson learned from concurrent programming is that concurrent programs cannot be made correct by testing. Trial and error development widely used in sequential program development is not acceptable in a concurrent programming environment any more. The solution rather is that the language must ensure correctness as pointed out by Brinch Hansen [Brinch Hansen, 1977, pp.xii].

This paper introduces only the main concepts of the proposed language to give an overall view but is not intended to be a full language definition. Since the language tools presented in this paper are based on the concepts of Dijkstra's guarded commands and Brinch Hansen's distributed processes, these concepts are discussed here in some details with special care given to the features used in this proposal but an interested reader should consult original papers, too. However, knowledge of the Prolog as well as the principles of OOP is assumed.

1. Concurrent programming

Operating systems gave the main inspiration for concurrent programming. The operating system of a computer supervises many simultaneous activities. Some basic problems to be solved in case of simultaneously executing and interacting programs are the protection of critical resources from simultaneous use (e.g. a lineprinter) and the interaction among concurrently executing processes like synchronization and communication. Therefore, language tools for concurrent programming were developed from the early operating systems.

Various tools have been developed for synchronization and communication. The first general tool was the semaphore due to E.W.Dijkstra [Dijkstra, 1968]. Although, the semaphore is a general and useful tool, its use is dangerous because the compiler can never check the correct use of the semaphore. Several more structured tools have been developed but unfortunately the semaphore is still popular and widely used today.

The conditional critical region due to C.A.R. Hoare is another general language tool that can be used both for synchronization and communication [Hoare, 1972]. Although, the concept is elegant and powerful, the circumstances at that time made the concept rejected [Brinch Hansen, 1982, pp.47]. The so-called monitor construct due to C.A.R. Hoare and P. Brinch Hansen made a compromise between elegance of the language and economy of CPU time and become the standard structured tool for synchronization and communication.

Supporting the programming of distributed systems two elegant language concepts have appeared at about the same time in 1978: the Communicating Sequential Processes concept (CSP) of C.A.R. Hoare and the Distributed Processes concept (DP) of P. Brinch Hansen. Both concepts apply the guarded commands construction of E.W. Dijkstra to make control of non-determinism aroused from concurrency. Although, the constructions of guarded commands do not consider a parallel programming environment, the CSP and the DP adapted it to parallel environment so that the guards refer to common shared variables. Shared variables in CSP are channels and in DP they are variables of a module that are only accessible from other modules through procedures supplied by the module. Note that such a module in DP is called a distributed process. The Guarded Commands concept is a very useful language tool in case of concurrent programs because the constructions of guarded commands have a non-deterministic semantics and this feature is quite necessary in a concurrent environment where interactions are inherently non-deterministic.

1.1 The concept of guarded mechanisms

The term guarded mechanism is used, in this paper, to refer to both the concept of guarded commands of Dijkstra [Dijkstra, 1975] and its applications in other language concepts. The concept of guarded commands probably comes from the desire not to overspecify a problem. In other words, give the executor of the program less control and leave it the choice. Guarded commands can control the execution of harmoniously cooperating programs. From the declarative point of view, the guarded commands construction is a means to express or to allow non-deterministic behaviour of programs. Procedurally, it means that the next step of the computation is not fully determined by the program text. The guarded commands construction in the thread of executions means that at those points the processor or interpreter is free to choose a next step from a set of available possibilities.

An example used by Dijkstra for demonstrating the elegance and power of the concept is the greatest common divisor of numbers. It is solved according to Euclid's algorithm. Let's see the program for three numbers: X, Y and Z.

```
x,y,z:= X,Y,Z;
do x>y -> x := x - y
 | y>z -> y := y - z
 | z>x -> z := z - x
od
```

The repetitive constructor synchronizes the repeated execution of three computation processes. At each repetition one of the processes proceeds. If none of the processes can continue, it means that the task is done and the repetitive construction terminates. At termination

```
x = y = z = (GCD of [X,Y,Z]).
```

The program above does not state more of the algorithm than necessary: carry on traversing the state space at one of the possible next steps. It does not specify the priority of the steps, since according to the algorithm there is none: the next computation step in the sequence is not determined by the program text. Constructions formed from guarded commands allow non-determinism in the sequence of computational steps.

The characteristics of guarded commands construction suit the non-deterministic nature of concurrent systems. Although Dijkstra did not consider concurrent computation when he introduced his concept (actually he did it later, see EDW508 and EDW554 [Dijkstra, 1982]), others adapted it in concurrent programming environment (see CSP and DP).

1.2 Distributed Processes

The language concept of DP introduces a special process paradigm. Usually a process is regarded as a single thread of predetermined control but DP processes are non-deterministic and interleaved multithreads. From another point of view, a DP process is a collection of private and shared variables, public handling procedures of the shared variables and optionally an initial goal. This kind of collection is called a distributed process. Therefore, a DP process is an encapsulated active module with well defined interface towards the outside. Shared variables can only be used from other modules through public procedures.

A DP program is a two level system. The inner level is a conventional concurrent system with interleaved execution and conditional critical regions [Hoare, 1972]. A public procedure can be called without restriction but the execution of such procedures might be postponed until a certain condition is fulfilled on the internal state space of the called module. Inside a module, that is called a process in DP, there exists a world of communicating sequential processes.

On the outermost level entities communicate with each other through remote procedure calls. Some entities have a potency to issue remote procedure calls, and others are capable of accepting such calls. Entities are labelled and labels are used in remote procedure calls. The caller addresses the target entity in the communication but not vice versa. The caller suspends every activation until the other entity performs a requested remote procedure call issued by the caller.

A module definition in DP consists of three main parts: variable declarations, procedure declarations and an initial goal.

```
process <name>
<own variables>
<common procedures>
<initial statements>
```

Accessing common variables from a procedure is only allowed from guarded regions. A guarded region is the combination of the conditional critical region of Hoare and the guarded commands of Dijkstra. Common variables are used in expressions of guard parts and can be modified only inside guarded mechanisms. Two forms of guarded regions are defined, called when statement and cycle statement respectively:

```
when B1:S1 | B2:S2 | ... | Bn:Sn end
cycle B1:S1 | B2:S2 | ... | Bn:Sn end
```

A when statement corresponds to Dijkstra's alternative constructor but the execution of a when statement is suspended rather than the program aborted whenever none of the guards yield true. The reason is that, since we are in a concurrent programming environment and guards may refer to common variables, there is a chance that other processes will modify the state space while a when statement is suspended and it can be resumed later. A cycle statement is an endless repetition of a when statement. The cycle statement never terminates.

DP is a static language in the sense that all the DP processes are created at program initialization time and they never terminate.

A simple and representative example is a definition of a bounded buffer module in DP. The module contains variables for storing data elements and for administration reasons, and defines procedures by which the contents of the buffer can be accessed. The defined procedure calls are the only operations on the buffer.

```
process buffer; s: seq[n]char; head,tail: int
proc send(c:char) when
  (head-tail mod n) < n: s[head]:= c; head:= head+1 end
proc rec(#c:char) when
  (head-tail mod n) > n: c:= s[tail]; tail:= tail+1 end
begin head:= 0; tail:= 0 end
```

The above defined buffer can be used by other modules in the following way:

```
call buffer.send('a')          call buffer.rec(v)
```

These procedure calls may be suspended by the buffer module if the status of the buffer does not fulfil the preconditions of the operations.

2. Towards a synthesis: Distributed Prolog Processes

The language concept of Distributed Prolog Processes (DPP) is an extended version of the concept of DP. However, it is merged with logic programming (Prolog) and a modularization concept is adapted to it. To make the program more flexible and introduce some dynamic scope, the naming of objects is eliminated from the concept but ports and many-to-one channels are introduced to substitute naming.

To achieve distributed execution, a DPP program is built from a set of interacting objects. An object is an autonomous entity that has its own goal and capability to evolve. Objects have port interfaces by which they may be connected to each other. Ports are represented by Prolog predicates while the behaviour of the objects are defined by Prolog programs. The predicates representing output and input ports play the role of built-in predicates and goals in the Prolog programs respectively. Interface predicates of objects are connected with many-to-one type channels. A channel is a conceptual entity. It can connect one or more output ports to a single input port. Whenever an object's Prolog program uses a predicate representing the output port it activates another Prolog evaluation thread in the scope of the object which is connected to the output port by its input port. This is called a remote predicate call which is the only form of communication between the objects. During the evaluation of an output predicate, i.e. a remote predicate call, the issuing object is suspended. An object can interact with only those objects that are connected to it by channels: dynamic scope.

Prolog clauses can be grouped into modules. Modules can be combined using already defined modules into new ones. The allocation of modules to newly created objects as well as the configuration of objects can be described in a Prolog like notation. In the standard Prolog part of the modules, a form of guarded commands is used to prepare for non-deterministic behaviour of interacting objects. Prolog backtracking is still used to achieve exhaustive search even through guarded parts, in certain circumstances.

The concepts of DPP result in a two level language. In the outer level, Prolog objects are defined by modules; collections of Prolog clauses and goals. Prolog objects come into being as the result of interpreting module specification clauses. Interaction between objects is possible through so-called remote resolution, an adapted form of remote procedure call of DP.

Inside an object, multiple Prolog evaluation threads coexist and communicate by a generalized form of Prolog data base manipulations. Beside multiple Prolog threads, objects can usually have sub-objects to make up an object hierarchy. One of the main problems a concurrent system faces is the non-deterministic behaviour of asynchronous processes at interaction points. The concept of guarded commands is adapted to the Prolog environment to solve this problem.

2.1 Introducing modularity into Prolog

The most apparent problem with Prolog is that it lacks modularity at the language level. Although many Prolog implementations allow modularity in some way, it is unsuitable for parallel evaluation. Two kinds of modularity are to be distinguished in a program: (1) modularity in the specification and (2) modularity in the arrangement of acting components. The former gives module hierarchy (textual arrangement), while the latter gives object hierarchy (arrangement of entities). In this paper, a sharp distinction is made between the two. The mapping between modules and objects is given in a Prolog like way too.

Modules form hierarchies in two different senses. A hierarchy of (1) module definitions and (2) a hierarchy of entities behaving according to a particular module specification (objects). A module hierarchy is analogous to classification and inheritance of OOP. Entities are analogous to objects in OOP.

2.2 Modules in the specification

A structured description of the architecture of the objects are achieved by introducing a module concept. A module in DPP is a collection of (1) standard Prolog clauses that are valid only in the textual scope of the module (module body), (2) Pseudo-Prolog clauses and (3) an initial goal sequence. A module is divided into a head and a body part. The module head provides an outer interface built from Prolog predicates.

```
module definition
<set of Pseudo-Prolog clauses>    % for configuration of objects
module body
<set of Pseudo-Prolog clauses>    % for classification
<set of Prolog clauses>               % for specifying behaviour
<single Prolog goal sequence>         % for own activity
end module definition.
```

A. The module head

The head might contain Pseudo-Prolog clauses only. They are just like ordinary Prolog clauses but '::--' is used as the sign of the conditional and terms can only have predicate arguments which are like variables. The clauses give name to the module and determine the structure of the object that will be created from the module. The interface of the module is also determined by the Pseudo-Prolog clauses. A Pseudo-Prolog clause consists of a head and a body.

```
<Pseudo-Head> ::-- <Pseudo-Body>
```

The Pseudo-Head is a single Prolog term with arguments as predicates. A predicate argument serves as a kind of variable. It is only a term in the scope of the module but can be used as a goal. Mapping of term variables and predicate definitions is done by module instantiation, so the meaning of the term depends on the environment where the term will be used as a predicate (polymorphism). Predicate arguments are represented by their functor and arity, augmented by their type. The type of the predicate arguments can be input (?), output (^) and dynamic fact (*). The Pseudo-Body is built from similar terms separated by the sign '//'. The functor of the Pseudo-Head gives also the name of the module. The Pseudo-Clause without Pseudo-Body is a Pseudo-Fact. It means that an object created from that module behaves according to the Prolog clauses of the module body. For example, a Pseudo-Clause specifying a buffer looks like a fact resulting an atomic module:

```
buffer(^emptyp/0,^emptyv/0,^fullp/0,^fullv/0,?put/1,?get/1,*store/1).
```

The parameters augmented by upper arrows (^) define terms which are used inside the module as goals but may not have matching local clause definitions. The parameters augmented by question marks (?) define terms for which the module defines clauses, so they can be used as predicates from outside the module. These may be activated in unpredictable times of the evaluation. However, if activated, they represent an additional thread of Prolog program, i.e. another interpreter, in the scope of an object specified by the module. The third kind of parameters are facts which are added to the dynamic data base of an object defined by the module, possibly overwriting existing default local data base clauses. Therefore, the full buffer module is as follows:

```
module definition
buffer( ^emptyp/0,^emptyv/0,^fullp/0,^fullv/0, ?put/1,?get/1, *st/1).
module body
emptyp.      emptyv.      fullp.       fullv.        st(B-B).
put(M) :- emptyp, retract(st(B-[M|B1])), assert(st(B-B1)), fullv.
get(M) :- fullp, retract(st([M|B]-B1)), assert(st(B-B1)), emptyv.
end module.
```

A Pseudo-Clause with a body is called a Pseudo-Rule. It means that an object created from that module behaves according to the Prolog clauses of the module body and additionally it has sub-objects represented by terms in the Pseudo-Body. Sub-objects are active simultaneously with the encapsulating object. Argument names in Pseudo-Rule have a local scope as usual with Prolog arguments and the new predicate names introduced represent many-to-one channels. For example, a bounded buffer can also be built from the module clauses of the buffer and two semaphore sub-objects (the semaphore module is defined later):

```
module definition
bounded_buffer( ?put/1, ?get/1, *bound(B) ) ::--
        buffer(^ewait/0,^esgnl/0,^fwait/0,^fsgnl/0,?put/1,?get/1,nil)
//      semaphore( ?ewait/0, ?esgnl/0, *count(B) )
//      semaphore( ?fwait/0, ?fsgnl/0, *count(0) ).
end module.
```

This means that the bounded buffer module is composed from three already defined modules making use of their external interfaces. The bounded buffer can inherit those predicates from buffer and semaphore modules what they define as interface predicates. In the scope of the outer module the interface terms are bound together. This means that whenever emptyp/0 is used in module buffer, it matches with the clause 'wait' in an object defined by the semaphore module (see later), since the channel ewait/0 connects them.

B. The module body

The body of a module might contain (1) Pseudo-Facts, (2) Prolog clauses and (3) a Prolog goal sequence.

A Pseudo-Fact in the module body means that the interface predicates of the module associated with the Pseudo-Fact are valid in the module body and can be used as a built-in predicate. Pseudo-Facts can be asserted and deleted dynamically. They serve for classification in the usual sense.

Prolog clauses, as usual, define facts and rules valid in the world of the module. Based on these facts and rules questions can be put up in this environment. Each object can have an initial question and more questions may appear (input predicates) during the lifetime of the object. Each question including the initial one involves a Prolog evaluation thread in the scope of the object. These threads are evaluated by the object in an interleaved manner.

The clauses of the module body are a mixture of ordinary Prolog clauses and guarded clauses. The form and semantics of guarded clauses are defined later in this paper.

2.3 Modules in execution

A DPP program consists of connected objects. Objects are created from module definitions and have their own state spaces and goals. The state space in Prolog programs can be represented at least in two ways: (1) in logical variables and (2) in clauses of the data base. The objects of DPP apply the second solution for representing the state space. Objects form a hierarchy: nested objects. Objects on the same level only can communicate directly. Note that textual modules are not objects but only specifications of possible objects. Objects are only introduced into the system at module head clauses if a clause has a body.

2.3.1 Adding parallelism

Two kinds of resolution are used in DPP. Resolution of module clauses at the outer level, and resolution of Prolog clauses at the inner level. The former is a Pseudo-Prolog while the latter is a normal Prolog level.

A. The outer level: a Pseudo-Prolog

Resolution of module clauses is a kind of meta-resolution. A program is defined by a set of modules. They consist of module clauses and an initial module goal. During resolution, the initial goal is matched with the heads of the available module clauses. If a matching one is found, the goal is substituted with the body of the clause and the predicate variables are bound. If a clause has no body, the goal is substituted with the empty body. In this way, an object tree is built. Each node hides a complete Prolog world with its local facts, rules and goals.

Parallelism is introduced at module clause resolution. Note that DPP realizes coarse grain parallelism, since it is the parallelism of objects which can be large sequential programs themselves. During module clause resolution, whenever a module clause head matches with a goal during module unification, as many AND objects come into being as many goals the module clause body has. Each AND object interprets a Prolog program and behaves according to the corresponding module definition. If the module clause consists of alternative module clauses, either it may result in OR parallelism or backtracking can be used to investigate alternative object configurations. For example:

```
bounded_buffer( ?put/1, ?get/1, *bound(B) ) ::--
       buffer(^ewait/0,^esgnl/0,^fwait/0,^fsgnl/0,?put/1,?get/1,*st/1)
//      semaphore( ?ewait/0, ?esgnl/0, *count(B) )
//      semaphore( ?fwait/0, ?fsgnl/0), *count(0) ).
bounded_buffer( ?put/1, ?get/1, *bound(B) ) ::--
       bbuffer( ?put/1, ?get/1, *bound(B) ).
```

If the module clause is a fact, it will result in an atomic object, otherwise it is a composite object. Module clause resolution, therefore, results in an object network. Objects are behaving according to the module definitions from which they were created.

B. The inner level: a separate Prolog world with interface

The inner level is a Prolog world. However, the resolution rule of Prolog clauses is modified because of the concurrent environment. The module from which an object is created has a set of internal Prolog clauses (module body) and a set of predicates at the module interface. During the inner level resolution the interpreter first tries to match a goal with a predicate from the module interface, but if no match is found, the search continues in the local data base. If a matching with a module interface predicate occurs, it may result in a remote resolution procedure. It means that the resolution continues in the scope of the object which defines the same predicate in its interface and is connected with a path to the present object. If there is no path, however, the alternative clauses defined in the same module are also searched.

It is semantically very important that during remote resolution, the caller object's Prolog program is suspended, since this feature makes scheduling possible. If the module from which an object is created contains accepting type interface predicates, the object is capable of evaluating those predicates at request. A request is issued by another object and results in an additional Prolog program in the scope of an object, which executes multiple Prolog programs in an interleaved manner.

Note that new Prolog evaluation threads are only introduced in a DPP program at module clause resolution when an object is created with an initial goal. At remote resolution, a sequential Prolog execution thread flows through the objects in the system only, since the caller is suspended and the target starts a new Prolog thread in the meantime.

2.3.2 Concurrency and non-determinism

In a DPP program two kinds of non-determinism may arise: (1) non-deterministic behaviour of concurrently acting objects and (2) search non-determinism usual in Prolog programs. The former is called *don't care non-determinism* and the latter *don't know non-determinism* in the literature. In contrast to GHC languages, the guarded mechanism applied by DPP is a means for both don't care and don't know non-determinism.

Non-deterministic behaviour of parallel objects is handled by help of guarded mechanisms applied in a sequential part of their definitions. Guarded mechanisms are represented by Prolog clauses of the form:

```
head :- (guard1 | body1) # (guard2 | body2) # ... # (guardn | bodyn).
```

where both the guard and the body part is a bracketed Prolog goal sequence. The guard may contain data base inquiries on the data base served for communication of concurrent Prolog evaluation threads within the scope of the same object. The sign '#' separates alternative clause bodies (OR definitions). The sign '|' separates the guard and the body part of a guarded clause. At guarded mechanisms, the usual LRDF strategy of Prolog machines is changed into a true non-deterministic search strategy. It means that any of the alternatives with true guard part can be chosen for forward execution, however, alternatives can be selected at backtracking too, until an exhaustive search is achieved. If no guards yield true, the execution of the Prolog program is delayed until at least one guard becomes true.

Backtracking is used to make exhaustive search for solutions. Although the original definition of the guarded commands ensures that if a guard is true, the corresponding mechanism is guarantied to terminate (see the weakest precondition [Dijkstra, 1975]), from the Prolog nature of the program (see don't know non-determinism) backtracking is allowed in DPP. If backtracking reaches a guarded mechanism, an alternative guard can be used and the execution continues forward, and if no guard is true, the current evaluation thread is suspended until at least one guard is true. The usual Prolog tool, the cut can be used to make a guarded mechanism deterministic like the commitment mechanism in GHC languages.

2.3.3 Deadlock detection and backtracking

Backtracking is allowed in DPP to substitute OR parallelism. Normally, backtracking occurs in a normal Prolog program if the interpreter arrives to a fail leaf of the search tree. Since a DPP program consists of many search trees (see multiple threads), a form of distributed deadlock detection and backtracking mechanism must be applied.

In the scope of an object, there is no need for any distributed deadlock detection between Prolog threads. If a thread exploring a single search tree arrives to a fail leaf, backtracking starts as in normal Prolog programs and either finds a Prolog choice point or arrives to the latest guarded choice. At a guarded choice point either another true guard can be chosen or the thread becomes suspended until at least one guard is true.

The distributed deadlock detection is applied in object level. Since the program consists of nested objects, distributed deadlock detection is layered according to the structure of the objects. Each object detects the deadlock state of its child objects and reports its own deadlock state to the parent object. Distributed backtracking is started at the uppermost level. The purpose of distributed backtracking is to substitute OR parallelism of objects. Alternative object configurations can search for solutions through backtracking.

The exact mechanism of distributed deadlock detection and backtracking is subject of a forthcoming paper and not dealt with here.

3. Example: a Bounded Buffer

The problem of producers and consumers is a well-known problem in concurrent programming. In concurrent programming environment, acting components may produce data elements and other components consume them. However, their harmonious cooperation should be established. The producer should not produce much more data than the consumer can consume and the consumer should be delayed when there is nothing to consume. A bounded buffer inserted into the communication path solves this problem. Note that several producers can produce elements into the same buffer and several consumers can consume from the same buffer. The following module specifies such a buffer:

```
module_definition
bounded_buffer( ?put/1, ?get/1, *bound/1 ).
module_body
bound(1).
length(Len, List) :- len(Len, List, 0), !.
len([], L, L).
len([_|T], L, Count) :- NC is Count+1, len(T, L, NC).
replace(M, M1) :- retract(M), assert(M1).
put(M) :- (store(L-[M|T]), length(Len, L), bound(B), Len < B |
              replace(store(L-[M|T]), store(L-T)).
get(M) :- (store(L-T), length(Len, L), Len > 0 |
              replace(store(L-T), store([M|L]-T)).
end module.
```

The module defines an object which accepts two operations: 'put/1' and 'get/1'. The object is a passive one. It does not have an own goal. Both the put/1 and the get/1 operation might be delayed, if the status of the buffer does not make it possible to complete them. Both operations are synchronized and the get/1 transfers a result item back to the caller.

Another possible definition is to compose the buffer from other modules. A module describing a semaphore is used:

```
module definition
semaphore( ?wait/0, ?signal/0, *counter/1 ).
module body
replace(M, M1) :- retract(M), assert(M1).
counter(1).
wait :- (counter(C), C > 0) | (C1 is C-1,
  replace(counter(C), counter(C1))).
```

```
signal :- counter(C), C1 is C+1, replace(counter(C), counter(C1)).
end module.
```

The module defines an object which accepts two operations: wait/0 and signal/0. An initial data base element can be placed in the local scope of the object: counter/1. If the third list is empty, the default clause is used but, if given, it replaces the default one. Making use of the semaphore module an alternative bounded buffer can be built as follows:

```
module definition
buffer( ?put/1, ?get/1, *bound(M) ) ::--
      semaphore( ?empty_wait/0, ?empty_signal/0, *counter(M) )
//    semaphore( ?full_wait/0, ?full_signal/0, *counter(0) ).
module body
st(B-B).
bound(1).
replace(M, M1) :- retract(M), assert(M1).
put(M) :- empty_wait, replace(st(B-[M|B1]), st(B-B1)), full_signal.
get(M) :- full_wait, replace(st([M|B]-B1), st(B-B1)), empty_signal.
end module.
```

The bounded buffer definition is composed from two objects behaving according to the semaphore module definition. The module contains additional local clauses. Note that the module does not contain any initial goal, so it is a passive one.

4. Comparison and assessment

The DPP concept merges DP with Prolog. This feature makes it possible for Prolog programmers to make use of the system easily. DPP makes it possible to write the bulk of the program in standard Prolog.

The module concept defined here lies the closest to SCOOP [Vaucher, et al., 1988], while the parallelism is similar to POOL-T [America, 1987], although DPP has arrived to these concepts from a different approach: see DP and modular programming. The main advantage of DPP compared with both SCOOP and POOL-T is that it identifies objects by position rather than by names, since DPP defines port interface on the modules and ports of objects are connected by channels. Furthermore, the access rights of objects are well defined, what is a critical issue in concurrent programming. Also note that the DPP unifies the OOPL's message sending and messages of concurrent parts of the program.

The present approach is different from existing concurrent Prolog proposals and implementations in the way it achieves explicit parallelism in Prolog programs. Most of the existing concurrent logic programming languages with explicit parallelism use stream communication through logical variables between AND parallel processes and apply guarded Horn clauses with commitment mechanism allowing don't care non-determinism only (Guarded Horn Clause Languages: Concurrent Prolog [Shapiro, 1986], GHC [Ueda, 1986], Parlog [Clark and Gregory, 1986]). Furthermore, limited OR parallelism is applied at evaluating alternative guards. The present solution, however, introduces AND parallel Prolog objects (Prolog worlds) with remote resolution between objects and a form of data base communication combined with guarded mechanisms between Prolog evaluation threads of a single object. The applied guarded mechanism is a combination of don't know non-determinism and don't care non-determinism. Another distinguishing feature is that existing languages, with a few exceptions (Delta Prolog [Pereira, et al., 1986], CS-Prolog [Futó and Kacsuk, 1989], Strand [Foster and Taylor, 1989], Amoeba-Prolog [Shizgal,1990], are primely oriented to shared memory machines but the features proposed here make execution of programs on distributed machines possible.

Conclusion

Parts of the concepts described here resulted from the experiment of CS-Prolog implementation techniques [Kacsuk, et al., 1989]. More investigation should be carried out, however, with respect to backtracking through guarded constructions and termination questions of objects.

The concepts seem to be useful in the field of concurrent programming but the concurrent logic applications are to be exploited in the future. Nevertheless, the concepts described here form a good starting point for further investigations.

References

America, P., "*POOL-T: A Parallel Object-Oriented Language*," in **Object-Oriented Concurrent Programming**, Massachusetts Institute of Technology, 1987, pp.199-220.

Brinch Hansen, P., **The Architecture of Concurrent Programs**, Prentice Hall, 1977.

Brinch Hansen, P., "*Distributed Processes: A Concurrent Programming Concept*," Comm. ACM 21, 11, pp.932-941, Nov. 1978.

Brinch Hansen, P., **Programming a Personal Computer**, Prentice Hall, 1982.

Clark, K.L., Gregory, S., "*PARLOG: parallel programming in logic*," ACM TOPLAS 8, 1, (1986), pp.1-49.

Dijkstra, E.W., "*Guarded Commands, Nondeterminancy, and Formal Derivation of Programs*," Comm. ACM 18, 8, pp.453-457, Aug. 1975.

Dijkstra, E.W., **Selected Writings in Computing: A Personal Perspective**. Springer-Verlag, 1982.

Foster, I, Taylor, S., **Strand: New Concepts in Parallel Programming**. Prentice-Hall, Englewood Cliffs, N.J. (1989).

Futó, I., Kacsuk, P., "*CS-Prolog on multitransputer systems*," Microprocessors and Microsystems, 13, 2, March 1989, pp.103-112.

Hoare, C.A.R., "*Towards a Theory of Parallel Programming*," in **Operating Systems Techniques**, Ed: C.A.R. Hoare and R.H. Perrot, Academic Press, 1972, pp.61-71.

Hoare, C.A.R., "*Communicating Sequential Processes*," Comm. ACM 21, 8, pp.666-677, Aug. 1978.

Kacsuk, P., Futó, I., Ferenczi, Sz., "*Implementing CS-Prolog on a communicating process architecture*," Journal of Microcomputer Applications, 13, (1990), pp.19-41.

Pereira, L.M., Monteiro, L., Cunha, J., Aparicio, J.N., "*Delta Prolog: A Distributed Backtracking Extension with Events*," Third International Conference on Logic Programming, London, UK, July 1986, Proceedings, Also in **Lecture Notes in Computer Science** 225, Springer-Verlag, 1986.

Shapiro, E., "*Concurrent Prolog: a progress report*," IEEE Computer, 19, 8, (1986), pp.44-58.

Shizgal, I., "*The Amoeba-Prolog System*," The Computer Journal, 33, 6, Dec. 1990, pp.508-517.

Ueda, K., "*Guarded Horn Clauses*," In **Logic Programming '85, Lecture Notes in Computer Science** 221, Springer-Verlag, Heidelberg 1986, pp.168-179.

Vaucher, J, Lapalme, G., Malenfant, J., "*SCOOP, Structured Concurrent Object Oriented Prolog*," European Conference on Object-Oriented Programming, Oslo, Norway, August 15-17, 1988, Proceedings;
Lecture Notes in Computer Science 322, Springer-Verlag, 1988, (eds:S.Gjessing, K.Nygaard)

FROM PARLOG TO POLKA IN TWO *EASY* STEPS

Andrew Davison
Dept. of Computing, Imperial College
London SW7 2BZ, UK
February 1991

ABSTRACT

In our opinion, the Object Oriented Programming (OOP) and concurrent Logic Programming (LP) paradigms offer complimentary functionality, which taken as a whole is more expressive than either separately. For this reason, the two object oriented extensions to the concurrent LP language Parlog discussed here support both paradigms.

The simpler extension is called *Parlog++*, which combines Parlog with the basic object oriented features of encapsulation, data hiding and message passing. It is a subset of a more complex language called *Polka*, which also supports multiple inheritance and self communication.

Since Polka combines two paradigms, it has the drawback that a potential user will need to be familiar with both programming approaches before being able to use the language. However, the learning barrier can be reduced by initially using the simpler language Parlog++. This has many benefits, not least being that it enables process-based programs to be written as objects. Once the programmer is familiar with Parlog++ it is considerably easier to move over to Polka which offers more OOP functionality. These two 'easy' steps are illustrated in this paper.

1. INTRODUCTION

The essential relationships between Parlog, Parlog++ and Polka can be represented pictorially as in figure 1.

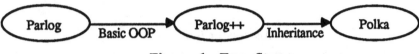

Figure 1: Two Stages.

The main advantage of an object oriented extension to Parlog is the introduction of the object oriented paradigm as a guiding principle for writing programs. This offers a simple but powerful model for representing programs as computational entities which communicate with each other via message passing. Message passing allows each class to have a well defined interface, and polymorphism means that the interface is flexible. In addition, the use of inheritance and self communication increases the reusability of programs and makes rapid prototyping of classes straightforward.

The OOP constructs in Parlog++ and Polka remove one of the weaknesses of Parlog : its lack of structure, especially for 'programming in the large', due to its reliance on the predicate as the main structuring mechanism. Module constructs could be utilised instead, but OOP offers more than just encapsulation. To be as useful, a module system must specify how communication is to take place between modules, and how modules can be reused in different applications. However, the main difference between conventional modules and classes is that a class is a collection of operations and *state*, which allows a programmer to deal with state changes within the object oriented formalism.

The utility of OOP begs the question of why Parlog is being extended at all, rather than being replaced by a language such as C++ or Smalltalk. One answer is that most OOP languages do not have a formal semantics, but this can be remedied by basing object oriented features inside a LP framework.

Another reason for using Parlog is that it can easily define relationships between entities – an ability which is missing from conventional OOP languages. Retaining predicates in Parlog++ and Polka may seem unnecessary because the same functionality can be achieved with objects alone. However, there are numerous cases where an algorithm can be more clearly stated as a predicate, rather than as an object. For instance, the encoding of the relationships in an object oriented database as predicates is much closer to their usual mathematical specification, and arguably simpler than in a language which only supports the object paradigm. The 'objects only' approach is typified by Vulcan [Kahn et al. 1987] and A'UM [Yoshida and Chikayama 1987].

Key advantages of Parlog for OOP include the availability of simple ways to specify concurrency, synchronisation, and state change. In addition, the logical variable can be used for such things as assignment, testing, data access, data construction, and various other forms of parameter passing. Parlog also offers committed choice non-determinism which may, in certain situations, be more suitable for programming simulations and computer systems than the don't know non-determinism found in Prolog. In such systems, it should not be possible to undo an action once taken, except by explicitly doing so, which is not the case with the backtracking mechanism. Guards are also useful, since they allow messages to be tested in arbitrarily complex ways before they are accepted.

In section 2, Parlog++ will be described and a simple 'keyStore' class defined. 'keyStore' will also be the basis of the Polka examples, which will utilise inheritance and self communication to generalise the program so that it can be reused in a number of ways. More information on Polka, including its meta level functionality, can be found in [Davison 1989].

Parlog will not be described; excellent descriptions have appeared in many places [Conlon 1989; Gregory 1987].

2. PARLOG++

A fruitful way of viewing Parlog++ is as a kind of 'higher level' Parlog, by which we mean that it offers a set of syntactic constructs on top of Parlog for the concise representation of common Parlog programming styles. Interestingly, it was found that the types of program which would benefit most from extra syntactic operations were those that used processes, and it was a short step from this to the work of Shapiro and Takeuchi, which equates an object to a process [Shapiro and Takeuchi 1983]. This relationship is the central idea behind the implementation of every language which combines OOP and concurrent LP, although it must be extended to deal with more advanced OOP features such as multiple inheritance and self communication. More importantly, the relationship between the OOP and concurrent LP elements of the language must be defined, such as whether both objects and predicates will be present. Essentially, Shapiro and Takeuchi's model is most useful as a way of viewing concurrent LP processes as objects.

All Parlog++ classes have a similar structure, which can be summarised by the following informal BNF :

```
parlog++  < class name >
< variable declarations >
clauses
    < clauses >
[
code
    < predicates >
]
end .
```

A complete, and more formal, BNF can be found in the [Davison 1989].

The words in bold are Parlog++ keywords, the phrases in angled brackets are informal descriptions of the text that should be placed there, while square brackets around text denotes that it is optional.

The BNF above shows how a Parlog++ class starts with the keyword parlog++, its name, and then declares its variables. These variables define streams and states for the class which may be visible to the user or invisible (local to the class). The clauses section contains the clauses for manipulating the variables; each clause is separated by either the Parlog parallel search operator '.' or the sequential search operator ';'. By using '.', OR-parallelism is supported.

A clause has the form :

< input messages > ⇒ < actions > < separator >

The actions to the right of the '⇒' will be a sequence of calls to Parlog predicates or Parlog++ operations separated by Parlog's parallel conjunction ',' or sequential conjunction '&', with the possibility of a guard. By using ',', AND-parallelism is supported.

A clause is chosen by input pattern matching its message part against messages arriving on the input stream. If the match is successful and any guard part also succeeds, then the clause body is executed.

The code section is optional – it contains Parlog predicates which are visible only to the class.

The class definition finishes with the **end** keyword.

2.1. A 'keyStore' class

The 'keyStore' Parlog++ class will be able to store and retrieve terms of the form:

```
entry( < key >, < email address > )
```

<key> is itself a term, containing three arguments :

```
key( < first name >, < surname >, < login name > )
```

Thus, a typical 'entry' term might be :

```
entry( key(andrew, davison, ad), ad@doc.ic.ac.uk )
```

The 'entry' store will be accessible only via two types of messages :

```
        insert( < entry > )
and     read( < entry >, < ok > )
```

'insert' messages will add entries to the 'keyStore' object, while 'read' messages will attempt to read entries. In a 'read', <entry> will be partially instantiated – its <key> argument will be fully bound, but the <email> value will be a variable. <email> will be bound by the object if an entry with the specified key is located, and the <ok> variable will be used to inform the sender of the success (or otherwise) of the read. In this simple example, there will be no means of removing entries.

The class is :

```
    parlog++    keyStore         Error ostream
            % input stream added automatically
    invisible    Store   state  <-  []
    clauses
        last   ⇒    Error :: last.        % terminate

        insert( entry(Key, Val))   ⇒        % insert entry
            insert_val( entry(Key, Val), Store, NewStore),
            Store becomes NewStore.

        read( entry(Key, Val), Ok)  ⇒           % read entry
            read_val(entry(Key, V), Store) :
                Ok = read(Key), Val = V           % successfully
```

```
          else
              Ok = no_such_entry(Key);          % unsuccessfully

      Msg      ⇒      Error :: Msg.              % reject message

code
   mode insert_val(?, ?, ^).          % insert value into list
   insert_val(Entry, [], [Entry]).
   insert_val(entry( key(F, S, L), Val),
      [entry(key(First, Surname, Login),Value)|Store],
      [entry(key(F, S, L), Val),
       entry(key(First, Surname, Login),Value)|Store]) <-
          S @< Surname : true;
   insert_val(Entry, [E|Store], [E|Store1])  <-
          insert_val(Entry, Store, Store1).


   mode read_val(?, ?).               % read value from list
   read_val(entry(Key,Val), [entry(Key,Value)|Store])  <-
          Val = Value;
   read_val(Entry, [E|Store]) <-
          read_val(Entry, Store).
end.
```

The class declares one visible output stream called Error, which will allow unrecognised messages to be output from the object. By default, there is also a visible input stream which enables the user to send messages to the object. In addition, an invisible state variable called Store is declared, which is initialised to [] when the object is first invoked. The difference between *invisible* and *visible* is that invisible variables will be hidden from the user at run time.

The class has four clauses – the first uses the Polka last keyword to signal that the input stream has closed and to automatically terminate the object after the actions to the right of the '=>' have been carried out. The actions in this case involve terminating the Error output stream which is achieved by sending a last message along Error by using the :: operation.

The second clause deals with 'insert' messages by calling insert_val/3, which is defined in the code section and so is invisible to the user. The predicate builds a list whose entries are sorted into alphabetical order by their surnames. This list becomes the new value for Store via the becomes operation. becomes does not carry out destructive assignment, and it is more accurate to think of it a replacement operator which only takes effect after the other operations in the clause have been carried out.

The third clause has an if-then-else format, dictated by the use of read_val/2 as a deep guard. If it succeeds, then the unbound < email > variable in the entry will be bound, which is signalled by binding Ok to :

```
read(Key)
```

The alternative is that the entry is not found and so Ok is bound to :

```
no_such_entry(Key)
```

The final clause is only executed if the others have failed to match against the input (or their guards have failed). This default nature is made possible by separating the clause from the others with a ';'. Note that the other clauses are separated by '.'s, so enabling them to be tried in parallel against incoming messages. The message form is a variable which will allow it to match against any message. This is then output on the error stream.

An invocation of 'keyStore' might be :

```
<- keyStore(Ins, Error),
   Ins = [ insert(entry(key(andrew,davison,ad),ad@doc.ic.ac.uk)),
           insert(entry(key(lee,naish,lnaish),lee@cs.mu.oz.au)),
           read(entry(key(lee,naish,lnaish), Email1), Ok1),
           read(entry(key(donald,peterson,don), Email2), Ok2),
           delete].
```

The call to 'keyStore' contains two arguments - the first is for the default input stream, the second for the error output stream. Since Store is defined as being invisible it will not appear as an argument. The input stream is sent five messages which are here represented by a list of 5 terms; in the actual Parlog++ system, messages can be sent interactively. However, this call does illustrate how a Parlog++ class can be treated as a Parlog goal, so permitting Parlog programs to use objects.

The bindings produced are :

```
Email1 = lee@cs.mu.oz.au
Ok1    = read(key(lee, naish, lnaish))

Email2 = _
Ok2    = no_such_entry( key(donald, peterson, don))

Error  = [delete]
```

It is worthwhile to consider what the Parlog++ 'keyStore' program offers over a Parlog version. Firstly, the use of an invisible variable and the code section help to enforce encapsulation and data abstraction, permitting the class to hide its internal representation. Another benefit is the way that the clauses of 'keyStore' show how a message spawns an action. Such conceptual aids greatly simplify the programming task, reducing programming errors and code size. Also, the use of typing ensures that class variables cannot be used incorrectly. For instance, it is not possible to send a message to a state variable.

A typical Parlog program for 'keyStore' would be based around a recursive predicate. In comparison, the Parlog++ class is shorter because its clauses are not recursive, and state values unused by a clause need not be mentioned.

A significant point is that Parlog is still present : AND- and OR- parallelism, unification and predicates can be employed. Of special interest is the use of back communication as a reply mechanism for messages. Parlog++ is a *combination* of two paradigms, as shown by the way that object oriented features such as encapsulation and message passing are utilised along side relations for searching and manipulating lists.

An important consideration is the overhead of using Parlog++ instead of Parlog. The compilation process is slightly longer since the Parlog++ code is translated into Parlog and then compiled. However, the Parlog code generated by the Parlog++ translator is, in most cases, indistinguishable from what a Parlog programmer would have written. In other words, there is no run time penalty to writing a program as a Parlog++ class.

There are a number of disadvantages to this program, which it shares with the Parlog version. Firstly, it is quite difficult to change the kinds of things stored. Admittedly, there is a certain amount of flexibility because of the use of logical variables and the absence of typing, but the representation is restricted by the way that the entry(Key, Val) structure is referred to by insert_val/3 and read_val/2. These predicates also fix the search strategy employed by the store - keys must be in ascending order by surname.

The type of flexibility implied by the preceding comments is quite difficult to achieve in Parlog or Parlog++, but is much easier in Polka. In the next section, a general purpose 'ordered_store' class is defined which neither specifies the structure of its entries nor how they are stored and retrieved. These issues will be dealt with by subclasses of 'ordered_store', as illustrated by one which specialises it to obtain the functionality of 'keyStore'.

3. POLKA

A Polka class looks very like a Parlog++ program, the main syntactic differences being that a class does not start with the **parlog++** keyword, and that there is some extra notation for carrying out inheritance and self communication. It is also unnecessary to declare a visible error output stream and to include clauses for dealing with termination and unrecognised messages.

3.1. A 'ordered_store' class

As with 'keyStore', this class can accept two types of messages :

insert(Val)   **and**   read(Val, Ok)

However, the allowable forms of Val are not specified, although Ok can be bound to :

read(Val)   **and**   no_such_entry(Val)

in a similar way to 'keyStore'. The class is defined as :

```
ordered_store          % Error ostream defined automatically
invisible     Store state <-  []
clauses
    insert(Val)    ⇒
        insert(Val, Store, NewStore),        % insert a value
        Store becomes NewStore.

    read(Val, Ok)    ⇒
        read(Val, Store, Ok).           % try to read a value


    % no last or 'error' clause required
code
    mode insert(?, ?, ^).                      % insert Val in list
    insert(Val, [], [Val]).
    insert(Val, [V|Vals], Vals1) <-
        self :: less(Val, V, Ans),                % is Val > V
        insert_ans(Ans, Val, V, Vals, Vals1).

    mode insert_ans(?, ?, ?, ?, ^).
    insert_ans(yes, Val, V, Vals, [Val,V|Vals]).     % Val < V
    insert_ans(no, Val, V, Vals, [V|Vals1]) <-       % Val ≥ V
        insert(Val, Vals, Vals1).
```

```
    mode read(?, ?, ^).                     % read Val from list
    read(Val, [], no_such_entry(Val)).      % cannot find Val
    read(Val, [V|Vals], Ok)   <-
        self :: equal(Val, V, Ans),         % is Val = V
        read_ans(Ans, Val, V, Vals, Ok).

    mode read_ans(?, ?, ?, ?, ^).
    read_ans(yes, Val, V, Vals, read(Val)). % is Val = V
    read_ans(no, Val, V, Vals, Ok) <-       % is Val ≠ V
        read(Val, Vals, Ok).
end.
```

'insert' messages are handled in the first clause by calling insert/3 which adds Val to Store, returning NewStore. insert/3 relies on a self message in its second clause :

```
        self :: less(Val, V, Ans)
```

This denotes that the object will send a 'less' message to itself, and Val will be tested to see if it is less than V. If it is, Ans is bound to yes, otherwise to no. The result is passed to insert_ans/5 which either continues the search through the list or inserts Val. In fact, the destination of the self message depends on the inheritance hierarchy to which 'ordered_store' belongs, but such a message is always sent to the class at the bottom of the hierarchy.

This approach means that the search mechanism (embedded in insert/3 and insert_ans/5) is separated from the criteria for inserting a value (which is handled by the 'less' message).

The second clause deals with 'read' messages by calling read/3. Its functionality depends on the self message :

```
        self :: equal(Val, V, Ans)
```

This will compare Val and V and bind Ans to yes if they are 'equivalent' in some sense, or bind it to no if they are not. As with insert/3, the result is passed to a subsidiary predicate which either extracts the element or continues the search.

This approach separates the criteria for selecting a value from the search, in a similar way to in insert/3. This means that 'ordered_store' never directly examines the structure of the elements which it stores - this is abstracted away by using self messages *and* logical variables (a good example of OOP and LP working together).

'ordered_store' would be called an *abstract class* by object oriented programmers, which signifies that it is of no real use on its own; it must be specialised for an actual application. In practice, this involves specifying the structure of a Store value, and also defining the meaning of 'less' and 'equal' messages.

3.2. The new 'keyStore' class

This class specialises 'ordered_store' in order to attain the same functionality as the 'keyStore' Parlog++ class. It inherits 'ordered_store' using an **inherits** declaration, which means that self messages sent from 'ordered_store' will be delivered to 'keyStore'. Thus, 'keyStore' should include clauses which can handle 'less' and 'equal' messages. The class is :

```
keyStore                % Polka version
inherits  ordered_store        % inheritance line
clauses
    less(entry(key(_, S, _), _),
       entry(key(_, Surname, _), _) , Ans)  ⇒
        S @< Surname : Ans = yes
        else  Ans = no.

    equal(entry(K, V), entry(Key, Val), Ans)    ⇒
        K == Key : Ans = yes, V = Val
        else  Ans = no.
end.
```

By inheritance, 'keyStore' can understand 'insert' and 'read' messages but, more importantly, it also contains clauses for handling 'less' and 'equal' messages.

The clauses specify that Store values are actually 'entry' terms and also state how they are to be compared. The first clause defines 'less' in terms of '@<', while the second defines 'equal' using '==' (the Parlog test unification construct).

An invocation of this 'keyStore' is similar to the Parlog++ version :

```
<- keyStore(Ins, Error).
```

In this example, inheritance is only being used to add 'insert' and 'read' clauses to 'keyStore' but, in general, it is also possible to redefine inherited clauses. This might be useful here as a means of changing the structure of the term returned by the < ok > variable of a 'read' message, so that it is closer to the format used in the original 'keyStore' class. Further details on this use of inheritance can be found in [Davison 1989].

4. SUMMARY

Two extensions to Parlog have been described : Parlog++ and Polka. Parlog++ supports a useful subset of the functionality of Polka which permits it to act as a 'bridge' between the concurrent LP features of Parlog and the full OOP power of Polka. However, Parlog++ is a useful language, allowing process-based programs to be written more succinctly. It supports a notation for encapsulation, data hiding, message passing and state updates which reinforces the programmer's view of a problem as a network of communicating objects. The main extra contributions of Polka are inheritance and self communication, both of which were examined in this paper. Together they permit a more general programming style where *abstract classes* (such as 'ordered_store') can be specialised in various ways. This technique is common in object oriented languages such as Smalltalk, but is neglected in concurrent LP.

REFERENCES

CONLON, T. 1989. *Programming in Parlog*, Addison Wesley, Reading, Mass.

DAVISON, A. 1989. "Polka : A Parlog Object Oriented Language", PhD thesis, Imperial College, November.

GREGORY, S. 1987. *Parallel logic programming in PARLOG*, Addison Wesley, Reading, Mass.

KAHN, K.M., TRIBBLE, D., MILLER, M.S., AND BOBROW, D.G. 1987. "Vulcan : Logical Concurrent Objects", In *Concurrent Prolog : Collected Papers*, E.Y. Shapiro (ed.), MIT Press, Cambridge, MA, Vol. 2, Chapter 30, pp.274-303.

SHAPIRO, E., AND TAKEUCHI, A. 1983. "Object Oriented Programming in Concurrent Prolog", *New Generation Computing* 1 (1983), pp.25-48.

YOSHIDA, K., AND CHIKAYAMA, T. 1987. "A'UM - Parallel Object-Oriented Language upon KL1", ICOT Technical Report : TR 335, Tokyo, Japan.

Precedences in Specifications and Implementations of Programming Languages

Annika Aasa

Programming Methodology Group, Dept. of Computer Sciences
Chalmers University of Technology
S-412 96 Göteborg, Sweden
E-mail: annika@cs.chalmers.se

Abstract

Although precedences are often used to resolve ambiguities in programming language descriptions, there has been no parser-independent definition of languages which are generated by grammars with precedence rules. This paper gives such a definition for a subclass of context-free grammars.

A problem with a language containing infix, prefix and postfix operators of different precedences is that the well-known algorithm, which transforms a grammar with infix operator precedences to an ordinary unambiguous context-free grammar, does not work. This paper gives an algorithm that works also for prefix and postfix operators. An application of the algorithm is also presented.

1 Introduction

Precedences are used in many language descriptions to resolve ambiguities. The reason for resolving ambiguities with precedences, instead of using an unambiguous grammar, is that the language description often becomes shorter and more readable. An unambiguous grammar that reflects different precedences of operators usually contains a lot of nonterminals and single productions. Consider for example an ambiguous grammar for simple arithmetic expressions and the unambiguous alternative.

$$
\begin{array}{lcl}
E & ::= & E \; + \; E \\
  & | & E \; - \; E \\
  & | & E \; * \; E \\
  & | & E \; / \; E \\
  & | & \text{int} \\
  & | & (\; E \;)
\end{array}
\qquad
\begin{array}{lcl}
E & ::= & E \; + \; T \\
  & | & E \; - \; T \\
  & | & T \\
\\
T & ::= & T \; * \; F \\
  & | & T \; / \; F \\
  & | & F \\
\\
F & ::= & \text{int} \\
  & | & (\; E \;)
\end{array}
$$

If the language also contains prefix and postfix operators, then the unambiguous grammar will be surprisingly large.

If a language has user-defined operators as for example ML [16] and PROLOG [19] it is also very convenient to use precedences. When a new operator is introduced, the grammar is augmented with a new production, and it is hard to imagine how a user should indicate where this production should be placed in an unambiguous grammar with different nonterminals.

When dealing with precedences, at least two questions arise. First, although precedences are used in many situations, there is no adequate definition of what it means for a production in a grammar to have higher precedence than another production. Precedences are only used to guide which steps a parser would take when there is an ambiguity in the grammar [3, 11, 20]. It is not always easy, given an ambiguous grammar and a set of disambiguating precedence rules, to decide if a parse tree belongs to the language.

The second question is if it is possible to transform a grammar with precedence rules to an ordinary context-free grammar. This is surprisingly complicated for grammars containing prefix and postfix operators of different precedences.

For a subclass of context-free grammars, we will give a parser-independent definition of precedences and an algorithm that transforms a grammar with precedences to an unambiguous context-free grammar.

2 Notation

We will only consider grammars with one nonterminal and in which every left-hand-side is either *atomic* or is of one of the following forms: A op A, A op, op A, where A is the nonterminal and op a terminal. Atomic left-hand-sides either consists solely of terminals (for example int) or is a nonterminal surrounded by terminals (for example (A)). We will sometimes use AE as a shorthand for *all* atomic left-hand-sides. Furthermore, all terminals must be distinct.

This kind of grammars generates languages with infix, prefix and postfix operators, but it is trivial to extend the ideas to grammars with distfix operators. The first restriction, only one nonterminal, is not as hard as it seems. In many language descriptions, precedences are used to resolve ambiguity in just one part of the language and that part can be described by a grammar with only one nonterminal. The same ideas of defining precedences can also be extended to more general grammars [2].

With precedence rules we mean both precedence and associativity rules. A grammar together with precedence rules will be called a *precedence grammar* and the notation is as follows.

$$
\begin{array}{llll}
E & ::= & \$\ E & 3 \\
  & | & E + E & 2 & \text{left associative} \\
  & | & \#\ E & 1 \\
  & | & \text{int} &
\end{array}
$$

The precedences are given as numbers together with the productions. For these simple grammars we can just as well say that it is the operators that have precedence. The precedence of an operator op will be denoted P(op). The operators can be divided into four kinds: left associative infix, right associative infix, prefix and postfix, and operators of different kinds are not allowed to have the same precedence. In the algorithm described in section 4 we suppose that there is only one operator of each precedence but it is trivial to extend the algorithm to handle more. We let the variable H range over precedence grammars.

We use the convention of precedence that a production with higher precedence has less binding power than one with lower precedence. Thus, for the usual arithmetic operators the addition operator + has higher precedence than the multiplication operator *. This convention of precedence is used for example in PROLOG [19] and OBJ [15]. In for example ML [16] and Hope [7] the opposite convention is used, and multiplication is said to have higher precedence than addition.

Since precedences have to do with structure we have to consider parse trees or syntax trees instead of strings when we talk about which language a precedence grammar defines. We will use syntax trees and we will for example picture the derivation

$$E \rightarrow E+E \rightarrow E*E+E \rightarrow E*E!+E \rightarrow^* 2*3!+4$$

as

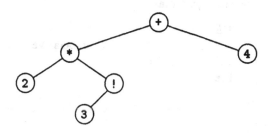

A *syntax tree for an operator* is a syntax tree with that operator as root.

3 Definition of precedence and associativity

One obvious question to ask is which language we define with a precedence grammar. The language is of course a subset of the language generated by the ambiguous grammar without precedence rules. The precedence rules throw away some parse trees. We will call the parse trees that should not be thrown away for *precedence correct*.

We think it is unsatisfactory to define the precedence correct trees in terms of a specific parsing method. A specification of a language should not involve a method to recognize it. If the language is defined by one parsing method it could be very hard to see if a parser that uses another method is correct.

We will define a predicate Pc_H which given a precedence grammar H defines the precedence correct trees. So, $Pc_H(t)$ holds if and only if the syntax tree t is correct according to the disambiguating rules in the grammar H. The predicate is defined in such a way that syntax trees built by an operator precedence parser [13] are precedence-correct. This is shown in [1]. I believe that the converse, i.e. that every precedence correct tree can be recognized by an operator precedence parser is true but I have not proved it.

Let us first make some reflections. A syntax tree with an infix operator as root has the following form.

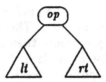

If it should be precedence correct, both the subtrees *lt* and *rt* must of course be precedence correct. Furthermore, there must be some requirements involving the precedence of the root operator. For languages with only infix operators it is enough to look at the precedences of the roots of the subtrees. They should be less than the precedence of the root. This is however not enough if the language also contains prefix and postfix operators. Consider the precedence grammar

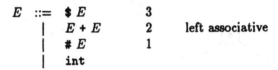

and the syntax trees

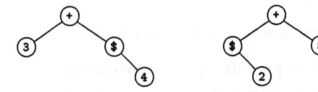

We want to consider the left syntax tree as precedence correct but not the syntax tree to the right. This illustrates that prefix operators with higher precedence than an infix operator must be allowed to occur in the right subtree.

Furthermore, consider the two syntax trees below, generated from the same grammar.

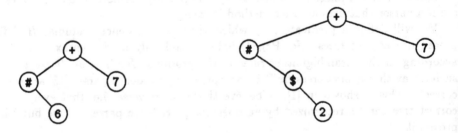

We want to consider the left syntax tree as precedence correct but not the syntax tree to the right. This illustrates that even if the precedence of the root operator of a subtree is less than the precedence of the whole tree the syntax tree need not be precedence correct.

To solve this problem we introduce two different kinds of precedence weight of a syntax tree, the left weight, Lw, and the right weight, Rw. Prefix operators have precedence only to the right, postfix operators only to the left and infix operators in both directions. The weights can depend both on the root operator and the weights of the subtrees, and we define them as follows.

Definition 1

$$Rw(AE) = 0 \qquad\qquad Lw(AE) = 0$$
$$Rw(t\ op) = 0 \qquad\qquad Lw(t\ op) = max(P(op), Lw(t))$$
$$Rw(op\ t) = max(P(op), Rw(t)) \qquad Lw(op\ t) = 0$$
$$Rw(lt\ op\ rt) = max(P(op), Rw(rt)) \quad Lw(lt\ op\ rt) = max(P(op), Lw(lt))$$

We can now give the definition of the predicate Pc_H that defines the precedence correct syntax trees.

Definition 2 *Given a precedence grammar H, the following rules define the predicate Pc_H, where Left, Right, Pre and Post respectively, denote the set of left associative infix operators, right associative infix operators, prefix operators and postfix operators.*

atomic expressions:

$$Pc_H(AE)$$

left associative infix operators:

$$\frac{op \in Left \quad Pc_H(lt) \quad Pc_H(rt) \quad Rw(lt) \leq P(op) \quad Lw(rt) < P(op)}{Pc_H(lt\ op\ rt)}$$

right associative infix operators:

$$\frac{op \in Right \quad Pc_H(lt) \quad Pc_H(rt) \quad Rw(lt) < P(op) \quad Lw(rt) \leq P(op)}{Pc_H(lt\ op\ rt)}$$

prefix operators:

$$\frac{op \in Pre \quad Pc_H(t) \quad Lw(t) < P(op)}{Pc_H(op\ t)}$$

postfix operators:

$$\frac{op \in Post \quad Pc_H(t) \quad Rw(t) < P(op)}{Pc_H(t\ op)}$$

An alternative way to define the precedence correct trees is to define which operators are allowed to occur in each subtree. To say that, we need a new definition which we also will use later.

Note that an operator with higher precedence can occur in a subtree of an operator with lower precedence. An example of this is the precedence correct syntax tree generated from the same precedence grammar as before.

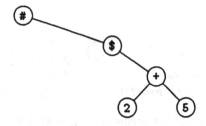

The prefix operator **$** *covers* the infix operator **+**.

Definition 3 *An occurrence of an operator in a syntax tree t is covered if it occurs in a subtree of an operator with higher precedence than itself. An occurrence of an operator is uncovered if it is not covered.*

Postfix operators can be in the left subtree of an infix node independent of their precedence but not in the right subtree. Analogously, prefix operators can be in the right subtree of an infix node independent of their precedence but not in the left subtree. The conclusion of this is that if a syntax tree *lt op rt* (where *op* is left associative) should be precedence correct both *lt* and *rt* must be precedence correct, all infix and prefix operators in *lt* with higher precedence than *op* must be covered and all infix and postfix operators in *rt* with higher or equal precedence than *op* must be covered.

4 Transformation to an unambiguous context-free grammar

Besides that it is interesting for theoretical reasons to know whether a precedence grammar can be transformed to an unambiguous context-free grammar, such an algorithm is sometimes needed in practice. For example, if we want to describe a language with a precedence grammar but parse the language with a parsing method that cannot handle precedence rules, then the algorithm is definitely needed. One such commonly used parsing method is recursive descent [9], and another is DCG [17]. It is not even obvious how to use precedence rules in Earley's algorithm [10] even if it is possible [2].

For grammars with only infix operators, there is a well-known algorithm that transforms them to ordinary context-free grammars by introducing one nonterminal for each precedence level [4]. But if the language also contains prefix and postfix operators, this method does not work. Consider the precedence grammar:

$$
\begin{array}{llll}
E & ::= & E\ ? & 4 \\
  & | & E + E & 3 \quad \text{left associative} \\
  & | & E\ ! & 2 \\
  & | & E * E & 1 \quad \text{left associative} \\
  & | & \texttt{int} &
\end{array}
$$

For this grammar the method of introducing one nonterminal for each precedence level does not work. Using the method naïvely would give the grammar:

$$
\begin{array}{lll}
E(4) & ::= & E(4)\ ? \quad | \quad E(3) \\
E(3) & ::= & E(3) + E(2) \quad | \quad E(2) \\
E(2) & ::= & E(2)\ ! \quad | \quad E(1) \\
E(1) & ::= & E(1) * E(0) \quad | \quad E(0) \\
E(0) & ::= & \texttt{int}
\end{array}
$$

But this grammar is incorrect since it does not generate *all* precedence correct syntax trees. For example, the syntax trees for the strings 7?+8 and 9+6?*8 are not derivable. Another attempt to construct a grammar from which precisely the precedence correct syntax trees are derivable is:

$$
\begin{aligned}
E(4) &::= E(3)\\
E(3) &::= E(3) + E(2) \mid E(2)\\
E(2) &::= E(1)\\
E(1) &::= E(1) * E(0) \mid E(0)\\
E(0) &::= \text{int} \mid E(2)\ ! \mid E(4)\ ?
\end{aligned}
$$

Here we have tried to incorporate the idea that a postfix operator forms an atomic expression. This grammar is not correct, either, since we can derive syntax trees which are *not* precedence correct. An example is the syntax tree for the string 3*4?.

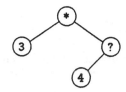

This illustrates that we must construct the grammar in such a way that, for every production $E ::= E_v * E_h$, it is not possible to derive a syntax tree for a postfix operator with higher precedence than $*$ from E_h. Syntax trees for postfix operators with lower precedence than $*$ should of course be derivable from E_h.

Let us now make some reflections about the syntax trees which should be derivable from the nonterminal E_v in the production $E ::= E_v * E_h$. It is harder because we sometimes want syntax trees for postfix operators with higher precedence than $*$ to be derivable from it and sometimes not. Consider the syntax tree:

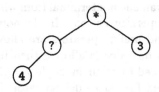

Since it is precedence correct, we must ensure that the syntax tree

is derivable from E_v in a production $E ::= E_v * E_h$. That is, we have shown that there must be at least one production $E ::= E_v * E_h$ such that we can derive syntax trees for postfix operators with higher precedence than $*$ from E_v. We will now show that we also must have productions $E ::= E_v * E_h$ such that we cannot derive such syntax trees from E_v. Consider the following syntax tree for the string 5+4?*3

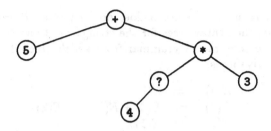

where the syntax tree above is a subtree in another syntax tree. This syntax tree is not precedence correct since ? has higher precedence than +. So in this case we must ensure that the syntax tree

is *not* derivable from E_v. Note that the occurrence of + is covered in the syntax tree

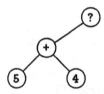

and that this syntax tree must be derivable from E_v. That is, if a postfix operator is allowed to occur in a syntax tree derived from a nonterminal, it can also cover other operators. The reasoning for the case where prefix operators are allowed to occur is analogous.

We have shown that we must have more than one nonterminal from which it is possible to derive a syntax tree with a specific infix operator as root. In the nonterminals there must be information about which postfix and prefix operators are allowed to occur in the left (right) subtree. If a postfix (prefix) operator is allowed then it can also cover other operators which otherwise are not allowed to occur in the left (right) subtree. So, the number of different nonterminals we need for each infix operator depends on how many postfix and prefix operators with higher precedence there are in the precedence grammar.

Our algorithm generates a grammar with nonterminals of the form $E(n, p, q)$ where the indices show which operators are allowed to occur in the syntax trees derived from the nonterminal. Before giving the algorithm we introduce some notation.

inl_i	the i:th left associative infix operator
inr_i	the i:th right associative infix operator
pre_i	the i:th prefix operator
$post_i$	the i:th postfix operator
$P_{pre}(n, p)$	the precedence of the p:th prefix operator with higher precedence than n
$P_{post}(n, q)$	the precedence of the q:th postfix operator with higher precedence than n.

Now we can define which operators must not occur in a syntax tree derived from $E(n, p, q)$.

1. An operator op for which $P(op) > max(P_{pre}(n, p), P_{post}(n, q))$.

2. An uncovered prefix operator op for which $P(op) > P_{pre}(n, p)$.

3. An uncovered postfix operator op for which $P(op) > P_{post}(n, q)$.

4. An uncovered infix operator op for which $P(op) > n$.

As mentioned earlier there must be more than one production for each operator, the number depends on the number of prefix and postfix operators with higher precedence. The algorithm generates the grammar by five rules which introduce the nonterminals $E(n, p, q)$ and the productions for them.

1. The rule for left associative infix operators.

$$E(P(inl_i)), p, q) \quad ::= \quad E(P(inl_i), 0, q)\ inl_i\ E(P(inl_i) - 1, p, 0)$$
$$|\quad E(P(inl_i) - 1, p, q)$$

> where $1 \leq i \leq$ number of left associative infix operators
> $0 \leq p \leq$ number of prefix operators with higher precedence than $P(inl_i)$
> $0 \leq q \leq$ number of postfix operators with higher precedence than $P(inl_i)$

2. The rule for right associative infix operators.

$$E(P(inr_i)), p, q) \quad ::= \quad E(P(inr_i) - 1, 0, q)\ inr_i\ E(P(inr_i), p, 0)$$
$$|\quad E(P(inr_i) - 1, p, q)$$

> where $1 \leq i \leq$ number of right associative infix operators
> $0 \leq p \leq$ number of prefix operators with higher precedence than $P(inr_i)$
> $0 \leq q \leq$ number of postfix operators with higher precedence than $P(inr_i)$

3. The rule for prefix operators.

$$E(P(pre_i)), p, q) \quad ::= \quad E(P(pre_i) - 1, p + 1, q)$$

> where $1 \leq i \leq$ number of prefix operators
> $0 \leq p \leq$ number of prefix operators with higher precedence than $P(pre_i)$
> $0 \leq q \leq$ number of postfix operators with higher precedence than $P(pre_i)$

4. The rule for postfix operators.

$$E(P(post_i)), p, q) \quad ::= \quad E(P(post_i) - 1, p, q + 1)$$

> where $1 \leq i \leq$ number of postfix operators
> $0 \leq p \leq$ number of prefix operators with higher precedence than $P(post_i)$
> $0 \leq q \leq$ number of postfix operators with higher precedence than $P(post_i)$

5. The rule for atomic expressions.

$$E(0, p, q) \quad ::= \quad AE$$
$$| \quad pre_i \ E(P(pre_i), p - i, 0) \qquad \text{where } 1 \le i \le p$$
$$| \quad E(P(post_j), 0, q - j) \ post_j \qquad \text{where } 1 \le j \le q$$

where $0 \le p \le$ number of prefix operators
$0 \le q \le$ number of postfix operators

The start symbol in the resulting grammar is the nonterminal $E(m, 0, 0)$ where m is the highest precedence.

Let us use the method to construct an unambiguous grammar for the language generated by the precedence grammar:

E	::=	E ?	4	
	|	$E + E$	3	left associative
	|	E !	2	
	|	$E * E$	1	left associative
	|	int		

For this grammar we have

$$inl_1 = * \qquad\qquad\qquad post_1 = !$$
$$inl_2 = + \qquad\qquad\qquad post_2 = ?$$

The rule for left associative infix operators yields the productions:

$$E(3, 0, 0) \quad ::= \quad E(3, 0, 0) + E(2, 0, 0) \quad | \quad E(2, 0, 0)$$

$$E(3, 0, 1) \quad ::= \quad E(3, 0, 1) + E(2, 0, 0) \quad | \quad E(2, 0, 1)$$

since $P(+) = 3$ and there is one postfix operator with higher precedence than + but no prefix operators, as well as the productions:

$$E(1, 0, 0) \quad ::= \quad E(1, 0, 0) * E(0, 0, 0) \quad | \quad E(0, 0, 0)$$

$$E(1, 0, 1) \quad ::= \quad E(1, 0, 1) * E(0, 0, 0) \quad | \quad E(0, 0, 1)$$

$$E(1, 0, 2) \quad ::= \quad E(1, 0, 2) * E(0, 0, 0) \quad | \quad E(0, 0, 2)$$

since $P(*) = 1$ and there are two postfix operators with higher precedence than * but no prefix operators. The rule for postfix operators yields the productions:

$$E(4, 0, 0) \quad ::= \quad E(3, 0, 1)$$

$$E(2, 0, 0) \quad ::= \quad E(1, 0, 1)$$

$$E(2, 0, 1) \quad ::= \quad E(1, 0, 2)$$

since $P(?) = 4$ and there are neither prefix operators nor postfix operators with higher precedence than ?, and since $P(!) = 2$ and there is one postfix operator with higher precedence than ! but no prefix operators. Finally the rule for atomic expression yields the productions:

$$E(0,0,0) \quad ::= \quad AE$$

$$E(0,0,1) \quad ::= \quad AE \quad | \quad E(2,0,0) \ !$$

$$E(0,0,2) \quad ::= \quad AE \quad | \quad E(2,0,1) \ ! \quad | \quad E(4,0,0) \ ?$$

since we have two postfix operators, ! and ?.

The resulting grammar contains some useless [5] nonterminals and a lot of single productions. These could easily be eliminated.

The correctness of the algorithm can be shown by proving that every precedence grammar, H, generates the same language, that is, the same set of syntax trees, as the grammar we obtain by applying the algorithm on H. This is done formally in [1].

5 Practical use of the algorithm

We have used the algorithm to implement an experimental language with user-defined distfix operators [1]. A new distfix operator is specified by the operator words and optionally with precedence and associativity. The parser is written in ML [16] and uses parser constructors due to Burge [6] and Fairbairn [12] (and Kent Petersson and Sören Holmström [18]). Using these parser constructors it is easy to write a parser given a grammar, since there are constructors that recognize terminal symbols, sequences, and alternatives and other constructors that introduce actions during the parsing.

The parser constructors construct a recursive descent parser and therefore the grammar must not be left recursive and it must express precedences of the involved operators.

In the parser for user-defined operators we use the rules in the algorithm described above. We have to do some changes in order to remove left recursion, and we never generate the entire grammar with all different nonterminals. Instead we see the rules as production schemas in a way that is similar to the hyper rules in two-level grammars [8], and instantiate the rules during the parsing. Hanson describes another technique for parsing expressions using recursive descent without introducing additional nonterminals [14].

Acknowledgements

I am very greatful to Kent Petersson for all advice, encouragement and help he has supplied. Thanks also to Mikael Rittri for his careful reading and valuable comments.

References

[1] Annika Aasa. Recursive Descent Parsing of User Defined Distfix Operators. Licentiate Thesis, Dept. of Computer Sciences, Chalmers University of Technology, S-412 96 Göteborg, Sweden, May 1989.

[2] Annika Aasa. Conctypes with precedences. In preparation, Dept. of Computer Sciences, Chalmers University of Technology, S-412 96 Göteborg, Sweden, 1991.

[3] A. V. Aho, S. C. Johnson, and J. D. Ullman. Deterministic Parsing of Ambiguous Grammars. *Communications of the ACM*, 18(8):441–452, August 1975.

[4] A. V. Aho, J. D. Ullman, and R. Sethi. *Compilers: Principles, Techniques, Tools*. Addison-Wesley Publishing Company, Reading, Mass., 1986.

[5] Roland Backhouse. *Syntax of Programming Languages, Theory and Practice*. Prentice Hall, 1979.

[6] W. H. Burge. *Recursive Programming Techniques*. Addison-Wesley Publishing Company, Reading, Mass., 1975.

[7] R. M. Burstall, D. B. McQueen, and D. T. Sannella. Hope: An Experimental Applicative Language. In *Proceedings of the 1980 ACM Symposium on Lisp and Functional Programming*, pages 136–143, Stanford, CA, August 1980.

[8] J. C. Cleaveland and R. C. Uzgalis. *Grammars for Programming Languages*. Elsevier North-Holland, 1977.

[9] A. J. T. Davie and R. Morrison. *Recursive Descent Compiling*. Ellice Horwood Limited, 1981.

[10] Jay Earley. An Efficient Context-Free Parsing Algorithm. *Communications of the ACM*, 13(2):94–102, February 1970.

[11] Jay Earley. Ambiguity and Precedence in Syntax Description. *Acta Informatica*, 4(2):183–192, 1975.

[12] Jon Fairbairn. Making Form Follow Function: An Exercise in Functional Programming Style. *Software-Practice and Experience*, 17(6):379–386, 1987.

[13] R. W. Floyd. Syntactic Analysis and Operator Precedence. *Journal of the ACM*, 10(3):316–333, 1963.

[14] David R. Hanson. Compact Recursive-descent Parsing of Expressions. *Software-Practice and Experience*, 15(12):1205–1212, 1985.

[15] J-P. Jouannaud K. Futatsugi, J. A. Goguen and J. Meseguer. Principles of OBJ2. In *Proceedings of the 12:th ACM Symposium on Principles of Programming Languages*, pages 52–66, 1985.

[16] R. Milner. Standard ML Proposal. *Polymorphism: The ML/LCF/Hope Newsletter*, 1(3), January 1984.

[17] Fernando C. N. Pereira and David H. D. Warren. Parsing as deduction. In *Proceedings of the 21'st Annual Meeting of the Association for Computational Linguistics*, pages 137 – 144, 1983.

[18] Kent Petersson. LABORATION: Denotationsemantik i ML, 1985. Dept. of Computer Sciences, Univ. of Göteborg and Chalmers Univ. of Tech, S-412 96 Göteborg, Sweden.

[19] Leon Sterling and Ehud Shapiro. *The Art of Prolog*. The MIT Press, Cambridge, Massachusetts, London, England, 1986.

[20] R.M. Wharton. Resolution of Ambiguity in Parsing. *Acta Informatica*, 6:387–395, 1976.

A randomized heuristic approach to register allocation

C. W. Keßler,[*] W. J. Paul, T. Rauber[†]

Computer Science Department, University Saarbrücken, Germany

Abstract

We present a randomized algorithm to generate contiguous evaluations for expression DAGs representing basic blocks of straight line code with nearly minimal register need. This heuristic may be used to reorder the statements in a basic block before applying a global register allocation scheme like Graph Coloring. Experiments have shown that the new heuristic produces results which are about 30% better on the average than without reordering.

1 Introduction

Register allocation is one of the most important problems in compiler optimizations. Among the numerous register allocation schemes proposed register allocation and spilling via graph coloring is generally accepted to give good results. But register allocation via graph coloring has the disadvantage of using a fixed evaluation order within a given basic block. This is the evaluation order given by the source program. But often there exists an evaluation order for the basic block that uses less registers. By using this order the global register allocation generated via graph coloring could be improved. The aim of this article is to achieve such an improvement by improving the evaluation order within the basic blocks.

We can represent a basic block by a directed acyclic graph (DAG); see Fig. 1 for an example. An algorithm that constructs a DAG for a given basic block is given in [1]. For the evaluation of DAGs the following results are known:

(1) If the DAG is a tree, the well–known algorithm of *Sethi* and *Ullman* (see [8]) generates an optimal evaluation in linear time (optimal means: uses as few registers as possible).

(2) The problem of generating an optimal evaluation for a given DAG is NP–complete (see [9]).

To generate a good evaluation order for a DAG that is not a tree, we have to find an heuristic to do this task. We present such an heuristic in the following sections.

The new heuristic uses a mix of several simple evaluation strategies that also include a randomized evaluation selection. These simple evaluation strategies are applied concurrently and the best evaluation generated is selected.

[*]research partially funded by the Leibniz program of the DFG
[†]research partially funded by DFG, SFB 124

The idea behind this approach is that there exists no uniform heuristic that generates good evaluations for every possible DAG. But most of the DAGs encountered in real programs belong to one of a few simple classes. For each of these classes there exists a simple algorithm that generates good, often optimal evaluations. By running these simple algorithms "in parallel" and choosing the best result we obtain a heuristic that copes with most of the DAGs encountered in real programs.

In section 2 we introduce the basic formalism and two simple depth–first–search (*dfs*) variants as evaluation strategies. In section 3 we present the randomized evaluation strategy, in section 4 a variant of the Labeling Algorithm (see [8]). In section 5 we put the pieces together and discuss the performance improvement reached by the reordering with respect to the original basic block.

2 Evaluating DAGs

We assume that we are generating code for a single processor machine with general purpose registers $\mathcal{R} = \{R_0, R_1, R_2 \ldots\}$ and a countable sequence of memory locations. The arithmetic machine operations are three-address-instructions of the following types:

$R_k \leftarrow R_i \, op \, R_j$	binary operation,	$op \in \{+, -, \times, \ldots\}$,
$R_k \leftarrow op \, R_i$	unary operation,	
$R_k \leftarrow \text{Load}(a)$	load register k from the memory location a, or	
$\text{Store}(a) \leftarrow R_k$	store the contents of register k into the memory location a,	

where $i \neq j \neq k \neq i$, $R_k, R_i, R_j \in \mathcal{R}$.

The following considerations are also applicable to the case $k = i$ or $k = j$. Our claim that the registers used by an operation must be mutually different makes the handling partially easier, but does not affect the validity of our results.

Definition: A *basic block* is a sequence of three–address–instructions that can only be entered via the first and only be left via the last statement.

A *directed graph* is a pair $G = (V, E)$ where V is a finite set of nodes and $E \subseteq V \times V$ is a set of edges. In the following let $n = |V|$ denote the number of nodes of the graph. A sequence of vertices $v_0, v_1, \ldots v_k$ with $(v_i, v_{i+1}) \in E$ is called a *path of length k from* v_0 to v_k. A *cycle* is a path from v to v. A directed graph is called *acyclic* if it contains no cycle (of length ≥ 1).

If $(w, v) \in E$, then w is called *operand* or *son* of v; v is called *result* or *father* of w, i. e. the edge is directed from the son to the father. A node which has no sons is a *leaf*, otherwise it is an *inner node*; in particular we call a node with two sons *binary* and a node with only one son *unary*. A node with no father is called *root* of G.

The *outdegree outdeg(w)* of a node $w \in V$ is the number of edges leaving w, i. e. the number of its fathers.

The data dependencies in a basic block can be described by a *directed acyclic graph (DAG)*. The leaves of the DAG are the variables and constants occurring as operands in the basic block; the inner nodes represent intermediate results. An example is given in Figure 1.

Let $G = (V, E)$ be a directed graph with n nodes. A mapping *ord*: $V \rightarrow \{1, 2, \ldots, n\}$ with

$$\forall (w, v) \in E: \quad ord(w) < ord(v)$$

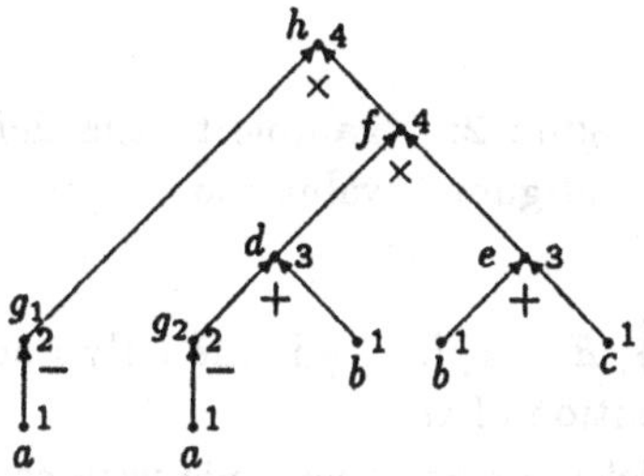

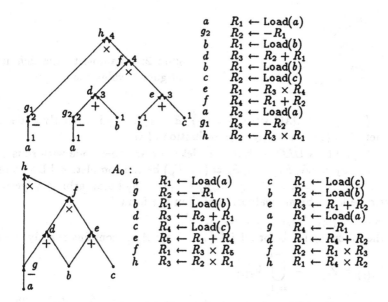

a	$R_1 \leftarrow \text{Load}(a)$
g_2	$R_2 \leftarrow -R_1$
b	$R_1 \leftarrow \text{Load}(b)$
d	$R_3 \leftarrow R_2 + R_1$
b	$R_1 \leftarrow \text{Load}(b)$
c	$R_2 \leftarrow \text{Load}(c)$
e	$R_1 \leftarrow R_3 \times R_4$
f	$R_4 \leftarrow R_1 + R_2$
a	$R_2 \leftarrow \text{Load}(a)$
g_1	$R_3 \leftarrow -R_2$
h	$R_2 \leftarrow R_3 \times R_1$

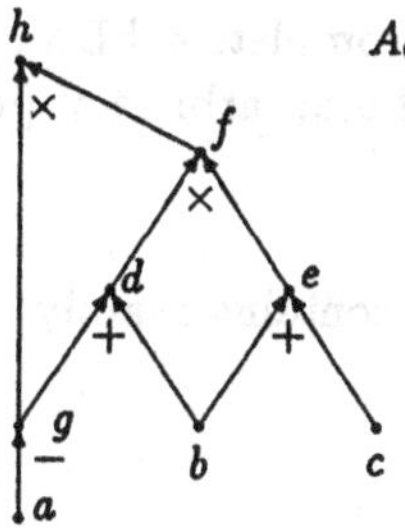

A_0 :

a	$R_1 \leftarrow \text{Load}(a)$	c	$R_1 \leftarrow \text{Load}(c)$
g	$R_2 \leftarrow -R_1$	b	$R_2 \leftarrow \text{Load}(b)$
b	$R_1 \leftarrow \text{Load}(b)$	e	$R_3 \leftarrow R_1 + R_2$
d	$R_3 \leftarrow R_2 + R_1$	a	$R_1 \leftarrow \text{Load}(a)$
c	$R_4 \leftarrow \text{Load}(c)$	g	$R_4 \leftarrow -R_1$
e	$R_5 \leftarrow R_1 + R_4$	d	$R_1 \leftarrow R_4 + R_2$
f	$R_1 \leftarrow R_3 \times R_5$	f	$R_2 \leftarrow R_1 \times R_3$
h	$R_3 \leftarrow R_2 \times R_1$	h	$R_1 \leftarrow R_4 \times R_2$

Figure 1: Example: At first the front end generates an expression tree for $(-a) \times ((-a + b) \times (b + c))$ and evaluates it according to the Labeling algorithm of *Sethi/ Ullman* with minimal register need 4. (The labels are printed at the right hand side of the nodes). The optimizer recognizes common subexpressions, constructs the DAG G and evaluates G in the original order resulting in an evaluation A_0. This reduces the number of instructions and hence the completion time of the basic block. However, now 5 instead of 4 registers are required for the new basic block since A_0 is not optimal. The better evaluation on the right hand side obtained by reordering A_0 needs only 4 registers, and that is optimal.

is called a *topologic order* of the nodes of G. It is well–known that for a directed graph a topological order exists iff it is acyclic. (see e. g. [5]).

Definition: (*Evaluation of a DAG G*) An evaluation A of G is a permutation of the nodes in V such that for all nodes $v \in V$ the following holds: If v is an inner node with sons $v_1, \ldots, v_k$ then v appears in A after v_i, $i = 1, ..., k$.

This means the evaluation A is *complete* and contains *no recomputations*, i. e. each node of the DAG appears exactly once in A. Moreover the evaluation is *consistent* because no node can be evaluated before all of his sons are. Thus each topological order of G is an evaluation, and vice versa.

Definitions: Let $G = (V, E)$ be a DAG. A DAG $S = (V', E')$ is called *subDAG* of G, if $V' \subseteq V$ and $E' \subseteq E \cap (V' \times V')$. — A subDAG $S = (V', E')$ of $G = (V, E)$ with root w is called *complete*, if:

$V' = \{v \in V : \exists \text{ path from } v \text{ to } w\}$ and

$E' = \{e \in E : e \text{ is an edge on a path from a node } v \in V' \text{ to } w\}$.

Definition: (*contiguous evaluation of a DAG G*)

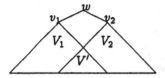

Figure 2: Example to the definition of a contiguous evaluation

(1) Let $G = (V, E)$ be a DAG with $V = \{v\}$, $E = \emptyset$, i. e. v is the only node, root and leaf. Then $A = (v)$ is a contiguous evaluation of G.

(2) Let $G = (V, E)$ be a DAG with root w, let w be an inner node with sons $v_1, \ldots, v_p$ ($p \geq 1$). Let $S_1 = (V_1, E_1), \ldots, S_p = (V_p, E_p)$ be the complete subDAGs of G with the roots $v_1, \ldots, v_p$. Let $\pi : \{1, ..., p\} \to \{1, ..., p\}$ be an arbitrary permutation. Furthermore let A be an evaluation of G in the form

$$A = (A_{\pi(1)}, A_{\pi(2)}, \cdots, A_{\pi(p)}, w),$$

where the following holds for all $j \in \{1, \ldots, p\}$: $A_{\pi(j)}$ contains exactly the nodes of

$$\tilde{V}_{\pi(j)} = V_{\pi(j)} - \bigcup_{i=1}^{j-1} V_{\pi(i)}$$

and A_j is a contiguous evaluation of $\tilde{G}_j = (\tilde{V}_j, E_j \cap (\tilde{V}_j \times \tilde{V}_j))$. Then A is a contiguous evaluation of G.

(3) Those are all contiguous evaluations of G.

From now on we suppose $p \leq 2$ because no node has more than two sons in our machine model.

Example: Consider Fig. 2: Let $G = (V, E)$ be a DAG with root w. w has the sons v_1 und v_2. Let $S_1 = (V_1, E_1)$ and $S_2 = (V_2, E_2)$ be the complete subDAGs of G with the roots v_1 resp. v_2. Let $V' = V_1 \cap V_2$. Let $\pi(1) = 1$ and $\pi(2) = 2$. Then $\tilde{V}_1 = V_1$ and $\tilde{V}_2 = V_2 - V'$. Let A_1 be a contiguous evaluation of $\tilde{G}_1$, and let A_2 be a contiguous evaluation of $\tilde{G}_2$. Then $A = (A_1, A_2, w)$ is a contiguous evaluation of G.

The advantage of a contiguous evaluation is the fact that it can be generated by simple algorithms (variations of *depth–first–search* (*dfs*)). In this paper we will restrict our attention to contiguous evaluations. This is already a heuristic because there are some DAGs for which a noncontiguous evaluation exists that uses less registers than every contiguous evaluation. However, in practice these cases seem to be rare. The smallest DAG of this kind we found so far has 14 nodes and is printed in Fig. 3.

Definitions: (cf. [9]) Let $num : \mathcal{R} \to \{0, 1, 2 \ldots\}$, $num(R_i) = i$ be a function that assigns a number to each register. — A mapping $reg : V \to \mathcal{R}$ is called a (consistent) *register allocation* for A if for all nodes $u, v, w \in V$ the following holds: If u is a son of w, and v appears in A between u and w, then $reg(u) \neq reg(v)$.

$$m(A) = \min_{reg \text{ is reg. alloc. for } A} \left\{ \max_{v \text{ appears in } A} \{num(reg(v)) + 1\} \right\}$$

is called the *register need* of the evaluation A. — An evaluation A for a DAG G is called *optimal* (w. r. to its register need $m = m(A)$) if for all evaluations A' of $Gm(A') \geq m(A)$. In general there will exist several optimal evaluations for a given DAG. — In this paper we always use the word "optimal" with respect to the register need.

Sethi proved in 1975 ([9]) that the problem of computing an optimal evaluation for a given DAG is NP–complete. Assuming $P \neq NP$ we expect an algorithm with nonpolynomial run time. Unfortunately, this problem often occurs in compiler construction and should

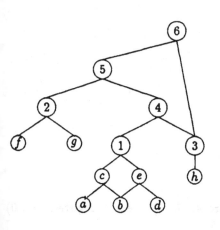

Figure 3: This DAG can be evaluated nonconti-guously (in the order $a, b, c, d, e, 1, f, g, 2, h, 3, 4, 5, 6$) using 4 registers, less than each contiguous evaluation: An evaluation using 4 registers must evaluate the subDAG with root 1 first. A conti-guous evaluation can do this only by evaluating the left son of node 6, the right son of node 5 and the left son of node 4 first. In order the evaluation to be contiguous nodes h and 3 must be evaluated after 1 and thus the values of nodes 3 and 4 must be held in two registers. But in order to evaluate nodes f, g and 2, three more registers are requi-red, thus the contiguous evaluation uses at least five registers altogether.

be solved fast. We will present a heuristic to produce fairly good evaluations in linear time.

Let *new_reg()* be a function which returns an available register and marks it busy, and let *regfree(reg)* be a function which marks the register *reg* free again. (For more details see [3]).

We apply *dfs*–variations to evaluate the given DAG contiguously. The crucial point in *dfs* is the order in which the sons of the binary nodes are visited. If we always visit the left son first, we obtain *left first search* (*lfs*):

Let $G = (V, E)$ be a DAG.
Set *visited(v)* = FALSE for all $v \in V$.
(1) **function** *lfs*(**node** v)
    (* evaluate the subDAG induced by node $v \in V$ *)
    begin
(2) *visited(v)* $\leftarrow$ TRUE;
(3) **if** v is not a leaf
(4) **then if not** *visited(lson(v))* **then** *lfs(lson(v))* **fi**
(5)   **if not** *visited(rson(v))* **then** *lfs(rson(v))* **fi**
    fi
(6) *reg(v)* $\leftarrow$ *new_reg()*; print(v, *reg(v)*);
(7) **if** v is not a leaf
(8) **then if** *lson(v)* will not be used any more **then** *regfree(reg(lson(v)))* **fi**
(9)   **if** *rson(v)* will not be used any more **then** *regfree(reg(rson(v)))* **fi**
    fi
    end *lfs*;

We get the information whether a node will be used later from a reference counter *ref(v)* for each node $v \in V$, which is initialized with *outdeg(v)* at the beginning of *lfs* and decremented when using v as operand. If *ref(v)* = 0, v will not be needed any more; the register containing the result of v can be marked free (line 8/9).

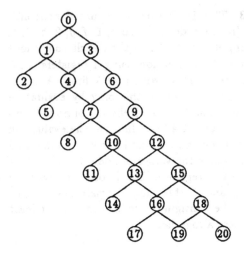

Figure 4: *lfs*(0) needs 9 registers, *rfs*(0) only 4.

lfs–evaluation of the DAG in Fig. 4:

node	2	5	8	11	14	17	19	16	13	10	7	4	1	20	18	15	12	9	6	3	0
in reg.	0	1	2	3	4	5	6	7	5	4	3	2	1	0	8	6	7	5	4	3	2

rfs–evaluation of the DAG in Fig. 4:

node	20	19	18	17	16	15	14	13	12	11	10	9	8	7	6	5	4	3	2	1	0
in reg.	0	1	2	3	0	1	2	3	0	1	2	3	0	1	2	3	0	1	2	3	0

Table 1: Two examples for an evaluation of the DAG in Fig. 4

We observe that each node v will be held in a register from its evaluation point until the last reference to v is reached. Thus, the register need m results from the highest marked register number plus one, as in the definition above.

We obtain *rfs*() by swapping lines (4) and (5).

new_reg() and *regfree*() have constant run time. So *lfs*() has run time $O(n)$ since *lfs*(v) is called exactly once for each node $v \in V$.

lfs() (or *rfs*()) might return a very bad evaluation when the DAG has a certain structure (see Fig. 4 and Tab. 1). For this reason we try to modify these algorithms in the next section.

3   Random first search

Lemma 1 *A unary node has no influence on a contiguous evaluation generated by a dfs–variation.*

That is evident since for a unary node u with son v, *dfs*(u) has no other choice than to call *dfs*(v).

Definition: (*tree–node*)
(1) Each leaf is a tree–node.
(2) An inner node is a tree–node iff all its sons are tree–nodes and none of them has outdegree > 1.

(3) Those are all tree–nodes.

Definition: (*label*)
  (1) For any leaf v is $label(v) = 1$.
  (2) For a unary node v is $label(v) = max\{label(son(v)), 2\}$.
  (3) For a binary node v is
      $label(v) = max\{3, \; max\{label(lson(v)), label(rson(v))\} + \; q\}$
      where $q = 1$ if $label(lson(v)) = label(rson(v))$, and 0 otherwise.

The Labeling–algorithm of *Sethi* and *Ullman* (see [8]) generates optimal evaluations for a tree with labels:

---

(1) **function** *labelfs*(**node** v)
    (∗ generates an optimal evaluation for the subtree with root v ∗)
    begin
(2) **if** v is not a leaf
(3) **then if** $label(lson(v)) > label(rson(v))$
(4)   **then** $labelfs(lson(v))$; $labelfs(rson(v))$
(5)   **else** $labelfs(rson(v))$; $labelfs(lson(v))$
      fi
    fi
(6) $reg(v) \leftarrow new\_reg()$; $print(v, reg(v))$;
(7) **if** v is not a leaf **then** $regfree(reg(lson(v)))$; $regfree(reg(rson(v)))$ **fi**
    end *labelfs*;

Definition: A *decision node* is a binary node which is not a tree–node.
It is clear that in a tree there are no decision nodes. In general, for a DAG let d be the number of decision nodes and b be the number of *binary tree–nodes*. Then $k = b + d$ denotes the number of all binary nodes of the DAG.

Lemma 2 *For a tree T with one root and b binary nodes there exist exactly 2^b different contiguous evaluations.*

Proof: For each binary node there are 2 possibilities to select the order in which the subtrees are evaluated. $\square$

Lemma 3 *For a DAG with one root and k binary nodes there exist at most 2^k different contiguous evaluations.*

Proof: Each contiguous evaluation of a DAG with one root is characterized by the permutations π applied at each node (see the definition of a contiguous evaluation). Thus, if the inner nodes $v_1, v_2, \ldots$ of the DAG have indegrees (the number of sons) $d_1, d_2, \ldots$, then there are at most $\prod_i (d_i!)$ contiguous evaluations. Since $d_i \leq 2$ for all i the lemma follows. $\square$

Lemma 4 *Let G be a DAG with d decision nodes and b binary tree–nodes which form t (disjoint) subtrees $T_1, \ldots, T_t$; in T_i there are b_i binary tree–nodes, $i = 1 \ldots t$, with $\sum_{i=1}^{t} b_i = b$. Then the following is true:*
If we fix an evaluation A_i for T_i then there remain at most 2^d different contiguous evaluations for G.

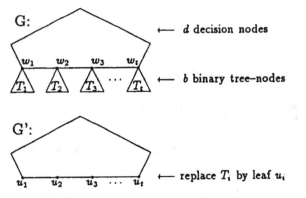

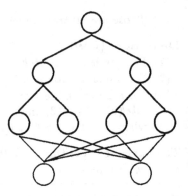

A DAG with $n - 2$ decision nodes

To the proof of lemma 4

Figure 5:

Proof: If we replace in G each subtree T_i with root w_i (see Fig. 5) by a leaf u_i, $i = 1, \ldots, t$, we obtain a reduced DAG G' with d decision nodes and 0 binary tree–nodes. According to lemma 3 there are at most 2^d different contiguous evaluations A' for G'. By replacing u_i in A' by the fixed evaluation A_i we get for each contiguous evaluation A' of G' exactly one contiguous evaluation A of G. □

Corollary 5 *If we evaluate all the tree–nodes in the DAG G by labelfs(), there remain at most 2^d different contiguous evaluations for G.*

Let $v_1, \ldots, v_d$ be the decision nodes of a DAG G and let $\beta = (\beta_1, \ldots, \beta_d) \in \{0, 1\}^d$ be a bitvector. Now we enumerate the 2^d different $\beta \in \{0, 1\}^d$ and start $dfs(root)$ with each β such that the following holds:
$\beta_i = 0$ iff in the call $dfs(v_i)$ the left son of v_i should be evaluated first, $1 \leq i \leq d$.
By doing this we obtain *all* (up to 2^d) possible contiguous evaluations for G provided that we use a fixed contiguous evaluation for the tree–nodes of G.
Unfortunately, the algorithm induced by that still might have exponential run time since a DAG with n nodes can have up to $d = n - 2$ decision nodes (e. g. consider a binary tree with $n - 2$ nodes; by adding two new nodes and $n - 1$ edges as given in Fig. 5, we get a DAG with $n - 2$ decision nodes).
Of course we do not want to invest exponential run time if we have a lot of decision nodes. Often a heuristic solution suffices[1]. This suggests to throw coins in order to generate several random bitvectors β and to hope that at least one of the evaluations computed by this procedure has a register need close to the optimum.

Algorithm *randomfs*: We generate a fixed number zv of random bitvectors (with $prob(\beta_i = 1) = 1/2$) and apply $dfs()$ to each β. Among the computed evaluations we select one with the least register need.
The run time of *randomfs* is $O(zv \cdot n)$ according to the discussion of *lfs*.
Of course, if $zv \geq 2^d$ we have enough time to enumerate all possible 2^d bitvectors, i. e. we simulate a binary counter on the $\beta = (0\ldots0000), (0\ldots0001), \ldots, (1\ldots1111)$. This procedure surely gives an optimal contiguous evaluation for G.

---

[1]In particular if we would like to evaluate vector DAGs (the nodes represent vectors of a certain length L) by vector processors, one or two additional registers do not matter very much, see [3]. The details will be given in [4].

The advantage of this method lies in the fact that the quality of the generated solution can be controlled by zv (zv may be passed as parameter to the compiler). That is why we are interested in the questions how good the computed evaluation is on the average and what size zv should have in order to get sufficiently good results.

We want to illustrate this problem for a special example: Consider the DAG of Fig. 4. It is easy to see that *randomfs* can generate an optimal contiguous evaluation with a register need of 4 only if at the decision nodes 0, 3, 6, 9 and 12 the right son is always visited first. The probability for the subDAG with the root 15 being evaluated first is $p = 1/2^5 = 1/32$, about 3%. The probability to find at least one optimal evaluation among zv possibilities is

$$ 1 - \left(\frac{31}{32} \right)^{zv} $$

in this example. If we wish that probability being over 90%, we conclude

$$ zv \geq \frac{\log 0.1}{\log 31 - \log 32} \approx 72.5, $$

for a probability of 50% we need $zv \geq 22$, and so on.

Of course we might be satisfied if the generated evaluation would require five instead of four registers. For the average register need with given zv we have found the following results for our example by experiments:

zv	0	1	3	5	6	7	8	10	12	15	18	20	30	50
reg	9	8.4	7.7	6.7	6.3	5.6	5.6	5.6	5.6	5.6	5.0	5.0	4.9	4.4

We can see that already for a relatively small size of zv, e.g. 10, a fairly good average register need is scored. Of course the improvement of the evaluation quality decreases for increasing zv, and the probability for the same bitvector being chosen twice certainly increases for increasing zv, e. g. for our example DAG with $d = 13$ decision nodes (thus 8192 possible bitvectors) the probability of at least one bitvector occuring several times is over 50% already for $zv = 107$.

Certainly these computations are limited to our example DAG above; a more general discussion of zv may be a subject of further research. For the present we will choose zv with respect to the run time of *randomfs*.

4 *labelfs* — another heuristic

It is possible to compute labels for all nodes of the DAG according to the formula of *Sethi/Ullman* for trees given above. In general a label–controlled evaluation of a DAG which is not a tree will not be optimal (otherwise $P = NP$). However, *labelfs* often scores fairly good results even for DAGs:

Start with $visited(v) = $ FALSE for all $v \in V$.

(1) function *labelfs*(node v)
    (* evaluate the subDAG induced by node v *)
    begin
(2) $visited(v) \leftarrow$ TRUE;

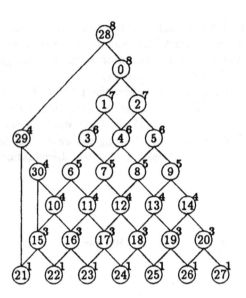

Figure 6: An extended 7–pyramid as counterexample to the assumption of *labelfs* giving always optimal evaluations: *labelfs*(28) allocates 11 registers, *rfs*(28) eight, *lfs*(28) nine registers. The label values are printed on the right top at each node.

(3) **if** v is not a leaf
(4) **then if** $label(lson(v)) > label(rson(v))$
(5a)   **then if not** $visited(lson(v))$ **then** $labelfs(lson(v))$ **fi**;
(5b)   **if not** $visited(rson(v))$ **then** $labelfs(rson(v))$ **fi**
(6a)   **else if not** $visited(rson(v))$ **then** $labelfs(rson(v))$ **fi**;
(6b)   **if not** $visited(lson(v))$ **then** $labelfs(lson(v))$ **fi**
      fi
  fi
(7) $reg(v) \leftarrow new\_reg()$; $print(v, reg(v))$;
(8) **if** v is not a leaf
(9) **then if** $lson(v)$ will no longer be used **then** $regfree(reg(lson(v)))$ **fi**
(10)   **if** $rson(v)$ will no longer be used **then** $regfree(reg(rson(v)))$ **fi**
  fi
  end;

For the DAG of Fig. 4 *labelfs* gives an optimal evaluation (4 registers).
But we give a counterexample where *labelfs* does not the best (Fig. 6, Tab. 2).
So it seems sensible to unify all heuristics considered so far in a combination called **V4** which applies all algorithms one after another and chooses the best evaluation generated.

5 V4 and its performance

V4 starts successively *lfs*, *rfs*, *labelfs* and *zv* instances of *randomfs*.
For test compilations (set $zv = 0$) we can stop after *lfs* since for testing programs any correct evaluation is sufficient. For the final optimizing compilation of a program we choose a great value of zv. The run time is $O((zv + 3) \cdot n)$ according to the discussions above.
We use V4 to improve the register need of Graph Coloring. Graph Coloring (*Chaitin*, 1981, see [2]) allocates the registers by coloring the nodes of a so–called register inter-

rfs–evaluation of the counterexample:

node	27	26	20	25	19	14	24	18	13	9	23	17	12	8	5	22
in reg.	0	1	2	0	3	1	2	4	0	3	1	5	2	4	0	3

node	16	11	7	4	2	21	15	10	6	3	1	0	30	29	28
in reg.	6	1	5	2	4	0	7	3	6	1	5	2	1	3	0

labelfs–evaluation of the counterexample:

node	21	22	15	23	16	10	24	17	11	6	25	18	12	7	3	26
in reg.	0	1	2	3	4	1	5	6	3	4	7	8	5	6	3	4

node	19	13	8	4	1	27	20	14	9	5	2	0	30	29	28
in reg.	9	7	8	5	6	3	10	3	4	3	4	3	4	1	0

Table 2: Two examples for evaluating the DAG of Fig. 6: The decisive difference between the *labelfs*– and the *rfs*–evaluation in this case is the fact that *labelfs* visits the left son first if the label values of both sons are equal. Of course we may formulate *labelfs* such that with equal labels always the right son is preferred, but then the reflection of the DAG of Fig. 6 at the vertical axis would be a counterexample to *labelfs* too.

random DAG No.	1	2	3	4	5	6	7	8	9	10
number of nodes	128	80	43	135	100	53	32	96	98	115
reg. need with V4	27	16	8	27	21	11	7	23	22	25
reg. need with GC	31	20	15	37	23	16	12	32	26	32
random DAG No.	11	12	13	14	15	16	17	18	19	20
number of nodes	130	55	125	32	81	5	97	79	50	119
reg. need with V4	28	11	25	8	21	3	18	18	12	25
reg. need with GC	35	16	36	14	26	4	25	25	15	34

Table 3: A series of tests with 20 randomly constructed DAGs: V4 always improved the register need of the original evaluation for the basic block (GC stands for "Graph Coloring without reordering by V4"); here we obtained the average ratio GC / V4 $\approx$ 1.38.

ference graph (RIG) which must be constructed from a fixed evaluation A (this is the evaluation given in the original basic block). Two nodes (symbolic registers, here identical with the DAG nodes) interfere (thus they are connected in the RIG by an edge) if they are active simultaneously in A, i.e. they cannot be assigned to the same physical register (same color). The coloring can be computed by a linear–time heuristic (applied here) or via *backtracking* (where exponential run time is possible). The number of different colors (*chromatic number*) needed for A corresponds to the register need m.

In order to show the advantages of the new heuristic we apply V4 with $zv = 10$ to randomly constructed DAGs with 30 to 150 nodes (average ca. 80), see Tab. 3. The result is surprisingly clear:

For the original evaluation of the basic block about 1/3 more registers are needed on the average than for the evaluation returned by V4. The improvement achieved by V4 might even be increased by choosing a greater zv.

This observation can only be explained by the fact that before the reordering we have *one* fixed evaluation A_0 (just the one which is given by the random construction of the test DAG), and in general this A_0 is noncontiguous. The probability for exactly this evaluation having a very low register need is rather small. On the other hand, V4 examines here $zv + 3 = 13$ (mostly) different evaluations, and only one of them must have a lower register need than A_0 to improve the result.

6   Final remarks

We are rather pleased with the results returned by V4, so we use it for the code optimizer in the implementation of a compiler for vector PASCAL which is being developed at our institute. For more details see [7]. The next step in that optimizer, the adaption of a computed evaluation of a vector DAG to a special vector processor, is described in [3] and will be presented in a later paper.

References

[1] Aho, A.V., Sethi, R., Ullman, J.D.: *Compilers: Principles, Techniques, and Tools.* Addison–Wesley (1986)

[2] Chaitin, G.J. et al.: *Register allocation via coloring.* Computer Languages Vol. 6, 47–57 (1981)

[3] Keßler, C.W.: *Code–Optimierung quasiskalarer vektorieller Grundblöcke für Vektorrechner.* Master thesis (1990), Universität Saarbrücken.

[4] Keßler, C.W., Paul, W.J., Rauber, T.: *Scheduling Vector Straight Line Code on Vector Processors.* Submitted to: First International Conference of the Austrian Center for Parallel Computation, Sept. 30 – Oct. 2, 1991, Salzburg (Austria).

[5] Mehlhorn, K.: *Data Structures and Algorithms 2: Graph Algorithms and NP–Completeness.* (1984)

[6] Paul, W.J., Tarjan, R.E., Celoni, J.R.: *Space bounds for a game on graphs.* Math. Systems Theory 10, 239–251 (1977)

[7] Rauber, Thomas: *An Optimizing Compiler for Vector Processors.* Proc. ISMM International Conference on Parallel and Distributed Computing and Systems, New York 1990, Acta press, 97–103

[8] Sethi, R., Ullman, J.D.: *The generation of optimal code for arithmetic expressions.* J. ACM, Vol. 17, 715–728 (1970)

[9] Sethi, R.: *Complete register allocation problems.* SIAM J. Comput. 4, 226–248 (1975)

Generating Efficient Code From Data-Flow Programs *

Nicolas Halbwachs , Pascal Raymond
IMAG - LGI (U.A. CNRS 398)
B.P.53X, 38041 Grenoble, France

Christophe Ratel
Merlin Gerin - SES
38050 Grenoble, France

Abstract

This paper presents the techniques applied in compiling the synchronous data-flow language LUSTRE. The most original technique consists in synthesising an efficient control structure, by simulating the behavior of boolean variables at compile-time. Here, the techniques are explained on a small subset of LUSTRE.

1 Introduction

Many authors [Kah74,McG82,PP83,AW85] have advocated the advantages of data-flow languages, mainly due to their mathematical soundness, the ease of formal program construction and transformation and the absence of side effects. However, no such language is actually used, mainly because no good compilers exist for standard machines. The absence of assignment and control structures makes it difficult to produce efficient code from a data-flow program.

We have argued elsewhere [BCHP86,CPHP87] that the declarative style allowed by data-flow languages makes them especially suitable for a class of real time programs: such programs can be found in domains (automatic control, signal processing, hardware simulation,...) where traditional tools are declarative (differential equations, operator networks,...). However, in these domains, the need for efficient code is even more important than in usual programming. On the other hand, in order to adapt data-flow languages to real time programming, we have proposed that they be given a synchronous interpretation: each operator receives all of its inputs at the same time, and all the operators synchronously respond to their inputs. Such an interpretation restricts the behavior of a program in such a way that the production of efficient code becomes possible, through a static synthesis of control structures, as done by the compiler of the imperative synchronous language ESTEREL [BG85]. The synchronous data-flow language LUSTRE [CPHP87] and its compiler are based on these ideas.

This paper deals with this compilation process, illustrated on a very simple language, which is a subset of LUSTRE and is introduced in section 2. In section 3, we show the simplest way for compiling a program in this language, as a single loop. Section 4 explains how a program may be partially simulated, and turned into a finite automaton representing the control skeleton of the object code: this automaton is obtained by a static simulation of the boolean variables involved in the program: in our language, boolean variables are used to model what is usually expressed through control structures in imperative languages. However, experience shows that this exhaustive simulation often produces an automaton far from minimal; this phenomenon, which is less apparent in the case of an imperative language like ESTEREL, comes from the fact that the declarative style encourages the definition of variables regardless of their effective use: one writes equations which are assumed to be always true; there are no control structures to indicate that some variables are only intermittently used, and the fact that they are always computed creates many irrelevant states. We shall therefore

---

*This work was partially supported by ESPRIT-BRA Project "SPEC", by PRC-C³ (CNRS) and by a contract from Merlin-Gerin

concentrate on a kind of "demand driven simulation", which consists of computing, in a given state, only those boolean variables which are involved in the computation either of the output or of the next state. In [BFH90], we have proposed a general algorithm for generating a minimal automaton from a finite state program. Section 5 presents the adaptation of this algorithm for compiling our language. In conclusion, we shall give some experimental results about the implementation of the method, which has been done in the compiler LUSTRE-V3.

2 A very simple data-flow language

2.1 Informal presentation

For the sake of simplicity, we shall illustrate the compilation process on a small subset of LUSTRE. As usual in the data-flow approach, this language operates over infinite sequences of values: any variable or constant is equated to the sequence of values it takes during the execution of the program, as the operators are intended to operate globally over sequences. The synchronous interpretation consists of considering that a program has a cyclic behavior, consisting of computing at the n-th cycle the n-th value of each variable (this strict synchrony may be released in full LUSTRE). A program is a set of variable definitions, expressed by a system of equations. The equation "X=E", where X is a variable representing the sequence $(x_1, x_2, \ldots, x_n, \ldots)$ and E is an expression representing the sequence $(e_1, e_2, \ldots, e_n, \ldots)$, means that $x_n = e_n$, for any positive integer n.

Expressions are built using constants (infinite constant sequences), variables and operators. Usual arithmetic, boolean and conditional operators are extended to pointwisely operate over sequences, and are hereafter referred to as "data operators". For instance, the expression "if X>=Y then X-Y else Y-X" represents the sequence whose n-th term is the absolute difference of the n-th terms of the sequences represented by X and Y.

In addition to those data operators, the language only contains one specific operator: the "pre" operator (for "previous"). If E is an expression representing the sequence $(e_1, e_2, \ldots, e_n, \ldots)$, and initial is a constant of the same type as E, then pre(E,initial) represents the sequence $(initial, e_1, e_2, \ldots, e_{n-1}, \ldots)$.

We are now able to define non-trivial sequences, by means of recursive definitions. For instance, the equation "N = pre(N + 1, 0)" defines N to be the sequence of naturals, and "C = not pre(C, false)" defines C to be the alternating sequence $(true, false, true, false, \ldots)$.

2.2 Formal semantics

Let us define the operational semantics of our language, which will be the basis of the compilation process. We define a memory σ to be a function from identifiers to values, and an history η to be an infinite sequence of memories. Given an history of input variables of a program, the semantics will provide the whole history of all variables. The semantics are described by means of structural inference rules [Plo81], which define the following predicates:

- $\eta \vdash eqs : \eta'$ which means "given the input history η, the program consisting of the system of equations eqs defines the global history η' "

- $eq \xrightarrow{\sigma} eq'$ which means "the right-hand side of the equation eq evaluates as $\sigma(X)$, where X is its left-hand side identifier, and eq will be later on considered as eq' ". If this predicate holds, eq will be said *compatible* with σ.

- $\sigma \vdash e \xrightarrow{v} e'$ which means "the expression e evaluates as v in the memory σ, and will be later on considered as e'"

Here are the rules, which are commented below:

$$\frac{eqs\xrightarrow{\sigma}eqs' \quad , \quad \eta \vdash eqs : \eta'}{\sigma_{in}.\eta \vdash eqs : \sigma.\eta'} \quad (1) \qquad \frac{eq\xrightarrow{\sigma}eq' \quad , \quad eqs\xrightarrow{\sigma}eqs'}{(eq; eqs)\xrightarrow{\sigma}(eq'; eqs')} \quad (2) \qquad \frac{\sigma \vdash E\xrightarrow{\sigma(X)}E'}{X=E\xrightarrow{\sigma}X=E'} \quad (3)$$

$$\frac{\sigma \vdash E_0\xrightarrow{v_0}E_0' \quad , \quad \sigma \vdash E_1\xrightarrow{v_1}E_1'}{\sigma \vdash E_0 \text{ or } E_1\xrightarrow{v_0 \vee v_1}E_0' \text{ or } E_1'} \quad (4) \qquad \frac{\sigma \vdash E\xrightarrow{w}E'}{\sigma \vdash \text{pre}(E,v)\xrightarrow{v}\text{pre}(E',w)} \quad (5)$$

(1) Rule of programs: If the program eqs is compatible with the memory σ and must be later on be considered as eqs', and if eqs' transforms the history η into η', then eqs transforms the history $\sigma_{in}.\eta$ into $\sigma.\eta'$, where σ_{in} denotes the restriction of σ to the inputs variables of the program, and ".". denotes the concatenation on histories.

(2) Systems of equations: A system of equations is compatible with a memory σ if and only if each equation is compatible with σ. Each equation is rewritten for further evaluation.

(3) Equations: An equation $X = E$ is compatible with σ if and only if its right-hand side evaluates as $\sigma(X)$ in σ. Its right hand-side is rewritten for further evaluation.

(4,5) Expressions: Any n-ary data operator op evaluates always in the same way: The rule 4 concerns the boolean operator or (rules for other data operators are similar). There is a special rule (5) for the operator pre: it always evaluates as it second operand, and the current value of its first operand is "stored" by the rewriting, to be returned at the next evaluation.

For instance, we can apply these rules to get the two first steps of the behavior of the "program" C = not pre(C, false) (we note $[v/C]$ the memory associating v to C):

$$\frac{\dfrac{[true/C] \vdash \texttt{C}\xrightarrow{true}\texttt{C}}{[true/C] \vdash \texttt{pre(C, false)}\xrightarrow{false}\texttt{pre(C, true)}}}{\dfrac{[true/C] \vdash \texttt{not pre(C, false)}\xrightarrow{true}\texttt{not pre(C, true)}}{\texttt{C = not pre(C, false)}\xrightarrow{[true/C]}\texttt{C = not pre(C, true)}}}$$

$$\frac{\dfrac{[false/C] \vdash \texttt{C}\xrightarrow{false}\texttt{C}}{[false/C] \vdash \texttt{pre(C, true)}\xrightarrow{true}\texttt{pre(C, false)}}}{\dfrac{[false/C] \vdash \texttt{not pre(C, true)}\xrightarrow{false}\texttt{not pre(C, false)}}{\texttt{C = not pre(C, true)}\xrightarrow{[false/C]}\texttt{C = not pre(C, false)}}}$$

3 Computation order and single loop

The simplest code we can generate from a program written in our language consists of a global infinite loop computing all the variables of the program, in suitable order. The existence of such an order must be considered first: The current value of a variable may only depend on its past values, since we don't intend to give sense to implicit definitions like X=X*X-1. The existence of a computation order comes down to the irreflexivity of some dependency relation among variables. Several dependency relations can be considered, which lead to different acceptance criteria; here are two examples:

- *Semantic dependency:* An expression E in a program eqs *semantically depends* on a variable X if and only if there exists a sequence of memories $(\sigma_0,\ldots,\sigma_n)$, a memory σ and a value w such that $eqs\xrightarrow{\sigma_0}eqs_0\xrightarrow{\sigma_1}\cdots\xrightarrow{\sigma_n}eqs_n$, $\sigma \vdash E_n\xrightarrow{v}E'$ and $\sigma[w/X] \not\vdash E_n\xrightarrow{v}E'$, where E_n is the expression such that $\sigma_0 \vdash E\xrightarrow{v_0}E_0\ldots\sigma_n \vdash E_{n-1}\xrightarrow{v_n}E_n$ and $\sigma[w/X]$ denotes the memory $\lambda Y.$ if $Y \equiv X$ then w else $\sigma(Y)$.

The semantic dependency is the most precise we can define: A variable X semantically depends on Y if and only if there is a behavior of the program leading to a situation where the value of X can depend of the current value of Y. The problem is that deciding whether a variable semantically depends on itself is generally undecidable. So, we need a stronger acceptance criterion.

- *Syntactic dependency:* In a given program *eqs*, a variable X, defined by the equation X = E *directly depends* on a variable Y if and only if Y appears outside any pre operator in E. Let us define the *syntactic dependency* relation to be the transitive closure of that direct dependency. Deciding of the irreflexivity of this relation is straightforward. Of course, an acceptance criterion based on the syntactic dependency will reject some meaningful programs, like the following:

      X = if C then Y else Z ; Y = if C then Z else X

since X syntacticly depends on C,Y,Z and X (since Y depends on X).

If the program does not contain dependency loops, any total order compatible with the dependency order is a suitable computation order. The choice of a particular order can influence the size of necessary memory. Generally speaking, the result of a pre operator must be stored in an auxiliary variable, in order to be available at the next cycle. However, this auxiliary memory can often be saved: Consider an expression pre(X,*v*), and assume that all the variables depending on that expression can be ordered before X. Then, there is no need of auxiliary memory for storing the previous value of X, since X itself can be used for that; for instance the "program" C = not pre(C,false) can be translated into

```
C:= true;                          C:= true; PRE_C:= C;
loop                 instead of    loop
    C := not C;                        C := not PRE_C; PRE_C:= C;
end                                end
```

More generally, if a variable X depends on the previous value of Y, X should be computed before Y, if possible. Let us consider an other example, which will be used throughout the remainder of the paper. It has been chosen for the purpose of illustrating the presented techniques rather than for its intuitive meaning (more realistic examples can be found in the bibliography [CPHP87,HCRP91]):

```
program EX
    input i:  bool; output n:  int; local x, y, z:  bool;
    n = if pre(x,true) then 0 else pre(n,0) + 1
    x = if pre(x,true) then false else z
    y = if pre(x,true) then pre(y,true) and i else pre(z,true) or i
    z = if pre(x,true) then pre(z,true)
            else (pre(y,true) and pre(z,true)) or i
end
```

In this program, the only constraint induced by the syntactic dependency relation is that x must be computed after z. Moreover,

(1)   since n depends on pre(x,true), it should be computed before x;
(2)   since y depends on pre(x,true) and pre(z,true), it should be computed before x and z;
(3)   since z depends on pre(x,true) and pre(y,true), it should be computed before x and y;

Now, since there is a contradiction between rules (2) and (3) above, a memory is necessary, for instance for storing the previous value of z. We get the following code:

```
n := 0; z := true; y := true; x := false; pre_z := z;
loop
    read(i);
    n := if x then 0 else n + 1;
    z := if x then z else (y and z) or i;
    y := if x then y and i else pre_z or i;
    x := if x then false else pre_z;
    pre_z := z; write(n)
end
```

or, by factorizing the conditional:

```
n:=0; z:=true; y:=true; x:=false; pre_z:=z;
loop
    read(i);
    if x then n:=0; y:=(y and i); x:=false;
    else n:=n+1; z:=(y and z) or i; y:=pre_z or i; x:=pre_z;
    endif;
    pre_z:=z; write(n)
end
```

4 "Data driven" control synthesis

The translation proposed in the previous section appears to often produce a quite inefficient code. As a matter of fact the single loop is a poor control structure. It thus appears that some more sophisticated control structure must be derived, in order to avoid irrelevant tests. For instance, in our example program, one could take advantage of the knowledge that, at a given cycle, x takes the value false in order to avoid testing the previous value of x at the next cycle. The idea is thus to execute a different code at the next cycle, according to the state of the variables at the current cycle.

Moreover such a control structure is suggested by our semantics. According to this semantics, an execution cycle of a program consists of two steps:

1. the current values of the variables are computed

2. the program is rewritten into a new program, which summarizes all the information necessary for executing the next cycle.

A program is therefore an automaton, whose states are the different forms that the program can take as it is rewritten. These forms only differ on the constant values appearing as second operands of the pre operators. These values will be called *state variables*.

For instance, consider the program C = not pre(C, false). From the semantic rules, we have (see the example derivation § 2.2):

$$C = \text{not pre}(C, \text{false}) \xrightarrow{[true/C]} C = \text{not pre}(C, \text{true})$$

and

$$C = \text{not pre}(C, \text{true}) \xrightarrow{[false/C]} C = \text{not pre}(C, \text{false})$$

So, this program corresponds to an automaton with two states, drawn on Fig. 1

Of course, the automaton may have infinitely many states (since state variables can be numbers), so the idea is to fold it onto a finite automaton which is the wanted control structure. An easy way to perform this finite folding consists of considering only boolean state variables, thus leaving integer computations be performed at runtime: they will be merely placed onto the transitions of the automaton.

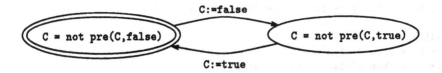

Figure 1: The automaton of the program C = not pre(C, false)

So, the control structure is built by simulating the behavior of boolean variables. Let us illustrate this procedure on our example program EX. Let eqs_0 be the system of equations defining its boolean variables (this system will be the initial state of the automaton):

```
eqs₀ :x = if pre(x,true) then false else z
       y = if pre(x,true) then pre(y,true) and i else pre(z,true) or i
       z = if pre(x,true) then pre(z,true)
              else (pre(y,true) and pre(z,true)) or i
```

In this state, the definition of n reduces to n:=0. From the semantic rules, we have two possible rewritings, according to the value of the input i:

- $eqs_0 \xrightarrow{\sigma_1} eqs_1$, where $\sigma_1 = [true/i, false/x, true/y, true/z]$ and

```
eqs₁ :x = if pre(x,false) then false else z
       y = if pre(x,false) then pre(y,true) and i else pre(z,true) or i
       z = if pre(x,false) then pre(z,true)
              else (pre(y,true) and pre(z,true)) or i
```

- $eqs_0 \xrightarrow{\sigma_2} eqs_2$, where $\sigma_2 = [false/i, false/x, false/y, true/z]$ and

```
eqs₂ :x = if pre(x,false) then false else z
       y = if pre(x,false) then pre(y,false) and i else pre(z,true) or i
       z = if pre(x,false) then pre(z,true)
              else (pre(y,false) and pre(z,true)) or i
```

Now, two new states (eqs_1 and eqs_2) have been reached. They must be considered in turn:

- In the state corresponding to eqs_1, we have n:=n+1, since the previous value of x is *false*. Whatever be the value of i in σ_3, we get $eqs_1 \xrightarrow{\sigma_3} eqs_0$, provided $\sigma_3(x) = \sigma_3(y) = \sigma_3(z) = true$.

- eqs_2 has two possible rewritings, according to the value of the input i:

  $eqs_2 \xrightarrow{\sigma_4} eqs_0$, where $\sigma_4 = [true/i, true/x, true/y, true/z]$

  $eqs_2 \xrightarrow{\sigma_5} eqs_3$, where $\sigma_5 = [false/i, false/x, true/y, false/z]$ and

```
eqs₃ :x = if pre(x,false) then false else z
       y = if pre(x,false) then pre(y,true) and i else pre(z,false) or i
       z = if pre(x,false) then pre(z,false)
              else (pre(y,true) and pre(z,false)) or i
```

Considering all the states in turn, we get:

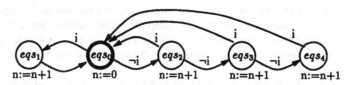

Figure 2: The control automaton of the program EX

- eqs_3 has two possible rewritings, according to the value of the input i:

$$eqs_3 \xrightarrow{\sigma_4} eqs_0$$

$eqs_3 \xrightarrow{\sigma_6} eqs_4$, where $\sigma_6 = [false/i,\ false/x,\ false/y,\ false/z]$ and

```
eqs4 : x = if pre(x,false) then false else z
       y = if pre(x,false) then pre(y,false) and i else pre(z,false) or i
       z = if pre(x,false) then pre(z,false)
                else (pre(y,false) and pre(z,false)) or i
```

- eqs_4 has two possible rewritings, according to the value of the input i:

$$eqs_4 \xrightarrow{\sigma_4} eqs_0 \qquad eqs_4 \xrightarrow{\sigma_6} eqs_4$$

All the reached states have been processed. In all states eqs_3, eqs_4, eqs_5, since the previous value of x is false, we get n:=n+1. The resulting automaton is drawn in Fig 2. A possible code would be the following:

```
EQSO : n:=0 ; write(n); read(i); if i then goto EQS1 else goto EQS2;
EQS1 : n:=n+1 ; write(n); read(i); goto EQSO;
EQS2 : n:=n+1 ; write(n); read(i); if i then goto EQSO else goto EQS3;
EQS3 : n:=n+1 ; write(n); read(i); if i then goto EQSO else goto EQS4;
EQS4 : n:=n+1 ; write(n); read(i); if i then goto EQSO else goto EQS4;
```

Such a method was first introduced for the compilation of ESTEREL [BG85] and is used by the LUSTRE-V2 compiler [CPHP87,Pla88]. Efficient algorithms exist for it, since it is similar to the method for building an accepting automaton for a regular expression [Brz64,ASU86,BS87]. The resulting code appears to be minimal, in some sense, with respect to the execution time.

5 "Demand driven" control synthesis

Experience with ESTEREL shows that the resulting automaton is generally small (while theoretically exponential), and close to the minimal one (with respect to bisimulation, the usual minimization criterion over automata). However, it is not the case with a declarative language. As a matter of fact, the automaton built from our example is not minimal (eqs_2, eqs_3 and eqs_4 are equivalent states).

The problem is that the declarative style encourages the definition of variables without care of their effective use: variables are defined at each cycle, but if they are only intermittently used, the automaton obtained by the above procedure may contain many equivalent states, which differ only in the values of state variables which affect neither the current outputs nor the future behavior of the program. Notice that the size of the automaton does not affect the execution time of the code (as far as this code does not exceed the memory size) but a tremendous extension of the code size is clearly unrealistic. A possible solution would be to apply a standard minimization algorithm [PT87,Fer90], which still supposes there is enough memory to store the automaton. A better solution consists of integrating the minimization into the generation process. So, we have designed a algorithm [BFH90] for generating directly the minimal automaton, by computing state variables according to a "demand-driven" strategy: A variable is computed in a state if and only if its value is necessary for computing either the output or the next state. In this section, we illustrate the use of this algorithm on our toy language.

The basic idea is the following: We want to avoid distinguishing between states, when they differ on state variables which are used neither in the current output nor in the future behavior of the program. Hence, we shall deal with classes of equivalent states, characterized by boolean formulas. The method is inspired from the standard minimization of automata, whose principles are the following:

- Two states are considered equivalent as long as they haven't been shown to be different;

- Two states are different if and only if, in response to the same inputs, either they produce different outputs, or they lead to states which have been shown to be different.

In a similar way, our demand-driven generation of the automaton will progressively refine a partition of the set of states. The initial partition consists of one class, i.e., all the states are considered equivalent. Processing a class C consists of generating the code for its outputs and computing its successor classes. Whenever this computation involves some unknown state information ι, the class C is split into two subclasses: the subset C_0 of states for which ι is false, and the subset C_1 of those for which ι is true. Then, predecessor classes of C are asked for the value of ι, and requested to choose their successor among C_0 and C_1.

In order to make clear the state information needed for computing an expression of the program, we shall put the expression into a normal form:

Definitions: An *instantaneous expression* is an expression which contains neither a pre operator applied to a boolean operand, nor any boolean variable which is not an input. An expression E is in *normal form* if and only if either it is an instantaneous expression, or it is of the form:

```
if pre(E_1,v_1) then F_1
else if pre(E_2,v_2) then F_2
else ...
else if pre(E_{n-1},v_{n-1}) then F_{n-1}
else F_n
```

where all the F_i are syntacticly different instantaneous expressions. It can be easily shown that any expression of the language can be put into such a form.

Now, let us apply the demand-driven generation to our example program EX: We shall consider the program written as follows, where boolean state variables are noted v_x, v_y, v_z:

```
program EX
    input i: bool; output n: int; local x, y, z: bool;
    n = if pre(x,v_x) then 0 else pre(n,0) + 1
    x = if pre(x,v_x) then false else z
    y = if pre(x,v_x) then pre(y,v_y) and i else pre(z,v_z) or i
    z = if pre(x,v_x) then pre(z,v_z)
                       else (pre(y,v_y) and pre(z,v_z)) or i
end
```

State classes will be characterized by predicates on v_x, v_y, v_z. We start with one class C, characterized by the formula *true*, in which we try to compute the output n. The expression defining n, "if pre(x,v_x) then 0 else pre(n,0) + 1", is already in normal form, which shows that the code for n depends on the state information v_x. So C is split into two subclasses: C_0, where the value of v_x is false, and C_1 where the value of v_x is true. So, in C_0 (resp. C_1), pre(x,v_x) is false (resp.

true) and the code for the output is n:=n+1 (resp. n:=0). Since in the initial form of the program, v_x is written true, the initial state belongs to C_1, which is (until now) the only clearly accessible class.

Let us process C_1: The code for the output is n:=0. Let us compute the next class, i.e, answer the question: Will v_x be true or false at the next cycle? The answer depends on the value of x in C_1, so, we have to compute x. The expression defining x, if pre(x,v_x) then false else z, must be put in normal form. We get:

x = if pre(x,v_x) then false else if pre(y and z,v_y and v_z) then true else i

In C_1, where v_x is true, it evaluates to false. So the next class is C_0, which is thus accessible, and must be processed.

In C_0, the code for the output is n:=n+1. For computing the next class, we have also to compute x, which evaluates to

if pre(y and z,v_y and v_z) then true else i

So, the next class depends on the state information (v_y and v_z). So, C_0 is split into

- C_{00} where both v_x and (v_y and v_z) are false;

- C_{01} where v_x is false and where (v_y and v_z) is true.

The predecessor class of C_0, C_1 is requested to choose between these two subclasses, that is to compute the value of (y and z), the normal form of which is:

if pre((x and y and z) or (not x and not(y and z)),
                (v_x and v_y and v_z) or (not v_x and not (v_y and v_z)))
then i
else if pre(x and not(y and z),v_x and not(v_y and v_z))
then false else true

In C_1, this expression reduces to

if pre(y and z, v_y and v_z) then i else false

So, the choice of the successor class of C_1 among C_{00} and C_{01} depends on the state information (v_y and v_z), which is unknown in C_1. So, C_1 is split into:

- C_{10}, where v_x is true and (v_y and v_z) is false;

- C_{11}, where both v_x and (v_y and v_z) are true.

The initial state belongs to C_{11}, which is the only accessible class, since until now, C_1 doesn't have any predecessor.

From the definition of C_{11}, (v_y and v_z) is true in C_{11}, so the value of (y and z) is i, and the choice of the successor class of C_{11} depends on the value of i: if i is false, the next class will be C_{00}, otherwise it will be C_{01}. As a consequence, both C_{00} and C_{01} are accessible classes, and must be processed.

In these classes, we already know that the code for the output is n:=n+1 (since we consider subclasses of C_0), so only the successor class must be chosen, among C_{00}, C_{01}, C_{10} and C_{11}, according to the values of x and (y and z):

- In C_{00}, both v_x and (v_y and v_z) are false, so both x and (y and z) evaluate to i. So if i is false the successor class is C_{00}, otherwise it is C_{11}.

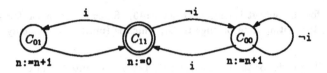

Figure 3: Minimal automaton of EX

- In C_{01}, v_x is false and $(v_y$ and $v_z)$ is true. From their normal forms, both x and $(y$ and $z)$ evaluate to true. So the successor class is C_{11}.

All the accessible classes have been processed, so the generation is complete, and we are left with the automaton of Fig. 3. Compared with the "data-driven" strategy applied in section 4, notice that the three equivalent states eqs_2, eqs_3, eqs_4 have not been distinguished: they all belong to the class C_{00}. Notice also that our algorithm only deals with accessible classes: The class C_{01}, which is not accessible, has not been considered for splitting. This makes the distinction with standard minimization algorithms.

6 Implementation and experimentation

As mentioned before, all the methods presented here have been implemented in the LUSTRE compilers. In this section, we shall discuss the experience gained from this implementation, and particularly from the LUSTRE-V3 compiler. This compiler is written in C++, and translates a LUSTRE program into a sequential C program.

Control synthesis: The synthesis of the control structure presented in section 4 improves the code performances of a rate ranging from 10% to 50%, according to the proportion of boolean computations involved in the source program. For instance, a digital watch written in LUSTRE and compiled into an automaton gives rise to 41 states, and runs about 20% faster than the same program compiled as a single loop. Practically, its stopwatch counts the 1/100 second (on a SUN3/60) in the former case, but not in the later. For a highly boolean program, like the example EX considered in the paper, the improvement in speed is about 50%. The counterpart of the improvement in speed is the code size, which is often large. However, the code is produced in a particular form, where actions are stored in table, so as to avoid copying their code in each state of the automaton. This way of producing the code [PS87] is common with the ESTEREL compiler. Moreover, their are several options for choosing state variables, so as to let the user manage the compromise between the speed and the code size.

"Demand driven" synthesis: The interest of minimizing the automaton during its generation appeared from experience. As a matter of fact, in many cases, the "data driven" method described in section 4 produces a lot of equivalent states, or cannot succeed in producing the automaton because of these irrelevant states, and over all, because of irrelevant transitions between equivalent states. One can be easily convinced that the number of transitions grows as the square of the number of equivalent states. This explains why the demand-driven algorithm runs generally faster, since it deals only with the minimal number of states and transitions. The demand-driven algorithm involves symbolic computation of boolean formulas. An efficient decision procedure has been implemented by means of Bryant's "Binary Decision Diagrams" [Bry86].

The following tables compare the generation times (on SUN 4) and the results of four compiling strategies, applied to our example EX and to the digital watch program:

- the single loop generation

- the data-driven generation alone

- the data-driven generation followed by a minimization by ALDEBARAN [Fer90], an automaton minimizer based on the Paige-Tarjan algorithm [PT87]

- the demand-driven generation

Times are given in second, and code sizes in Kbytes.

Program EX	single loop	data-driven	data-driven + minimization	demand-driven
Compilation time	0.5	0.5	1.2	0.3
Number of states	1	5	3	3
Number of transitions	1	11	5	5
Size of the generated code	1.56	0.89	0.85	0.85
Average reaction time	$29\mu s$	$15\mu s$	$15\mu s$	$15\mu s$

Program WATCH	single loop	data-driven	data-driven + minimization	demand-driven
Compilation time	1.3	8.1	41.5	6.8
Number of states	1	81	41	41
Number of transitions	1	1163	474	342
Size of the generated code	7.88	30.94	19.15	16.51
Average reaction time	$820\mu s$	$590\mu s$	$590\mu s$	$578\mu s$

Aknowledgements: We are indebted to Gérard Berry for the idea of synthesizing the control structure of a program as a finite automaton. John Plaice wrote the compiler LUSTRE-V2 which implements the data-driven automaton generation. Many ideas about the ordering of computations along the transitions of the automaton are also due to him. Christian Berthet and Jean-Christophe Madre taught us how to implement a boolean decision procedure by means of "binary decision diagrams", without which the demand-driven algorithm would not work so well.

References

[ASU86] A. Aho, R. Sethi, and J. Ullman. *Compilers : Principles, Techniques and Tools*. Addison-Wesley, 1986.

[AW85] E. A. Ashcroft and W. W. Wadge. LUCID, *the data-flow programming language*. Academic Press, 1985.

[BCHP86] J-L. Bergerand, P. Caspi, N. Halbwachs, and J. Plaice. Automatic control systems programming using a real-time declarative language. In *IFAC/IFIP Symp. 'SOCOCO 86*, may 1986.

[BFH90] A. Bouajjani, J. C. Fernandez, and N. Halbwachs. Minimal model generation. In *Workshop on Computer-Aided Verification*, june 1990.

[BG85] G. Berry and G. Gonthier. *The synchronous programming language* ESTEREL, *design, semantics, implementation*. Technical Report 327, INRIA, 1985. to appear in Science of Computer Programming.

[Bry86]   R. E. Bryant. Graph-based algorithms for boolean function manipulation. *IEEE Transactions on Computers*, C-35(8), 1986.

[Brz64]   J. A. Brzozowski. Derivative of regular expressions. *JACM*, 11(4), 1964.

[BS87]   G. Berry and R. Sethi. From regular expressions to deterministic automata. *TCS*, 25(1), 1987.

[CPHP87]   P. Caspi, D. Pilaud, N. Halbwachs, and J. Plaice. LUSTRE: a declarative language for programming synchronous systems. In *14th ACM Symposium on Principles of Programming Languages*, january 1987.

[Fer90]   J. C. Fernandez. An implementation of an efficient algorithm for bisimulation equivalence. *Science of Computer Programming*, 13(2-3), may 1990.

[HCRP91]   N. Halbwachs, P. Caspi, P. Raymond, and D. Pilaud. The synchronous dataflow programming language LUSTRE. *Proceedings of the IEEE, Special Issue on Synchronous Programming*, To appear, 1991.

[Kah74]   G. Kahn. The semantics of a simple language for parallel programming. In *IFIP 74*, North Holland, 1974.

[McG82]   J. R. Mc Graw. The VAL language: description and analysis. *TOPLAS*, 4(1), january 1982.

[Pla88]   J. A. Plaice. *Sémantique et compilation de LUSTRE, un langage déclaratif synchrone.* Thesis, Institut National Polytechnique de Grenoble, 1988.

[Plo81]   G. D. Plotkin. *A structural approach to operational semantics.* Lecture Notes, Aarhus University, 1981.

[PP83]   N.S. Prywes and A. Pnueli. Compilation of nonprocedural specifications into computer programs. *IEEE Transactions on Software Engineering*, SE-9(3), may 1983.

[PS87]   J. A. Plaice and J-B. Saint. The LUSTRE-ESTEREL portable format. 1987. Unpublished Report, INRIA, Sophia Antipolis.

[PT87]   R. Paige and R. Tarjan. Three partition refinement algorithms. *SIAM J. Comput.*, 16(6), 1987.

On the borderline
between grammars and programs

C. H. A. Koster
Informatics department
University of Nijmegen
The Netherlands

J. G. Beney
Informatics department
INSA de Lyon
France

Abstract

We describe some of the engineering considerations and trade-offs in the design of a new Compiler Description Language, CDL3. The language is based on Extended Affix Grammars, where the affix rules are used to define tree types. The execution model is deterministic and depth-first, except that part of the work can be delayed until a second pass over the implicit parse tree. It is checked statically whether the program can indeed be executed in two passes without backtracking. A simple module structure allows separate compilation in a safe way.

As a running example, we treat the translation of a block-structured language.

1 Introduction

Programmers have always [17] been fascinated by the close kinship between grammars and related software, such as parsers, interpreters and compilers. This fascination led on the one hand to tools for the automatic generation of software from grammars (Compiler Compilers, Meta-compilers, Environment Generators) and on the other hand to the development of syntactic formalisms amenable to the automatic generation of software, including the various forms of two-level grammars (like W-Grammars [20], Attribute Grammars [10] and Affix Grammars [12]).

In the work on meta-compilation based on two-level grammars two different directions can be distinguished [4]:

- (*grammatical specification*) On the one hand, some researchers [15, 9] want to allow maximum freedom in the writing of grammars and therefore in the order of evaluation of the attributes. They strive for minimal well-formedness conditions that permit this evaluation, even at the price of multiple traversals of a parse tree and nondeterminism.

- (*grammatical programming*) On the other hand, the restriction to L-attributed grammars made in e.g. [11, 1, 2, 14] leads to efficient context-sensitive parsers, where the attribute evaluation can guide the parsing and without the need to build a parse tree. But since there is only one deterministic left-to-right pass, in case another order of evaluation is needed the programmer must still explicitly build an abstract tree that is later used to traverse the same structure.

For describing and implementing programming languages, which is our aim, a deterministic formalism is to be preferred. But many problems (such as forward referencing) require more than one pass over the parse tree. Some attributes are to be calculated using information that comes later in the source text. For that purpose part of the attitube evaluation may have to be delayed. For many applications, a two-pass approach is sufficient: to collect information in the first pass, and use it in the second. That is why we chose to base our language on 2-pass L-attributed Affix Grammars.

Another major difference is in the choice of a second grammatical level defining the semantic objects. In W-Grammars and Affix Grammars, the domains of the meta-variables are specified by means of Context-Free meta-grammars, which also serve as a typing system. Various effective affix representations have been explored: strings [15], trees [2], sets [13] and lattices [18]. For reasons of expressivity as well as efficiency, we choose to interpret the affix domains as trees, similar to the functor structures in PROLOG, rather than as strings.

A third choice is to be made between top-down and bottom-up analysis. The construction of the table for a bottom-up parser must take the whole grammar into account, while a top-down parser is easily divided into several sub-grammars that do not need to be recompiled when only one is modified. Also a top-down parser can more readily be extended with parameters (affixes) than a table-driven one. That is why we chose to implement a two-pass variant of recursive descent parsing.

2 Grammar and program

The goal of the CDL3 project is, as the acronym suggests, to design a new Compiler Description Language. In order to combine a high degree of expressiveness and formal tractability with efficient implementation, we choose a notation based on the syntactic formalism of Extended Affix Grammars (EAGs) [19, 15] while adhering to well-understood principles of Software Engineering.

2.1 The language

A CDL3 program can be seen as a sugared version of an EAG, which is quite similar to PROLOG extended with a meta-grammar and without the logical variable, all affix values being ground. We shall describe the notation only superficially, since we are mainly interested in bringing out the design decisions; details and examples follow in later sections.

A program consists of meta-rules (see 2.3), which serve as *type-declarations*, and syntax rules playing the role of *procedures*. The order of these declarations is free.

A procedure can have affixes as *input-* (= inherited) and *output-* (= synthesized) parameters, as indicated by the *direction* signs > in its heading, e.g.:

```
ACTION p (>INPAR, OUTPAR>):
```
Procedures may be *generic* in the sense that any number of procedures can have the same name provided they differ in name and/or type of parameters.

A CDL3-program is L-attributed: the *flow-of-information* within a procedure body is strictly left-to-right. No information can be passed from one (failed) alternative to the next.

The *control-structures*, which are of syntactic origin, are those which have been made familiar by PROLOG: The semicolon as a nondeterministic OR-operator between alternatives and the comma as a MacCarthy AND-operator within an alternative. Right-recursion is used to express repetition. The execution is deterministic and sequential (but an alternative can have a *second pass*, see 4).

An alternative may end in an *enclosed group*, a group of alternatives between brackets, that can be seen as an anonymous procedure. This allows procedure bodies to be left-factored.

A procedure may end with one or more WHERE-clauses, which serve as *refinements*. The refinements of a procedure are implicitly parametrized with the same parameters as the procedure itself. This convention eliminates some stereotypical parameters.

Another kind of stereotypical parameters can be seen in examples like

$$p\ (>\!X_1, X_3\!>):$$
$$a\ (X_1, X_2),\ b\ (X_2, X_3).$$

As a shorthand for such pairs of parameters, winding their way through the calls, we also admit *transient* parameters, written

$$p\ (>\!X\!>):\ a(X),\ b(X).$$

At a transient parameter position we also allow a *pair of affixes* separated by a |-sign (a special kind of comma). As an example, the call p(0|OFFSET) calls p with 0 as input parameter and OFFSET as output parameter.

2.2 Grammatical programming

By these notational conventions (and a few more pragmatic decisions, e.g. concerning built-in operations), the grammatical formalism has turned into a programming language. The programming style resulting from the abovementioned control structures gives meaning to the hackneyed term "structured programming". The refinements and procedures support Top-Down programming. In conjunction with a simple module system (see 5), the generic procedures support Bottom-Up programming.

In fact, various researchers have introduced the notion of *grammatical programming* [3, 8] to describe programming based on syntactic structure.

Grammatical programming can be seen as the exploitation of some homomorphism: between syntax and input, (abstract) syntax and output, input and output (the Jackson method), input and computation (interpreters) or even between different computations and one same datastructure. A formal basis for this observation can be found in [7].

2.3 The meta-level

The meta-rules of a CDL3-program are CF rules, with affixes (written in capital letters) as the meta-nonterminals and small-letter words as meta-terminals. They serve to define datastructures and the domains of the affixes.

In W-Grammars and EAGs, the values of the attributes are considered as strings over the meta-alphabet which are concatenated. This approach leads to concise formulations, but any ambiguity in the meta-grammar will necessitate backtracking.

A way to avoid this backtracking as well as some dynamic patern-matching is to represent such meta-values as trees, that is to keep track of the way those values were built, and to require that each meta-alternative is marked by a different terminal affix.

To be more precise, the meta-rules of the CDL3-program form a Context-Free Grammar in (near) Greibach Normal form, every alternative (except maybe for the last one) starting with a distinct terminal symbol (its *marker*). Only the last meta-alternative is allowed to have no marker.

As an example, we describe the environment structure for a simple block-structured language by the following meta-rules:

```
RANGE:: empty; RANGE DEF.
```
A RANGE is in fact a linear list of DEFs. There is no predefined list-type or tree-type.

```
DEF:: where IDF has OFFSET in BNO.
```
In the example language to be described, the affixes BNO and OFFSET will denote a static block number and the offset of a variable within that block, respectively.

```
BNO, OFFSET:: INT.
```
The affixes BNO and OFFSET are introduced as synonyms for INT. Another way to introduce a synonym for an affix is to index it with an integer — so OFFSET1 is also a synonym for OFFSET.

```
IDF:: TEXT.
```
The types INT and TEXT are the only predefined types.

Trees can be built and decomposed by means of *guards*. A guard is a confrontation between an affix and one of its productions (an *affix expression*), or between two affixes with the same domain. The guards serve as elementary actions and tests.

Four kinds of guards can be distinguished:

`[BNO = BNO1]`	*equality* of (ground) affix values
`[RANGE =: RANGE DEF]`	*split* combined with conformity test
`[where IDF has OFFSET in BNO =: DEF]`	*join*
`[RANGE =: RANGE1]`	*assignment* to RANGE1

As a further extension, we allow affix expressions to occur at both inherited and derived positions, standing for implicit guards. Letting E stand for some affix expression, A for an affix of which it is a production and p for a procedure having one A parameter, in a procedure heading

- p(>E): is interpreted as p(>A): with [A =: E] inserted at the beginning of every alternative of p, and

- p(E>): as p(A>): with [E =: A] inserted at the end of every alternative.

Similarly, in a procedure call

- p(E) is interpreted as [E =: A], p(A) for an input-parameter, and

- p(E) as p(A), [A =: E] for an output-parameter.

The same sugaring conventions apply to the components of a transient parameter. By these conventions, most guards may remain implicit.

3 Static Semantic Checks

We want to distinguish statically between *effects* of the execution of a program, i.e. desired side effects, and *defects*, i.e. spurious side effects. To that end we consider the program as a grammar and require it to be LL(1) (and therefore deterministic).

Each procedure is classified according to two binary criteria:

- (has it an effect) Does an execution of the procedure cause a change in state (of the input, the output or the variables global to the procedure)?

- (does it yield a control value) Can the execution of the procedure fail (and thereby influence the flow of control) or does it always succeed?

The programmer specifies a type for each procedure, as follows:

PREDICATE   has effect, may fail
ACTION   has effect, always succeeds
TEST   no effect, may fail
FUNCTION   no effect, always succeeds.

This specification is checked on the one hand against the body of the procedure, and on the other hand against the use made of the procedure in the program.

Consider a production rule of a CF grammar satisfying the LL(1) restriction

$$r : m_1, m_2, ..., m_n; \ other \ alternatives.$$

and an associated recursive descent recognition procedure

PREDICATE r: m_1, m_2, ..., m_n; other alternatives.

where each of the members m_i of the CF grammar similarly has a corresponding recognition procedure m_i. Since the grammar satisfies the LL(1) restriction, none of the *other alternatives* has a starter in common with $m_1, m_2, ..., m_n$. Assume none of the m_i produces empty (all of the m_i are predicates, in the sense given above).

The predicate m_1 can only succeed upon recognizing one of the starters of m_1. Consequently, successful return from m_1 indicates that none of the other alternatives could have succeeded — even if one of $m_2, ..., m_n$ fails. There is therefore no point in backtracking to the other alternatives if m_1 succeeds but $m_2, ..., m_n$ fails. We will therefore insist (and verify) that $m_2, ..., m_n$ must succeed. Any later failure is forbidden.

More generally, an alternative may have a 'point of no return', viz. the first member which is a predicate or action; before this point there can be any number of tests or functions, after it only functions or actions. Any failure after this point leads to a *defect* — a side effect upon failure. A defect can be removed by backtracking explicitly, or by the introduction of an *error alternative*, as in

```
PRED is parameter list(PLIST>):
  is open symbol,
    ( is parameter list(PLIST),
      ( is close symbol;
        error("missing close symbol"))
    [nil =: PLIST]).
```

In CDL3, defects are forbidden!

A defect corresponds to a place where the program might have to backtrack. A program without defects will never need to backtrack.

Experience from CDL2 [6] shows that this semantic consistency check (together with a number of checks based on the directions of parameters) is very effective both in catching deep errors and in imposing a rigorous discipline on the use of side effects.

3.1 Example: a simple language

We want to recognize a simple block-structured language with the following CF syntax:

```
program: block.
block: "begin", units, "end".
units: unit, units; unit.
unit: application; definition; block.
application: "apply", identifier.
definition: "define", identifier.
```

By turning the rules of this grammar into procedures and suitably extending them with affixes we obtain a recursive descent parser, which builds for each block a local symbol table (RANGE) as a result of parsing.

```
R:: RANGE.
ACTION program:
  block(0).
ACT block(>BNO):
  should be token("begin"),
    units(BNO, 0|OFFSET, empty|RANGE),
      should be token("end").
ACTION units(>BNO, >OFFSET>, >R>):
  application, units tail;
  definition(BNO, OFFSET, R), units tail;
  block(BNO+1), units tail
WHERE units tail:
  is token(";"), units; +.
```

The refinement units tail is implicitly parametrized with the affixes BNO, OFFSET and R of the procedure units. The + indicates the empty alternative (which always succeeds) and - will be used to indicate an alternative that always fails.

```
PRED application:
  is token("apply"),
    ( is identifier(IDF);
      error("identifier expected")).
PRED definition(>BNO, >OFFSET|OFFSET+1>, >R>):
  is token("define"),
      ( is identifier(IDF), [R where IDF has OFFSET in BNO =: R];
      error("identifier expected")).
```

4 The second pass

An important property of Attribute Grammars (and Affix Grammars) is the freedom they give in the order of evaluation of components and their attributes, a freedom however which may have to be bought at the expense of a rather expensive mechanism for multipass or delayed evaluation. All kinds of subclasses (n-pass, n-visit, etc.) have been introduced to allow more efficient evaluation models.

In CDL3 we have chosen to support two-pass evaluation. Some actions from an alternative can be delayed until a second pass over the execution tree (the *second-pass elements*). These actions are meant to evaluate and use attributes that could not be known before the end of the first pass (the *second-pass parameters*). After the first pass is completed, the second pass of each successful procedure is executed, starting at the root and proceeding in *postfix* order: the second passes of all children are executed in textual order before the second pass of the parent. The second pass of a procedure can access both its first- and second-pass parameters, as well as the second pass output-parameters of its children.

The second pass shows its usefulness in many compilation problems where it obviates the need for an explicit abstract tree.

This execution model can also be seen as a form of partial evaluation of the parse tree, where the second pass builds the subtrees that were left unevaluated during the first pass. Since such a delayed subtree might itself have a second pass, this notion could be applied recursively, but we shall not pursue this possibility here.

Of course some compiler tasks imply a form of comparison of subtrees. In this case a data structure must be built that represents the relevant abstract tree. However, many interesting transductions can be described without introducing such an intermediate data structure.

More complicated transductions can be achieved by composing two-pass transducers using piping.

4.1 Adding context conditions ...

We now want to introduce checks on the context conditions, for which the information is available on completion of the first pass. To this end, some of the procedures are extended with a second-pass affix ENV denoting an environment composed of ranges. second-pass elements.

    ENV:: conc ENV RANGE; only RANGE.
The second pass elements are separated from the first-pass elements by the sign /. The additions are shown in bold face.

    ACTION program:
        block(0 / empty).
The outermost block stands an empty environment.

    ACTION block(>BNO / >ENV):
        should be token("begin"),
            units(BNO, 0|OFFSET, empty|RANGE / conc ENV RANGE),
                should be token("end").
Notice that the units of a block stand in an environment composed in the second pass out of the surrounding environment and the declarations belonging to that block.

```
ACTION units(>BNO, >OFFSET>, >R> / >ENV):
  application (/ ENV), units tail;
  definition(BNO+1, OFFSET, R), units tail;
  block(BNO+1 / ENV), units tail
WHERE units tail:
  is token(";"), units; + .
PRED application( / >ENV):
  is token("apply"),
    ( is identifier(IDF),
        ( found(IDF, ENV, OFFSET, BNO);
          error("missing definition for "+IDF);
      error("identifier expected")).
PRED definition(>BNO, >OFFSET|OFFSET+1, >R>):
  is token("define"),
    ( is identifier(IDF) ,
        ( found(IDF, R, OFFSET1, BNO1),
            error("multiple definition for "+IDF);
          [R where IDF has OFFSET in BNO =:R] );
      error("identifier expected")).
TEST found(>IDF, >conc ENV RANGE, OFFSET>, BNO>):
  found(IDF, RANGE, OFFSET, BNO);
  found(IDF, ENV, OFFSET, BNO).
TEST found(>IDF, >only RANGE, OFFSET>, BNO>):
  found(IDF, RANGE, OFFSET, BNO).
TEST found(>IDF, >RANGE DEF, OFFSET>, BNO>):
  [DEF =: where IDF has OFFSET in BNO];
  found(IDF, RANGE, OFFSET, BNO) ).
TEST found(>IDF, >empty, OFFSET>, BNO>): - .
```

4.2 ... and code generation

Finally we shall introduce code generation, in the form of a transduction of the input to a sequence of abstract instructions of the form

reserve bn, size	reserve size words for block no bn
apply bn, offs	apply the identifier with offset offs and block no bn
unreserve bn, size	free the storage of this block.

To that end we introduce metarules describing the output

```
CODE:: empty; CODE INSTR.
```

```
INSTR:: reserve BNO OFFSET; apply BNO OFFSET; unreserve BNO OFFSET.
```

We extend some of the previous procedures.

```
ACTION program:
  block(0/ empty , empty| CODE).
```

```
ACTION block(>BNO/ >ENV , >CODE| CODE unreserve BNO OFFSET>):
  should be token("begin"),
    units(BNO, 0|OFFSET, empty|RANGE/ conc ENV RANGE ,
      CODE reserve BNO OFFSET| CODE),
      should be token("end").
ACTION units(>BNO, >OFFSET>, >R>/ >ENV , >CODE>):
  application(/ ENV , CODE), units tail;
  definition(BNO+1, OFFSET, R), units tail;
  block(BNO+1/ ENV , CODE), units tail
WHERE units tail:
  is token(";"), units; + .
PRED application(/ >ENV , >CODE>):
  is token("apply"),
    ( is identifier(IDF),
        ( found(IDF, R, OFFSET, BNO) ,
          [CODE apply BNO OFFSET =: CODE];
          error("missing definition for "+IDF);
        error("identifier expected")).
```

In a number of simple steps we have achieved the required transduction, performing syntax analysis, static semantics and generation of a nontrivial translation in two passes, without explicitly constructing an abstract tree as an intermediary.

5 Modules

A major concern in writing large grammars (as in writing large programs) is the large number of attributes to be passed explicitly from one part of the grammar to others. This communication overhead, resulting from the explicit applicative character of grammars, has to be overcome in order to obtain a practical notation for programming. Uncritically introducing a notation for global variables on top of the grammatical formalism is unsatisfactory, both from a formal viewpoint and from Software Engineering considerations.

In CDL3, global variables can only be introduced in separate modules. A *module* is considered as an abstract data type that has precisely one instance. Since there is only one instance of it, the module need never be passed explicitly as a parameter.

A module consists of one structured object (an affix expression comprizing all its global variables) and the definitions of procedures to handle this object. The global variables are invisible transient parameters to all the procedures exported by the module and recursively of the procedures that call those procedures. Their introduction again serves to get rid of stereotypical parameters, and gain some security in the bargain.

5.1 Preludes and postludes

A module may have two special procedures, the *prelude* and the *postlude* of the module, which will be called respectively at the begining and at the end of the execution of the program.

The introduction of preludes and postludes allows to factor out the initializations and finalizations from the program. The programmer need not worry globally about propagating and calling initializations and finalizations, and can concentrate on the main algorithm (*separation of concerns*).

5.2 Example of a module

As a simple example of a module, just to show the notation used, we will provide lexical analysis procedures for the previous program.

```
MODULE lexico = buffered LINE.
```

The module has one global variable, named LINE.

```
DEFINES IDF,
        PRED is token(>TEXT),
        PRED should be token(>TEXT),
        PRED is identifier(IDF>),
        TEST is end of source.
```

The interface specifies all exported procedures and types, as well as the modules which will be used (none, in this case).

```
LINE, IDF:: TEXT.

PRELUDE read buffer.
ACTION read buffer:
  read line(LINE).
TEST is end of source:
  [LINE =: ""] .
PRED is token(>TEXT):
  is prefix(TEXT, LINE), layout.
```

The test **is prefix (>TEXT1, >TEXT2>)** is a built-in text operation.

```
ACTION should be token(>TEXT):
  is prefix(TEXT, LINE), layout;
  error(TEXT + " expected).
ACTION layout:
  is prefix(" ", LINE), layout;
  is prefix("\n", LINE), read buffer, layout;
  +.

L, D, T:: TEXT.
PRED is identifier(L+T>):
  is letter(L), rest identifier(T), layout.
PRED is letter(L>):
    split(LINE, L, LINE1), before("a", L), before(L, "z"),
      [LINE1 =: LINE] .
```

The predefined FUNCTION **split(>TEXT, CHAR>, TEXT1>)** splits the first character off a text.

```
PRED is digit(D>):
    split(LINE, D, LINE1), before("0", D), before(D, "9"),
      [LINE1 =: LINE] .
```

6 Implementation status

A compiler for CDL3 has been implemented by J. Beney, generating C code which is compatible with most C compilers on PC's and UNIX workstations. Work still has to be invested in optimization and support tools. In the present state, the compiler is efficient enough to compile itself in a reasonable time on a PC.

7 Conclusion

Two principles underlie the design of CDL3: the exploitation of the borderline between grammars and programs, and the exploitation of the benefits of voluntary restrictions.

By the first principle, the advantages of the interpretation of programs as grammars (static semantic checks, the defect-philosophy) are combined with Software Engineering considerations (support of structured programming, modular programming, preludes and postludes, as well as the use of special tools and environments).

According to the second principle, we strive neither for maximal efficiency nor for maximal generality, but aim for expressiveness of notation and checkability of properties of programs combined with quite good efficiency, at the price of small and well-understood restrictions in generality (LL(1) base grammar, second level in Greibach Normal Form). Disregarding these restrictions, a CDL3 program can easily be embedded in a richer formalism (such as PROLOG or EAGs), mainly by turning | and / into comma's and eliminating some syntactic sugar. But then we loose the advantages afforded by the restrictions, in particular checkability.

We hope to have struck the right balance, resulting in a compact notation for the efficient development of secure syntax-based transducers, such as compilers, interpreters and man-machine interfaces.

References

[1] J. G. Beney, L. Frecon, *Langage et Systeme d'Ecriture de Traducteurs*, RAIRO Informatique Vol. 14, No. 4, pp 379-394, 1980.

[2] J. G. Beney, J. F. Boulicaut, *An Affix-Based Compiler Compiler designed as a Logic Programming System*, Proceedings of the CC'90, Schwerin, October 1990.

[3] P. Y. Cunin, M. Griffiths, J. Voiron *Comprendre la Compilation*, Lecture Notes in Computer Science, Springer-Verlag, 1980.

[4] P. Deransart, M. Jourdan, B. Lohro, *A survey on Attribute Grammars*, Lecture Notes in Computer Science 323, Springer-Verlag, 1988.

[5] P. Deransart, M. Jourdan (Eds), *Attribute Grammars and their Applications*, Lecture Notes in Computer Science 461, Springer-Verlag, 1990.

[6] H. Feuerhahn, C. H. A. Koster, *Static Semantic Checks in an Openended Language*, In: P. G. Hibbard and S. A. Schuman (Eds.), *Constructing Quality Software*, North Holland Publ. Cy., 1978.

[7] M. Fokkinga, J. Jeuring, L. Meertens, E. Meijer, *Translation of Attribute Grammars into Catamorphisms*, to appear in The Squiggolist, 1991.

[8] E. C. R. Hehner, S. A. Silverberg, *Programming with Grammars: an Exercise in Methodology-directed Language Design*, the Computer Journal, Vol 26 no 3, 1983.

[9] M. Jourdan, D. Parigot, *The FNC-2 System: Advances in Attribute Grammar Technology*, RR-834 INRIA April 1988.

[10] D. E, Knuth, *Semantics of Context-Free Languages*, Mathematical System Theory, Vol. 2, No. 2, pp 127-145, june 1968 and Vol. 5, pp 95-96, may 1971.

[11] C. H. A. Koster, *A Compiler Compiler*, Report MR127/71, Mathematical Centre, Amsterdam, 1971.

[12] C. H. A. Koster, *Affix Grammars*, In: J. E. L. Peck(Ed.), ALGOL 68 Implementation, North-Holland Publishing Co., Amsterdam 1971.

[13] C. H. A. Koster, *Affix Grammars for Natural Languages*, Proceedings Summer School on Attribute Grammars and their Applications, Prague, June 1991.

[14] J. Lewi, K. De Vlaminck, J. Huens, M. Huybrecht, *Project LILA User's Manual*, Report CW7, Applied Mathematics and Programming Division, Katholieke Universiteit Leuven, 1977.

[15] H. Meijer, *PROGRAMMAR: a Translator Generator*, Ph.D. Thesis, University of Nijmegen, 1986.

[16] H. Meijer, *The Project on Extended Affix Grammars at Nijmegen*, In: [5].

[17] H. H. Metcalfe, *A Parameterized Compiler based on Mechanical Linguistics*, Planning Research Corporation, Los Angeles, Calif. 1963.

[18] M. P. G. Moritz, *Description and Analysis of Static Semantics by Fixed-Point Equations*, Ph.d. Thesis, University of Nijmegen, 1989.

[19] D. A. Watt, *Analysis-oriented two-level Grammars*, Ph.D. thesis, University of Glasgow, 1974.

[20] A. van Wijngaarden et al. (Eds.), *Revised Report on the Algorithmic Language ALGOL68*, Acta Informatica 5, pp. 1–236, 1975; also SIGPLAN Notices 12(5), pp. 5–70, 1977.

Efficient Incremental Evaluation of Higher order Attribute Grammars

Harald Vogt, Doaitse Swierstra and Matthijs Kuiper

Department of Computer Science, Utrecht University
P.O. Box 80.089, 3508 TB Utrecht, The Netherlands
E-Mail: {harald,doaitse,matthys}@cs.ruu.nl

Abstract

This paper presents a new algorithm for the incremental evaluation of Ordered Attribute Grammars (OAGs), which also solves the problem of the incremental evaluation of Ordered *Higher order* Attribute Grammars (OHAGs). Two new approaches are used in the algorithm.

First, instead of caching all results of semantic functions in the grammar, all results of visits to trees are cached. There are no attributed trees, because all attributes are stored in the cache. Trees are build using hash consing, thus sharing multiple instances of the same tree and avoiding repeated attributions of the same tree with the same inherited attributes. Second, each visit computes not only synthesized attributes but also *bindings* for subsequent visits. Bindings, which contain attribute values computed in one visit and used in subsequent visits, are also stored in the cache. As a result, second and further visits get a subtree containing only all necessary earlier computed values (the bindings) as a parameter.

The algorithm runs in $O(|\text{Affected}| + |\text{paths\_to\_roots}|)$ steps after modifying subtrees, where paths_to_roots is the sum of the lengths of all paths from the root to all modified subtrees, which is almost as good as an optimal algorithm for first order AGs, which runs in $O(|\text{Affected}|)$.

1 Introduction

Attribute grammars describe the computation of attributes: values attached to nodes of a tree. The tree is described with a context free grammar. Attribute computation is defined by semantic functions. AGs are used to define languages and form the basis of compilers, language-based editors and other language based tools. For an introduction and more on AGs see: [Deransart, Jourdan and Lorho 88, WAGA 90].

Higher Order AGs [Vogt, Swierstra and Kuiper 89, TC90, Swierstra and Vogt 91] remove the artificial distinction between the syntactic level (context free grammar) and the semantic level (attributes) in attribute grammars. This strict separation is removed in two ways: First, trees can be *used* directly as a value within an attribute equation. Second, a part of the tree can be *defined* by attribution. Trees used as a value and trees defined by attribution are known as non-terminal attributes (NTAs).

It is known that the (incremental) attribute evaluator for Ordered AGs [Kastens 80, Yeh 83, Reps and Teitelbaum 88] can be trivially adapted to handle Ordered Higher Order AGs [Vogt, Swierstra and Kuiper 89]. The adapted evaluator, however, attributes each instance of equal NTAs separately. This leads to nonoptimal incremental behaviour after a change to a NTA, as can be seen in the recently published algorithm of [TC90]. Our evaluation algorithm handles multiple occurrences of the same NTA (and the same subtree) efficiently in $O(|\text{Affected}| + |\text{paths\_to\_roots}|)$ steps, where paths_to_roots is the sum of the lengths of all paths from the root to modified subtrees.

The new incremental evaluator can be used for language-based editors like those generated by the Synthesizer Generator ([Reps and Teitelbaum 88]) and for minimizing the amount of work for restoring semantic values in tree-based program transformations. The new algorithm is based on the combination of the following four ideas:

- The algorithm computes attribute values by using visit functions. A visit function takes as first parameter a tree and some of the inherited attributes of the root of the tree. Part of the returned values are some of the synthesized attributes of the root of the tree. Our evaluator consists of visit functions that recursively call each other to attribute the tree.

- As in [TC90]'s algorithm, trees are build using hash consing for trees. In our algorithm this is the only representation for trees, thus multiple instances of the same tree will be shared.

  Because many instantiations of a NTA may exists, each with its own attributes, attributes are no longer stored with the tree, but in a cache. In a normal incremental treewalk evaluator a partially attributed tree can be considered as a cache. Attributes needed by a visit and computed by a previous visit are not recomputed but are found in the partially attributed tree.

- A call to a visit function corresponds to a visit in a visitsequence of an Ordered HAG. Instead of caching the results of *semantic functions*, as was done in [Pugh 88], the results of *visit functions* are cached. This is more efficient because a cache hit of a visit function means that this visit to (a possible large) tree can be skipped. Furthermore, a visit function may return the results of several semantic functions at a time.

- Although the above idea seems appealing at first sight, a complication is the fact that attributes computed in an earlier visit may have to be available for later visits when necessary.

  Bindings contain attribute values computed in one visit and used in future visits to the same tree: Each visit function computes synthesized attributes *and* bindings for subsequent visits. Bindings computed by earlier visits are passed as an extra parameter to visit functions.

The visit functions can be implemented in any imperative or functional language. Furthermore, visit functions have no free variables and can therefore be viewed as supercombinators [Hughes 82].

Efficient caching is partly achieved by efficient equality testing between parameters of visit functions, which are trees, inherited attributes and bindings. Therefore, hash consing for trees and bindings is used, so testing for equality between trees and between bindings is reduced to a fast pointer comparison. Furthermore, several space optimizations for bindings are possible.

Although the computation of these bindings may appear to be cumbersome, they have a considerable advantage when evaluating incrementally. They contain precisely the information on which visits do depend and no more.

The remainder of this article is structured as follows. Section 2 presents an informal definition and example. Section 3 defines visit functions and bindings. A space optimization for bindings is presented in section 4. We discuss visit function caching in section 5. Section 6 presents the results of a simulation of our algorithm for incremental evaluation for a larger example. The incremental behaviour of the algorithm is discussed in section 7. Section 8 presents the conclusions.

2 Informal definition and example

First, visitsequences from which the visit functions will be derived are presented and illustrated by an example. Then bindings and visit functions for the example will be shown. Finally, incremental evaluation will be discussed.

2.1 Visit(sub)sequences

In [Vogt, Swierstra and Kuiper 89] the equivalent of OAGs ([Kastens 80]) for HAGs, the so called Ordered Higher order Attribute Grammars (OHAGs), are defined. An OHAG is characterized by the existence of a total order on the defining attribute occurrences in a production p. This order induces a fixed sequence of computation for the defining attribute occurrences, applicable in any tree production p occurs in. Such a fixed sequence is called a *visitsequence* and will be denoted by VS(p). The following introduction to visit-sequences for a HAG is almost literally taken from [Kastens 80].

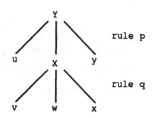

Figure 1: A structure tree.

This evaluation order is the starting point for the construction of a flexible and efficient attribute evaluation algorithm. It is closely adapted to the particular attribute dependencies of the AG. The principle is demonstrated here. Assume that an instance of X is derived by

$$S \Rightarrow uYy \rightarrow_p uvXxy \rightarrow_q uvwxy \Rightarrow s.$$

The corresponding part of the structure tree is shown in Figure 1. An attribute evaluation algorithm traverses the structure tree using the operations "move down to a descendant node" (e.g. from K_y to K_x) or "move up to the ancestor node" (e.g. from K_x to K_y). During a visit of node K_y some attributes defined in production p are evaluated according to semantic functions, if p is applied at K_y. In general several visits to each node are needed before all attributes are evaluated. A local tree walk rule is associated with each p. It is a sequence of four types of moves: move up to the ancestor, move down to a certain descendant, evaluate a certain attribute and evaluate followed by expansion of the labeled tree by the value of a non-terminal attribute. The last instruction is specific for a HAG.

VS(p) is split into *visitsubsequences* VSS(p,v) by splitting after at each "move up to the ancestor" instruction in VS(p). The attribute grammar in Figure 2 is used in the sequel only to demonstrate visitsubsequences, bindings and visit functions. Section 6 contains a larger example.

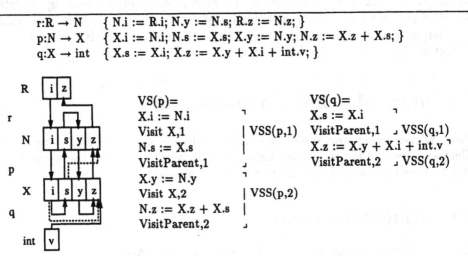

$$r:R \rightarrow N \quad \{ N.i := R.i;\ N.y := N.s;\ R.z := N.z;\ \}$$
$$p:N \rightarrow X \quad \{ X.i := N.i;\ N.s := X.s;\ X.y := N.y;\ N.z := X.z + X.s;\ \}$$
$$q:X \rightarrow int \quad \{ X.s := X.i;\ X.z := X.y + X.i + int.v;\ \}$$

Figure 2: An example AG (top), the dependencies (left) and visitsequences (right). The dashed lines indicate dependencies on an attribute defined in the first visit and used in the second visit. VS(r) is omitted.

2.2 Visit functions for the example grammar

The evaluator is obtained by translating each visitsubsequence VSS(p,v) into a *visit function visit_N_v* where N is the left hand side of p.

All visit functions together form a functional attribute evaluator program. We use a Miranda-like notation [Turner 1985] for visit functions. Because the visit functions are strict, which results in explicit scheduling of the computation, visit functions could also be easily translated into Pascal or any other non-lazy imperative language.

The first parameter in the definition of $visit\_N\_v$ is a *pattern* describing the subtree to which this visit is applied. The first element of the pattern is a marker, a constant which indicates the applied production rule. The other elements are identifiers representing the subtrees of the node. Following the functional style we will have one set of visit functions for each production with left hand side N.

All other arguments, *except* the last, of $visit\_N\_v$ represent the inherited attributes used in VSS(p,v). Before we discuss the results of a visit function, consider the grammar in Figure 2 again. The inherited attribute X.i and the synthesized attribute X.s in Figure 2 are also used in the second visit to N and X but passed to or computed in the first visit.

Therefore, every $visit\_N\_v$ not only computes synthesized attributes but also *bindings* (inherited and synthesized attributes *computed* in $visit\_N\_v$ and *used* in subsequent visits to N). So $visit\_N\_v$ computes also a list of $(max_N - v - 1)$ bindings, one for each subsequent visit (here max_N is the maximum number of visits to N). The bindings *used* in $visit\_N\_v+i$ but *computed* in $visit\_N\_v$ are denoted by $binds\_N^{v \to v+i}$.

The last argument of $visit\_N\_v$ is a list of bindings for $visit\_N\_v$ computed in earlier visits $1 \ldots (v-1)$ to N. The bindings themself are lists containing attribute values and further bindings. Both lists are constructed using hash consing. Elements of a list are addressed by projection, e.g. $binds\_N^{i \to v}.1$ is the first element of the list.

We now turn to the visit functions for the visitsubsequences VSS(p,v) and VSS(q,v) of the example grammar. We will put a box around attributes that are returned in a binding. In the example this concerns $\boxed{X.i}$ and $\boxed{X.s}$. The first visit to N will return the synthesized attribute $N.s$, and a binding list $binds\_N^{1 \to 2}$ containing the later needed $X.s$ together with $binds\_X^{1 \to 2}$. The binding list $binds\_N^{1 \to 2}$ is denoted by $[\boxed{X.s}, binds\_X^{1 \to 2}]$.

```
visit_N_1 (p [X]) N.i = (N.s, binds_N^{1→2})
        where X.i = N.i
              (X.s, binds_X^{1→2}) = visit_X_1 X X.i
              N.s = X.s
              binds_N^{1→2} = [X.s, binds_X^{1→2}]
```

In the above definition (p [X]) denotes the first argument: a tree at which production p is applied, with one son, denoted X. The second argument is the inherited attribute i of N. The function returns the synthesized attribute s and binding for $X.s$ together with the bindings from the first visit to subtree X. Function $visit\_N\_2$ does not return a binding because it is the *last* visit to a N-tree. Here the projections on $binds\_N^{1 \to 2}$ can be made implicit by replacing $[binds\_N^{1 \to 2}]$, the last parameter of $visit\_N\_2$, by $[[\boxed{X.s}, binds\_X^{1 \to 2}]]$.

```
visit_N_2 (p [X]) N.y [binds_N^{1→2}] = N.z
        where X.y = N.y
              binds_X^{1→2} = binds_N^{1→2}.2
              X.z = visit_X_2 X X.y [binds_X^{1→2}]
              X.s = binds_N^{1→2}.1
              N.z = X.z + X.s
```

The other visit functions have a similar structure.

$visit\_X\_1$ (q int) X.i = (X.s, $binds\_X^{1 \to 2}$)
  where X.s = X.i
    $binds\_X^{1 \to 2}$ = [$\boxed{\texttt{X.i}}$]

$visit\_X\_2$ (q int) X.y [$binds\_X^{1 \to 2}$] = X.z
  where $\boxed{\texttt{X.i}}$ = $binds\_X^{1 \to 2}$.1
    X.z = X.y + X.i + int.v

We have chosen the order of definition and use in the *where* clause in such a way that the visit functions could be also defined in an imperative language. A *where* clause contains three kinds of definitions: (1), assignments and visits from the corresponding VSS(p,v). (2), lookups of attributes and bindings in bindings (for example in $visit\_N\_2\ binds\_X^{1 \to 2}$ is selected from $binds\_N^{1 \to 2}$). And (3), definitions for returned bindings. The precise definition of visit functions and bindings is given in section 3.

2.3 Incremental evaluation

After a tree T is modified into T', in our model with hash consed trees, T' shares all unmodified parts with T. To evaluate the attributes of T and T' the *same* visit function $visit\_R\_1$ is used, where R is the root non-terminal. Note that tree T' is totally rebuild before $visit\_R\_1$ is called.

The incremental evaluator automatically skips unchanged parts of the tree because of cache-hits of visit functions, provided the inherited attributes have not changed. Hash consing for trees and bindings is used to achieve efficient caching, for which fast equality tests are essential. Because separate bindings for each visit are computed, for example $visit\_N\_1$ and $visit\_N\_4$ could be recomputed after a subtree replacement, but $visit\_N\_\{2,3\}$ could be found in the cache and skipped. Some other advantages are illustrated in Figure 3, in which the following can be noted:

- NTA1 and NTA2 are defined by attribution, indicated by boxes (□).

- Multiple instances of the same (sub)tree, for example a multiple instantiated NTA, are *shared* by using hash consing for trees (Trees T2 and T'2).

- Those parts of an attributed tree derived from NTA1 and NTA2 which can be reused after NTA1 and NTA2 change value are *identified automatically* because of the hash consing for trees and cached visit functions (Trees T3 and T4 in (b)). This holds also for a subtree modification in the initial parse tree (Tree T1).

- Because trees T1, T3 and T4 may be attributed the same in (a) and (b) they will be skipped after the subtree modification and the amount of work which has to be done in (b) is $O(|\text{Affected T'2}| + |\text{paths\_to\_roots}|)$ steps, where paths_to_roots is the sum of the lengths of all paths from the root to all subtree modifications (NEW, X1 and X2).

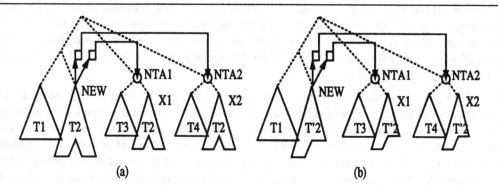

Figure 3: A subtree modification at node NEW induces subtree modifications at node X1 and X2 in the trees derived from NTA1 and NTA2. In this example we suppose that all instantiated trees T1, T2, T'2, T3 and T4 are attributed the same in (a) and (b).

3   Visit functions and bindings

We now turn to the definition of visit functions and bindings.

Let p be a production of the form $p:N \rightarrow \dots X \dots$. Let VS(p) be the visitsequence for p. Let max_N be the maximum number of visits to N. Let $VSS(p,1) \dots VSS(p,max_N)$ be the visitsubsequences in VS(p).

VSS(p,v) is translated into the visit function $visit\_N\_v$:

$$
visit\_N\_v \; (p\;[X])\; inh_v^N \; [binds\_N^{1\rightarrow v},\; \dots,\; binds\_N^{(v-1)\rightarrow v}] =
$$
$$
(syn_v^N,\; binds\_N^{v\rightarrow v+1},\; \dots,\; binds\_N^{v\rightarrow max_N})
$$
where Lines from 1) to 3).
  1) The assignments and visits in VSS(p,v).
  2) Lookups of attributes and bindings computed in earlier visits.
  3) Definitions for the returned bindings.

inh_v^N are the available inherited attributes needed in and not available in visits before VSS(p,v). syn_v^N are the synthesized attributes computed in VSS(p,v). The elements 1) to 3) are defined as follows. 1) is just copying from VSS(p,v). In 1) a Visit X,w is translated into

$$
(syn_w^X,\; binds\_X^{w\rightarrow w+1},\; \dots,\; binds\_X^{w\rightarrow max_X}) =
$$
$$
visit\_X\_w \; X \; inh_w^X \; [binds\_X^{1\rightarrow w},\; \dots,\; binds\_X^{(w-1)\rightarrow w}]
$$

When X is a non-terminal attribute, the variable defining X is used as the first argument pattern for $visit\_X\_w$.

There are three kinds of lookups in 2): Inherited attributes, synthesized attributes and bindings. The lookup method is the same for all, so we will only describe the method for an inherited attribute here. Let N.inh be an inherited attribute of N which is used in $visit\_N\_v$ but *not* defined in $visit\_N\_v$. Then, the lookup N.inh = $binds\_N^{e \rightarrow v}.f$ is added, for the appropriate e and f.

In 3) the bindings results of $visit\_N\_v$ are defined. Recall that the $binds\_N^{v \rightarrow v+i}$ are defined in terms of the visitsequence of production p. $binds\_N^{v \rightarrow v+i}$ is defined as a list containing those inherited attributes of N and synthesized attributes of sons of N used in $visit\_N\_v$ and in $visit\_N\_v+i$ (denoted by $inout\_N_p^{v \rightarrow v+i}$) plus the bindings of sons of N computed by visits to N and used in subsequent visits to those sons of N in $visit\_N\_v+i$ (denoted by $binds\text{-}sons\_N_p^{v \rightarrow v+i}$). For example $binds\_N^{1 \rightarrow 2}$ in the example visit functions in section 2 is

$$binds\_N^{1 \rightarrow 2} = [inout\_N_p^{1 \rightarrow 2}, binds\text{-}sons\_N_p^{1 \rightarrow 2}] = [X.s, binds\_X^{1 \rightarrow 2}],$$

where during execution the value of $binds\_X^{1 \rightarrow 2}$ will be [X.s, [X.i]]. $inout\_N_p^{v \rightarrow v+i}$ and $binds\text{-}sons\_N_p^{v \rightarrow v+i}$ are defined as follows:

$$inout\_N_p^{v \rightarrow v+i} = \{a \mid (a \in N.inh \cup X.syn) \wedge (a \in VSS(p,v)) \wedge (a \in VSS(p,v+i)) \}$$

$$binds\text{-}sons\_N_p^{v \rightarrow v+i} = \{ binds\_X^{w \rightarrow j} \mid (visit\_X\_w \in VSS(p,v)) \wedge (w < j < max_X) \\ \wedge (visit\_X\_j \in VSS(p,v+i)) \}$$

The following theorem holds for the above defined functional program.

Theorem 3.1 *Let HAG be a well-defined Ordered Higher Order Attribute grammar, and let S be a structure tree of HAG. The execution of the above defined functional program for HAG with input S terminates and attributes the tree S correctly. Furthermore, no attributes are evaluated twice.*

4 Optimizations for bindings

Several optimizations that reduce the number of occurrences of values in $binds\_N^{v \rightarrow v+i}$ are possible. We only mention that some $binds\_N^{v \rightarrow v+i}$ may always be empty, which may be deduced statically from the grammar.

5 Visit function caching

This section describes the implementation of function caching used for caching the visit functions of the functional evaluator and was inspired upon [Pugh 88]. A *hash table* is used to implement the cache. A single cache is used to store the cache results for all functions. Tree T, labeled with root N, is attributed in visit v by calling *visit_N_v* T inherited_attributes bindings

The result of this function is uniquely determined by the function-name, the arguments of the function and the bindings. The visit functions can be cached as follows:

```
cached_apply(visit_N_v, T, args, binds) =
    index := hash(visit_N_v, T, args, binds)
    ∀ <function, tree, arguments, bindings, result> ∈ cache[index] do
        if function = visit_N_v and EQUAL(tree,T)
            and EQUAL(arguments,args) and EQUAL(bindings,binds)
        then return result
    result := visit_N_v T args binds
    cache[index] := cache[index] ∪ {<visit_N_v, T, args, binds, result>}
    return result
```

To implement visit function caching, we need efficient solutions to several problems. We need to be able to

- compute a hash index based on a function name and an argument list. For a discussion of this problem, see [Pugh 88] for more details.

- determine whether a pending function call matches a cache entry, which requires efficient testing for equality between the arguments (in case of trees and bindings very large structures!!) in the pending function call and in a candidate match.

The case of trees and bindings in the last problem is solved by hash consing for trees and bindings.

6 A large example

Consider the higher order AG in Figure 4, which describes the mapping of a structure consisting of a sequence of defining identifier occurrences and a sequence of applied identifier occurrences onto a sequence of integers containing the index positions of the applied occurrences in the defining sequence. For example, the sentence **let a,b,c in c,c,b,c ni** is mapped onto the sequence [3,3,2,3]. The following can be noted in Figure 4:

- The attribute $\overline{ENV}$ is a non-terminal (higher order) attribute. The tree structure is built using the *constructor functions* update and empty, which correspond to the respective productions for $\overline{ENV}$. The attribute $\overline{ENV}$ is instantiated (i.e. a copy of the tree is attributed) in the occurrence of the first production of APPS, and takes the rôle of a semantic function.

- Note that there may exist many instantiations of the $\overline{ENV}$-tree, all with different attributes. There thus does not any longer exist an one-to-one correspondence between attributes and abstract syntax trees.

$$
\begin{array}{ll}
p_0(): & \text{ROOT} \rightarrow \text{let DECLS in APPS ni} \\
p_1(): & \text{DECLS} \rightarrow \text{DECLS ident} \\
p_2(): & \text{DECLS} \rightarrow \text{EMPTY\_decls} \\
p_3(): & \text{APPS} \rightarrow \text{APPS ident } \overline{ENV} \\
& \overline{ENV} := \text{APPS}_0.\text{env} \\
& \overline{ENV}.\text{param} := \text{ident.id} \\
& \text{APPS}_1.\text{env} := \text{APPS}_0.\text{env} \\
& \text{APPS}_0.\text{seq} := \text{APPS}_1.\text{seq} ++ [\overline{ENV}.\text{index}] \\
p_4(): & \text{APPS} \rightarrow \text{EMPTY\_apps} \\
\underline{\text{update}(\_,\_,\_)}: & \text{ENV} \rightarrow \text{ident number ENV} \\
\underline{\text{empty}()}: & \text{ENV} \rightarrow \text{EMPTY}
\end{array}
$$

Figure 4: The higher order AG, only the attribution rules for $p_3()$ are shown.

Figure 5.a shows the tree for the sentence **let a,b,c in c,c,b,c ni** which was attributed by a call to

visit_ROOT_1 (p_0 (DEF(DEF(DEF(DEF EMPTY_decls a) b) c))
               (USE(USE(USE(USE(USE EMPTY_apps c) c) b) c)))

Incremental reevaluation after removing the declaration of c is done by calling

visit_ROOT_1 (p_0 (DEF(DEF(DEF EMPTY_decls a) b))
               (USE(USE(USE(USE(USE EMPTY_apps c) c) b) c)))

The resulting tree is shown in Figure 5.b, note that only the APPS-tree will be totally revisited, the first visits to the DECLS and ENV trees generate cache-hits and further visits to them are skipped. Simulation shows that in this example 75% of all visit-function calls and tree-build calls which have to be computed in 5.b if there was no cache are found in the cache build up by 5.a when using caching. So 75% of the "work" was saved!

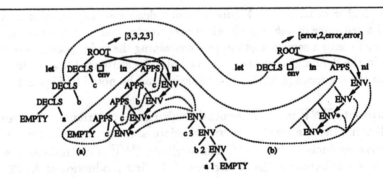

Figure 5: The tree before (a) and after removing c (b) from the declarations in **let a,b,c in c,c,b,c ni**. The * indicate cache-hits looking up c. The dashed lines denote sharing.

7   Incremental evaluation performance

In this section the performance of the functional evaluator with respect to incremental evaluation is discussed. The goal is to prove that the derived incremental evaluator recomputes in the worst case a number of semantic function calls bounded by $O(|\text{Affected}|)$. Here Affected is the set of attribute instances in the tree which contain a different value, together with the set of attribute instances newly created after a subtree modification.

This desire can be only partly fulfilled; it will be shown that the worst case boundary is given by $O(|\text{Affected}|+|\text{paths\_to\_roots}|)$. Here paths_to_roots are all nodes on the path to the initial subtree modification and on the paths to the root nodes of induced subtree modifications in trees derived from NTAs. The paths_to_roots part cannot be omitted because the reevaluation starts at the root of the tree and ends as soon all replaced subtrees are either reevaluated or found in the cache.

Let VIS be the mapping from a HAG to visit functions as discussed in section 3. Let T be a tree consistently attributed according to a HAG. Suppose T was attributed by VIS(HAG)(T). Let T' be the tree after a subtree modification and suppose T' was attributed by VIS(HAG)(T').

Theorem 7.1 *Let* Affected_Applications *be the set of function applications that need to be computed and will not be found in the cache when using VIS(HAG)(*T'*) with function caching for visits and hash consing for trees. Then, Affected_Applications is* $O(|\text{Affected}| + |\text{paths\_to\_roots}|)$*.*

8   Conclusions and future work

A new algorithm for the incremental evaluation of HAGs was presented. Two new approaches are succesfully combined. First, the results of visit functions are cached instead of results of semantic functions. Second, bindings are used containing attribute values computed in earlier visit functions and used by subsequent visit functions visiting the same tree.

Several space-optimizations for bindings are possible. Sharing and efficient caching is achieved by hash consing for trees and bindings.

The resulting algorithm runs in $O(|\text{Affected}| + |\text{paths\_to\_roots}|)$ steps after subtree modifications, where paths_to_roots is the sum of the lengths of all paths from the root to all subtree modifications, which is almost as good as an optimal algorithm for first order AGs (which runs in $O(|\text{Affected}|)$).

There are several other ways to improve the performance of which we will mention two here. First, it is possible to split the attribute equations in the visitsubsequences into independent parts. Then, several independent visit functions for one visitsubsequence can be derived. As a consequence only those parts of the attribute equations will be recomputed of which the input has changed. Second, it is possible to compute not only bindings for future visits but also those parts of the tree which will be actually visited by those future visits. So several future visits might be skipped during incremental evaluation when those parts of the tree where those future visits depend on have not changed.

References

[Deransart, Jourdan and Lorho 88] Deransart, P., M. Jourdan and B. Lorho. *Attribute Grammars: Definitions, Systems and Bibliography*. LNCS 323, Springer Verlag, Aug. 1988.

[WAGA 90] Deransart, P., M. Jourdan (Eds.). *Attribute Grammars and their Applications*. Proceedings of the International Workshop on Attribute Grammars and their Applications (WAGA), LNCS 461, Paris, September 19-21, 1990.

[Hughes 82] Hughes, R.J.M. *Super-combinators: A New Implementation Method for Applicative Languages*. In Proc. ACM Symp. on Lisp and Functional Progr., Pittsburgh, 1982.

[Hughes 85] Hughes, R.J.M. *Lazy memo functions*. In Proc. Conference on Functional Progr. and Comp. Architecture, Nancy, pages 129–146, LNCS 201, Springer Verlag 1985.

[Kastens 80] Kastens, U. *Ordered Attributed Grammars*. Acta Informatica, 13, pages 229–256, 1980.

[Pugh 88] Pugh, W.W. *Incremental Computation and the Incremental Evaluation o Functional Programs*. Tech. Rep. 88-936 and Ph.D. Thesis, Dept. of Computer Science, Cornell Univ., Ithaca, N.Y., Aug. 1988.

[Reps and Teitelbaum 88] Reps, T. and T. Teitelbaum. *The Synthesizer Generator: A System for Constructing Language-Based Editors*. Springer-Verlag, NY, 1988.

[Swierstra and Vogt 91] Swierstra, S.D. and H.H. Vogt. *Higher Order Attribute Grammars*. In the proceedings of the International Summer School on Attribute Grammars, Applications and Systems, (To Appear), Prague, June 4-13, 1991.

[TC90] Teitelbaum, T. and R. Chapman. *Higher-Order Attribute Grammars and Editing Environments*. ACM SIGPLAN '90 Conference on Programming Language Design and Implementation, White Plains, New York, pages 197-208, June, 1990.

[Turner 1985] Turner, D.A. *Miranda: A non-strict functional language with polymorphic types*. In J. Jouannaud, editor, Funct. Progr. Lang. and Comp. Arch., pages 1-16 Springer, 1985.

[Vogt, Swierstra and Kuiper 89] Vogt, H.H., S.D. Swierstra and M.F. Kuiper. *Higher Order Attribute Grammars*. ACM SIGPLAN '89 Conference on Programming Language Design and Implementation, Portland, Oregon, pages 131-145, June, 1989.

[Yeh 83] Yeh, D. *On incremental evaluation of ordered attributed grammars*. BIT, 23:308 320, 1983.

PROFIT: A System Integrating
Logic Programming and Attribute Grammars

Jukka Paakki

Department of Computer Science, University of Helsinki
Teollisuuskatu 23, SF-00510 Helsinki, Finland

Abstract

PROFIT is a system and a programming language where Prolog is extended with concepts needed in compiler writing applications. The special facilities include a deterministic error-recovering form of definite clause grammars (DCGs), functions as conditional equations, and functional terms modelling inherited and synthesized attributes of attribute grammars. PROFIT supports a multi-paradigm method of writing compilers in a compact and natural way. Most notably PROFIT can be used to express one-pass compilation within the framework of logical one-pass attribute grammars, a proper superset of L-attributed grammars.

1. Introduction

Finding a suitable methodology for implementing programming languages has been one of the most popular and important topics in the history of computing research. The idea of automatically translating programs into executable machine code was introduced in the early 1950's. Since then a variety of methods and formalisms have been devised to make language implementation (especially compilation) a professional discipline. This research has, indeed, succeeded to the extent that compiler construction is often nominated as one of the most mature areas within the current software engineering scene (see e.g. [Sha90]).

The current practice of producing production compilers, however, does not fully utilize the advanced principles developed. Commercial compilers are usually implemented in a general-purpose algorithmic language following (at least in the front-end) a recursive descent style. This procedural approach is rather low-level with respect to the application domain, a fact that makes hand-written compilers hard to create and maintain. A more scientific and application-oriented approach is followed in formalisms designed especially for programming language specification and automatic generation of compiler components from such specifications. Especially attribute grammars [Knu68], providing both syntactic and semantic facilities, and associated compiler writing systems (see e.g. [DJL88]) have been promoted as a higher-level alternative to the algorithmic compiler writing methodology.

Despite their evident conceptual merits and a number of textbooks (e.g., [WaG84], [ASU86]) introducing the idea, attribute grammars and systems based on them have not reached a common popularity among language implementors. One of the main reasons to this unfortunate situation has certainly been the fact that the developed metalanguages, while being strong in their formal properties, are quite primitive with respect to paradigmatic concepts available in modern programming languages. In other words, the research on attribute grammars has concentrated mainly on the theoretical issues of the formalism and neglected the development of metalanguages as practical software tools. It is not until recently that also the metalanguages have been considered an important subject of conceptual evolution. One of the current directions in developing attribute-grammar-based metalanguages is to introduce into them mechanisms from well-established programming paradigms. An overview of the state of research in this area is given in [Paa91a].

We present PROFIT (PROlog dialect For Implementing Translators), a system for compiler writing. The main objective of PROFIT is to provide a simple notion of *logic attribute grammars*, a paradigm founded on the integration of concepts from logic programming with concepts from attribute grammars. In the following, we address the approach by first discussing its methodological motivation. Thereafter, PROFIT is described by concentrating on its two centerpieces: a deterministic error-recovering form of Prolog's definite clause grammars (DCGs) modelling the context-free grammar component of attribute grammars, and functional terms modelling the attribute component. We also briefly present the implementation of PROFIT and finally discuss its applicability.

2. Foundations

The logic programming paradigm has been promoted as providing a relevant framework for compiler writing (see e.g. [War80], [CoH87]). Indeed, some logic programming characteristics, such as DCGs [PeW80] for parsing and unification for tree pattern matching, are powerful and expressive in this application area. The practical merits of logic programming in compiler writing have not been studied in-the-large. Usually the techniques are demonstrated with trivial toy examples, and only a few complete experiments have been reported. These include the compilers for Estelle [Mon89] and for Edison [Paa91b], both written in Prolog.

Our experiment in compiler writing [Paa91b] showed that the general-purpose facilities of Prolog and their backtracking implementation are not good enough in practice. In this experiment we produced a compiler for Edison [Bri82] using three different approaches: a procedural approach (using Berkeley Pascal [JGH83] as the implementation language), an attribute-grammar-based approach (using GAG [KHZ82] and PGS [KIM89]), and a logic approach (using C-Prolog [Per84] and Quintus Prolog [Qui86]). The main methodological problems with the Prologian version were found to be the purely syntactic nature of unification (imposing an unnatural evaluation of semantic values) and the restricted facilities for symbol processing. On the performance side the conventional implementation of DCGs made the compiler absolutely too inefficient for real use.

On the other hand, the Edison experiment showed the potential of the dual declarative-operational nature of Prolog. The conclusion was that we can well develop a compiler writing tool with a flexible multi-paradigm metalanguage by integrating into the logic programming basis some useful concepts from procedural programming and from attribute grammars.

This view has been discussed in [Paa89] where the general guidelines behind PROFIT are given. Since then, we have proceeded on extending Prolog in particular with two features we felt the most important ones when advancing Prolog towards a compiler writing language, namely with practical DCGs and with functional facilities. By these we can solve a compilational problem in a style that is both declarative and procedural as a mixture of attribute grammars and logic programming (Prolog).

3. Practical DCGs

Since the seminal paper [PeW80], definite clause grammars have been a widely used logic programming formalism for expressing the syntactic structure of languages. When considering DCG rules as "productions", DCGs can be seen as a logic counterpart to context-free grammars. From a broader perspective, when considering the predicate arguments as "attributes" and the embedded procedure calls as "semantic rules", DCGs can also be considered as attribute grammars. Hence, DCGs are a most natural choice for implementing the parser-driven phase of a compiler written in Prolog.

Besides their relation with context-free and attribute grammars, a major reason to the success of DCGs is that they can be translated into Prolog in a straightforward way. That is why a number of Prolog implementations provide a DCG preprocessor as an acquisition. The conventional implementation scheme, however, introduces severe problems when considering the use of DCGs in practical compiler writing:

(1) the DCG preprocessor generates backtracking parsers,

(2) the preprocessor does not check the properties of the underlying
     context-free grammar (such as left recursion or circularity),

(3) no checking of syntax errors is provided by the parser, and

(4) scanning is not interleaved with parsing.

In our Edison experiment we circumvented some of these shortcomings explicitly, for instance by using a most artificial syntactic definition of Edison. Still, the backtracking nature of the parser generated from the DCG imposed an exponential worst-case time complexity on the Edison compiler.

Because the DCG *notation* is, besides well-known, also quite compact and elegant we decided to retain it in PROFIT and only change the *implementation*. We considered the most serious drawback of the conventional DCG implementation be its total lack of the syntactic error concept. That is why our DCG facility is based on an error-recovering implementation that follows the tradition of parser generators built around attribute grammars. A DCG generates a deterministic top-down parser, with built-in syntactic error recovery based on the combination of *panic mode* and *phrase level* methods, as described e.g. in [WeM80]. The context-free grammar underlying a PROFIT DCG must be of class LL(1). Hence, our implementation concentrates on the problems (1) - (3) mentioned above.

The DCG facility is provided in PROFIT in three alternative forms. First, the conventional implementation can still be utilized. Second, a top-down recursive descent parser can be generated (corresponding to an "LL(1) parser program"), as presented in [Paa90b]. And third, a DCG metainterpreter can be used (corresponding to a "table-driven LL(1) parser"), as presented in [PaT90]. For making it easier to transform an underlying context-free grammar into an LL(1) form, the system also provides grammar transformations for removing left recursion, for left-factoring, and for eliminating useless productions.

Our alternative DCG implementations produce parsers that always succeed. When recognizing a syntax error in the input, the parser gives an error message, finds a suitable synchronization point in the remaining stream, and continues in normal mode until the end of input is reached.

4. Functional facilities

The logic programming paradigm is based on using relations to express computation. This works well within certain type of applications, but in most cases part of the solution must be obtained by algorithmic means, using expressions or (more generally) *functions*. That is why several studies have been carried out on methods for integrating functions into logic programming (for an overview, see e.g. [DeL86]).

Although functions as a mathematical concept are a special case of relations (and thus a logic language in a sense already has a functional language as subset), the explicit separation of functions from relations has strong pragmatic motivations. A significant number of Prolog programs tend to be algorithmic [Dra87], i.e., they express deterministic computations from ground input values to ground output values. Thus a functional style is more suitable for such programs than a relational one. Using a functional notation for the algorithmic parts of a program makes it more readable and concise. Moreover, functions are more efficient to implement than relations, thanks to the deterministic nature.

The most typical functional feature needed in many practical situations is arithmetic. Prolog supports arithmetic computations with the built-in evaluator predicate *is* and with a number of machine oriented arithmetic operations. This solution, however, is rather clumsy since it makes it necessary to use explicit calls for *is*, as well as temporary variables holding the subvalues of computations. Methodologically the solution is unsatisfactory because it introduces a deviation from the logical basis of Prolog. These disadvantages apply to other kinds of algorithmic predicates as well.

The compiler writing methodology is largely founded upon algorithms. This fact is reflected also in attribute grammars: therein, the semantic aspects of a language are expressed with *semantic functions*. Therefore, the relation-based logic programming languages fall short in this application area, unless they provide some form of functional facilities as well.

As an example of a typical compilation situation, we give a fragment of code generation in Prolog, extracted from the compiler given in [War80]:

```
encodesubexpr(expr(Op,Expr1,Expr2),N,D,
    (Expr2code;instr(store,Addr);Expr1code;instr(Opcode,Addr))) :-
    complex(Expr2),
    lookup(N,D,Addr),
    encodesubexpr(Expr2,N,D,Expr2code),
    N1 is N+1,
    encodesubexpr(Expr1,N1,D,Expr1code),
    memoryop(Op,Opcode).
```

This piece of the compiler generates code for expressions having the abstract form expr(Op,Expr1,Expr2).

The original notation is unnecessarily verbose and unaesthetical. It can be written in a cleaner form using functional terms:

```
encodesubexpr(expr(Op,Expr1,Expr2),N,D,
     Expr2code;instr(store,lookup(N,D));Expr1code;
     instr(memoryop(Op),lookup(N,D)))) :-
  complex(Expr2),
  encodesubexpr(Expr2,N,D,Expr2code),
  encodesubexpr(Expr1,N+1,D,Expr1code).
```

We have marked the functional (sub)terms, i.e., calls to the functions lookup, memoryop, and +. In essence, the terms underlined represent synthesized information, and the term **in boldface** represents inherited information. In the attribute grammar terminology, "encodesubexpr is a nonterminal with four attributes, the first three (source code, temporary name, symbol table) being inherited and the last one (target code) being synthesized". The argument terms of the encodesubexpr occurrences thus represent attribute values of the corresponding nonterminal occurrences, such that those values that are not direct copies are given explicitly as functional terms.

We have based the design of PROFIT's functional features on our Edison experiment. An analysis of the Edison compiler showed that functional expressions were needed essentially just on the term level, as predicate arguments. When used in the body of a clause, such a functional term is expected to behave like an *inherited attribute*: its value is needed when executing the associated predicate, and thus the term shall be available when entering the predicate. When used as an argument of the head, a functional term corresponds to a *synthesized attribute*: its value is not needed until after returning from the associated predicate, and it is typically evaluated in the predicate's body. Such dual expressions (inherited, synthesized) are most common in compilers, and that is why they should be provided in some form in a compiler writing system.

The provision of inherited and synthesized attribute values in a simple form is the fundamental principle behind the amalgamation of functions into PROFIT. The nature of the functional facilities is more precisely as follows:

1. In addition to conventional constructor terms of Prolog, PROFIT provides *functional terms* having the syntactic form

$$@f(a_1,...,a_n), n >= 0,$$

where f is a *function* with arity n, and the a_is are the *arguments* of the functional term. An argument may be any constructor or functional term. A term is considered functional also if any of its arguments is functional.

2. Let a goal g in the body of a clause for p have a functional term $@f(a_1,...,a_n)$ as argument:

$$p(...) :- ..., g(...,@f(a_1,...,a_n), ...), ...$$

Now $@f(a_1,...,a_n)$ is called an *inherited functional term*. When g becomes the current goal in the execution of p, $@f(a_1,...,a_n)$ is evaluated by calling the function f with arguments $a_1, ..., a_n$ and replaced with the value returned by f *before* unifying goal g with a clause. This applies also in case the inherited functional term is of the form $t(..., @f(...), ...)$.

3. Let the head of a clause for g have a functional term as its argument:

$$g(..., @f(a_1,...,a_n), ...) :- g_1(...),...,g_m(...).$$

Now $@f(a_1,...,a_n)$ is called a *synthesized functional term*. When the head is unified with the current goal $g(...,b,...)$ where b is the term in the position corresponding to $@f(a_1,...,a_n)$, the unification of b and $@f(a_1,...,a_n)$ will immediately succeed, but b and $@f(a_1,...,a_n)$ are not actually bound until *after* having executed the body $g_1(...),...,g_m(...)$. Then b will be unified with the evaluated value of $@f(a_1,...,a_n)$, obtained by calling the function f with arguments $a_1,...,a_n$. When calling f, the arguments have their current values according to the execution of $g_1(..),...,g_m(..)$. This applies also in case the synthesized functional term is of the form $t(...,@f(...),...)$.

4. The functions are deterministic in the sense that they must never fail in returning a value. In case the function is unable to return a proper value, an error message is given, and execution is aborted. Note that by definition the actual value of a synthesized functional term must always unify with the corresponding term in the current goal, otherwise execution will be aborted.

We do not try to remove the deficiences of Prolog, such as incompleteness or incorrectness. We feel that the shortcomings of Prolog are well-known and well-studied to the extent that they can be avoided and even utilized, if considered necessary. An important aspect of using PROFIT is to understand the operational semantics of the additional features it introduces, and to apply them in a suitable manner.

In our setting functions are deterministic objects that always return a single output value (or an error message), computed from the input arguments. An important requirement is that of *information flow discipline*: when executing a function, information may not flow back to the caller of the function through the arguments. In other words, a function may not update its arguments. A suitable equational execution mechanism for such behaviour is reduction (i.e. [EmY87]) where two terms t_1 and t_2 in an equation $t_1 = t_2$ are matched by term rewriting such that no substitutions are introduced in the subterms of t_1. Unlike in most related languages based on reduction (e.g. Funlog [SuY86]), we do not require that t_1 (i.e., an input argument) is a ground term. This requirement is, in our opinion, too restricting since the use of incomplete arguments is in many cases useful for producing incomplete results to be completed in a later phase of the execution (see Chapter 6).

Since we want to make predicates available in function declarations, the functions are realized in PROFIT as conditional equations of the form

$$F := T :- G_1, ..., G_n.$$

where F is the function heading, T the output value (term), and the G_is are guards (predicates). When executed, this function will return the value T, provided that all the guards succeed. Several declarations can be given for F.

A conventional requirement in term rewriting systems is that the headings of two separate declarations shall not match. Since we are not aiming at completeness, we have applied a more pragmatic variant of the restriction: in case the headings of several function declarations match, the first one whose guards succeed is selected. Another pragmatic relaxation to conventional practices is that T may contain free logical variables not contained in F. This solution is sensible since in our concept of conditional equations a term rewriting system is extended with a predicate logic system, in the form of predicates G_i. Now the free logical variables of T can be provided by the external logic system. Since our implementation scheme for functions is translation into relations (see Chapter 5), the *constructor discipline* [ODo85] is applied: the heading F cannot contain functional terms.

As mentioned, functional terms are given using the notation @f(...) which makes them even syntactically different from the constructors; @ is called the *evaluation operator*. This also is a pragmatic choice since otherwise the syntactical similarity of constructor and functional terms would make it harder to identify the functional extensions made to a conventional Prolog program. Moreover, this makes overloading of functors possible. Some frequently used functional operators are predefined, such as @+, @-, @/, and @*. In order to make the notation of nested functional terms more compact, a shorthand is provided: $@(f_1(f_2(...f_n(...))))$ binds the evaluation operator implicitly down to the level i such that $f_1, f_2, ..., f_i$ have been declared as functions, but f_{i+1} has not; thus the functors $f_{i+1}, ..., f_n$ are considered as constructors. For instance 1@+2@+3@*4 can be written as @(1+2+3*4).

As an example, we give in PROFIT an attribute grammar taken from [ASU86], constructing an abstract syntax tree for an arithmetic expression. We employ both the DCG notation and the functional facilities. Note that the underlying context-free grammar must be LL(1).

```
exp(T2)       --> term(T1) , exp2(T1,T2).
exp2(T1,T3) --> ['+'] , term(T2) , exp2(@mknode('+',T1,T2),T3).
exp2(T1,T3) --> ['-'] , term(T2) , exp2(@mknode('-',T1,T2),T3).
exp2(T,T)     --> [].
term(T)       --> ['('] , exp(T) , [')'].
term(@mkleaf(num,N)) --> [num(N)].

mknode(Op,T1,T2) := plus(T1,T2)  :- Op=='+'.
mknode(Op,T1,T2) := minus(T1,T2) :- Op=='-'.
mkleaf(num,V)     := number(V).
```

Due to the way functional terms have been used, the single argument of both exp and term can be regarded as a synthesized attribute, the first argument of exp2 as an inherited attribute, and the second argument of exp2 as a synthesized attribute.

5. Implementation

Since we have decided to integrate compiler writing facilities into the logic programming basis, the programming language PROFIT is a direct extension of Prolog. Accordingly, the implementation scheme of PROFIT is translation into ordinary Quintus Prolog [Qui86] where the system is running. Hence, all the Prolog mechanisms can still be used, and the special PROFIT features are smoothly included into a logic program by a preprocessor.

The translation of the DCG facility into Prolog follows a conventional deterministic, error-recovering, top-down, recursive descent implementation of LL(1) grammars. The context-free grammar giving rise to the recursive descent parser is obtained from the DCG by considering each predicate symbol in it as a nonterminal symbol and each self-standing list as a terminal symbol (an empty list representing the empty string), and by stripping off the embedded procedure calls. The production to be applied is selected on the basis of the lookahead symbol, and a synchronization action between the parser and the input is made both at entry and at exit of a nonterminal. For details, we refer to [Paa90b].

The translation of the functions of PROFIT into Prolog [Nyk90] is done according to the principles given e.g. in [EmY87]: a function $f(a_1, ..., a_n)$ is expressed as a predicate $f(a_1, ..., a_n, r)$ where r is the result of the function. This translation retains the semantics of functions since constructor discipline is followed. Information flow discipline of functions is checked at runtime of the generated Prolog program. This is done by introducing a call for predicate *freeze* (see e.g. [StS86]) for each argument of the function before calling the corresponding predicate. This guarantees that the predicate cannot update the function arguments (*freeze* transforms a term into a ground form). After executing the function, the arguments are returned into their original form using the inverse predicate, *melt*. Each function declaration is translated into a Prolog clause. If the result value is given as a functional term, its evaluation is resolved by inserting the corresponding predicate calls after the guard part of the generated clause, otherwise the result is inserted directly as the result argument of the predicate. In addition to clauses that correspond to function declarations, default clauses are generated for error reporting. Explicit cuts are inserted at the end of the clauses, in order to make them deterministic in the style of functions.

An inherited functional term in a call *C* is translated into Prolog by substituting the term in *C* with a new variable *V*, and by inserting a call to the corresponding predicate immediately before *C*, with variable *V* as the last argument in the call. If the inserted predicate call still contains functional terms, the same is repeated recursively; consecutive and nested inherited functional terms are evaluated left-to-right, bottom-up.

A synthesized functional term is translated by substituting the term in a clause's head with a new variable, and by inserting the call(s) for the corresponding predicate(s) after the body of the clause (left-to-right, bottom-up). This strategy achieves the delayed synthesized behaviour of the terms.

Lazy evaluation is a traditional scheme for the evaluation of functional expressions such that an expression is not evaluated until its value is actually needed. This form of evaluation is useful in particular in connection with infinite data structures and streamized programming. Since the concept of lazy reduction fits well in our implementation strategy, PROFIT provides *lazy functions* in addition to the normal, "eager" functions described so far.

The principle of lazy evaluation in PROFIT is akin to the method presented in [Nar85]: a (typically recursive) function is evaluated in a lazy fashion by transforming the function into a data structure, and by evaluating this data structure using a special *reduce* predicate. Now the number of evaluations is controlled implicitly by Prolog's unification mechanism, and the evaluation will be stopped as soon as a proper value has been generated.

For instance, the Fibonacci numbers can be defined in PROFIT as follows:

```
fibonacci(N) := 1 :- N =< 1.
fibonacci(N) := @(fibonacci(N-1) + fibonacci(N-2)) :- N > 1.
```

and a lazy list of Fibonacci numbers as follows:

```
lazy fiblist(N) := [@fibonacci(N) | @fiblist(N+1)].
fcall(X,@X).
```

Now the query

```
?- fcall(fiblist(0),X).
```

produces a potentially infinite list of numbers from Fibonacci(0):

```
X = [1 | fiblist(1)] ;
X = [1,1 | fiblist(2)] ;
...
X = [1,1,2,3,5,8,13,21,34,55,89 | fiblist(11)] ;
...
```

6. Discussion

Because PROFIT is a proper superset of Prolog, the logical part of the language has exactly the same properties as Prolog. All the known troublespots of Prolog are thus included in PROFIT as well, such as incompleteness, negation as failure, cuts, etc. Moreover, since the implementation strategy is translation into Prolog, even the functional part inherits a number of these semantically problematic features and also loses potential efficiency of functional determinism. Hence, in efficiency PROFIT is not stronger than Prolog.

The main contribution of PROFIT is the conceptual level it provides. By using PROFIT as the implementation language, a number of algorithmic aspects can be solved in a natural way and embedded flexibly within a logic (Prologian) frame. Inherited and synthesized functional terms, as well as lazy and eager functions are conceptually powerful tools for expressing deterministic computations and infinite data structures in a controlled fashion. The compile-time and runtime error messages make the functional part of PROFIT more protected against simple errors than the corresponding languages are.

Since the model for PROFIT's functional terms has been taken from attribute grammars, the language is most suitable for expressing compilation-oriented computations, especially when using the DCG notation. Because of the deterministic top-down, left-to-right execution mechanism underlying, PROFIT's DCGs are close to one-pass attribute grammars, especially L-attributed grammars ([LRS74], [Boc76]). In notation PROFIT comes close to extended attribute grammars [WaM83], a mixture of attribute and affix grammars [Kos71]. There are some differences, however, between PROFIT and conventional one-pass attribute grammars. PROFIT is rather liberal with respect to attributes because of its general purpose objectives: an argument position of a predicate p can be represented as an inherited functional term in one occurrence of p, and as a synthesized functional term in another occurrence of p.

Other generalizations to the conventional (strict) attribute grammar notion are provided by logical variables and lazy evaluation. Using these, one can express and implement (temporarily) incomplete and infinite data structures, and even circular attribution rules. These facilities raise the expressive power of PROFIT beyond one-pass attribute grammars, even though the actual execution would flow in principle as a one-pass process. One can characterize that PROFIT includes conventional one-pass attribute grammars as a subset since they can be mapped directly to a deterministic PROFIT program using functional terms. More exactly, thanks to logical variables associated with the attribute grammar formalism, PROFIT is a suitable language for expressing *logical one-pass attribute grammars* [Paa90a], a superset of L-attributed grammars.

In a logical one-pass grammar, one can make use of a restricted form of right-to-left attribute dependencies within a one-pass evaluation scheme. The formalism is based on employing logical variables to achieve delayed evaluation of attribute instances, and it is most useful in compactly expressing "passive" data values (such as target code of a compiler). When the underlying context-free grammar is LL(1) (as in PROFIT's DCGs), logical one-pass evaluation can be carried out completely during parsing. The definition [Paa90a] gives an exact semantics of logical one-pass grammars and of those PROFIT programs that correspond to such grammars.

PROFIT makes it possible to write a compiler in a more natural and readable way than by doing it in pure Prolog. An analysis of the Edison compiler written in Prolog [Paa91b] shows that a significant number of inherently algorithmic primitives had been expressed in an unnatural way using predicates: about 12% of the predicates in the compiler program would have been more natural given as functions. About 20% of the clauses contained functional arguments in the body, and about 8% in the head. When applying functional facilities, the total length of the compiler decreased from 2400 lines to 2100 lines. (It can be noted that the Pascal version of the compiler contains 4800 lines and the attribute grammar specification 4500 lines). Similar results were obtained when rewriting a tutorial compiler [War80] in a mixed relational/functional style [Pöy89]; we believe that such figures would be quite typical in conventional compilers written in Prolog.

Our multi-pass Edison compiler could be transformed into a one-pass compiler, obtained by expressing it as a logical one-pass attribute grammar in PROFIT. In that case the whole compiler would essentially be a DCG with functional terms as arguments, corresponding to a syntax-directed translation scheme. The conceptual power of logical one-pass attribute grammars can be demonstrated by showing how forwarded procedure calls of Edison can be compiled in PROFIT, as "2" in:

```
pre proc p                      "1"
proc q begin  p end             "2"
post proc p begin ... end       "3"
```

The DCG rule for procedure calls would have the following pattern:

```
proc_call(Env, [ACode,instance(...),proccall(@find(N,Env))]) -->
    [id(N)], arg_part(ACode).
...
find(Name,Env) := Address :-    Env = [proc(Name,Address)|_].
find(Name,Env) := @find(Name,R)     :- Env = [_|R].
find(Name,Env) := 0                 :- error(...).
```

For simplicity, the symbol table Env is here represented as a list, an entry for a procedure being a pair proc(name, address of code). At point "1", the pair proc(p,A) is inserted in the symbol table, the uninstantiated variable A representing the code address of p. (Note that it is unknown at the moment.) At point "2" the call for p generates the incomplete instruction proccall(A). Finally, at point "3" the code address of p (say 100) becomes known, and the symbol table entry for p is completed into proc(p,100). Simultaneously, the unification mechanism used for attribute evaluation *implicitly* backpatches the incomplete call instruction into proccall(100). Hence, explicit backpatching of target code can be avoided, and the DCG-attribute-grammar retains a simple form. Note that such completion of the undefined code address would necessarily require an additional explicit backpatching operation within a conventional (strict) one-pass evaluation scheme. (Note, however, that this solution, while being conceptually elegant, is rather space-consuming since it requires the target code to be kept in memory.)

PROFIT's multi-paradigm features can be employed in compilation by solving problems of different nature with different techniques. DCGs and the functional facilities can be used to express those aspects that are inherently deterministic in practice, such as parsing and symbol table operations. Logical variables can be utilized in a logical one-pass fashion to represent temporarily incomplete values, such as the target code. Finally, the normal backtracking predicates of Prolog can be used e.g.

in code optimization to find optimal target sequences using backtracking. We feel that this selection of various computation mechanisms within a common frame is the most important practical merit of the approach of integrating logic programming and attribute grammars.

At the moment, the PROFIT system is incomplete in many respects. The topics for further development include the provision of general purpose standard predicates and functions (especially for symbol table management), and the integration of scanning and parsing in the DCG facility. Also, PROFIT could be developed more towards a compiler writing system based mainly on the concept of logical one-pass attribute grammars. (Currently, PROFIT can be characterized as merely an implementation language of such grammars since their properties are not analyzed.)

PROFIT shares some ideas with other systems and languages. A number of languages, such as EqLog [GoM86] and Funlog [SuY86], integrate functional and logic primitives. Alternative DCG implementations are proposed e.g. in [MTH83] and in [Nil86]. General perspectives on the use of the logic programming paradigm in compiler writing are discussed e.g. in [Abr87] and in [BrP89].

The conceptual relationship between attribute grammars and logic programming has been studied e.g. in [DeM85] and in [DeM89]. PROFIT applies these formal results in practice by operationally integrating concepts from both attribute grammars and from logic programming into a single language. Other languages/systems that can be considered as a synthesis of the two paradigms are presented e.g. in [Att89], in [Hen89], and in [RiL89].

Acknowledgements. I wish to thank Kari Pöysä and Matti Nykänen for their contribution in the design and implementation of PROFIT. The comments of the anonymous referees have helped in improving the presentation.

References

[Abr87] Abramson H.: Towards an Expert System for Compiler Development. Technical Report 87-33, Department of Computer Science, University of British Columbia, 1987.

[ASU86] Aho A.V., Sethi R., Ullman J.D.: Compilers - Principles, Techniques, and Tools. Addison-Wesley, 1986.

[Att89] Attali I.: Compiling Typol with Attribute Grammars. In: [PLI88], 252-272.

[Boc76] Bochmann G.V.: Semantic Evaluation from Left to Right. Communications of the ACM 19, 2, 1976, 55-62.

[Bri82] Brinch Hansen P.: Programming a Personal Computer. Prentice-Hall, 1982.

[BrP89] Bryant B.R., Pan A.: Rapid Prototyping of Programming Language Semantics Using Prolog. In: Proc. of IEEE COMPSAC '89, Orlando, Florida, 1989, 439-446.

[CoH87] Cohen J. Hickey T.J.: Parsing and Compiling Using Prolog. ACM Transactions on Programming Languages and Systems 9, 2, 1987, 125-163.

[DeL86] DeGroot D., Lindstrom G.: Logic Programming - Functions, Relations, and Equations. Prentice-Hall, 1986.

[DeM85] Deransart P., Maluszynski J.: Relating Logic Programs and Attribute Grammars. Journal of Logic Programming 2, 2, 1985, 119-155.

[DeM89] Deransart P., Maluszynski J.: A Grammatical View of Logic Programming. In: [PLI88], 219-251.

[DJL88] Deransart P., Jourdan M., Lorho B.: Attribute Grammars - Definitions, Systems and Bibliography. Lecture Notes in Computer Science 323, Springer-Verlag, 1988.

[Dra87] Drabent W.: Do Logic Programs Resemble Programs in Conventional Languages? In: Proc. of the 1987 IEEE Symposium on Logic Programming, San Francisco, California, 389-396.

[EmY87] van Emden M.H., Yukawa K.: Logic Programming with Equations. Journal of Logic Programming 4, 4, 1987, 265-288.

[GoM86] Goguen J.A., Meseguer J.: EQLOG: Equations, Types and Generic Modules for Logic Programming. In: [DeL86], 295-363.

[Hen89] Henriques P.R.: A Semantic Evaluator Generating System in Prolog. In: [PLI88], 201-218.

[JGH83] Joy W.N., Graham S.L., Haley C.B., McKusick M.K., Kessler P.B.: Berkeley Pascal User's Manual, Version 3.0. Computer Science Division, Department of Electrical Engineering and Computer Science, University of Berkeley, 1983.

[KHZ82] Kastens U., Hutt B., Zimmermann E.: GAG: A Practical Compiler Generator. Lecture Notes in Computer Science 141, Springer-Verlag, 1982.

[KIM89] Klein E., Martin M.: The Parser Generating System PGS. Software - Practice and Experience 19, 11, 1989, 1015-1028.

[Knu68] Knuth D.E.: Semantics of Context-Free Languages. Math. Systems Theory 2, 1968, 127-145.

[Kos71] Koster C.H.A.: Affix Grammars. In: Algol68 Implementation (J.E.Peck, ed.), North-Holland, 1971, 95-109.

[LRS74] Lewis P.M., Rosenkrantz D.J., Stearns R.E.: Attributed Translations. Journal of Computer and System Sciences 9, 3, 1974, 279-307.

[Mon89] Monin J.-F.: A Compiler Written in Prolog: The Véda Experience. In: [PLI88], 119-131.

[MTH83] Matsumoto Y., Tanaka H., Hirakawa H., Miyoshi H., Yasukawa H.: BUP: A Bottom-Up Parser Embedded in Prolog. New Generation Computing 1, 1983, 145-158.

[Nar85] Narain A.: A Technique for Doing Lazy Evaluation in Logic. In: Proc. of the 1985 IEEE Symposium on Logic Programming, Boston, Massachusetts, 261-269.

[Nil86] Nilsson U.: Alternative Implementation of DCGs. New Generation Computing 4, 2, 1986, 383-399.

[Nyk90] Nykänen M.: Implementing Functional Prolog Terms (in Finnish). Report C-1990-17, Department of Computer Science, University of Helsinki, 1990.

[ODo85] O'Donnell M.J.: Equational Logic as a Programming Language. The MIT Press, 1985.

[Paa89] Paakki J.: A Prolog-Based Compiler Writing Tool. In: Proc. of the Workshop on Compiler Compiler and High Speed Compilation, Berlin (GDR), 1988. Report 3/1989, Academy of the Sciences of the GDR, 1989, 107-117.

[Paa90a] Paakki J.: A Logic-Based Modification of Attribute Grammars for Practical Compiler Writing. In: Proc. of the 7th Int. Conference on Logic Programming (D.H.D.Warren, P.Szeredi, eds.), Jerusalem, 1990. The MIT Press, 1990, 203-217.

[Paa90b] Paakki J.: A Practical Implementation of DCGs. In: Proc. of the Third Int. Workshop on Compiler Compilers, Schwerin, 1990. Report 8/1990, Akademie der Wissenschaften der DDR, 1990, 249-257.

[Paa91a] Paakki J.: Paradigms for Attribute-Grammar-Based Language Implementation. Report A-1991-1, Department of Computer Science, University of Helsinki, 1991.

[Paa91b] Paakki J.: Prolog in Practical Compiler Writing. The Computer Journal 34, 1, 1991, 64-72.

[PaT90] Paakki J., Toppola K.: An Error-Recovering Form of DCGs. Acta Cybernetica 9, 3, 1990, 211-221.

[Per84] Pereira F. (ed.): C-Prolog User's Manual, Version 1.5. EdCAAD, Department of Architecture, University of Edinburgh, 1984.

[PeW80] Pereira F.C.N., Warren D.H.D.: Definite Clause Grammars for Language Analysis - A Survey of the Formalism and a Comparison with Augmented Transition Networks. Artificial Intelligence 13, 1980, 231-278.

[PLI88] Proc. of the Int. Workshop on Programming Languages Implementation and Logic Programming (PLILP '88), Orléans, 1988. Lecture Notes in Computer Science 348 (P.Deransart, B.Lorho, J.Maluszynski, eds.), Springer-Verlag, 1989.

[Pöy89] Pöysä K.: Extending Prolog with Functional Features (in Finnish). Report C-1989-71, Department of Computer Science, University of Helsinki, 1989.

[Qui86] Quintus Computer Systems Inc.: Quintus Prolog Reference Manual, Version 6, 1986.

[RiL89] Riedewald G., Lämmel U.: Using an Attribute Grammar as a Logic Program. In: [PLI88], 161-179.

[Sha90] Shaw M.: Prospects for an Engineering Discipline of Software. IEEE Software, Nov. 1990, 15-24.

[StS86] Sterling L., Shapiro E.: The Art of Prolog. The MIT Press, 1986.

[SuY86] Subrahmanyam P.A., You J.-H.: FUNLOG: A Computational Model Integrating Logic Programming and Functional Programming. In: [DeL86], 157-198.

[WaG84] Waite W.M., Goos G.: Compiler Construction. Springer-Verlag, 1984.

[WaM83] Watt D.A., Madsen O.L.: Extended Attribute Grammars. The Computer Journal 26, 2, 1983, 142-153.

[War80] Warren D.H.D.: Logic Programming and Compiler Writing. Software - Practice and Experience 10, 2, 1980, 97-125.

[WeM80] Welsh J., McKeag M.: Structured System Programming. Prentice-Hall, 1980.

Towards a Meaning of LIFE

Hassan Aït-Kaci*
hak@prl.dec.com

Andreas Podelski†
anp@litp.ibp.fr

Abstract

LIFE (Logic, Inheritance, Functions, Equations) is an experimental programming language proposing to integrate three orthogonal programming paradigms proven useful for symbolic computation. From the programmer's standpoint, it may be perceived as a language taking after logic programming, functional programming, and object-oriented programming. ¿From a formal perspective, it may be seen as an instance (or rather, a composition of three instances) of a Constraint Logic Programming scheme due to Hvhfeld and Smolka refining that of Jaffar and Lassez.

We start with an informal overview demonstrating LIFE as a programming language, illustrating how its primitives offer rather unusual, and perhaps (pleasantly) startling, conveniences. The second part is a formal account of LIFE's object unification seen as constraint-solving over specific domains. We build on work by Smolka and Rounds to develop type-theoretic, logical, and algebraic renditions of a calculus of order-sorted feature approximations.

> ... the most succint and poetic definition: 'Créer, c'est unir' ("To create is to unify"). This is a principle that must have been at work from the very beginning of life.
>
> KONRAD LORENZ, *Die Rückseite des Spiegels.*

1 Introduction

As a denominating substantive, 'LIFE' means Logic, Inheritance, Functions, and Equations. This appellation also designates an experimental programming language designed after these four precepts for specifying data structures and computations. As for what LIFE means as a programming language, it is the purpose of this document to present the foundation for a declarative semantics for LIFE as a specific instantiation of a Constraint Logic Programming scheme with a particular constraint language. In fact, LIFE's computation structure can be factorized into a composition of sublogics that are each expressively more convenient and pragmatically more efficient for some limited purposes. We have been able to find a class of interpretations of approximation structures adequate to represent LIFE's objects (LIFE terms), and we describe three syntactic representations thereof: (1) a type-theoretic term language; (2) an algebraic language; and, (3) a logical (clausal) language. All three admit semantics over LIFE's approximation structures. We call these structures Order-Sorted Feature (*OSF*-)interpretations. We show the mutual syntactic and semantic equivalence of the three representations. This allows us to shed some light on, and reconcile, three common interpretations of multiple inheritance as (1) set inclusion; as (2) algebraic endomorphisms; and, (3) as logical implication.

---

*Digital Equipment Corporation, Paris Research Laboratory, 85 avenue Victor Hugo, 92500 Rueil-Malmaison, France.

†L. I. T. P., Universiti de Paris 7, 2 place Jussieu, 75251 Paris Cedex 05, France.

Our approach centers around the notion of an *OSF*-algebra. This notion was already used implicitly in [AK84, AK86] to give a semantics to ψ-terms. Gert Smolka's work on Feature Logic [Smo88, Smo89] made the formalism emerge more explicitly, especially in the form of a "canonical *OSF*-graph algebra," and a recent publication paper uses it in the context of showing an undecidability result for semiunification of cyclic structures was [DR90]. The latter do not consider order-sorted graphs and focus only on features. The former considers both the order-sorted and the unsorted case but does not make explicit the syntactic and semantic mappings between the alge-
.braic, the logical, and the type-theoretic views. However, the logics considered in [Smo88, Smo89] are richer than we consider here, allowing explicit negation and quantification. Naturally, all these extensions can be considered in our framework as will be shown in a forthcoming report.

There are many benefits to seeing LIFE's constraints algebraically, especially if the view is in complete and natural coincidence with logical and implementational views. One nice outcome of this approach is our understanding that sorted and labeled graph-structures used in our implementation of LIFE's approximation structures form a particularly useful *OSF*-algebra which happens to be a canonical interpretation (in the sense of Herbrand) for the satisfiability of *OSF*-clauses. This is important as there is no obvious initiality result, our setting having no values but only approximations. (As a matter of fact, we show that the *OSF*-graph algebra is *final* in the category of *OSF*-algebras with *OSF*-morphisms.) Indeed, approximation chains in *OSF*-algebras can very well be infinitely strictly ascending (getting better and better...); and this is the case of our version presented here—*all* approximation chains are non Noetherian! We do not care, as only "interesting" approximations, in a sense to be made rigorously precise, are of any use in LIFE.

With this generalizing insight, we can give a crisp interpretation of Life's approximation structures as principal filters in *OSF*-interpretations for the information-theoretic approximation ordering ($\sqsubseteq$) derived from the existence of (*OSF*-)endomorphisms. Thereby, they may inherit a wealth of lattice-theoretic results such as that of being closed under *joins* ($\sqcup$), or equivalently, set-intersection ($\cap$) in the type interpretation (Ψ) with the inclusion ordering ($\subseteq$), conjunction (&) in the logical interpretation (Φ) with the implication ordering ($\succeq$), and graph-unification ($\wedge$) in the canonical (graph) interpretation $\mathcal{G}$ with the (graph) approximation ordering.

We have much to say about LIFE. This document is the first of a sequence of several that will give a full account of the complete meaning of LIFE. The purpose of this document is therefore an informal introduction to the most salient features of LIFE and the layout of a semantic basis to serve as a foundation for interpretation structures of LIFE. The second report has already been completed [AKP91b]. There, we detail an operational semantics for functional reduction in LIFE and show it in complete concordance with the three abstract semantics of *OSF*-constraints developed here. In particular, one will find there the treatment of functions as approximation structures with an operational meaning, and the declarative accounts of (1) suspension/resumption of functional reduction based on mixing unification and matching [AKP91b], as well as that of (2) automatic deamonic constraint triggering. We are pleased by these results as they amount to giving declarative specifications to a behavior of programs where explicit control annotations (*e.g.*, "wait until something happens") seem hard to avoid. Nevertheless, we show in [AKP91b] that processes sharing information can communicate by refining or expecting information to specific approximation thresholds. In particular, this can be given correct and automatic operational semantics by ways of non-deterministic constraint normalization rules which stall on insufficient information.

This article is organized as follows. We first make an informal show of LIFE's basic features. We hope by this to convince the reader that a unique functionality of programming is available to a LIFE user. We do this by way of small yet (pleasantly) startling examples. Following that, we give the essentials of the three formal *OSF*-constraint systems and their interpretations.

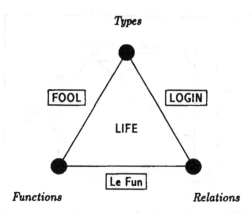

Types

FOOL   LOGIN

LIFE

Functions   Le Fun   *Relations*

Figure 1: The LIFE molecule

2 LIFE, Informally

LIFE is a trinity. The function-oriented component of LIFE is directly derived from functional programming languages with higher-order functions as first-class objects, data constructors, and algebraic pattern-matching for parameter-passing. The convenience offered by this style of programming is one in which expressions of any order are first-class objects and computation is determinate. The relation-oriented component of LIFE is essentially one inspired by the Prolog language [CM84, O'K90]. Unification of first-order patterns used as the argument-passing operation turns out to be the key of a quite unique and hitherto unusual *generative* behavior of programs, which can construct missing information as needed to accommodate success. Finally, the most original part of LIFE is the structure-oriented component which consists of a calculus of type structures—the ψ-calculus [AK84, AK86]—and accounts for some of the (multiple) inheritance convenience typically found in so-called object-oriented languages.

An algebra of term structures adequate for the representation and formalization of frame-like objects is given a clear notion of subsumption interpretable as a subtype ordering, together with an efficient unification operation interpretable as type intersection. Disjunctive structures are accommodated as well, providing a rich and clean pattern calculus for both functional and logic programming.

Under these considerations, a natural coming to LIFE has consisted thus in first studying pairwise combinations of each of these three operational tools. Metaphorically, this means realizing edges of a triangle (see Figure 1) where each vertex is some essential operational rendition of the appropriate calculus. LOGIN [AKN86] is simply Prolog where first-order constructor terms have been replaced by ψ-terms, with type definitions [AKN86]. Its operational semantics is the immediate adaptation of that of Prolog's SLD resolution. Le Fun [AKNL87, AKN89] is Prolog where unification may reduce functional expressions into constructor form according to functions defined by pattern-oriented functional specifications. Finally, FOOL is simply a pattern-oriented functional language where first-order constructor terms have been replaced by ψ-terms, with type definitions.

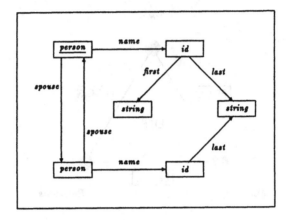

Figure 2: A commutative functional diagram

LIFE is the composition of the three with the additional capability of specifying arbitrary functional and relational constraints on objects being defined.

The next subsection gives a very brief and informal account of the calculus of type inheritance used in LIFE (ψ-calculus). The reader is assumed familiar with functional programming and logic programming.

2.1 ψ-Calculus: computing with types

The ψ-calculus consists of a syntax of structured types called ψ-terms together with subtyping and type intersection operations. Intuitively, as expounded in [AKN86], the ψ-calculus is a convenience for representing record-like data structures in logic and functional programming more adequately than first-order terms do, without loss of the well-appreciated instantiation ordering and unification operation.

Let us take an example to illustrate. Let us say that one has in mind to express syntactically a type structure for a *person* with the property, as expressed for the underlined symbol in Figure 2, that a certain functional diagram commutes.

The syntax of ψ-terms is one simply tailored to express as a term this kind of approximate description. Thus, in the ψ-calculus, the information of Figure 2 is unambiguously encoded into a formula, perspicuously expressed as the ψ-term:

$$X : person(name \Rightarrow id(first \Rightarrow string,$$
$$last \Rightarrow S : string),$$
$$spouse \Rightarrow person(name \Rightarrow id(last \Rightarrow S),$$
$$spouse \Rightarrow X)).$$

It is important to distinguish among the three kinds of symbols which participate in a ψ-term expression. Thus, we assume given a set S of sorts or *type constructor symbols*, a set $\mathcal{F}$ of *features*, or *attributes symbols*, and a set V of *variables* (or *coreference tags*). In the ψ-term above, for example, the symbols *person, id, string* are drawn from S, the symbols *name, first, last, spouse* from $\mathcal{F}$, and the symbols X, S from V. (We capitalize variables, as in Prolog.)

A ψ-term is either *tagged* or *untagged*. A tagged ψ-term is either a variable in V or an expression of the form $X : t$ where $X \in V$ and t is an untagged ψ-term. An untagged ψ-term is either *atomic* or *attributed*. An atomic ψ-term is a type symbol in S. An attributed ψ-term is an expression of the form $s(\ell_1 \Rightarrow t_1, \ldots, \ell_n \Rightarrow t_n)$ where $s \in S$ and is called the ψ-term's *principal* type, the ℓ_i's are mutually distinct attribute symbols in $\mathcal{F}$, and the t_i's are ψ-terms ($n \geq 0$).

Variables capture coreference in a precise sense. They are coreference tags and may be viewed as typed variables where the type expressions are untagged ψ-terms. Hence, as a condition to be *well-formed*, a ψ-term must have all occurrences of each coreference tag consistently refer to the same structure. For example, the variable X in:

$$person(id \Rightarrow name(first \Rightarrow string,$$
$$last \Rightarrow X : string),$$
$$father \Rightarrow person(id \Rightarrow name(last \Rightarrow X : string)))$$

refers consistently to the atomic ψ-term *string*. To simplify matters and avoid redundancy, we shall obey a simple convention of specifying the type of a variable at most once and understand that other occurrences are equally referring to the same structure, as in:

$$person(id \Rightarrow name(first \Rightarrow string,$$
$$last \Rightarrow X : string),$$
$$father \Rightarrow person(id \Rightarrow name(last \Rightarrow X)))$$

In fact, since there may be circular references as in $X : person(spouse \Rightarrow person(spouse \Rightarrow X))$, this convention is necessary. Finally, a variable appearing nowhere typed, as in $junk(kind \Rightarrow X)$ is implicitly typed by a special universal type symbol $\top$ always present in S. This symbol will be left invisible and not written explicitly as in $(age \Rightarrow integer, name \Rightarrow string)$, or written as the symbol ₀. In the sequel, by ψ-term we shall always mean well-formed ψ-term.

Similarly to first-order terms, a subsumption pre-order can be defined on ψ-terms which is an ordering up to variable renaming. Given that the set S is partially-ordered (with a greatest element $\top$), its partial ordering is extended to the set of attributed ψ-terms. Informally, a ψ-term t_1 is subsumed by a ψ-term t_2 if (1) the principal type of t_1 is a subtype in S of the principal type of t_2; (2) all attributes of t_2 are also attributes of t_1 with ψ-terms which subsume their homologues in t_1; and, (3) all coreference constraints binding in t_2 must also be binding in t_1.

For example, if *student* < *person* and *paris* < *cityname* in S then the ψ-term:

$$student(id \Rightarrow name(first \Rightarrow string,$$
$$last \Rightarrow X : string),$$
$$lives\_at \Rightarrow Y : address(city \Rightarrow paris),$$
$$father \Rightarrow person(id \Rightarrow name(last \Rightarrow X),$$
$$lives\_at \Rightarrow Y))$$

is subsumed by the ψ-term:

$$person(id \Rightarrow name(last \Rightarrow X : string),$$
$$lives\_at \Rightarrow address(city \Rightarrow cityname),$$
$$father \Rightarrow person(id \Rightarrow name(last \Rightarrow X))).$$

In fact, if the set S is such that *greatest lower bounds* (GLB's) exist for any pair of type symbols, then the subsumption ordering on ψ-term is also such that GLB's exist. Such are defined as the *unification* of two ψ-terms. A detailed unification algorithm for ψ-terms is given in [AKN86]. Consider for example the poset displayed in Figure 3 and the two ψ-terms:

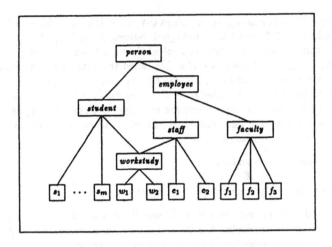

Figure 3: A lower semi-lattice of sorts

$$X : student(advisor \Rightarrow faculty(secretary \Rightarrow Y : staff,$$
$$assistant \Rightarrow X),$$
$$roommate \Rightarrow employee(representative \Rightarrow Y))$$

and:

$$employee(advisor \Rightarrow f_1(secretary \Rightarrow employee,$$
$$assistant \Rightarrow U : person),$$
$$roommate \Rightarrow V : student(representative \Rightarrow V),$$
$$helper \Rightarrow w_1(spouse \Rightarrow U)).$$

Their unification (up to tag renaming) yields the term:

$$W : workstudy(advisor \Rightarrow f_1(secretary \Rightarrow Z : workstudy(representative \Rightarrow Z),$$
$$assistant \Rightarrow W),$$
$$roommate \Rightarrow Z,$$
$$helper \Rightarrow w_1(spouse \Rightarrow W)).$$

Last in this brief introduction to the ψ-calculus, we explain type definitions. The concept is analogous to what a global store of constant definitions is in a practical functional programming language based on the λ-calculus. The idea is that types in the signature may be specified to have attributes in addition to being partially-ordered. Inheritance of attributes from all supertypes to a subtype is done in accordance to ψ-term subsumption and unification. For example, given a simple signature for the specification of linear lists $S = \{list, cons, nil\}$ with $nil < list$ and $cons < list$, it is yet possible to specify that $cons$ has an attribute $tail \Rightarrow list$. We shall specify this as:

$$list := \{nil; cons(tail \Rightarrow list)\}.$$

¿From which the partial-ordering above is inferred.

As in this $list$ example, such type definitions may be recursive. Then, ψ-unification $modulo$ such a type specification proceeds by unfolding type symbols according to their definitions. This is done by need as no expansion of symbols need be done in case of (1) failures due to order-theoretic clashes (*e.g.*, *cons(tail ⇒ list)* unified with *nil* fails; *i.e.*, gives ⊥); (2) symbol subsumption (*e.g.*, *cons* unified with *list* gives just *cons*), and (3) absence of attribute (*e.g.*, *cons(tail ⇒ cons)* unified with *cons* gives *cons(tail ⇒ cons)*). Thus, attribute inheritance may be done "lazily," saving much unnecessary expansions.

In LIFE, a basic ψ-term denotes a functional application in FOOL's sense if its root symbol is a defined function. Thus, a *functional expression* is either a ψ-term or a conjunction of ψ-terms denoted by $t_1 : t_2 : \ldots : t_n$. An example of such is *append(list, L) : list*, where *append* is the FOOL function defined as:

> $list := \{[]; [@|list]\}.$
>
> $append([], L : list) \Rightarrow L.$
> $append([H|T : list], L : list) \Rightarrow [H|append(T, L)].$

This is how functional dependency constraints are expressed in a ψ-term in LIFE. For example, in LIFE the ψ-term *foo(bar ⇒ X : list, baz ⇒ Y : list, fuz ⇒ append(X, Y) : list)* is one in which the attribute *fuz* is derived as a list-valued function of the attributes *bar* and *baz*. Unifying such ψ-terms proceeds as before modulo suspension of functional expressions whose arguments are not sufficiently refined to be provably subsumed by patterns of function definitions.

As for relational constraints on objects in LIFE, a ψ-term t may be followed by a *such-that* clause consisting of the logical conjunction of literals $\ell_1, \ldots, \ell_n$. It is written as $t \mid \ell_1, \ldots, \ell_n$. Unification of such relationally constrained terms is done modulo proving the conjoined constraints. We will illustrate this very intriguing feature with two examples: prime.life (Section 2.5) and quick.life (Section 2.4). In effect, this allows specifying *daemonic constraints* to be attached to objects. Such a "deamon-constrained" object's specified (renamed) relational and (equational) functional formula is normalized by LIFE, its proof being triggered by unification at the object's creation time.

LIFE's type definitions and deamonic constraints are not treated formally here but in a forthcoming report. There, one will also find correct declarative and operational semantics of LIFE with this convenience, and more.

We give next some LIFE examples.

2.2 Order-sorted logic programming: happy.life

Peter, Paul and Mary are students, and students are persons.

```
student := {peter;paul;mary}.
student <| person.
```

Grades are good grades or bad grades. A and B are good grades, while C, D and F are bad grades.

```
grade := {goodgrade;badgrade}.
goodgrade := {a;b}.
badgrade := {c;d;f}.
```

Goodgrades are good things.

```
goodgrade <| goodthing.
```

Every person likes herself. Every person likes every good thing. Peter likes Mary.

```
likes(X:person,X).
likes(person,goodthing).
likes(peter,mary).
```

Peter got a C, Paul an F and Mary an A.

```
got(peter,c).
got(paul,f).
got(mary,a).
```

A person is happy if s/he got something that s/he likes, or, if s/he likes something that got a good thing.

```
happy(X:person) :- got(X,Y),likes(X,Y).
happy(X:person) :- likes(X,Y),got(Y,goodthing).
```

To the query 'happy(X:student)?' LIFE answers X = mary (twice—see why?), then gives X = peter, then fails. (It helps to draw the sort hierarchy order diagram.)

2.3 Passive constraints: lefun.life

```
p(X, Y, Z) :- q(X, Y, Z, Z), r(X, Y).

q(X, Y, X+Y, X*Y).
q(X, Y, X+Y, (X*Y)-14).

r(3, 5).
r(2, 2).
r(4, 6).
```

Upon a query 'p(X,Y)?' the predicate p selects a pair of expressions in X and Y whose evaluations must unify, and then selects values for X and Y. The first solution selected by predicate q sets up the residual equation (or *residuation*, or *suspension*) that $X + Y = X * Y$ (more precisely that both $X + Y$ and $X * Y$ should unify with Z), which is not satisfied by the first pair of values, but is by the second. The second solution sets up $X + Y = (X * Y) - 14$ which is satisfied by $X = 4, Y = 6$.

The next two examples show the use of higher-order functions such as map:

```
map(@, []) => [].
map(F, [H|T]) => [F(H)|map(F,T)].

inc_list(N:int, L:list, map(+(N),L)).
```

To the query 'inc_list(3,[1,2,3,4],L)?' LIFE answers L = [4,5,6,7].

In passing, note the built-in constant @ as the primeval LIFE object (formally written ⊤) which approximates anything in the universe.

Note that it is possible, since LIFE uses ψ-terms as a universal object structure, to pass arguments to functions by keywords and obtain the power of partial application (currying) in all arguments (as opposed to λ-calculus which requires left-to-right currying). For example of an (argument-selective) currying, consider the (admittedly pathological) LIFE program:

```
curry(V) :- V = G(2=>1), G = F(X), valid(F), pick(X), p(sq(V)).
```

```
sq(X) => X*X.

twice(F,X) => F(F(X)).

valid(twice).

p(1).

id(x) => x.

pick(id).
```

What does LIFE answer when 'curry(V)?' is the query? The relation curry is the property of a variable V when this variable is the result of applying a variable function G to the number 1 as its second argument. But G must also be the value of applying a variable function F to an unknown argument X. The predicate valid binds F to twice, and therefore binds V to twice(X,1). Then, pick binds X to the identity function. Thus, the value of G, twice(X), becomes twice(id) and V becomes now bound to 1, the value of twice(id,1). Finally, it must be verified that the square of V unifies with a value satisfying property p.

2.4 Functional programming with logical variables: quick.life

This is a small LIFE module specifying (and thus, implementing) C.A.R. Hoare's "Quick Sort" algorithm functionally. This version works on lists which are not terminated with ☐ (nil) but with uninstantiated variables (or partially instantiated to a non-minimal list sort). Therefore, LIFE makes difference-lists *bona fide* data structures in functional programming.

```
q_sort(L,order => 0) => undlist(dqsort(L,order => 0)).

undlist(X\Y) => X | Y=☐.

dqsort(☐) => L\L.
dqsort([H|T],order => 0)
   => (L1\L2) : where
                   ((Less,More) : split(H,T,(☐,☐),order => 0),
                    (L1\[H|L3]) : dqsort(Less,order => 0),
                    (L3\L2)     : dqsort(More,order => 0)).
where => ©.

split(©,☐,P) => P.
split(X,[H|T],(Less,More),order => 0) =>
        cond(O(H,X),
             split(X,T,([H|Less],More),order => 0),
             split(X,T,(Less,[H|More]),order => 0)).
```

The function dqsort takes a regular list (and a parameterized comparison boolean function 0) into a difference-list form of its sorted version (using Quick Sort). The function undlist yields a regular form for a difference-list. Finally, notice the definition and use of the (functional) constant where which returns the most permissive approximation (©). It simply evaluates its arguments (*a priori* unconstrained in number and sorts) and throws them away. Here, it is applied to three arguments at (implicit) positions (attributes) 1 (a pair of lists), 2 (a difference-list), and 3 (a difference-list). Unification (denoted ":") takes care of binding the local variables Less, More, L1,

L2, L3, and exporting those needed for the result (L1, L2). The advantage (besides perspicuity and elegance) is performance: replacing **where** with **@** inside the definition of dqsort is correct but keeps around three no-longer needed argument structures at each recursive call.

Here are some specific instantiations:

```
string_sort(L:string_list) => q_sort(L, order => string_less).
```

```
natural_sort(L:int_list) => q_sort(L, order => '<').
```

such that to the query:

```
L = string_sort(["is","This","sorted","lexicographically"])?
```

LIFE answers:

```
L = ["This","is","lexicographically","sorted"].
```

2.5 High-school math specifications: prime.life

A prime number is a positive integer whose number of proper factors is exactly one. This can be expressed in LIFE as:

```
posint := I:int | I>0=true.
```

```
prime := P:posint | number_of_factors(P) = one.
```

where:

```
number_of_factors(N:posint)
   => cond(N<=1,
           undefined,
           factors_from(N,2)).
```

```
factors_from(N:int,P:int)
   => cond(P*P>N,
           one,
           cond(R:(N/P)==int_part(R),
               many,
               factors_from(N,P+1))).
```

```
posint_stream => {1;1+posint_stream}.
```

```
list_all_primes :- write(posint_stream:prime), nl, fail.
```

As for **@**, the dual built-in constant **undefined** is the final LIFE object (formally written ⊥) and is approximated by anything in the universe. Operationally, it just causes failure equivalent to that due to an inconsistent formula. Any object that is not a non-strict functional expression (such as cond) in which undefined occurs will lead to failure (⊥ as an object or the inconsistent clause as a formula). Also, LIFE's functions may contain infinitely disjunctive objects such as streams. For instance, posint_stream is such an object (a 0-ary function constant) whose infinitely many disjuncts are the positive integers enumerated from 1. Or, if a limited stream is preferred:

```
posint_stream_up_to(N:int)
   => cond(N<1,
```

```
          undefined,
          {1;1+posint_stream_up_to(N-1)}).

list_primes_up_to(N:int)
   :- write(posint_stream_up_to(N):prime), nl, fail.
```

3 Formal LIFE

This section makes up the second part of this paper and sets up formal foundations upon which to build a full semantics of LIFE. The gist of what follows is the construction of a logical constraint language for LIFE type structures with the appropriate semantic structures. (All the proofs can be found in [AKP91a].)

3.1 Syntax

We shall consider domains which are coherently described by classifiers (*i.e.*, partially-ordered sorts) and whose elements may be functionally related with one another through attributes (*i.e.*, labels or features). Thus, our specific signatures comprise the symbols for sorts and features and regulate their intended interpretation.

Definition 1 (OSF-Signature) *An order-sorted feature signature (or simply OSF-signature) is a tuple* $\langle S, \leq, \wedge, \mathcal{F} \rangle$ *such that:*

▷ S *is a set of sorts containing the sorts* $\top$ *and* $\bot$;

▷ $\leq$ *is a decidable partial order on* S *such that* $\bot$ *is the least and* $\top$ *is the greatest element;*

▷ $\langle S, \leq, \wedge \rangle$ *is a lower semi-lattice* ($s \wedge s'$ *is called the greatest common subsort of sorts* s *and* s');

▷ $\mathcal{F}$ *is the set of feature symbols.*

In what follows, let $\langle S, \leq, \wedge, \mathcal{F} \rangle$, be some fixed *OSF*-signature. We will also refer to $\mathcal{V}$, a countably infinite set of *variables* that we always denote as capitalized and possibly subscripted letters.

3.1.1 OSF-terms

We now introduce the syntactic objects that we intend to use as type formulae to be interpreted as subsets of the domain of an *OSF*-algebra.

Definition 2 (OSF-Terms) *An order-sorted feature term (or, OSF-term)* ψ *is an expression of the form:*

$$\psi = X : s(\ell_1 \Rightarrow \psi_1, \ldots, \ell_n \Rightarrow \psi_n) \tag{1}$$

where X *is a variable in* $\mathcal{V}$, s *is a sort in* S, $\ell_1, \ldots, \ell_n$ *are features in* $\mathcal{F}$, $n \geq 0$, *and* $\psi_1, \ldots, \psi_n$ *are OSF-terms.*

Note that the equation above includes $n = 0$ as a base case. That is, the simplest *OSF*-terms are of the form $X : s$. We call the variable X in the above *OSF*-term the *root* of ψ (noted $Root(\psi)$), and say that X is "sorted" by the sort s and "has attributes" $\ell_1, \ldots, \ell_n$. The set of variables occurring in ψ is given by $Var(\psi) = \{X\} \cup \bigcup_{j \leq n} Var(\psi_j)$.

Example 1 The following is an example of the syntax of an *OSF*-term.

$$X : person(name \Rightarrow N : \top(first \Rightarrow F : string),$$
$$name \Rightarrow M : id(last \Rightarrow S : string),$$
$$spouse \Rightarrow P : person(name \Rightarrow I : id(last \Rightarrow S : \top),$$
$$spouse \Rightarrow X : \top)).$$

Note that, in general, an *OSF*-term may have redundant attributes (*e.g.*, *name* above), or the same variable sorted by different sorts (*e.g.*, X and S above).

The syntax of *OSF*-term allows some to be in a form where there is apparently ambiguous or even implicitly inconsistent information. For instance, in the *OSF*-term of Example 1, it unclear what the attribute *name* could be. Or, if *string* and *number* are two sorts such that $string \wedge number = \perp$, what the *ssn* attribute is for the *OSF*-term $X : \top(ssn \Rightarrow string, ssn \Rightarrow number)$, and whether indeed such a term's denotation is empty or not. The following notion is useful to this end.

Definition 3 (ψ-term) *A normal OSF-term ψ is of the form $\psi = X : s(\ell_1 \Rightarrow \psi_1, \ldots, \ell_n \Rightarrow \psi_n)$ where:*
  ▷ *there is at most one occurrence of a variable Y in ψ such that Y is the root variable of a non-trivial OSF-term (i.e., different than $Y : \top$);*
  ▷ *s is a non-bottom sort in S;*
  ▷ *$\ell_1, \ldots, \ell_n$ are pairwise distinct features in $\mathcal{F}$, $n \geq 0$,*
  ▷ *$\psi_1, \ldots, \psi_n$ are normal OSF-terms.*
We call Ψ the set that they constitute.

Example 2 One could verify easily that this *OSF*-term is a ψ-term:

$$X : person(name \Rightarrow id(first \Rightarrow string,$$
$$last \Rightarrow S : string),$$
$$spouse \Rightarrow person(name \Rightarrow id(last \Rightarrow S),$$
$$spouse \Rightarrow X))$$

Intuitively (and, as we will see later, also formally) this *OSF*-term always "denotes" exactly the same set as the one of Example 1.

Given an arbitrary *OSF*-term ψ, it is natural to ask whether there exists a ψ-term ψ' with the same "denotation" in every *OSF*-interpretation $\mathcal{A}$. We shall see that the normalization procedure in the next subsection allows either to determine whether an *OSF*-term denotes the empty set or produce an equivalent ψ-term form for it. Moreover, the same procedure is used to compute the GLB $\psi_1 \wedge \psi_2$ (*i.e.*, unification) of two *OSF*-terms ψ_1 and ψ_2. (The subsumption problem is treated in [AKP91b].)

3.1.2 OSF-clauses

An alternative syntactic presentation of the information conveyed by *OSF*-terms can be given using logical means as an *OSF*-term can be translated into a constraint formula bearing the "same" meaning. This is particularly useful as proof-theoretic procedures such as constraint normalization can be devised in the form of semantics preserving simplification rules, and thus allow the effective use of types as constraints formulae in a constraint logic programming context.

Definition 4 (*OSF*-Constraint) *An order-sorted feature constraint (OSF-constraint) is an expression of either of the forms:*
  ▷ $X : s$

▷ $X \doteq Y$

▷ $X.\ell \doteq Y$

where X and Y are variables in V, s is a sort in S, and ℓ is a feature in $\mathcal{F}$. An order-sorted feature clause (OSF-clause) $\phi_1 \& \dots \& \phi_n$ is a finite, possibly empty conjunction of OSF-constraints $\phi_1, \dots, \phi_n$ $(n \geq 0)$.

One may read the three atomic forms of OSF-constraints as, respectively, "X lies in sort s," "X is equal to Y," and "Y is the feature ℓ of X." The set $Var(\phi)$ of (free) variables occurring in an OSF-clause ϕ is defined in the standard way.

Example 3 The following is an example of the syntax of an OSF-clause:

$$
\begin{aligned}
X : person \ \& \ &X.name \ \doteq N \ \& \ N : \top \ &\& \ N.first \ \doteq F \ \& \ F : string \\
\& \ &X.name \ \doteq M \ \& \ M : id \ &\& \ M.last \ \doteq S \ \& \ S : string \\
\& \ &X.spouse \doteq P \ \& \ P : person \ &\& \ P.name \ \doteq I \ \& \ I : id \\
& & \& \ I.last \ \doteq S \ \& \ S : \top \\
& & \& \ P.spouse \doteq X \ \& \ X : \top.
\end{aligned}
$$

Given an OSF-clause ϕ and variables X, Y, Z occurring in ϕ, we say that Y is reachable from X if: $X \equiv Y$; or, $X.l \doteq Z$ is a constraint in ϕ and Y is reachable from Z.

Definition 5 (Rooted OSF-Clauses) A rooted OSF-clause ϕ_X is an OSF-clause ϕ together with a distinguished variable X (called its root) such that every variable Y occurring in ϕ is reachable from X.

Definition 6 (Solved OSF-Constraints) An OSF-clause ϕ is called _solved_ if for every variable X, ϕ contains:

▷ at most one sort constraint of the form $X : s$, with $\bot < s$;

▷ at most one feature constraint of the form $X.l \doteq Y$ for each l; and,

▷ only one occurrence of the variable X, if X occurs in an equality constraint of the form $X \doteq Y$.

We call Φ the set of all OSF-clauses in solved form, and Φ_R the subset of Φ of rooted solved OSF-clauses.

Proposition 1 (Normalization of OSF-clauses) Given an OSF-clause ϕ, choosing non-deterministically and applying any applicable rule among the four transformation rules below until none applies will always terminate in a solved OSF-clause, or a `fail` constraint of the form $X : \bot$.

$$(\textit{Inconsistent Sort}) \qquad \frac{\phi \ \& \ X : \bot}{X : \bot}$$

$$(\textit{Sort Intersection}) \qquad \frac{\phi \ \& \ X : s \ \& \ X : s'}{\phi \ \& \ X : s \wedge s'}$$

$$(\textit{Feature Decomposition}) \qquad \frac{\phi \ \& \ X.\ell \doteq Y \ \& \ X.\ell \doteq Y'}{\phi \ \& \ X.\ell \doteq Y \ \& \ Y \doteq Y'}$$

$$(\textit{Variable Elimination}) \qquad \frac{\phi \ \& \ X \doteq Y}{\phi[X/Y] \ \& \ X \doteq Y} \qquad \text{if } X \in Var(\phi)$$

Example 4 The normalization of the *OSF*-clause given in Example 3 leads to the solved *OSF*-clause which is the conjunction of the equality constraint $N \doteq M$ and the following solved *OSF*-clause:

$$X : person \ \& \ X.\,name \ \doteq N \ \& \ N : \top \qquad \& \ N.\,first \ \doteq F \ \& \ F : string$$
$$\& \ M.\,last \ \doteq S \ \& \ S : string$$
$$\& \ X.\,spouse \doteq P \ \& \ P : person \ \& \ P.\,name \ \doteq I \ \& \ I : id$$
$$\& \ I \,.\,last \ \doteq S \ \& \ S : \top$$
$$\& \ P\,.\,spouse \doteq X \ \& \ X : \top.$$

3.1.3 OSF-graphs

We will now introduce the notion of *order-sorted feature graph* (*OSF*-graph) which is closely related to that of normal *OSF*-terms, and to that of rooted solved *OSF*-clauses. The exact syntactic and semantic mutual correspondence between these three notions is to be established precisely.

Definition 7 (OSF-Graphs) *An OSF-graph $G = (N, E, \lambda_N, \lambda_E, X)$ is a directed labeled graph, together with a distinguished node called its root $X \in N$ such that:*
  ▷ *each node of G is denoted by a variable X, i.e., $N \subseteq \mathcal{V}$;*
  ▷ *each node X of G is labeled by a non-bottom sort s, i.e., $\lambda_N(N) \subseteq \mathcal{S} - \{\bot\}$;*
  ▷ *each (directed) edge (X, Y) of G is labeled by a feature, i.e., $\lambda_E(E) \subseteq \mathcal{F}$;*
  ▷ *no two edges outgoing from the same node are labeled by the same feature, i.e., if $\lambda_E((X, Y)) = \lambda_E((X, Z))$, then $Y = Z$ (G is deterministic);*
  ▷ *every node lies on a directed path starting at the root (G is connected).*

Let us call $D^{\mathcal{G}}$ the domain of all *OSF*-graphs.

Example 5 The *OSF*-graph depicted in Figure 2 has the six nodes X (the root), N, F, S, P and I; the variable names denoting the nodes are usually omitted in a graphical representation (where they are semantically redundant, in a sense that will become clear in Section 8).

3.2 Syntactic bijections

Effective syntactic translations between normal *OSF*-terms, *OSF*-graphs, and solved rooted *OSF*-clauses:

$$\phi : \Psi \to \Phi_R \qquad \psi_\phi : \Phi_R \to \Psi$$
$$G : \Psi \to D^{\mathcal{G}} \qquad \psi_G : D^{\mathcal{G}} \to \Psi$$

can be defined in such a way that,

Proposition 2 (Syntactic Bijections) *There is a one-one correspondence between normal OSF-terms OSF-graphs and rooted solved OSF-clauses syntactic mappings* $\psi : (\Phi_R + D^{\mathcal{G}}) \to \Psi$, *and* $G : \Psi \to D^{\mathcal{G}}$, *and* $\phi : \Psi \to \Phi_R$ *putting the syntactic domains* Ψ, $D^{\mathcal{G}}$, *and* Φ_R *in bijection. That is,*

$$Id_{\Phi_R} = \phi \circ \psi_\phi \quad \text{and} \quad \psi_\phi \circ \phi = Id_\Psi,$$
$$Id_\Psi = \psi_G \circ G \quad \text{and} \quad G \circ \psi_G = Id_{D^{\mathcal{G}}}.$$

We "overload" the notation of the mapping ψ ($= \psi_\phi + \psi_G$) to work either on rooted solved *OSF*-clauses or *OSF*-graphs. We say that the *OSF*-clause $\phi(\psi)$ is obtained from "dissolving" the (not necessarily normal) *OSF*-term ψ.

Example 6 Let ψ be the *OSF*-term of Example 1. Its dissolved form $\phi(\psi)$ is the *OSF*-clause of Example 3. Let ϕ be the solved form of this *OSF*-clause from Example 4 (without equality constraints). The rooted solved *OSF*-clause ϕ_X corresponds to the normal *OSF*-term in Example 2 and to the *OSF*-graph in Example 5 in the sense of the bijections above.

3.3 Semantics and OSF-orderings

The three kinds of syntactic objects have a meaning in a class of interpretation structures specified by a signature:

Definition 8 (OSF-Algebra) *An order-sorted feature algebra (or simply OSF-algebra) over the signature* $\langle S, \leq, \wedge, \mathcal{F} \rangle$ *is a structure* $\mathcal{A} = \langle D^{\mathcal{A}}, (s^{\mathcal{A}})_{s \in S}, (\ell^{\mathcal{A}})_{\ell \in \mathcal{F}} \rangle$ *such that:*

> ▷ $D^{\mathcal{A}}$ *is a non-empty set, called the domain of* $\mathcal{A}$ *(or, universe);*
> ▷ *for each sort symbol* s *in* S, $s^{\mathcal{A}}$ *is a subset of the domain; in particular,* $\top^{\mathcal{A}} = D^{\mathcal{A}}$ *and* $\bot^{\mathcal{A}} = \emptyset$;
> ▷ *the greatest lower bound (GLB) operation on the sorts is interpreted as the intersection; i.e.,* $(s \wedge s')^{\mathcal{A}} = s^{\mathcal{A}} \cap s'^{\mathcal{A}}$ *for two sorts* s *and* s' *in* S.
> ▷ *for each feature* ℓ *in* $\mathcal{F}$, $\ell^{\mathcal{A}}$ *is a total unary function from the domain into the domain; i.e.,* $\ell^{\mathcal{A}} : D^{\mathcal{A}} \mapsto D^{\mathcal{A}}$;

An *OSF*-algebra $\mathcal{A}'$ is a subalgebra of the *OSF*-algebra $\mathcal{A}$ if $\ell^{\mathcal{A}'} = \ell^{\mathcal{A}}$, for $\ell \in \mathcal{F}$, and $s^{\mathcal{A}'} \subseteq s^{\mathcal{A}}$, for $s \in S$. For any non-empty subset S of $D^{\mathcal{A}}$, the subalgebra generated by S, denoted $\mathcal{A}[S]$, is the least subalgebra of $\mathcal{A}$ containing S.—The set of all $\mathcal{A}$-valuations, *i.e.*, functions $\alpha : \mathcal{V} \mapsto D^{\mathcal{A}}$, is denotet $Val(\mathcal{A})$.

3.3.1 OSF-Terms

Let $\psi = X : s(l_1 \Rightarrow \psi_1, \ldots, l_n \Rightarrow \psi_n)$ be an *OSF*-term. Given the interpretation $\mathcal{A}$, the denotation $[\![\psi]\!]^{\mathcal{A},\alpha}$ of the *OSF*-term ψ under a valuation $\alpha : \mathcal{V} \mapsto D^{\mathcal{A}}$ $Dom(\alpha) \supseteq Var(\psi)$ is given by the following inductive definition:

$$[\![\psi]\!]^{\mathcal{A},\alpha} = \{\alpha(X)\} \cap s^{\mathcal{A}} \cap \bigcap_{1 \leq i \leq n} (l_i^{\mathcal{A}})^{-1}([\![\psi_i]\!]^{\mathcal{A},\alpha}). \tag{2}$$

Observe that this always specifies a singleton, or possibly the empty set. Generally, the meaning of an *OSF*-term ψ is understood as the set denoted by ψ when one abstracts from particular valuations. Then, all variables occurring in an *OSF*-term are implicitly existentially quantified at the term's outset. Thus, the *denotation* of the *OSF*-term ψ is defined as:

$$[\psi]^{\mathcal{A}} = \bigcup_{\alpha \in Val(\mathcal{A})} [\psi]^{\mathcal{A},\alpha} \tag{3}$$

For the denotation of several *OSF*-terms in the same context, they are viewed as a tupel constituting one *OSF*-term $\psi = (\psi_1,\ldots,\psi_n)$ (which is short for $\psi = (Z : \top(1 \Rightarrow \psi_1,\ldots,n \Rightarrow \psi_n)$, where $1,\ldots,n \in \mathcal{F}$ are interpreted as projections). This way, the consistent interpretation of variables shared between *OSF*-terms is ensured.

Definition 9 (Term Subsumption Ordering) *If ψ and ψ' are two OSF-terms, then (ψ is subsumed by ψ'),*

$$\psi \leq \psi' \quad iff \quad for \ all \ OSF\text{-}algebras \ \mathcal{A} : [\psi]^{\mathcal{A}} \subseteq [\psi']^{\mathcal{A}}.$$

The *OSF*-terms generalize first-order terms in many respects. In particular, if we see a first-order term as an expression denoting the set of all terms that it subsumes, then we obtain the special case where *OSF*-terms are interpreted as subsets of a free term algebra $\mathcal{T}(\Sigma, V)$, which can be seen naturally as a special *OSF*-algebra.

3.3.2 OSF-Clauses

Given $\mathcal{A}$, an *OSF*-algebra and $\alpha \in Val(\mathcal{A})$ such that $Dom(\alpha) \supseteq Var(\phi)$, then $\mathcal{A}, \alpha \models \phi$, is defined inductively as:

$$\mathcal{A}, \alpha \models X : s \quad \text{iff} \quad \alpha(X) \in s^{\mathcal{A}};$$
$$\mathcal{A}, \alpha \models X \doteq Y \quad \text{iff} \quad \alpha(X) = \alpha(Y);$$
$$\mathcal{A}, \alpha \models X.l \doteq Y \quad \text{iff} \quad l^{\mathcal{A}}(\alpha(X)) = \alpha(Y);$$
$$\mathcal{A}, \alpha \models \phi \ \& \ \phi' \quad \text{iff} \quad \mathcal{A}, \alpha \models \phi \ \text{and} \ \mathcal{A}, \alpha \models \phi'.$$

Definition 10 (Logical Implication Ordering) *If ϕ and ϕ' are two OSF-clauses, then (ϕ implies ϕ'), $\phi \succeq \phi'$ iff*

$$for \ all \ \mathcal{A}, \alpha \ s.t. \ \mathcal{A}, \alpha \models \phi, \ there \ exists \ \alpha' \ s.t. \ \mathcal{A}, \alpha' \models \phi'.$$

If ϕ_X and $\phi'_{X'}$ are two rooted OSF-clauses, then (ϕ_X implies $\phi'_{X'}$), $\phi_X \succeq \phi'_{X'}$, iff

$$for \ all \ \mathcal{A}, \alpha \ s.t. \ \mathcal{A}, \alpha \models \phi, \ there \ exists \ \alpha' \ s.t. \ \alpha(X) = \alpha'(X') \ and \ \mathcal{A}, \alpha' \models \phi'.$$

3.3.3 OSF-Graphs

Proposition 3 *The set $D^{\mathcal{G}}$ of all OSF-graphs forms an OSF-algebra structure.*

Let $\mathcal{G}$ be this *OSF*-graph algebra. In the interpretation $\mathcal{G}$, a sort $s \in \mathcal{S}$ denotes the set $s^{\mathcal{G}}$ of *OSF*-graphs G whose root is labeled by a sort s' such that $s' \leq s$. A feature $l \in \mathcal{F}$ denotes the function selecting the maximal subgraph rooted at the end of the edge labeled l if such an edge goes out of the root, or a "trivial graph" consisting of its root Z labeled $\top$, where Z is a "new" variable.—The ordering on *OSF*-graphs is the special case of the "algebraic approximation ordering" on the elements of any given *OSF*-algebra $\mathcal{A}$.

An *OSF*-algebra *homomorphism* $\gamma : \mathcal{A} \mapsto \mathcal{B}$ between two *OSF*-algebras $\mathcal{A}$ and $\mathcal{B}$ is a partial function $\gamma : D^{\mathcal{A}} \mapsto D^{\mathcal{B}}$ such that its domain is a subalgebra of $\mathcal{A}$ and:

$$\triangleright\ \gamma(\ell^A(d)) \;=\; \ell^B(\gamma(d)) \quad \text{for all}\quad d \in D^A;$$
$$\triangleright\ \gamma(s^A) \subseteq s^B.$$

Definition 11 (Algebraic Approximation Ordering) *A partial order $\sqsubseteq_A$ on the domain of a given an OSF-algebra A is defined by saying that, for two elements d and e in D^A, d approximates e,*

$$d \sqsubseteq_A e \quad \text{iff} \quad \gamma(d) = e \ \text{ for some endomorphism } \ \gamma : A \mapsto A.$$

The domain of γ in the definition above can be chosen as $A[d]$, the subalgebra of A generated by d.

3.4 Semantic transparency

3.4.1 Characterizing solutions: initiality vs. finality

Let ϕ be a solved *OSF*-clause and X a variable occurring in ϕ, but not in an equality constraint; we say that a conjunct in ϕ *constrains* the variable X if it has an occurrence of a variable which is reachable from X. One can thus construct the *OSF*-clause $\phi(X)$ which consists of all conjuncts constraining X. Clearly, $\phi(X)$ is the maximal subclause of ϕ rooted in X.—We also consider $\phi(X)$ as a rooted *OSF*-clause, with the implicit root variable X.—If X occurs in ϕ in the form $X \doteq Y$, one sets $\phi(X) = \phi(Y)$.

Definition 12 (Solution Algebra) *Let ϕ be an OSF-formula in solved-form, and $D^{\mathcal{G},\phi} = \{G(\phi(X)) \mid X \in Var(\phi)\}$. The subalgebra $\mathcal{G}[D^{\mathcal{G},\phi}]$ of the OSF-graph algebra $\mathcal{G}$ generated by $D^{\mathcal{G},\phi}$ is called the solution algebra of ϕ.*

Given an *OSF*-clause in solved form ϕ, any *OSF*-algebra A is said to be ϕ-*admissible* if there exists some A-valuation α such that $A, \alpha \models \phi$. It comes as no surprise that the initial solution algebra of any solved *OSF*-clause ϕ is ϕ-admissible, and so is any *OSF*-algebra containing it—$\mathcal{G}$, in particular.

Proposition 4 (Extracting Solutions) *The solutions of a solved OSF-clause ϕ in any ϕ-admissible OSF-algebra A are given by OSF-algebra homomorphisms from the initial solution algebra: For each $\alpha \in Val(A)$ such that $A, \alpha \models \phi$ there exists an OSF-algebra homomorphism $\gamma : \mathcal{G}[D^{\mathcal{G},\phi}] \mapsto A$ such that:*

$$\alpha(X) = \gamma(G(\phi(X))).$$

The *OSF*-graph associated with the rooted solved *OSF*-clause $\phi(X)$, $G(\phi(X))$, can be viewed as the maximally connected subgraph rooted in the node X of the (not rooted) graph which can be associated with the solved *OSF*-clause ϕ.

Corollary 1 (Weakly initial models) *The solution algebra of ϕ is weakly initial in* OSF(ϕ), *the full subcategory of ϕ-admissible OSF-algebras.*[1]

Proposition 5 (Extending Solutions) *Let A and B be two OSF-interpretations, and let $\gamma : A \mapsto B$ be an OSF-homomorphism between them. Let ϕ be any OSF-clause such that $A, \alpha \models \phi$ for some A-valuation α. Then, for any B-valuation β obtained as $\beta = \gamma \circ \alpha$ it is also the case that $B, \beta \models \phi$.*

---

[1]An object o is *weakly* initial (resp., final) in a category if there is at least one arrow $a : o \to o'$ (resp., $a : o' \to o$) for any other object o' in the category. Weakly initial (resp., final) objects are not necessarily mutually isomorphic. If the object o admits *exactly one* such arrow, it is initial (resp., final). Initial (resp., final) objects are necesarily mutually isomorphic.

Proposition 6 (Weak Finality of $\mathcal{G}$) *There exists a totally defined homomorphism γ from any OSF-algebra $\mathcal{A}$ into the OSF-graph algebra $\mathcal{G}$.*

In other words, the *OSF*-graph algebra $\mathcal{G}$ is a weakly final object in the category OSF of *OSF*-algebras with *OSF*-homomorphisms. Thus, we have the interesting situation that, if in the *OSF*-algebra $\mathcal{A}$ a solution $\alpha \in Val(\mathcal{A})$ of an *OSF*-clause ϕ exists, it is given by a homomorphism from the *OSF*-graph algebra $\mathcal{G}$ into $\mathcal{A}$ (more precisely, by a total homomorphism on the solution algebra of ϕ); and a solution of ϕ in $\mathcal{G}$ can be obtained as the image of α under a homomorphism from $\mathcal{A}$ into $\mathcal{G}$, and, thus, always exists. That is:

Corollary 2 (Canonicity of $\mathcal{G}$) *An OSF-clause is satisfiable iff it is satisfiable in the OSF-graph algebra.*

This canonicity result was originally proven proof-theoretically by Smolka [Smo88], and then by Dörre and Rounds [DR90], directly, for the case of feature graph algebras without sorts. It is obtained purely semantically as a simple corollary in our setting. That is, Proposition 6 gives the deep reason for the canonicity result.

Corollary 3 (Principal Canonical Solutions) *Every solved OSF-clause $\phi(X)$ is satisfiable, namely, in the OSF-graph algebra $\mathcal{G}$ under the valuation α given by $\alpha(X) = G(\phi(X))$. Moreover, α is a principal solution of ϕ in the OSF-algebra $\mathcal{G}$: The OSF-graph $\alpha(X)$ approximates the OSF-graph assigned to the variable X by any other solution of ϕ.*

3.4.2 Characterizing the denotations of OSF-terms

The *OSF*-graph associated to the normal *OSF*-term ψ, $G(\psi)$, is the principal element of ψ interpreted in the canonical *OSF*-algebra, $\mathcal{G}$. In particular, ψ has a non-empty denotation ("is consistent").

Proposition 7 (Interpretability of Canonical Solutions) *The denotation of the normal OSF-term ψ in any OSF-algebra $\mathcal{A}$ can be characterized by total homomorphisms from the subalgebra of $\mathcal{G}$ generated by $G(\psi)$ into $\mathcal{A}$:*

$$[\![\psi]\!]^{\mathcal{A}} = \{\gamma(G(\psi)) \mid \gamma : \mathcal{G}[G(\psi)] \mapsto \mathcal{A} \text{ is an OSF-algebra homomorphism}\}.$$

In particular, the denotation of ψ in $\mathcal{A}$ is empty, if such homomorphisms do not exist.

Corollary 4 (ψ-Types as Graph Filters) *The denotation of a normal OSF-term in the OSF-graph algebra is the set of all OSF-graphs which the corresponding OSF-graph approximates; i.e.,*

$$[\![\psi]\!]^{\mathcal{G}} = \{g \in D^{\mathcal{G}} \mid G(\psi) \sqsubseteq_{\mathcal{G}} g\}.$$

In lattice-theoretic terms, this result characterizes the canonical type denotation of a ψ-term as the principal approximation filter generated by its graph form.

3.4.3 Characterizing the OSF-orderings

Proposition 8 (Semantic Transparency of Orderings) *If the normal OSF-terms ψ, ψ', the OSF-graphs g, g' and the rooted solved OSF-clauses ϕ_X, ϕ'_X respectively correspond to one another through the syntactic mappings, then the following are equivalent statements:*

- $g \sqsubseteq_G g'$; "g is a graph approximation of g';"
- $\psi' \le \psi$; "ψ' is a subtype of ψ;"
- $\phi'_X \succeq \phi_X$; "ϕ is true of X whenever ϕ' is true of X;"
- $[\psi]^G \subseteq [\psi']^G$. "the set of graphs filtered by ψ is contained in that filtered by ψ'."

Strictly speaking, our *OSF*-orderings are preorders rather than orderings. It really does not matter, in fact. Recall that a preorder (reflexive,transitive) o is a "looser" structure than either an order (anti-symmetric preorder) or an equivalence (symmetric preorder). It may be tightened into an order by factoring over its underlying equivalence ($\equiv_o = o \cap o^{-1}$), its "symmetric core." Then, the quotient set over $\equiv_o$ is partially ordered by o. Hence, if we define, in all three frameworks, *equivalence* as the symmetric core ($\equiv_o$) of the corresponding preorder ($o = \sqsubseteq, \le, \succeq$), then Proposition 8 extends readily to these equivalence relations, and therefore the quotients are in order-bijection. It is really the respective syntactic object modulo the equivalence relation that we are interested in, in order to describe our intuition of "approximation objects" and multiple inheritance among those.

Proposition 9 (Computing the GLB of two *OSF*-terms) *If the GLB of the two OSF-terms ψ_1 and ψ_2 is $\bot$, then the OSF-clause $\phi(\psi_1)$ & $\phi(\psi_2)$ normalizes to the* fail *constraint; otherwise (ψ_1 and ψ_2 are unifiable), it normalizes to a solved-form OSF-clause ϕ', and the GLB $\psi_1 \wedge \psi_2$ is represented by the normal OSF-term $\psi(\phi')$.*

Proposition 10 (Computing the LUB of two *OSF*-graphs) *Given two OSF-graphs G_1 and G_2, an OSF-graph approximated by both, G_1 and G_2, is computed by setting $G' = G(\phi(G_1)$ & $\phi(G_2))$, if it exists; moreover, G' is the principal OSF-graph with this property (i.e., approximating all other ones).*

3.5 Conclusion

The results of this section are used to describe formally the declarative and operational semantics of LIFE programs and prove that they coincide, in the type-theoretic, algebraic, or logical language [AKP91a]. The solutions of a LIFE query can be represented equivalently in either of these formalisms; *i.e.*, an answer can be given by a ψ-term, an endomorphism, or a solved *OSF*-clause. This answer does not say, however, that there exist solutions in *every* minimal model of the program. That is, the answer refers to the solutions in the *final* interpretation structure (and not the initial one, as in PROLOG). *If* solutions exist for some other model, then these can be homomorphically constructed from an *OSF*-algebra which is initial with respect to that answer.

Acknowledgements

We acknowledge first and foremost Gert Smolka for his enlightening work on feature logic and for mind-opening discussions. He pointed out that ψ-terms were solved formulae and he also devised explicitly the notion of feature algebras. Bill Rounds has also been a source of great inspiration. In essence, our quest for the meaning of LIFE has put their ideas and ours together. To these friends, we owe a large part of our understanding.

 Each of us authors has enjoyed tremendously the excitement of seeking together a meaning for LIFE and, of course, each shamelessly blames the other for all remaining mistakes...

References

[AK84]   Hassan Aït-Kaci. *A Lattice-Theoretic Approach to Computation Based on a Calculus of Partially-Ordered Types*. PhD thesis, University of Pennsylvania, Philadelphia, PA, 1984.

[AK86]   Hassan Aït-Kaci. An algebraic semantics approach to the effective resolution of type equations. *Theoretical Computer Science*, 45:293–351, 1986.

[AKN86]   Hassan Aït-Kaci and Roger Nasr. LOGIN: A logic programming language with built-in inheritance. *Journal of Logic Programming*, 3:185–215, 1986.

[AKN89]   Hassan Aït-Kaci and Roger Nasr. Integrating logic and functional programming. *Lisp and Symbolic Computation*, 2:51–89, 1989.

[AKNL87]   Hassan Aït-Kaci, Roger Nasr, and Patrick Lincoln. Le Fun: Logic, equations, and Functions. In *Proceedings of the Symposium on Logic Programming (San Francisco, CA)*, pages 17–23, Washington, DC, 1987. IEEE, Computer Society Press.

[AKP91a]   Hassan Aït-Kaci and Andreas Podelski. Towards the meaning of LIFE. PRL Research Report 11, Digital Equipment Corporation, Paris Research Laboratory, Rueil-Malmaison, France, 1991.

[AKP91b]   Hassan Aït-Kaci and Andreas Podelski. Functions as passive constraints in LIFE. PRL Research Report 13, Digital Equipment Corporation, Paris Research Laboratory, Rueil-Malmaison, France, 1991.

[CM84]   William F. Clocksin and Christopher S. Mellish. *Programming in Prolog*. Springer-Verlag, Berlin, Germany, 2nd edition, 1984.

[DR90]   Jochen Dörre and William C. Rounds. On subsumption and semiunification in feature algebras. In *Proceedings of the 5th Annual IEEE Symposium on Logic in Computer Science (Philadelphia, PA)*, pages 301–310, Washington, DC, 1990. IEEE, Computer Society Press.

[HS88]   Markus Höhfeld and Gert Smolka. Definite relations over constraint languages. LILOG Report 53, IWBS, IBM Deutschland, Stuttgart, Germany, October 1988. To appear in the Journal of Logic Programming.

[JL87]   Joxan Jaffar and Jean-Louis Lassez. Constraint logic programming. In *Proceedings of the 14th ACM Symposium on Principles of Programming Languages*, Munich, W. Germany, January 1987.

[O'K90]   Richard O'Keefe. *The Craft of Prolog*. MIT Press, Cambridge, MA, 1990.

[Smo88]   Gert Smolka. A feature logic with subsorts. LILOG Report 33, IWBS, IBM Deutschland, Stuttgart, Germany, May 1988.

[Smo89]   Gert Smolka. Feature constraint logics for unification grammars. IWBS Report 93, IWBS, IBM Deutschland, Stuttgart, Germany, November 1989. To appear in Journal of Logic Programming.

U-Log, An Ordered Sorted Logic with Typed Attributes

Paul Y Gloess

Université de Technologie de Compiègne

B.P. 649 - 60206 Compiègne - France - ℡: (33)44 23 44 63, Fax: (33)44 20 30 56, e-mail: gloess@frutc51.bitnet

U.R.A. C.N.R.S. N°817, Heuristique et Diagnostic des Systèmes Complexes

Abstract

We extend Aït-Kaci ψ-term theory by constraining the type of arguments through an "assorted" signature. A new glb is defined, based on a "filtration" function which terminates under certain conditions. We obtain a lower lattice of "filters" with a type semantics. We introduce an equivalence relation among filters and obtain a lower lattice of "equi-filters" which has a more natural partial order and semantics. We define a constructive semantics of a Prolog extension to filters, adapted from Huet's explanation of Prolog as a polymorphic type inference system. Inference is represented by three filters, thus allowing meta-reasoning.

1   Introduction and Motivations

Many efforts have been devoted to declarative object oriented extensions of Prolog, e.g.: LOGIN [Aït-Kaci 86]; LIFE [Aït-Kaci 88]; CIEL [Gandriau 89]; ALF [Mellender 88]. Conrad [Conrad 87, 88] and Gallaire [Gallaire 88] advocate the "type reasoning ability" of such extensions. Schmidt-Schauß [Schmidt-Schauß 89] has proposed to add term declarations to an ordered sorted logic. These approaches mostly rely upon extended unification. Andreoli [Andreoli 90] advocates a different approach, called "dynamic objects", which extends the syntax of clauses and Prolog inference.

Our starting point is Aït-Kaci ψ-term theory, which provides records (ψ-terms) with a type semantics and extends Prolog to them.

The first motivation of our work is that in the above theory, record arguments are unconstrained. In other words, the signature is only concerned with sort ordering. It does not specify type information about labels (used as keywords), unlike ordered multisorted signatures in algebra theory, or most record definition facilities provided by programming languages, e.g., LISP [Steele 84] or Pascal.

Smolka and Aït-Kaci have proposed a notion of "feature type hierarchies" which allows the specification of type constraints on record attribute [Smolka 89]. Their semantics rely upon the very general context of ordered multi-sorted equational theories [Kirchner 88]. The definition of these "feature type hierarchies" is rather complex, and equational unification [Jouannaud 90] not easy to implement. Dorre and Rounds [Dorre 90] show a similar approach, but do not cope with attribute inheritance.

Our work is aimed at the same goal, but does not use equational theory: it is mainly a simple extension of the original ψ-term theory. We add an *assortment* to the specification of a partial order among sorts: the assortment specifies attribute type constraints. We define a *filter* as a ψ-term that respects the assortment. The glb operation of the ψ-calculus is not an internal operation to the set of filters. We show that combining this glb with a function called *filtration* endows the set of filters with a lower lattice structure. Filtration preserves ψ-term semantics.

Filtration is a worthy operation in a Prolog context because it may yield ⊥ (interpreted as failure) where ordinary ψ-unification would have succeeded. It also optimizes the computation of a denotation of a ψ-term, as a constraint resolution mechanism. Filtration does not always terminate: we show some sufficient conditions of termination that are often met in practice (e.g., well-foundedness of the sort partial order). We exhibit a simple sufficient condition, called ∧-φ-finiteness of the signature, of termination, which applies to a class of polymorphic signatures.

We remark that the ψ-term partial order is not very natural in the context of our filters: we propose a new partial order among classes of semantically equivalent filters.

Finally, we apply our theory to the definition of a meta-circular constructive semantics for Prolog extended to filters, following Huet's explanation of Prolog as a polymorphic type inference system: it is based on three inference rules which are written as three filters.

By lack of space, most proofs are omitted: more details are available in [Gloess 91, 90].

2   From ψ-terms to Filters

Definitions and theorems belonging to Aït-Kaci ψ-term theory and used here are stated in Appendix 8.

In the original ψ-term theory, a signature is a triple $< \mathcal{S}, \leq, \mathcal{L} >$, where $\mathcal{S}$, the set of sorts, is partially ordered by $\leq$, and $\mathcal{L}$ is the set of labels (also called attributes). $\mathcal{S}$ is always assumed to have a minimum element, denoted $\perp$. It is sometimes assumed to have a maximum element, denoted $\top$, and to be a lower lattice whose glb operation is then denoted $\wedge$. We implicitly make this assumption whenever we use this operation (whether directly or not).

We introduce type constraints on sort attributes, by adding a fourth element φ called assortment:

Definition   **Assortment**

An assortment is a total function $\varphi : \mathcal{S} \times \mathcal{L} \to \mathcal{S}$ satisfying both conditions:

$$\forall s, t \in \mathcal{S} \quad \forall \ell \in \mathcal{L} \qquad \perp < s < t \ \& \perp < \varphi(t,\ell) \ \Rightarrow \ \perp < \varphi(s,\ell) \leq \varphi(t,\ell) \qquad (i)$$

$$\forall \ell \in \mathcal{L} \qquad\qquad\qquad \varphi(\perp,\ell) \ = \ \perp \qquad\qquad\qquad (ii)$$

The intent of this definition is to achieve "attribute inheritance" of common object oriented programming languages. Let us discuss conditions (i) and (ii) with the help of a concrete example.

Consider a *person* as a sort, and *:lives_in* as an attribute. The fact that a person lives in a *house* (another sort) may be regarded as type information, which we represent by

$$\varphi(person, :lives\_in) \ = \ house$$

We shall say that sort *person* owns attribute *:lives_in*, on the grounds that house $> \perp$. (In general we shall say that sort s owns attribute ℓ if and only if $\varphi(s,\ell) \neq \perp$.) On the other hand, the house sort does not own the *:lives_in* attribute (it would not make sense!). Therefore:

$$\varphi(house, lives\_in) \ = \ \perp$$

Consider a *student* as a kind of *person*, i.e., $\perp <$ student $<$ person. Condition (i) forces *student* to own the *:lives_in* attribute (this is "attribute inheritance"). It also asserts the fact that a student lives in a (kind of) house. Hence the value of $\varphi(student, :lives\_in)$ must be strictly greater than $\perp$ but less or equal to *house* e.g.,

$$\varphi(student, :lives\_in) \ = \ dormitory$$

provided that $\perp <$ dormitory $\leq$ house.

Condition (ii) states that the minimum sort $\perp$ has no attribute.

The specification of φ is equivalent to the specification of profiles for unary function symbols in the context of ordered multi-sortered signatures: $\varphi(s,\ell) = s'$ corresponds to the $\ell:s \to s'$ profile.

From now on, we consider only *assorted* signatures $\Sigma = < \mathcal{S}, \leq, \mathcal{L}, \varphi >$. A signature $< \mathcal{S}, \leq, \mathcal{L} >$ is an assorted signature for the permissive assortment constantly equal to $\top$ (except on $\perp$).

We are now interested in ψ-terms that do not violate the assortment φ. We call them Σ-filters or filters:

Definition   **Filter**

A Σ-filter is a ψ-term $< f:D \to \mathcal{S}, \equiv >$ such that:

$$\forall u :: \ell \in D \qquad\qquad \perp \ < \ f(u::\ell) \ \leq \ \varphi(f(u), \ell)$$

The above condition simply states that the tree f agrees with the assortment φ, for each branch labeled ℓ leading from node at occurrence u to subnode at u::ℓ. Note that ($\bot$) is the only filter (and the only ψ-term) containing the $\bot$ sort in one of its node (see Appendix 8 for our concrete syntax). It is often denoted $\bot$.

The lower bound of two filters (computed by the glb operation of the Ψ calculus) is not necessarily a filter, since it may violate the assortment φ. Example:

$$\mathcal{S} \;=\; \{a, b, a', b', f, g, h\} \qquad \mathcal{L} \;=\; \{1, 2\}$$

a'	< a	$\varphi(f,1)$	= a		
b'	< b	$\varphi(g,2)$	= b	ψ_1 =	(f 1 a)
h	< f	$\varphi(h,1)$	= a'	ψ_2 =	(g 2 b)
h	< g	$\varphi(h,2)$	= b'	$\psi_1 \wedge \psi_2$ =	(h 1 a 2 b) violates φ.

For convenience, we denote Ψ the set of ψ-terms on $<\mathcal{S}, \leq>$, $\mathfrak{C}_\Sigma$ the set of filters. Intermediate between these two sets, we denote Ψ_α the set of ψ-terms that are α-coherent, where $\alpha: \mathcal{S} \to \mathbb{N}$ is the sort arity defined by

$$\alpha(s) \;=\; \operatorname{card}(\{\ell \in \mathcal{L} \mid \varphi(s,\ell) \neq \bot\})$$

Precisely, a ψ-term is α-coherent if it does not violate the arity α, that is, if $\varphi(f(u),\ell) > \bot$ for each u::ℓ in the ψ-term domain. It is not requested that f(u::ℓ) be less or equal to $\varphi(f(u),\ell)$. It turns out that Ψ_α is stable by the ψ-term glb operation (as defined in Appendix 8), and is thus a lattice when $<\mathcal{S}, \leq>$ is.

In order to get a lower bound operator in $\mathfrak{C}_\Sigma$, we introduce filtration. Filtration operates on any ψ-term, whether α-coherent or not. It either does not terminate or produces a filter within a finite number of steps. Each step produces a new ψ-term by correcting all the φ violations present in the previous one. New violations may emerge from such corrections.

Définition   Filtering Sequence

Let $\psi = <f: D \to \mathcal{S}, \equiv> \in \Psi$. We call filtering sequence of ψ the sequence $(\psi_n)_{n \in \mathbb{N}}$ defined by:

$$\forall n \in \mathbb{N} \qquad \psi_n \;=\; (\bot) \qquad \text{if } \exists [u] \in D/\equiv f_n([u]) = \bot,$$
$$=\; <f_n: D \to \mathcal{S}, \equiv> \qquad \text{otherwise,}$$

where the sequence $(f_n)_{n \in \mathbb{N}}$ is recursively defined by:

$$f_0 \;=\; f$$
$$\forall [w] \in D/\equiv \qquad f_{n+1}([w]) \;=\; f_n([w]) \wedge \operatorname{Inf}(\{\varphi(f_n([u]),\ell) \mid \ell \in \mathcal{L}, u \in D, u::\ell \in [w]\})$$

Note that all ψ-terms in the sequence share the same domain and coreference relation, except if one of them collapses to ($\bot$), and hence all the subsequent ones. When the sequence is constant after a certain rank (that is, when $\psi_{n+1} = \psi_n$ for some n, whether $\psi_n = (\bot)$ or not), we say that it converges toward this constant value. This defines a partial function called filtration:

$$\text{filtration}_\Sigma: \; \Psi \;\to\; \mathfrak{C}_\Sigma$$

since the limit, when it exists, is necessarily a Σ-filter, as it does not violate φ.

Let us give a few examples. The signature is given graphically: vertical unlabeled line segments show inheritance; labeled arrows correspond to the assortment φ. By convention, we assume implicit inheritance of attribute type constraints. Similarly, we implicitly assume $\varphi(s,\ell) = \bot$ unless specified.

Example showing convergence in two steps:

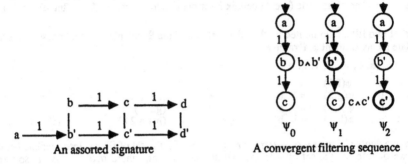

An assorted signature   A convergent filtering sequence

In general, convergence may occur in any number of steps, as one can easily imagine.

Example where sequence collapses to ($\bot$) in one step:

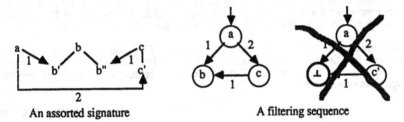

An assorted signature   A filtering sequence

The fact that the filtering sequence may collapse to ($\bot$) is precisely what makes filtration a worthy operation! As we shall see, filtration preserves type semantics, that is, the interpretation of a ψ-term after filtration is exactly the same as before. Therefore, both ψ_0 and ψ_1 represent the same set: $\emptyset$. But ψ_1 does it in a more obvious manner. It is of special importance to detect this case of semantic failure (equivalent to unification failure) as soon as possible in a Prolog context, in order to avoid undesirable inferences.

Even when filtration does not yield ($\bot$), it is useful because it nevertheless narrows the domains of variables contributing to the denotation of a ψ-term, thus optimizing its actual computation. One may regard filtration as a constraint resolution mechanism

Definition   Filtering Signature

An assorted signature $< \mathcal{S}, <, \mathcal{L}, \varphi >$ is said to be filtering if the filtering sequence of any ψ-term on $< \mathcal{S}, < >$ is convergent.

Before we study conditions of convergence of filtering sequences, let us state our main results about $\mathfrak{C}_\Sigma$ and its semantics.

Theorem   Filter Lower Lattice

If Σ is a filtering signature, then $< \mathfrak{C}_\Sigma, <, \underline{\wedge} >$ is a lower lattice for the $\underline{\wedge}$ glb operation defined by

$$\underline{\wedge}: \quad \mathfrak{C}_\Sigma \times \mathfrak{C}_\Sigma \quad \rightarrow \quad \mathfrak{C}_\Sigma$$

$$\psi_1 \underline{\wedge} \psi_2 \quad = \quad \text{filtration}_\Sigma(\psi_1 \wedge \psi_2)$$

where $\wedge$ is the Ψ glb operation.

Proof: Let $\psi = \psi_1 \wedge \psi_2$ be the glb (in Ψ) of two filters ψ_1 and ψ_2. Let $(\psi_n)_{n \in \mathbb{N}}$ be the filtering sequence and ψ' its limit, which is a filter. Let us show by recurrence that if some filter ψ'' is a common lower bound of ψ_1 and ψ_2 then it is a lower bound of any term ψ_n in the filtering sequence. Clearly $\psi'' \leq \psi_0$ since $\psi_0 = \psi_1 \wedge \psi_2$. Let us show that

$$\psi'' \leq \psi_n \;\Rightarrow\; \psi'' \leq \psi_{n+1}$$

If ψ_n is the limit, then $\psi_n = \psi_{n+1}$ hence this is true. Otherwise, by definition of the filtering sequence, ψ_n violates φ for some classes and labels and $\psi_{n+1} \leq \psi_n$ and ψ_{n+1} is the greatest lower bound (in Ψ) of ψ_n that does not violate φ on these classes and labels. Hence $\psi'' \leq \psi_{n+1}$, since $\psi'' \leq \psi_n$ and ψ'' does not violate φ at all. It follows that the limit ψ' is the glb of ψ_1 and ψ_2 in $\mathfrak{C}_\Sigma$.

In order to obtain coherent interpretations of filters, we must start with an interpretation of sorts and labels that respects the assortment φ, that is, a Σ-algebra. We call it a Σ-interpretation:

Definition Σ-Interpretation

A Σ-interpretation on a universe $\mathcal{U}$ is a pair of applications $< \mathfrak{I}_{\mathscr{S}}, \mathfrak{I}_{\mathscr{L}} >$:

$$\mathfrak{I}_{\mathscr{S}}: \qquad \mathscr{S} \;\rightarrow\; \mathscr{P}(\mathcal{U})$$
$$\mathfrak{I}_{\mathscr{L}}: \qquad \mathscr{L} \;\rightarrow\; \mathcal{U} \rightarrow \mathcal{U}$$

such that $\mathfrak{I}_{\mathscr{S}}: < \mathscr{S}, \leq, \wedge, \perp, ^\top > \rightarrow < \mathscr{P}(\mathcal{U}), \subset, \cap, \varnothing, \mathcal{U} >$ is a homomorphism and such that $\mathfrak{I}_{\mathscr{L}}$ satisfies the conditions:

$$\begin{array}{lll} \forall s \in \mathscr{S} & \perp < \varphi(s,\ell) \;\Rightarrow\; \mathrm{DEF}(\mathfrak{I}_{\mathscr{L}}(\ell)) \supset \mathfrak{I}_{\mathscr{S}}(s) & \text{(i)} \\ \forall s \in \mathscr{S} & \perp < \varphi(s,\ell) \;\Rightarrow\; \mathfrak{I}_{\mathscr{L}}(\ell)(\mathfrak{I}_{\mathscr{S}}(s)) \subset \mathfrak{I}_{\mathscr{S}}(\varphi(s,\ell)) & \text{(ii)} \end{array}$$

Because our definition is compatible with the definition given by Aït-Kaci (we have only added conditions (i) and (ii)), we can apply the same construction for extending $\mathfrak{I}_{\mathscr{S}}$ to Ψ. We view this construction as solving a finite system of constraints associated with a ψ-term (see Appendix 8).

Theorem   Filter Type Semantics

Let Σ be a filtering signature. Let $< \mathfrak{I}_{\mathscr{S}}, \mathfrak{I}_{\mathscr{L}} >$ be any Σ-interpretation of universe $\mathcal{U}$. Let $\mathfrak{J}_{\mathscr{S}}$ be the ψ-extension of $\mathfrak{I}_{\mathscr{S}}$. Then

$$\mathfrak{J}_{\mathscr{S}}: \qquad < \mathfrak{C}_\Sigma, \leq, \wedge, \perp, ^\top > \;\rightarrow\; < \mathscr{P}(\mathcal{U}), \subset, \cap, \varnothing, \mathcal{U} >$$

is an homomorphism.

Proof: Immediate from Aït-Kaci type semantics result and compatibility of filtration with type semantics (hint: consider sets of constraints associated with successive elements of a filtering sequence).

We have thus obtained a type semantics for $\mathfrak{C}_\Sigma$, provided that filtration$_\Sigma$ is a total function.

3   Convergence of Filtering Sequences

We study conditions under which the filtering sequence of a ψ-term converges. We first look at the ψ-term structure itself, regardless of the assorted signature Σ. We then focus on the nature of Σ, looking at sufficient conditions of termination of all ψ-terms.

We define the depth of a ψ-term as the upper bound in $\mathbb{N} \cup \{\omega\}$ of the length of each occurrence in its domain. We denote $|\psi|$ the depth of ψ. A ψ-term ψ is cyclic if and only if $|\psi| = \omega$. Similarly, we define the depth of a subset L of $\mathscr{L}^*$ as the upper bound of the length of its elements and use the notation $|L|$ as well: this applies to coreference classes $[u] \in D/\equiv$ of a given ψ. Filtering operates deeper and deeper within a ψ-term:

Proposition   **Filtering Depth**

Let $(f_n)_{n \in N}$ be the sequence of functions $D \to \mathscr{S}$ associated with a ψ-term ψ in the definition of a filtering sequence. Then

$$\forall n \in N \; \forall u \in D \qquad f_{n+1}([u]) < f_n([u]) \;\Rightarrow\; |[u]| > n$$

Hence, the filtering sequence of any acyclic ψ-term converges: thus, there is no problem of termination if we restrict ourselves to finite terms.

We now study signatures. Recall that in this context the set of sorts is assumed to be a lower lattice. A well founded signature is a signature whose inheritance partial order is well founded, i.e., there is no infinite strictly decreasing sequence of sorts $s_0 > s_1 > \ldots$. We have:

Theorem   **Well Founded Implies Filtering**

Any well founded signature is filtering.

Proof: Let ψ be a ψ-term. If the filtering sequence collapses to $\perp$, nothing remains to be proved. Otherwise, consider the $(f_n)_{n \in N}$ sequence of functions associated with ψ in the definition of filtering sequence. Let $[u_1], \ldots, [u_k]$ be finite set of coreference classes of ψ. Since $(f_n)_{n \in N}$ is decreasing by definition, each of the k sequences $(f_n([u_1]))_{n \in N}, \ldots, (f_n([u_k]))_{n \in N}$ is decreasing in $\mathscr{S}$ but cannot strictly decrease for ever. Hence these k sequences are constant after a certain rank computed as the maximum of k ranks. Consequently the limit of the filtering sequence is reached.

This result is of practical interest since $\mathscr{S}$ is usually finite, which implies its partial order is well founded. Another condition achieves the same effect: *finitely labeled* signatures, that is, signatures whose all sorts except a finite number own no label at all, are filtering (similar proof). Signatures with a finite set of sorts are obviously finitely labeled.

Neither well founded nor finitely labeled signatures comprehend the case of polymorphic sorts. Consider the simple polymorphic signature obtained by adding the infinitely decreasing sequence of parametric sorts:

$$list(s) > list(list(s)) > list^n(s) > \ldots$$

to some finite signature Σ with at least one sort $s > \perp$. Add one label, say *:car*, with

$$\varphi(list^{n+1}(s), \mathrm{:car}) \;=\; list^n(s)$$

It is neither well founded nor finitely labeled. Therefore these conditions are too drastic if one is interested in polymorphism. Note that well-foundedness is nevertheless requested in [Smolka 89]. We have tried a few ideas to cover polymorphism. One of them was suggested by Aït-Kaci and seems related to some notions in [Smolka 89] although there is no obvious mapping.

A sort s is said to be *principal for* a label ℓ if it owns ℓ and if no other upper bound of s owns ℓ. A sort s is said *principal* if it is principal for at least one label. A signature Σ is said to be *principally assorted* if for each label ℓ, each strictly ascending chain $s_0 < s_1 < \ldots$ of sorts such that s_0 owns ℓ contains a sort s_n principal for ℓ. In other words, there is no ascending chain owning the same attribute ℓ for ever. A signature is said to be *finitely attributed* if it is principally assorted and contains a finite number of principal sorts.

Note the difference between *finitely labeled* and *finitely attributed*! Finitely attributed signatures are a more general notion, compatible with polymorphism. The above polymorphic signature, and Σ_2 (see Appendix 9) as well, are finitely attributed, provided that Σ is finitely attributed, but they are not finitely labeled.

Unfortunately, there is a (quite simple) finitely attributed signature which is not filtering.

Example: Let $\mathscr{S}$ be $Z-$, the set of zero or negative integers, augmented with $\perp$ and $^\top$. Let $\mathscr{L} = \{-\}$. Consider the simple assortment:

$$\forall z \in Z- \qquad \varphi(z,-) = z-1$$
$$\varphi(^\top,-) = \varphi(\bot,-) = \bot$$

The cyclic $(:= X (z - X))$ ψ-term violates φ, for any sort $z \in Z-$. The filtering sequence diverges, since the sort z is replaced with z-1, z-2, ..., endlessly. Yet this signature is finitely attributed: it has only one principal sort which is 0, and there is no infinite ascending chain.

Another idea is based on the natural extension φ^* of φ as defined by:

$$\varphi^*: \qquad \mathcal{S} \times \mathcal{L}^* \rightarrow \mathcal{S}$$
$$\varphi^*(s, \ell u) = \varphi^*(\varphi(s,\ell), u)$$
$$\varphi^*(s, (\ )) = s$$

Definition   Radius

We call radius of a sort s (resp. signature Σ) the ordinal $\mathcal{R}(s)$ (resp. $\mathcal{R}(\Sigma)$) defined by

$$\mathcal{R}(s) = \sup(\{|u| \ | \ u \in \mathcal{L}^*, \varphi^*(s, u) > \bot \})$$
$$\mathcal{R}(\Sigma) = \sup(\{\mathcal{R}(s) | s \in \mathcal{S}\})$$

There are polymorphic signatures, such as Σ_2, whose radius $\mathcal{R}(s)$ is finite for each sort s. Unfortunately, we have the negative result in response to a previous conjecture [Gloess 90]:

Proposition   Finite Radius CounterExample

There are non filtering signatures Σ with finite radius $\mathcal{R}(\Sigma)$.

Proof: Consider the following signature with only one label, ℓ, and infinite set of sorts consisting of two infinitely decreasing sequences $(s_n)_{n \in N}$, $(t_n)_{n \in N}$. Precisely, Σ is defined as the least signature satisfying

$$s_{n+1} < s_n$$
$$t_{n+1} < t_n$$
$$s_{n+1} < t_n$$
$$\varphi(s_n, \ell) = t_n$$

The simple ψ-term $(:= X (s_0 \ell X))$ has a divergent filtering sequence $((:= X (s_n \ell X)))_{n \in N}$.

Turning away from this negative result, we now propose a simple and rather barren condition that is applicable to certain polymorphic signatures, such as Σ_2 given in Appendix 9:

Definition and Proposition $\wedge$-φ-Finiteness Implies Filtering

A signature Σ is $\wedge$-φ-finite if for each finite set S of sorts and each finite set L of labels, the least superset of S stable by $\wedge$ and all applications $(s \mapsto \varphi(s,\ell))$ for $\ell \in L$ is finite.

Any $\wedge$-φ-finite signature is filtering.

In other words, starting with a finite set of sorts and a finite set of labels, we cannot generate an infinite set of sorts by repeated application of the glb operation or φ limited to the specified labels. Note that well founded signatures are $\wedge$-φ-finite. The converse is untrue!

We conjecture that the converse of the above proposition is true, that is: Is any filtering signature necessarily $\wedge$-φ-finite? This necessary condition seems much harder to prove!

4   Toward a More Natural Partial Order

Shifting from convergence problems back to the semantics of our filters, we observe that the partial order inherited from Ψ does not always reflect the subset ordering of filter denotations. Consider for instance a signature with the following sorts and assortment:

student	<	person	φ(person, :lives_in) =	house
dormitory	<	house	φ(student, :lives_in) =	dormitory

We would like to have

$$\text{(student)} \quad < \quad \cdot\text{(person :lives\_in (house))}$$
$$\text{(student)} \quad < \quad \text{(person :lives\_in (dormitory))}$$

although neither of these relations hold according to the definition of the partial order in Ψ. This would nevertheless be natural since we know that the set of students is a subset of the set of persons who live in a house (or even in a dormitory), no matter what Σ-interpretation we use.

Note that if we replace the left member with the filter (student :lives_in (dormitory)) in either relation, we get it right in either case according to Ψ partial order! The trouble is that we also have

$$\text{(student :lives\_in (dormitory))} \quad < \quad \text{(student)}$$

although the denotations are really equal, because the apparent ":lives_in (dormitory)" constraint is implicit from the signature.

This example suggests that we consider filters that only differ from each other by partial completion using the assortment to be equivalent with respect to a "natural order". We shall call equi-filters such equivalence classes. Accordingly, we replace $\mathfrak{G}_\Sigma$ with its quotient by this equivalence relation, and define a partial order among classes, as well as a glb operation. We shall finally obtain a lower lattice with a type semantics.

Definition φ-completion

Let $\psi = \langle f{:}D \rightarrow \mathscr{S}, \equiv \rangle$ and $\psi' = \langle f'{:}D' \rightarrow \mathscr{S}, \equiv' \rangle$ two ψ-terms. We say that ψ' completes ψ and write $\psi \overset{\approx}{\rightarrow} \psi'$ if ψ' is obtained from ψ by adding a single coreference class of the form $[u{::}\ell]$, for a label ℓ such that $\varphi(f(u),\ell) > \perp$, and defining f' on that new class by

$$f'(u{::}\ell) \quad = \quad \varphi(f(u),\ell)$$

Example: The drawing below illustrates this notion, assuming $\varphi(c,\ell) = d > \perp$.

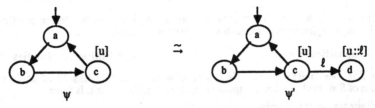

Completion relation is compatible with ψ-semantics and filtration. Its inverse (called *trimming*) is confluent and noetherian. Hence each ψ-term has a normal form obtained by iterated and exhaustive trimming. We shall denote $\simeq$ the reflexive and transitive closure of the union of $\overset{\approx}{\rightarrow}$ and its inverse: it is an equivalence relation. We call equi-filter any equivalence class in the quotient $\mathfrak{G}_\Sigma/\simeq$.

Theorem   **Equi-Filter Lower Lattice and Type Semantics**

$< \mathfrak{S}_\Sigma / \simeq, \tilde{\leqslant}, \tilde{\wedge}, \{\bot\}, \{^\top\} >$ is partially ordered by the relation $\tilde{\leqslant}$ defined below. It is a lower lattice for the $\tilde{\wedge}$ glb operation defined below, with $\{\bot\}$ and $\{^\top\}$ for bottom and top elements. We denote $\tilde{\psi}$ the class of ψ. By definition

$$\tilde{\psi}_1 \tilde{\leqslant} \tilde{\psi}_2 \quad \Leftrightarrow \quad \exists \psi'_1 \simeq \psi_1 \; \exists \psi'_2 \simeq \psi_2 \;\; \psi'_1 \leqslant \psi'_2 \tag{i}$$

$$\tilde{\psi}_1 \tilde{\wedge} \tilde{\psi}_2 \;\; = \;\; \tilde{\psi} \quad \text{with } \psi = \psi_{1 \wedge} \psi_2 \tag{ii}$$

Let $< \mathfrak{I}_\mathcal{G}, \mathfrak{I}_\mathcal{L} >$ be any Σ-interpretation of universe $\mathcal{U}$. Let $\mathfrak{I}$ be the ψ-extension of $\mathfrak{I}_\mathcal{G}$. Then $\mathfrak{I}$ is constant on each equi-filter and defines an homomorphism:

$$\mathfrak{I}: \;\; < \mathfrak{S}_\Sigma / \simeq, \tilde{\leqslant}, \tilde{\wedge}, \{\bot\}, \{^\top\} > \;\; \rightarrow \;\; < \mathcal{P}(\mathcal{U}), \subset, \cap, \varnothing, \mathcal{U} > \;\; .$$

The proof of this result is not obvious: it requires several technical lemmas, just to show that $\tilde{\leqslant}$ is a well defined partial order, and that $\tilde{\wedge}$ is a well defined operation. We conjecture that the partial order defined by (i) is exactly the subset partial order of the denotations:

$$\tilde{\psi}_1 \tilde{\leqslant} \tilde{\psi}_2 \quad \Leftarrow \quad \forall < \mathfrak{I}_\mathcal{G}, \mathfrak{I}_\mathcal{L} > \text{a } \Sigma\text{-interpretation whose } \mathfrak{I}_\mathcal{G} \text{ extends to } \mathfrak{I}, \qquad \mathfrak{I}(\psi_1) \subset \mathfrak{I}(\psi_2$$

the converse of which has been established as part of our theorem. If this were true, it would provide a syntactic and effective characterization of semantic inclusion.

5   Prolog Extension Semantics

In our extension, a Prolog program is a set of special filters called "clauses". A *clause* has three attributes:

$$\varphi(\text{clause}, \text{:head}) \;\; = \;\; ^\top \;\; = \;\; \varphi(\text{clause}, \text{:body})$$
$$\varphi(\text{clause}, \text{:id}) \;\; = \;\; \text{string}$$

This formalization of clauses as filters makes it possible to use the filter partial ordering in our statement of Prolog inference, in lieu of the classical subsumption partial ordering. Our semantics relies upon an adaptation of Huet "principal type theorem", and a slight change of vocabulary.

Given a program P and a question q, the semantics of $< P, q >$ is defined as the set of maximal filters $r \leqslant q$ such that there exists a proof tree of r. We call these filters r maximal answers to the question q. Following Huet's notation, we write $P \models T{:}r$ to mean that T is a proof tree of r whose nodes are clauses in P (or clause identifiers). According to the principal type theorem, for each proof T on the program P of some type, there is a maximal type (i.e., maximal answer) r such that $P \models T{:}r$.

It is straightforward that our semantics of $< P, q >$ is obtained as the set of all $r \wedge q \neq \bot$, where r is a maximal type for some proof tree T on P. It is interesting to see that Prolog inference itself, that is, the formation of proof trees and answers, may be described using three clause filters, where "$\models$" is a new sort equiped with two attributes (corresponding to the proof tree or forest and the response) and *pair* and *null* are obvious sorts used to represent lists:

(*clause*	:id	"Derivation of a proof tree using a clause."
	:head	($\models$:proof (*proof* :top $\mathcal{O}$:forest TS) :response R)
	:body	(*pair* :car (*clause* :head R :body HS :id $\mathcal{O}$)
		:cdr (*pair* :car ($\models$:proof TS :response HS) :cdr (*null*))))
(*clause*	:id	"Derivation of the empty forest."
	:head	($\models$:proof (*null*) :response (*null*))
	:body	(*null*))
(*clause*	:id	"Derivation of the non empty forest."
	:head	($\models$:proof (*pair* :car T :cdr TS) :response (*pair*:car H :cdr HS))
	:body	(*pair* :car ($\models$:proof T :response H)
		:cdr (*pair* :car ($\models$:proof TS :response HS) :cdr (*null*))))

This metacircular implementation of extended Prolog inference demonstrates the expressive power of this extension: it can effectively be used for explanation purposes as shown in Appendix 10. It can be altered in order to test different inference schemes in the way of Metalog [Dincbas 79].

6   Conclusion

We have defined an extension of Aït-Kaci ψ-term theory which allows the specification of type constraints on term arguments. This is done by adding a function called assortment to the partially ordered set of sorts and set of labels of a ψ-term signature. The assortment must only satisfy two simple conditions that merely correspond to attribute inheritance and attribute type inheritance in object oriented programming. These conditions are much simpler than those required in the algebraïc context used by other authors.

Filters are ψ-terms obeying the constraints specified by the assortment. The set of filters is a lower lattice with a type semantics, for a new glb operation which is a combination of the ψ-term glb and a function called filtration. It is well defined under some conditions, one of which is compatible with certain polymorphic sorts. It is easy to implement as a simple variation of ψ-term unification, without requiring equational unification. Filtration preserves ψ-term type semantics: the practical interest of performing filtration during unification is to detect failure that would otherwise go undetected.

We have observed that the partial order among filters (as inherited from ψ-terms) does not always reflect the semantic order of set inclusion. A more natural partial order arises in the set of so-called equi-filters, which is also a lower lattice with a type semantics.

We have then adapted Huet's explanation of Prolog as a polymorphic type inference system to provide a constructive semantics of a Prolog extension to filters. Extended Prolog inference can be completely represented by three filters, thus yielding a reflexive Prolog suitable for meta-reasoning.

Our theory, which we call a unified object logic model, is a contribution to the efforts devoted to the problem of mixing the object and Prolog paradigms.

7   References

[Aït-Kaci 88] Aït-Kaci & P. Lincoln: "LIFE, A Natural Language for Natural Language", MCC Technical Report Number ACA-ST-074-88, Austin, February 1988.
[Aït-Kaci 86] Aït-Kaci & R. Nasr: "LOGIN: A Logic Programming Language with Built-in Inheritance", Journal of Logic Programming 3(3), pp. 187-215, 1986.
[Andreoli 90] J.M. Andreoli & R. Pareschi, "Linear Objects: Logical Processes with Builtin Inheritance", in 9th Conference on Logic Programming, Jerusalem, Israel, 1990.
[Conrad 88] T. Conrad, "Equator: A Many-Sorted PROLOG Based on Equational Unification", pp. 171-183, Actes du 7ème Séminaire de Programmation en Logique, CNET, Mai 1988.
[Conrad 87] T. Conrad, "Termes Typés et Termes Globaux", pp. 119-130, Actes du 6ème Séminaire de Programmation en Logique, CNET, Mai 1987.
[Dincbas 79] M. Dincbas, "Le Système de Résolution de Problèmes Metalog", Rapport 3146/Deri, C.E.R.T. Toulouse 1979.
[Dorre 90] J. Dorre & W.C. Rounds, "On Subsumption and Semi-Unification in Feature Algebras", in Proc. of the Fifth Symposium on Logic in Computer Science, 1990.
[Gallaire 88] H. Gallaire, "Multiple Reasoning Styles in Logic Programming", in Proceedings of FGCS'88 Conference (ICOT), Tokyo, 1988.
[Gandriau 89] M. Gandriau & C. Massoutie, "Classes et Types: Aides à la Programmation Logique", pp. 57-69, Actes du 8ème Séminaire de Programmation en Logique, CNET, Mai 1989.
[Gloess 91] P.Y. Gloess, "U-Log, an Ordered Sorted Logic with Typed Attributes (Extended Version)", Report N°91/12/DI, Université de Technologie de Compiègne, June 1991.
[Gloess 90] P.Y. Gloess, "Contribution à l'Optimisation de Mécanismes de Raisonnement dans des Structures Spécialisées de Représentation des Connaissances", Thèse de Doctorat d'Etat, Université de Technologie de Compiègne, 22 Janvier 90.
[Huet 86] G. Huet, "Deduction and Computation", Rapport de Recherche INRIA N°513, Avril 1986.
[Jouannaud 90] J-P. Jouannaud & C. Kirchner, "Solving Equations in Abstract Algebras: a Rule-Based Survey of Unification", L.R.I. Research Report N°561, Université d'Orsay ParisXI, March 1990.
[Kirchner 88] C. Kirchner, "Order-Sorted Equational Unification", Rapport de Recherche INRIA N°954, Décembre 1988.
[Mellender 88] F. Mellender, "An Integration of Logic and Object-Oriented Programming", pp. 181-185, SIGPLAN Notices, Vol. 23, N°10, 1988.

[Schmidt-Schauß 89] M. Schmidt-Schauß, "Computational Aspects of an Ordered Sorted Logic with Term Declarations", Lectures Notes in Artificial Intelligence, Vol. 395, Springer-Verlag, ISBN 3-540-51705-7 and 0-387-517-05-7, 1989.
[Smolka 89] G. Smolka & H. Aït-Kaci, "Inheritance Hierarchies: Semantics and Unification", to appear in Journal of Symbolic Computation, Special Issue on Unification, C. Kirchner, Ed., March 1989.
[Steele 84] G.L. Steele Jr., "Common LISP: The Language, Digital Press", ISBN 0-932376-41-X, 1984.

8   Appendix: Basic Notations and Elements of ψ-Term Theory

A *signature* is a triple $< \mathscr{S}, \leq, \mathscr{L} >$ where $\mathscr{S}$, the set of sorts, is partially ordered by $\leq$, and $\mathscr{L}$ is the set of labels, also called attributes. In the sequel, we assume Σ is such a signature. $\mathscr{L}^*$ is the monoïd of finite sequences of labels, denoted u, or $(\ell_1 \dots \ell_n)$, and called occurrences. We denote $^\wedge$ the operation of concatenation, and use $u::\ell$ as an abbreviation for $u^\wedge(\ell)$. We denote u/v, when it exists, the unique occurrence w such that $u = v^\wedge w$. These operations extend naturally to $\mathscr{P}(\mathscr{L}^*)$.

A *tree domain* is a non empty subset D of $\mathscr{L}^*$ which is stable by prefixing, i.e.: $u^\wedge v \in D \Rightarrow u \in D$. A tree domain is finitely branched if for each $u \in D$, there is a finite number of occurrences of the form $u::\ell$. A *tree* is an application $f: D \to \mathscr{S}$. A tree is finitely branched if its domain is finitely branched. For each $u \in D$, f/u designates the tree $D/u \to \mathscr{S}$ defined by $f/u(v) = f(u^\wedge v)$ for each $v \in D/u$; f/u is called subtree of f at occurrence u. A *regular* tree is a tree which has a finite number of subtrees.

A *ψ-term* is a pair $< f, \equiv >$ of a regular and finitely branched tree $f: D \to \mathscr{S}$ and an equivalence relation $\equiv$ in D, called coreference, such that: $D/\equiv$ is finite; f is constant on each coreference class (denoted [u] for $u \in D$); $\equiv$ is compatible with concatenation to the right: if $u::\ell \in D$ and $u \equiv v$ then $v::\ell \in D$ and $u::\ell \equiv v::\ell$. A concrete syntax may be used to denote ψ-terms. We write $(s\ \ell_1\ t_1 \dots \ell_n\ t_n)$ rather than $s(\ell_1 \Rightarrow t_1, \dots, \ell_n \Rightarrow t_n)$, and $(:= X\ t)$ rather than $X:t$ to express coreference.

We denote Ψ the set of ψ-terms. It is partially ordered. Let $\psi = < f: D \to \mathscr{S}, \equiv >$ and $\psi' = < f': D' \to \mathscr{S}, \equiv' >$ be two ψ-terms. The partial order is defined by:

$$\psi \leqslant \psi' \quad \Leftrightarrow \quad \psi = (\perp) \quad \text{or} \quad \{D' \subset D \text{ and } [\equiv' \subset \equiv] \text{ and } [\forall u \in D'\ f(u) \leqslant f'(u)]\}$$

$< \mathscr{S}, \leq >$ may be considered as a subset of $< \Psi, \leq >$ by identifying sort s with ψ-term (s). If $\mathscr{S}$ is a lower lattice with $\wedge$ glb operation, this glb extends to Ψ. Let $\psi_1 = < f_1: D_1 \to \mathscr{S}, \equiv_1 >$ and $\psi_2 = < f_2: D_2 \to \mathscr{S}, \equiv_2 >$ be two ψ-terms. Their glb $\psi = < f: D \to \mathscr{S}, \equiv >$ is defined as follows:

$$\forall n \in \mathbf{N} \qquad D_{(n)} \quad = \quad \text{DEF}(\equiv_{(n)}) \tag{i}$$

$$\equiv_{(0)} \quad = \quad (\equiv_1 \cup \equiv_2)^+ \tag{ii}$$

$$\forall n \in \mathbf{N} \qquad \equiv_{(n+1)} \quad = \quad (\equiv_{(n)} \cup v_{(n)})^* \tag{iii}$$

$$\forall n \in \mathbf{N} \qquad v_{(n)} \quad = \quad \{ <u_1^\wedge w, u_2^\wedge w> \mid u_1 \equiv_{(n)} u_2,$$
$$(u_1^\wedge w) \in D_{(n)} \text{ or } (u_2^\wedge w) \in D_{(n)} \} \tag{iv}$$

$$D \quad = \quad \cup \{D_{(n)} \mid n \in \mathbf{N}\} \quad = \quad \text{DEF}(\equiv) \tag{v}$$

$$\equiv \quad = \quad \cup \{\equiv_{(n)} \mid n \in \mathbf{N}\} \tag{vi}$$

$$\forall u \in D \qquad f(u) \quad = \quad \text{Inf}(\{f_1(v) \wedge f_2(v) \mid v \in D_1 \cup D_2, u \equiv v\}) \tag{vii}$$

Note that (vii) is slightly incorrect: we implictly assume that f_1 and f_2 have been extended to $D_1 \cup D_2$, by setting $f_1(v)$, respectively $f_2(v)$, to $^\top$ whenever $f_1(v)$, respectively $f_2(v)$, is undefined.

Let $< \mathscr{I}_\mathscr{S}, \mathscr{I}_\mathscr{L} >$ be any Σ-algebra, that is, $\mathscr{I}_\mathscr{S}$ interprets sorts as subsets of a universe $\mathscr{U}$, with $\leqslant$ mapped into $\subset$, $\wedge$ into $\cap$, $\perp$ into $\varnothing$, $^\top$ into $\mathscr{U}$, and $\mathscr{I}_\mathscr{L}$ associates each label with a partial function from $\mathscr{U}$ to $\mathscr{U}$. Then $\mathscr{I}_\mathscr{S}$ can be homomorphically extended to Ψ. The extension, $\mathscr{I}_\mathscr{S}: \Psi \to \mathscr{P}(\mathscr{U})$ can be defined, on each $\psi \in \Psi$ by solving a set of constraints $\mathscr{C}$ associated with ψ as follows. We consider each coreference class in ψ as the name of a variable that will participate in $\mathscr{C}$. For each variable x, associate the constraint "$x \in \mathscr{I}_\mathscr{S}(f(x))$"; to each pair of variables of the form x=[u] and y=[u::ℓ], associate the constraint "$\mathscr{I}_\mathscr{L}(\ell)(x) \in \mathscr{I}_\mathscr{S}(y)$". Call $\mathscr{C}$ the conjunction of these constraints. A solution of $\mathscr{C}$ assigns a value (in $\mathscr{U}$) to each variable so that $\mathscr{C}$ is true. Let z denote the [( )] variable: then $\mathscr{I}_\mathscr{S}(\psi)$ is by definition the set of values assigned to z by the solutions of $\mathscr{C}$. (Note that Aït-Kaci equivalent definition differs in its formulation.)

9 Appendix: An Example of Polymorphic Sorts Built as Filters

We start with an assorted signature $\Sigma_1 = <\mathcal{S}_1, \leq_1, \mathcal{L}_1, \varphi_1>$. We add the sorts *pair* and *null* to $\mathcal{S}_1$ and obtain $\mathcal{S}_2$. We add the labels :car and :cdr, obtaining $\mathcal{L}_1$. The partial order and assortment are extended to $\leq_2$ and φ_2 by the declarations

$$
\begin{array}{llll}
null & <_2 \ ^\top & \varphi_2(pair, \text{:car}) & = \ ^\top \\
\bot & <_2 \ null & \varphi_2(pair, \text{:cdr}) & = \ ^\top \\
pair & <_2 \ ^\top & & \\
\bot & <_2 \ pair & &
\end{array}
$$

We obtain Σ_2. Define $\Sigma = <\mathfrak{C}, \leq, \mathcal{L}_2, \varphi>$ as the assorted signature where $\mathfrak{C}$ is the subset of $\mathfrak{C}_{\Sigma_2}$ consisting of *uncoreferenced* filters, that is, filters whose coreference relation is the identity relation. Such filters are really finite trees, and do not need the := coreference operator in their concrete syntax representation. The partial order $\leq$ is simply inherited from the ψ-term partial order in $\mathfrak{C}_{\Sigma_2}$. It is a lower lattice if this was true for Σ_1. The assortment $\varphi: \mathfrak{C} \times \mathcal{L}_2 \to \mathcal{S}$ is well defined as an extension of φ_2 by:

$$
\begin{array}{llll}
\varphi[(s \ \ell_1 t_1 \dots \ell_n t_n), \ell] & = & t_k & \text{if } \ell = \ell_k \text{ for some k with } 1 \leq k \leq n, \\
& = & (s') & \text{with } s' = \varphi_2(s, \ell) \text{ otherwise.}
\end{array}
$$

10 Appendix: A Session with U-Log Meta-Interpreter

After entering the reflexive definition of U-Log in U-Log (see Section 5), and the clauses

```
(clause :id   "app1"            (clause :id   "app2"
        :head (app () *1 *1)            :head (app (*e . *1) *m (*e . *r))
        :body ( ))                      :body ((app *1 *m *r)))
```

corresponding to the classical definition of *append* in Prolog.(U-Log permits positional syntax as well as keyworded syntax, and the use of LISP syntax for lists or pairs, which is more legible than nested *pair* and *null* expressions. We have abbreviated *append* into *app*), we obtain the following session (abbreviating :forest into :fo, :response into :re, :proof into :pr for legibility):

*;;; What is the proof tree of (app (1 2) (3) *r)?*
```
(? (⊨ :re (app (1 2) (3) *r) :pr *t))
((⊨ :PR (PROOF :OP "APP2"
                   :FO ((PROOF :OP "APP2" :FO ((PROOF :OP "APP1" :FO NIL)))))
      :RE (APP (1 2) (3) (1 2 3)))))
```

;;; What can be proven with proof tree (proof :op "app2" :fo ((proof :op "app1" :fo nil))))?
```
(? (⊨ :pr (proof :op "app2" :fo ((proof :op "app1" :fo nil)))) :re *r))
((⊨ :PR (PROOF :OP "APP2" :FOREST ((PROOF :OP "APP1" :FO NIL)))
      :RE (APP (*E) *R (*E . *R))))
```

;;; What can be proven with what proof tree?
```
(? (⊨ :pr *t :re *r))
((⊨ :PR (PROOF :OP "APP1" :FO NIL) :RE (APP NIL *L *L)))
((⊨ :PR (PROOF :OP "APP2" :FO ((PROOF :OP "APP1" :FO NIL)))
      :RE (APP (*E) *R (*E . *R))))
((⊨ :PR (PROOF :OP "APP2" :FO ((PROOF :OP "APP2" :FO ((PROOF :OP "APP1" :FO NIL)))))
      :RE (APP (*E1 *E2) *R (*E1 *E2 . *R))))
...
```

Acknowledgements

We thank Hassan Aït-Kaci and Gérard Huet for their preçious advice and helpful remarks on an earlier version of this work. We thank Jacques Carlier for helpful discussions and for providing a counter-example to our "finite radius" conjecture. We thank French D.R.E.T. for financial support through grant N°89.34.323.00.470.75.01. We thank GRAPHAEL for its support.

Compilation of Predicate Abstractions in Higher-Order Logic Programming

Weidong Chen
Computer Science & Engineering
Southern Methodist University
Dallas, TX 75275-0122
wchen@csvax.seas.smu.edu

David Scott Warren
Computer Science Department
SUNY at Stony Brook
Stony Brook, NY 11794-4400
warren@sbcs.sunysb.edu

Abstract

We explore higher-order logic programming and its relationship to computation in predicate calculus. The framework is based upon a logic of untyped λ-calculus which has a general model-theoretic semantics and whose equality theory corresponds to α-equivalence. The focus of the paper is on computing with predicate abstractions that are formalized by equivalence axioms with respect to a notion called *top reduction*. It is shown that, under certain conditions, computation with predicate abstractions can be compiled into predicate calculus and all most general answers are still preserved.

1 Motivation

The nature and the role of higher-order logic programming have been investigated in several different directions. In [8], D. Miller and G. Nadathur took a full-fledged approach, and developed a formal proof-theoretic basis for higher-order logic programming using Church's simple theory of types. One of the novel features is the replacement of first-order terms with typed λ-terms and a corresponding notion of higher-order unification [7]. However, the expressive power does not come for free since higher-order unification in typed λ-calculus is known to be undecidable in general [6].

In contrast, D.H.D. Warren showed informally [11] that certain uses of predicate variables and abstractions can be translated into efficient Prolog programs, and thus such extensions of Prolog are not needed. From a computational point of view, it indicates that there is no significant overhead for having these convenient features. Unfortunately, questions remain as to what is exactly the semantics that warrants such a translation, and whether the translation always preserves the semantics. A semantic account seems to be lacking also in [5].

If there is a spectrum of higher-order logic programming systems with regard to expressive power, the above two approaches may serve as the extreme cases in terms of manipulations of λ-terms. Our objective is to explore useful alternatives in between, and derive a more tractable framework for higher-order logic programming.

Our approach has been to extend predicate calculus with higher-order features in a step by step manner. In [3], we developed a logic of abstraction-free λ-calculus, called HiLog. HiLog treats predicates, functions, and atomic formulas as first-class objects, and yet it has a direct translation into predicate calculus.

Next, we extend HiLog with λ-abstractions to include all terms in untyped λ-calculus. This immediately raises the issue of higher-order unification. It turns out that there are two alternative views of λ-conversions. The equational view is such that terms are *equal* if they are convertible to each other. The resulting equality theory would be λ-equivalence, with respect to which the problem of unification seems still open for untyped λ-calculus. The other view, which we call *relational*, is that terms may related to each other by λ-conversions, but they are not necessarily equal. This opens up the possibilities of choosing a more tractable equality theory, such as α-equivalence, and incorporating other conversions in a more controlled manner.

Our approach is based upon the relational view of λ-conversions. In [4], we developed a logic of untyped λ-calculus, called L_α, whose equality theory corresponds to α-equivalence. L_α provides a foundational framework for investigating other λ-conversions in a systematic, controlled way. In particular, this paper explores predicate abstractions that are formalized by equivalence axioms (not equality axioms) in L_α with respect to a restricted notion of reduction, called *top reduction*. We show that under certain conditions, computing with predicate abstractions can be compiled into predicate calculus while preserving all most general answers.

Predicate abstractions provide a mechanism for anonymous predicates. They can be used to avoid proliferation of named auxiliary predicates that are used only once or twice. Consider a generic predicate definition:

list(A)([]).
list(A)([X|L]) :- A(X), list(A)(L).

(We follow the convention of variables beginning with upper case letters and other symbols starting with lower case letters.) With predicate abstractions, we may simply call list(λX.X=<100)(L) without introducing a named predicate for X=<100. In solving a subgoal with an abstraction as the predicate, e.g., (λX.X=<100)(Y), β-reduction is applied to simplify it into Y =< 100.

Anonymous predicates are also useful for encapsulation and module definitions. For example, the following query contains a local predicate definition Link:

?- Link = λ(X,Y).exists(λZ.and(edge(X,Z),edge(Z,Y))), Link(N1,N2).

A syntactically more pleasing notation may be

?- Link(N1, N2) where { Link(X, Y) :- edge(X, Z), edge(Z, Y).}

Because of space limitations, many details have been omitted; they can be found in [4]. Section 2 presents a brief description of the logic of α-equivalence, called L_α, and its encoding in predicate calculus. Section 3 incorporates predicate abstractions into L_α, and describes a compilation technique for certain programs and queries. We conclude with a discussion of computational strategies for arbitrary programs and queries.

2 L_α: A Logic of Alpha Equivalence

The alphabet contains, in addition to parentheses, connectives and quantifiers, the following disjoint sets of symbols: a countable set of parameters and a denumerable set of variables.

Terms are inductively defined as follows:

- A parameter or a variable is a term;

- An application $t(t_1, \cdots, t_n)$ or an abstraction $\lambda(X_1, ..., X_n).t$ is a term

where $n \geq 1$, $t, t_1, \cdots, t_n$ are terms, and $X_1, ..., X_n$ are distinct variables. *Atomic formulas* are simply terms. Complex formulas are inductively defined from atomic ones using connectives and quantifiers in the usual way.

L_α extends λ-calculus [2] with multi-arity applications and abstractions. Nevertheless, concepts in λ-calculus, can be easily adapted to L_α.

Definition 2.1 *A domain structure* $\mathbf{D}$ *is a quadruple* $< U, I, \mathcal{F}, \mathcal{S} >$, *where*

- U *is a nonempty domain of universe;*

- $I(s) \in U$ *for every parameter symbol s;*

- $\mathcal{F}$ *is a function from U to* $\prod_{k=1}^{\infty}[U^k \rightarrow U]$;

- $\mathcal{S}$ *is a function from* $\bigcup_{k=1}^{\infty}[U^k \rightarrow U]$ *to U*

For each $u \in U$ and each $k \geq 1$, the k-th projection of $\mathcal{F}(u)$ is a k-ary function in $[U^k \rightarrow U]$, which we denote by $u_{\mathcal{F}}^{(k)}$.

Given a domain structure $\mathbf{D}$ and a variable assignment ν, ν can be extended to all terms as follows:

- $\nu(s) = I(s)$ for each parameter symbol s;

- $\nu(t_0(t_1, ..., t_n)) = \nu(t_0)_{\mathcal{F}}^{(n)}(\nu(t_1), ..., \nu(t_n))$;

- $\nu(\lambda(X_1, ..., X_n).t) = \mathcal{S}(f)$, where f is the n-ary function such that for every $d_1, ..., d_n \in U$, $f(d_1, ..., d_n) = \nu'(t)$, and ν' is the variable assignment such that $\nu'(X_i) = d_i$ for each $i(1 \leq i \leq n)$, and $\nu'(Y) = \nu(Y)$ for all other variables Y.

f is called the *function associated with* $\lambda(X_1, ..., X_n).t$ *under* ν.

Definition 2.2 *A semantic structure* $\mathbf{M}$ *is a quintuple* $< U, U_{true}, I, \mathcal{F}, \mathcal{S} >$, *where* $< U, I, \mathcal{F}, \mathcal{S} >$ *is a domain structure, and* $U_{true} \subseteq U$.

Given a semantic structure **M** and a variable assignment ν, let A be an atomic formula. A is *satisfied* by **M** under ν, denoted by $\mathbf{M} \models A[\nu]$, if and only if $\nu(A) \in U_{true}$. The meaning of complex formulas is defined in the standard way. A formula ϕ is *true* in **M** if $\mathbf{M} \models \phi[\nu]$ for every variable assignment ν. A formula is *valid* if it is true in every semantic structure.

Theorem 2.1 ([4]) *For any terms t and s, t and s are α-equivalent if and only if for every semantic structure* **M** *and variable assignment ν, $\nu(t) = \nu(s)$*

L_α can be encoded in predicate calculus. The encoding can be viewed as a generalization of D.H.D. Warren's method in [11]. Let $\mathcal{L}_\alpha$ be a language of L_α with a set $\mathcal{V}$ of variables and a set $\mathcal{C}$ of parameter symbols. We define a language $\mathcal{L}_P$ of predicate calculus with the set of variables $\mathcal{V}$, constants $\mathcal{C}$, a unique predicate symbol call, and for each $n \geq 1$, an $n+1$-ary function symbol apply_{n+1}. In addition, $\mathcal{L}_P$ contains denumerably many function symbols, which will be used to represent abstractions in L_α.

Let t be a predicate calculus term in $\mathcal{L}_P$, and $X_1, ..., X_n$ be distinct variables. A *simple abstraction* is of the form $\lambda(X_1, ..., X_n).t$. A simple abstraction $\lambda(X_1, ..., X_n).t$ is *most general* if each occurrence of a term in t that does not contain any $X_i (1 \leq i \leq n)$ is a distinct variable. Obviously, any simple abstraction can be written in the form $(\lambda(X_1, ..., X_n).t)\theta$, where $\lambda(X_1, ..., X_n).t$ is most general, and θ is a substitution in predicate calculus such that no $X_i (1 \leq i \leq n)$ appears in θ.

Let $\mathcal{N}$ denote a set of pairs of the form $< f, a >$, where f is a function symbol of arity k for some $k \geq 0$, and a is a most general simple abstraction with k distinct free variables. Intuitively, $\mathcal{N}$ records the function symbols that have been introduced for encoding simple abstractions so far.

Initially, let $\mathcal{N}$ be the empty set. We define inductively $\mathcal{N}$, and a transformation encode_t that encodes L_α terms as predicate calculus terms. The encoding is adapted from the encoding of HiLog in predicate calculus [3].

- $\mathsf{encode}_t(X) = X$, for each variable $X \in \mathcal{V}$;

- $\mathsf{encode}_t(s) = s$, for each parameter symbol $s \in \mathcal{C}$;

- $\mathsf{encode}_t(t_0(t_1, ..., t_n)) =$
  $\qquad \mathsf{apply}_{n+1}(\mathsf{encode}_t(t_0), \mathsf{encode}_t(t_1), ..., \mathsf{encode}_t(t_n));$

- for abstraction $\lambda(X_1, ..., X_n).t$, let $\lambda(X_1, ..., X_n).\text{encode}_t(t)$ be of the form $(\lambda(X_1, ..., X_n).t')\theta$, where

    - $\lambda(X_1, ..., X_n).t'$ is a most general simple abstraction with k distinct free variables $Y_1, ..., Y_k$ from left to right for some $k \geq 0$; and

    - θ is a substitution in predicate calculus for $Y_1, ..., Y_k$ such that no $X_i(1 \leq i \leq n)$ appears in θ.

Either there exists some pair $< f, a > \in \mathcal{N}$ such that f is a k-ary function symbol, and a and $\lambda(X_1, ..., X_n).t'$ are identical up to renaming of variables, or we introduce a new k-ary function symbol f, and add $< f, \lambda(X_1, ..., X_n).t' >$ to $\mathcal{N}$. In either case,

$$\text{encode}_t(\lambda(X_1, ..., X_n).t) = f(Y_1, ..., Y_k)\theta$$

We require that $\mathcal{L}_P$ contain all the new functions symbols in $\mathcal{N}$. For encoding of formulas,

- $\text{encode}_f(A) = \text{call}(\text{encode}_t(A))$, where A is an atomic formula.

encode_f can be extended inductively to complex formulas.

Theorem 2.2 ([4]) *For any L_α formula ϕ, ϕ is valid in L_α if and only if $\text{encode}_f(\phi)$ is valid in predicate calculus.*

The encoding of L_α implies that L_α has a sound and complete proof theory. Furthermore, the encoding encode_t is a one-to-one and onto mapping from terms in $\mathcal{L}_\alpha$ to terms in $\mathcal{L}_P$. It follows that encode_t preserves all unifiers, and most general unifiers, in particular.

3 Computing with Predicate Abstractions

L_α provides only α-conversion directly; no other λ-conversions are built in. While a user may define rules for reducing specific function applications, it is useful to build some conversions into a logic programming system. The advantage with L_α is that various conversions, depending upon the underlying application, may be incorporated by axioms. In the following, we consider predicate abstractions that are formalized by equivalence axioms of atomic formulas.

3.1 Axioms of Predicate Abstractions

Abstractions as predicates usually have an intended meaning. For example, when λ X.student(X) is used as a predicate as in (λ X.student(X))(john), β-reduction should be applied to reduce this to student(john). In some applications, predicates themselves have to be reduced. For instance, (λ X.(λ Y.parent(X, Y)))(mary)(john) should be reduced to (λ Y.parent(mary, Y))(john) before being reduced to parent(mary, john).

To formalize this intuitive idea, we introduce a restricted notion of β-reduction, called *top reduction*, denoted by $\xrightarrow{t}$. It is defined inductively as follows:

- $(\lambda(X_1, ..., X_n).s)(s_1, ..., s_n) \xrightarrow{t} s[s_1/X_1, ..., s_n/X_n]$;

- if $t \xrightarrow{t} t'$ and $t_1, ..., t_n$ are terms, then $t(t_1, ..., t_n) \xrightarrow{t} t'(t_1, ..., t_n)$.

The converse of $\xrightarrow{t}$ is called *top expansion*. Let $\xrightarrow{t}\hspace{-0.5em}\rightarrow$ denote the reflexive and transitive closure of $\xrightarrow{t}$, and $\xleftarrow{t}\rightarrow$ the reflexive, symmetric, and transitive closure of $\xrightarrow{t}$. Top reduction is deterministic and has the property that no reduction is carried out inside abstractions or arguments of applications.

For predicate abstractions, we introduce an axiom $t \equiv s$ for every pair of terms t and s such that $t \xrightarrow{t} s$. This axiom scheme is denoted by $\mathcal{PA}$.

3.2 Infinitary SLD-Resolution with $\mathcal{PA}$

We now consider logic programming with predicate abstractions. We assume that the notions of definite and negative clauses are exactly as in predicate calculus, except that atomic formulas are as in L_α. A *program* is a finite set of definite clauses. A *goal* is a conjunction of atomic formulas. A *query* is a negative clause.

Because of the additional axioms $\mathcal{PA}$, each program P can really be viewed as $P \cup \mathcal{PA}$, where each axiom is represented by two clauses. While $P \cup \mathcal{PA}$ is infinite, its semantics can still be characterized using the least model semantics [1], except that the Herbrand universe (or base) is the quotient by α-equivalence of the set of all closed terms in L_α. Let $\mathcal{M}(P)$ denote the least Herbrand model of $P \cup \mathcal{PA}$.

Computationally, unification with respect to α-equivalence has been studied in [9], and it satisfies the property of most general unifiers. SLD-resolution can also be extended to infinite programs. By using the encoding of L_α in predicate calculus, we have the following.

Theorem 3.1 *Let P be a program and Q a query in L_α. If T is an SLD-tree for Q with respect to $P \cup \mathcal{PA}$, then $\mathrm{encode}(T)$ is an SLD-tree for $\mathrm{encode}(Q)$ with respect to $\mathrm{encode}(P) \cup \mathrm{encode}(\mathcal{PA})$ in predicate calculus, and vice versa.*

The problem, of course, is the infinitary nature of SLD-trees here, which is due to the fact that top expansion is highly nondeterministic. Furthermore, the encoding is infinite, i.e., it requires infinitely many function symbols.

3.3 Rigid Programs and Non-floundered Queries

To avoid top expansions, we impose certain conditions on programs and queries. We say that an atomic formula A is *flexible* if it starts with a variable or if $A \xrightarrow{t} B$ for some B. An atomic formula is *rigid* if it is not flexible. A *rigid program* is a program in which the head of each clause is a rigid atomic formula.

With each rigid logic program P, we associate a function T_P over Herbrand interpretations. We define T_P such that $A \in T_P(I)$ iff there exists a variant of a clause in P

$$B : -B_1, \cdots, B_n. \ (n \geq 0)$$

and some closed substitution θ such that $A \xrightarrow{t} B\theta$ and $\{B_1\theta, \cdots, B_n\theta\} \subseteq I$. Notice that top expansion is not used since T_P requires only $A \xrightarrow{t} B$, and not $A \xleftarrow{t} B$. It turns out that the least fixed point of T_P is also a model of $\mathcal{PA}$.

Lemma 3.1 *Let P be a rigid program. Then $lfp(T_P) = T_P \uparrow \omega = \mathcal{M}(P)$.*

We modify SLD-resolution for rigid programs such that a computation rule selects only atomic formulas that do not start with a variable. If such a selection is not possible for a nonempty query, we say that the query is *floundered*.

Theorem 3.2 *Let R be a fixed computation rule, P a rigid program, and Q a non-floundered query of the form :- G. Then*

1. $\mathcal{M}(P) \models \exists(G)$ *if and only if there is an SLD-answer for G.*

2. *The SLD-tree for Q is failed if and only if* $\mathcal{M}(P) \models \neg\exists(G)$.

3. $\mathcal{M}(P) \models \forall(G\delta)$ *if and only if there is an SLD-answer for G more general than* δ*, where* δ *is a substitution.*

It turns out that for rigid programs and non-floundered queries, the encoding can be restricted to the finite set of terms in the program and the query, and still preserve all most general answers.

Let Σ be the finite set of terms occurring in a rigid program P and a non-floundered query Q. We assume that Σ is closed under subterms. Although the encoding encode_t has been defined for all terms in L_α, only terms Σ need to be transformed in practice, and therefore only a finite number of new function symbols need to be introduced. Let $\mathcal{L}_P$ be the resulting language of predicate calculus.

The encoding of Σ is sufficient in the sense that all terms that may appear in the SLD-tree for Q with respect to P can be represented by predicate calculus terms in $\mathcal{L}_P$. We define the closure $uni(\Sigma)$ of Σ under substitutions as follows:

- $\Sigma \subseteq uni(\Sigma)$;

- $t\{t_1/X_1, \cdots, t_n/X_n\} \in uni(\Sigma)$ if $t, t_1, \cdots, t_n$ are in $uni(\Sigma)$, and X_i's are distinct variables.

New terms will be generated by unification or top reduction in the process of computation, both of which involve only substitutions. Thus $uni(\Sigma)$ contains all possible terms that may appear in an SLD-tree. The encoding for Σ can be extended to all terms in $uni(\Sigma)$, by

$$\text{encode}_t(t\{t_1/X_1, \cdots, t_n/X_n\}) =$$
$$\text{encode}_t(t)\{\text{encode}_t(t_1)/X_1, \cdots, \text{encode}_t(t_n)/X_n\}.$$

Intuitively top reduction does not generate new abstractions. Therefore the encoding can be done at compile time for Σ.

The encoding of a program P is the same as before. However, for axioms $\mathcal{PA}$, we only need a finite number of them. Let $< h, a >$ be any pair where h is a k-ary function symbol introduced by the encoding for a most general simple abstraction a of the form $\lambda(X_1, ..., X_n).t$, which has distinct free variables $Y_1, ..., Y_k$ from left to right. We introduce a definite clause for h:

$$\mathsf{call}(\mathsf{apply}_{n+1}(h(Y_1, \cdots, Y_m), X_1, \cdots, X_n)) \text{ :- } \mathsf{call}(t).$$

which has the effect of performing top reduction. Let encode_t be the set of all clauses for new function symbols.

Theorem 3.3 *Let P be a rigid program and Q a non-floundered query of the form :- G in L_α. Then the SLD-tree for $\mathsf{encode}(G)$ with respect to $\mathsf{encode}(P) \cup \mathsf{encode}_t$ satisfies the following:*

1. *$\mathcal{M}(P) \models \exists(G)$ if and only if there is an SLD-answer for $\mathsf{encode}(G)$;*

2. *The SLD-tree is failed if and only if $\mathcal{M}(P) \models \neg\exists(G)$.*

3. *$\mathcal{M}(P) \models \forall(G\delta)$ if and only if there is an SLD-answer for $\mathsf{encode}(G)$ more general than $\mathsf{encode}(\delta)$.*

4 Discussion

We conclude with a discussion of a novel domain closure assumption, negation, and some issues for further investigation.

Top expansion is highly nondeterministic. It may also generate new abstractions, which presents problems for compilation into predicate calculus. The assumptions of rigid programs and non-floundered queries are necessary in order to avoid top expansion and still preserve all most general answers.

Another alternative is to impose a certain domain closure assumption such that the Herbrand universe will be constructed using only parameters and abstractions that appear in a program. This will restrict $\mathcal{PA}$ to a certain extent, and also avoid some nonsensible answers, such as $\lambda X.\mathsf{db}(\mathsf{john})$, in

```
db(john).
?- Group(david).
```

In L_α, for simplicity, no connectives or quantifiers are allowed inside abstractions. For logic programming applications, connectives and quantifiers can be defined using rules, e.g.,

```
not(X) :- ~ X.
and(X, Y) :- X, Y.
or(X, Y) :- X; Y.
exists(X) :- X(Y).
```

Any semantics that is applicable to all programs, such as well founded semantics [10], can be employed to provide a precise meaning for programs containing parameters not, and, or, exists.

This work suggests a general approach to incorporating various notions of λ-conversions by adding axioms on top of a basic logic of α-equivalence. It may provide a framework for studying various logic programming systems that lie between Prolog on the one hand, and systems such as λProlog on the other. There are also implementation issues, such as efficient handling of λ-terms and reductions, which need further investigation.

References

[1] K.R. Apt and M.H. van Emden. Contributions to the theory of logic programming. *JACM*, 29(3):841–862, July 1982.

[2] H.P. Barendregt. *The Lambda Calculus: Its Syntax and Semantics (Revised Edition)*. North-Holland Publishing Co., 1984.

[3] W. Chen, M. Kifer, and D.S. Warren. HiLog: A first-order semantics for higher-order logic programming constructs. In *Proc. North American Conference on Logic Programming*, October 1989.

[4] W. Chen and D.S. Warren. Predicate abstractions in higher-order logic programming. Manuscript, March 1991.

[5] M.H.M. Cheng, M.H. van Emden, and B.E. Richards. On Warren's method for functional programming in logic. In *Proc. ICLP*, pages 546–560, 1990.

[6] W.D. Goldfarb. The undecidability of the second-order unification problem. *Theoretical Computer Science*, 13:225–230, 1981.

[7] G.P. Huet. A unification algorithm for typed λ-calculus. *TCS*, 1:27–57, 1975.

[8] G. Nadathur and N. Miller. Higher-order horn clauses. *JACM*, 37(4), 1990.

[9] J. Staples and P.J. Robinson. Efficient unification of quantified terms. *Journal of Logic Programming*, 5:133–149, 1988.

[10] A. van Gelder, K.A. Ross, and J.S. Schlipf. Unfounded sets and well-founded semantics for general logic programs. In *Proc. 7th ACM Symp. on PODS*, 1988.

[11] D.H.D. Warren. Higher-order extensions to Prolog: are they needed? *Machine Intelligence*, 10:441–454, 1982.

Abstract Interpretation: A Kind of Magic

Ulf Nilsson

Department of Computer and Information Science
Linköping University
S–581 83 Linköping, Sweden

email: urn@IDA.LiU.SE

Abstract: *Magic sets and, more recently, magic templates are used in the field of deductive databases to facilitate efficient bottom up evaluation of database queries. Roughly speaking a top down computation is simulated by first transforming the program and then executing the new program bottom up. In this paper we give a new and very simple proof that this approach is equivalent to the collecting interpretation of the abstract interpretation framework of C. Mellish. As a side-effect we also prove that "bottom up" abstract interpretation based on the magic templates transformation is equally powerful as this particular abstract interpretation framework, but less powerful than other (more precise) abstract interpretation frameworks.*

1 Introduction

The fields of deductive databases and logic programming are closely related. In fact, many of the results from one discipline are directly applicable also in the other. The objective of this paper is to shed some light on the relationship between abstract interpretation of logic programs and the magic templates method.

The area of abstract interpretation was founded by P. Cousot and R. Cousot [CC77]. It is an attempt to provide a common framework for the great variety of data-flow analysis methods developed in the 70s. The ideas were adopted and adapted by e.g. the logic programming community to provide a basis for inferring run-time properties of logic programs. It was, for instance, used by C. Mellish [Mel87] to infer mode-information (information about how arguments are instantiated when procedure-calls go ahead and succeed). Two principal approaches to abstract interpretation of logic programs can be singled out: "top down" approaches which are based on SLD-resolution (for instance that of C. Mellish) and "bottom up" approaches based on some kind of fixed point semantics of logic programs (for instance, the work of K. Marriott and H. Søndergaard [MS89]).

One of the outstanding issues in the deductive database community is the notion of magic sets (e.g. [Ban86, BR87]) and more recently magic templates [Ram88]. The motivation behind this work is the insight that SLD-resolution with backtracking is inappropriate for databases implementations. The main reason being that it generates one tuple at a time while in database systems it is — for computational reasons — advantageous to compute complete relations by using e.g. operations of relational algebra. However, this problem can be solved by bottom up computations similar to the fixed point semantics of logic programs. Unfortunately, in a query-answering system, bottom up computations of logic programs tend to do a great deal of unnecessary work since very few of the facts obtained in the computation are actually needed to answer the query. In response to these two problems the magic sets (templates) methods were developed. The rough idea is that instead of executing the original program (bottom up) the program (and a query) are transformed into a new program which is executed bottom up. By this method, both of the above-mentioned problems are (at least partially) solved.

In this paper we try to relate the notions of abstract interpretation of logic programs and the method of magic templates. In fact, we show that the magic templates method is a particular instance of abstract interpretation. This result is to the best of our knowledge not published, however a similar observation was independently made by Kanamori [Kan90]. As a side-effect we also prove that bottom up abstract interpretation is equally powerful as the top down abstract interpretation of Mellish by exploiting the magic templates transformation. However, we also note that using this approach some precision has to be sacrificed.

The rest of this paper is organised accordingly: After a few notational conventions we give a brief account of the magic templates method in Section 3. The standard presentation is accompanied by an alternative formulation intended to somewhat bridge the gap between the two fields. In Section 4 we discuss the notion of abstract interpretation as presented by C. Mellish. In Section 5 we show that the magic templates method can be viewed as an instance of abstract interpretation. We also discuss the relation between bottom up and top down abstract interpretations.

2 Preliminaries

The notation and terminology is standard with a few small exceptions listed below. For reference see e.g. [Llo87] or [NM90].

In what follows we assume that a fixed alphabet Σ is given. By Atom we denote the set of all atoms which can be constructed using Σ.

By a *renaming* we mean a substitution of the form $\{x_1/y_1, \ldots, x_n/y_n\}$ where $y_1, \ldots, y_n$ is a permutation of $x_1, \ldots, x_n$. Renamings are sometimes referred to as *invertible* substitutions.

Let E be an expression (a term or a formula) and θ a renaming. The expression $E\theta$ is called a *variant* of E. This relation is an equivalence relation which can naturally be extended (point-wise) to substitutions.

Unification of terms and atoms is extended to tuples of terms and atoms in the obvious way. That is, a unifier of $A_1 \ldots A_n$ and $B_1 \ldots B_n$ is a substitution θ such that $A_1\theta \equiv B_1\theta, \ldots, A_n\theta \equiv B_n\theta$.

By the notation $\overline{\mathrm{mgu}}(A_1 \ldots A_n, B_1 \ldots B_n)$ we mean a most general unifier of the tuples $A_1 \ldots A_n$ and $B_1' \ldots B_n'$ where B_i' is a variant of B_i such that $var(B_1') \cap \ldots \cap var(B_n') \cap var(A_1 \ldots A_n) = \emptyset$. If $n = 0$ the result is the empty substitution. Note that the function is not well-defined since $B_1, \ldots, B_n$ may have many variants which satisfy the condition. This may be remedied using some additional machinery. For instance, by letting $\overline{\mathrm{mgu}}$

return an equivalence class of variants of the mgu. However to simplify the notation we nevertheless use this definition with the remark that it is at least well-defined modulo renaming.

By a *prefix* of a clause $A_0 \leftarrow A_1, \ldots, A_n$ we mean any clause $A_0 \leftarrow A_1, \ldots, A_m$ such that $0 \leq m \leq n$. If P is a program then let $\tilde{P}$ be the set of all prefixes of all clauses in P.

3 Magic Templates

In this section we review the notion of magic templates of Ramakrishnan [Ram88]. We first provide the standard definition and then try to shed some additional light on it by giving a syntactically different but essentially equivalent definition based on the standard operational semantics of Prolog programs (that is, SLD-resolution with a leftmost computation rule).

In contrast to most work in the field of deductive databases we do not impose any assumptions about the program being divided into an extensional and intensional part (or put alternatively, the extensional part is empty). However, all the results carry over also to the case when the separation is made.

Let $A \in Atom$ be of the form $p(t_1, \ldots, t_n)$. By $magic(A)$ we mean the new atom $magic\_p(t_1, \ldots, t_n)$ where we implicitly assume that the new predicate symbol does not appear in Σ.

Definition 3.1 Let P be a definite program. By $magic(P)$ we mean the smallest set of clauses such that:

- if $A_0 \leftarrow A_1, \ldots, A_n \in P$, then $A_0 \leftarrow magic(A_0), A_1, \ldots, A_n \in magic(P)$;

- if $A_0 \leftarrow A_1, \ldots, A_n \in P$, then $magic(A_i) \leftarrow magic(A_0), A_1, \ldots, A_{i-1} \in magic(P)$ for each $1 \leq i \leq n$.

■

Example 3.2 Consider the following program P:

```
path(X, Y) ← edge(X, Y).
path(X, Z) ← edge(X, Y), path(Y, Z).
edge(a, b).
edge(b, c).
edge(b, a).
```

Then $magic(P)$ is the program:

```
path(X, Y) ← magic_path(X, Y), edge(X, Y).
path(X, Z) ← magic_path(X, Z), edge(X, Y), path(Y, Z).
edge(a, b) ← magic_edge(a, b).
edge(b, c) ← magic_edge(b, c).
edge(b, a) ← magic_edge(b, a).
magic_edge(X, Y) ← magic_path(X, Y).
magic_edge(X, Y) ← magic_path(X, Z).
magic_path(Y, Z) ← magic_path(X, Z), edge(X, Y).
```

■

In order to answer the query $\leftarrow A$ the new program is extended with a seed (fact) magic(A) and the answers are computed bottom up using a function $T_P: \wp(\text{Atom}) \to \wp(\text{Atom})$ which generalises the standard immediate consequence operator of van Emden and Kowalski (e.g. [Llo87, EK76]):

$$T_P(x) = \{A_0\theta \mid \quad A_0 \leftarrow A_1, \ldots, A_n \in P, B_1, \ldots, B_n \in x \text{ and}$$
$$\overline{mgu}(A_1 \ldots A_n, B_1 \ldots B_n) = \theta \neq \perp\}$$

The answers are obtained as the smallest set $x \subseteq \text{Atom}$ satisfying $T_P(x) = x$ (that is, as the least fixed point of T_P). For definite programs such a set is known to exist and it can be obtained as the least upper bound of the iteration:

$$\emptyset, T_P(\emptyset), T_P(T_P(\emptyset)), \ldots$$

Depending on P it is sometimes possible to find the fixed point after a finite number of iterations. However, the procedure is always exhaustive in the sense that any atom in the least fixed point of T_P is found after a finite number of iterations. For definite programs without functors (so called datalog programs) the fixed point is always reached after a finite number of iterations.

Example 3.2 [Continued] When extended with the fact:

magic_path(X, Y).

The (bottom up) computation of magic(P) yields the following fixed point (modulo renaming) after five iterations:

edge(a, b).	path(a, c).	magic_edge(b, A).	magic_path(b, A):
edge(b, c).	path(a, a).	magic_edge(c, A).	magic_path(c, A).
edge(b, a).	path(b, b).	magic_edge(a, A).	magic_path(a, A).
	path(a, b).	magic_edge(A, B).	magic_path(A, B).
	path(b, c).		
	path(b, a).		

■

To shed some additional light on the magic templates transformation we consider the following (as we shall see in Example 4.1 somewhat incorrect) argument based on [Bry90]. Given a clause of the form:

$$A_0 \leftarrow A_1, \ldots, A_n$$

What does it take for a call to this clause to succeed? Informally speaking in order for a call A_0 to succeed there must first be a call which unifies with A_0. Secondly all the body literals $A_1, \ldots, A_n$ must be successfully instantiated. This may be formalized as follows:

$$succ(A_0) \leftarrow call(A_0), succ(A_1), \ldots, succ(A_n).$$

This is isomorphic to the first part of Definition 3.1. Also note that technically this new clause uses a different alphabet than the original program. In the new alphabet the predicate symbols of the old alphabet become functors and the only predicate symbols are call/1 and succ/1. The atoms of the new clause are said to be *embeddings* of the old atoms.

Next consider the body literal A_i in a clause of the form:

$$A_0 \leftarrow A_1, \ldots, A_{i-1}, A_i, A_{i+1}, \ldots, A_n$$

What does it take for an instance of the call A_i to go ahead given that Prolog's computation rule is being used? Informally there must be a call to A_0 and, secondly, the body literals $A_1, \ldots, A_{i-1}$ must succeed. That is:

$$call(A_i) \leftarrow call(A_0), succ(A_1), \ldots, succ(A_{i-1}).$$

This time the result is isomorphic to the second part of Definition 3.1. Finally we may specify that we want to call the program with the goal $\leftarrow A$ by the addition of the fact $call(A)$.

4 Abstract Interpretation

The ultimate aim of abstract interpretations [CC77] is to infer (run-time) properties of programs. This goal is usually attained indirectly by computing (or more generally approximating) the intended model of the program (typically some excerpt from all possible computation states of the program). Approximations are generally needed because of the non-computability of the intended model.

Following Cousot and Cousot [CC77] an abstract interpretation of a program P consists of:

- a complete lattice $\langle D; \sqsubseteq \rangle$ of possible models of P;

- a monotone function $f: D \rightarrow D$ which assigns one particular meaning to the programs. This meaning is usually given by means of the least fixed point of the function (i.e. the, wrt $\sqsubseteq$, least $x \in D$ satisfying $f(x) = x$).

Practically the user specifies two abstract interpretations — the first defines (or at least approximates) the intended model of the program. This interpretation is usually called a *base* or *collecting* or *static* interpretation and if no restrictions are imposed on the the program this model is normally non-computable. The objective of the second abstract interpretation thus is to effectively approximate the intended model of the program. Needless to say, for the approximation to be of any use certain relations must hold between the two interpretations.

In this paper we are concerned mainly with datalog programs and there will be no need for approximations. We are thus restricting our attention to the collecting interpretation and the intended model of the program.

A number of frameworks to support abstract interpretation of logic programs have emerged recently (e.g. [Bru90, JS87, MS88, Mel87, Nil90]). All of these have different collecting interpretations and consequently facilitate inference of different properties. Here we primarily focus our attention on that of C. Mellish [Mel87] which is an approach whose collecting interpretation is founded on SLD-resolution and therefore often referred to as a "top down" abstract interpretation. But we will also relate this top-down approach to frameworks founded on a fixed point semantics and therefore often referred to as "bottom up" abstract interpretation (cf. [MS88]).

4.1 "Top Down" Abstract Interpretation

In the abstract interpretation framework of C. Mellish the intended meaning of a program consists of two sets of atoms — Call and Succ. The former is the set of all possible procedure-calls given a set Init of initial subgoals and Prolog's computation rule. The latter is the succeeding instances of the procedure-calls. Call and Succ are often referred to as call and success pattern. In the framework of C. Mellish these two sets are defined by two recursive equations. Here we give the following simplified formulation:

$$Call = Init \cup \bigcup_{A_0 \leftarrow A_1, \dots, A_l \in \tilde{P}} \{A_l\theta \mid B_0 \in Call, B_1, \dots, B_{l-1} \in Succ \text{ and}$$
$$\overline{mgu}(A_0 \dots A_{l-1}, B_0 \dots B_{l-1}) = \theta \neq \perp\}$$

$$Succ = \bigcup_{A_0 \leftarrow A_1, \dots, A_n \in P} \{A_0\theta \mid B_0 \in Call, B_1, \dots, B_n \in Succ \text{ and}$$
$$\overline{mgu}(A_0 \dots A_n, B_0 \dots B_n) = \theta \neq \perp\}$$

Given a program P and a set Init of initial call patterns we thus have two recursive equations:

$$Call = call_P(Call, Succ)$$
$$Succ = succ_P(Call, Succ)$$

Finding a solution to these equations can alternatively be stated as finding a fixed point of the mapping $\Psi_P: \wp(Atom) \times \wp(Atom) \to \wp(Atom) \times \wp(Atom)$:

$$\Psi_P(x, y) = \langle call_P(x, y); succ_P(x, y)\rangle$$

Now it is easy to prove that $\wp(Atom) \times \wp(Atom)$ is a complete lattice (under componentwise set inclusion) and that Ψ_P is continuous (modulo renaming) and as a consequence monotone. Thus, $(\wp(Atom) \times \wp(Atom); \Psi_P)$ is an abstract interpretation which attempts to characterise the intended model of the program by means of the least fixed point of Ψ_P. It is well-known that such a fixed point exists and that it can be approximated by the least upper bound of the iteration:

$$(\emptyset, \emptyset), \Psi_P(\emptyset, \emptyset), \Psi_P(\Psi_P(\emptyset, \emptyset)), \dots, \Psi_P^n(\emptyset, \emptyset), \dots$$

Now question is whether Ψ_P actually characterises the intended model of the program. It turns out that this is not always the case as demonstrated by the following example:

Example 4.1 Let P be the program:

$$p(X) \leftarrow q(a).$$
$$p(X) \leftarrow q(X), r(X).$$
$$q(X).$$
$$r(X).$$

and Init $= \{p(A)\}$. Mellish's approach yields the sets:

$$\text{Call} = \{r(A), r(a), q(A), q(a), p(A)\}$$
$$\text{Succ} = \{p(a), r(a), r(A), p(A), q(A), q(a)\}$$

However, it is easy to see that there is no Prolog derivation whose first goal is $\leftarrow p(A)$ which contains the call pattern $r(a)$ and, similarly, there are no success patterns $r(a)$ and $p(a)$. Hence, the collecting interpretation of Mellish is in general not able to exactly characterise the intended meaning of programs. However, it has been shown that it provides a conservative approximation of the intended model, i.e. it produces a superset of all actual call and success patterns [Mel87]. ∎

4.2 "Bottom Up" Abstract Interpretation

A different and often simpler framework for abstract interpretation can be obtained by means of a fixed point semantics similar to that of M. van Emden and R. Kowalski [EK76] or M. Fitting [Fit85]. For instance, if P is built from the same alphabet as Atom then:

$$\langle \wp(\text{Atom}); T_P \rangle$$

is an abstract interpretation of P (the domain is ordered under set inclusion and T_P is monotone). The existence of top down and bottom up approaches obviously raises the question of their relative strength. One would perhaps be led to believe that for the purpose of inferring run time properties the former is better since it is founded on SLD-resolution while the latter is founded on a fixed point semantics. However, as we shall see in the next section, by using the magic templates transformation the bottom up approach yields a model which is isomorphic to the model produced in the framework of Mellish. On the other hand, since this model is only an approximation of the intended model it may also be argued that bottom up abstract interpretation (as described here) cannot be used to characterise the exact intended model of the program. Something which is possible in most "top down" frameworks (but notably not in that of Mellish).

5 The Magic of Abstract Interpretation

In this section we first show, for a given program P, the equivalence between C. Mellish's collecting interpretation of P and the result from a bottom up computation of of $\text{magic}(P)$. Thereafter we take advantage of the magic templates transformation and show that Mellish's intended model can be obtained through the bottom up abstract interpretation given above.

In order to compare the output of the operator $T_{\text{magic}(P)}$ which produces a set of embedded atoms (in one of the forms $\text{succ}(\ldots)$ or $\text{call}(\ldots)$) and the output of Ψ_P — which consists of two not necessarily disjoint sets of "ordinary" atoms — we introduce a bijective colouring function from pairs of atoms to the set of embedded atoms. Let $x, y \subseteq \text{Atom}$ and define $\text{colour}(x, y)$ to be $\{\text{call}(z) \mid z \in x\} \cup \{\text{succ}(z) \mid z \in y\}$.

Without lack of generality we may assume that the query given by the user is of the form $\leftarrow A$ where $A \in \text{Atom}$. For the magic templates method it means that $\text{magic}(P)$ is extended with the fact $\text{magic}(A)$ and for the abstract interpretation it means that $\text{Init} = \{A\}$.

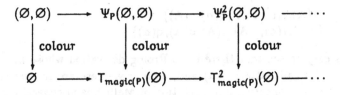

Figure 1: Isomorphism of Ψ_P and $T_{magic(P)}$

Theorem 5.1 If $x, y \subseteq$ Atom then $colour(\Psi_P(x, y)) = T_{magic(P)}(colour(x, y))$. ∎

Proof: The proof consists of two parts taking account of embedded atoms of the forms $call(\ldots)$ and $succ(\ldots)$. That is:

$$succ(A_0\theta) \in colour(\Psi_P(x, y))$$
iff
$$A_0 \leftarrow A_1, \ldots, A_n \in P \text{ and}$$
$$B_0 \in x, B_1, \ldots, B_n \in y \text{ and}$$
$$\overline{mgu}(A_0 \ldots A_n, B_0 \ldots B_n) = \theta$$
iff
$$succ(A_0) \leftarrow call(A_0), succ(A_1), \ldots, succ(A_n) \in magic(P) \text{ and}$$
$$call(B_0), succ(B_1), \ldots, succ(B_n) \in colour(x, y) \text{ and}$$
$$\overline{mgu}(call(A_0) \ldots succ(A_n), call(B_0) \ldots succ(B_n)) = \theta$$
iff
$$succ(A_0\theta) \in T_{magic(P)}(colour(x, y))$$

Similarly $call(A_i\theta) \in colour(\Psi_P(x, y))$ iff $call(A_i\theta) \in T_{magic(P)}(colour(x, y))$. ∎

Corollary 5.2 x is a fixed point of Ψ_P iff $colour(x)$ is a fixed point of $T_{magic(P)}$. ∎

In other words both operations have isomorphic least fixed points. We can even show that the iterations which lead to the isomorphic fixed points produce isomorphic intermediate results (cf. Figure 1):

Corollary 5.3 For all $n \geq 0$, $colour(\Psi_P^n(\emptyset, \emptyset)) = T_{magic(P)}^n(\emptyset)$. ∎

6 Conclusions

Corollaries 5.2 and 5.3 provide a very simple but yet strong equivalence between the collecting interpretation of C. Mellish and the bottom up computation of program resulting from the magic templates transformation. They not only state that the end-results (that is, the least fixed points) are equivalent but also that all the intermediate results in the computations of least fixed points are isomorphic.

Although providing a surprisingly simple but yet rigid proof, the final result of this paper may, in a sense, not seem very surprising. Recent results by F. Bry [Bry90] shows that many of the proposed evaluation schemes for logic programs — including magic templates and SLD-resolution with tabulation (OLDT-resolution [TS86]) — really are equivalent. Moreover the latter provides a basis for the abstract hybrid interpretation scheme of T. Kanamori and T. Kawamura [KK87]. This has been exploited by Kanamori [Kan90] who uses the *Alexander templates* method [RLK86] (a generalisation of the magic templates method) as a basis for an abstract interpretation scheme. Also C. Mellish uses similar techniques [Mel90]. By using partial evaluation of an *interpreter* for magic templates (much in the same style as that of F. Bry [Bry90]) Mellish is able to obtain a mode inference program which is then evaluated bottom up.

As a side-effect this work also demonstrates a (potential) weakness of the magic templates method. As shown in Example 4.1 the method may introduce overhead — call and success patterns which never appear in any real (top-down) derivation. The basic problem is that the success and call patterns are not coupled. It is quite possible that the magic templates method can be improved in this respect. In fact, the Alexander templates method which uses a more elaborate transformation than that of the magic templates method solves this problem. Likewise, most of the existing abstract interpretation frameworks do not suffer from the imprecision inherent in the framework of C. Mellish. In the magic templates method it is possible to circumvent the problem by using a subsumption-order instead of a renaming-order. However, such tests are fairly expensive. Moreover, this work also shows that the magic templates transformation is not really needed! Using the techniques of abstract interpretation is is possible to operate directly on the original program without any overhead.

The close relationship between the two fields may potentially lead to more efficient evaluation techniques both in the field of abstract interpretation and in the deductive database field. Techniques such as Jacobi's iteration strategy and the chaotic iteration strategy have been used both for data-flow analysis [Cou81] and for abstract interpretation of logic programs [Nil90].

Finally, recent results in the area of abstract interpretation of logic programs have indicated that "bottom up" abstract interpretation can replace traditional "top down" approaches by exploiting the magic templates method. Our results show that this is only true to a certain extent. If one is prepared to live with the imprecision which arises from such methods then "bottom up" abstract interpretation can be used to infer run-time properties. However, at least using the approach described here such a method is bound to give poorer results than more precise "top down" abstract interpretation frameworks. It should be noted that this need not be true when using more elaborate fixed point characterisations than that discussed in this paper.

For uniformity we have deliberately avoided dividing the program into an intensional and extensional part. It should be noted that the results can be adapted also to the case where the program is partitioned into two parts by a simple modification of the equations in Section 4.1.

Acknowledgments

I am grateful to Jan Małuszyński for his patient supervision. Thanks are also due to François Bry for introducing me to magic templates and to the anonymous referees. The work was partially supported by the Swedish National Board for Technical Development (STU and STUF) and by Esprit BRA 3020 "Integration".

References

[Ban86] F. Bancilhon et. al. Magic Sets and Other Strange Ways to Implement Logic Programs. In *Proc. of 5th ACM Symposium on Principles of Database Systems*, pages 1–15, 1986.

[BR87] C. Beeri and R. Ramakrishnan. On the Power of Magic. In *Proc of 6th Symposium on Principles of Database Systems*, pages 269–283, 1987.

[Bru90] M. Bruynooghe. *A Practical Framework for the Abstract Interpretation of Logic Programs*. To appear in J. Logic Programming, 1990.

[Bry90] F. Bry. Query Evaluation in Recursive Databases: Bottom-up and Top-down Reconciled. *Data and Knowledge Engineering*, 1990. To appear.

[CC77] P. Cousot and R. Cousot. Abstract Interpretation: A Unified Lattice Model for Static Analysis of Programs By Construction or Approximation of Fixpoints. In *Conf. Record of Fourth ACM Symposium on POPL*, pages 238–252, Los Angeles, 1977.

[Cou81] P. Cousot. Semantic Foundations of Program Analysis. In S. Muchnick and N.D. Jones, editors, *Program Flow Analysis*, pages 303–342. Prentice Hall, 1981.

[EK76] M. van Emden and R. Kowalski. The Semantics of Predicate Logic as a Programming Language. *J. of ACM*, 23(4):733–742, 1976.

[Fit85] M. Fitting. A Kripke-Kleene Semantics for Logic Programs. *J. of Logic Programming*, 2(4):295–312, 1985.

[JS87] N.D. Jones and H. Søndergaard. A Semantics-Based Framework for the Abstract Interpretation of Prolog. In S. Abramsky and C. Hankin, editors, *Abstract Interpretation of Declarative Languages*, pages 123–142. Ellis Horwood, 1987.

[Kan90] T. Kanamori. Abstract Interpretation Based On Alexander Templates. Technical Report TR-549, ICOT, 1990.

[KK87] T. Kanamori and T. Kawamura. Analyzing Success Patterns of Logic Programs by Abstract Hybrid Interpretation. Technical Report TR-279, ICOT, 1987.

[Llo87] J.W. Lloyd. *Foundations of Logic Programming*. Springer-Verlag, second edition, 1987.

[Mel87] C. Mellish. Abstract Interpretation of Prolog Programs. In S. Abramsky and C. Hankin, editors, *Abstract Interpretation of Declarative Languages*, pages 181–198. Ellis Horwood, 1987.

[Mel90] C. Mellish. Using Specialisation to Reconstruct Two Mode Inference Systems, 1990.

[MS88] K. Marriott and H. Søndergaard. Bottom-up Abstract Interpretation of Logic Programs. In *Proc. of Fifth International Conf/Symposium on Logic Programming*, Seattle, pages 733–748. MIT Press, 1988.

[MS89] K. Marriott and H. Søndergaard. Semantics-Based Dataflow Analysis of Logic Programs. In G. Ritter, editor, *Information Processing 89*, pages 601–605. North-Holland, 1989.

[Nil90] U. Nilsson. Systematic Semantic Approximations of Logic Programs. In *Programming Language Implementation and Logic Programming 90*, Lecture Notes in Computer Science 456, pages 293–306. Springer-Verlag, 1990.

[NM90] U. Nilsson and J. Małuszyński. *Logic, Programming and Prolog.* John Wiley & Sons, 1990.

[Ram88] R. Ramakrishnan. Magic Templates: A Spellbounding Approach to Logic Programming. In *Proc. of Fifth International Conf/Symposium on Logic Programming*, Seattle, pages 140–159. MIT Press, 1988.

[RLK86] J. Rohmer, R. Lescœur, and J.-M. Kerisit. The Alexander Method—A Technique For the Processing of Recursive Axioms in Deductive Databases. *New Generation Computing*, 4(3):273–285, 1986.

[TS86] H. Tamaki and T. Sato. OLD Resolution with Tabulation. In E Shapiro, editor, *Proc. of Third International Conf. on Logic Programming*, London, Lecture Notes in Computer Science 225, pages 84–98. Springer-Verlag, 1986.

Abstract interpretation for type checking

G.Filè and P.Sottero
Dep. of Pure and Applied Mathematics
University of Padova
Italy
email: mat01@unipad.infn.it

Abstract. The typed logic languages are more expressive than the usual untyped ones, but run-time type-checking is in general quite costly. Compile-time type checking is a classical application of the abstract interpretation paradigm. We describe a general abstract interpretation framework and inside it we develop two new methods for the compile-time type-checking of typed logic programs. The first method applies to a restricted class of programs (those that are type-preserving and use a finite number of types) and it detects the programs that need no type-checking at all. The second one applies to any program, but, in general, it only avoids part of the run-time type-checking.

Introduction

The addition of types to logic programs makes them more expressive and it also allows error detection at compile time. The first attempt of defining such a type system for logic languages is that of Mycroft and O'Keefe [MO'K84] and it is based on Milner's work on typing applicative languages [Mil78]. Later Hanus [Han89] has deepened the semantical aspects of this type system.

In all typed languages type-checking must be enforced and already Bakus et al. [Bak57] in the Fortran compiler and Naur [Nau65] in the Algol compiler describe compile time type-checking techniques. In the present paper we use the technique of abstract interpretation for performing compile-time type-checking of typed logic languages. In [MO'K84] it is stated that the typed programs and goals satisfying the following two conditions are type-correct: terminology of [Han89] is used,

(a) all the type-declarations of the functor symbols are **type-preserving**, i.e., all the type variables present in the types of the arguments are also in the type of the result;

(b) all clauses satisfy the **genericity condition**, i.e., the head of the clause has type equal (modulo renaming) to the type of the corresponding predicate symbol,

Unfortunately these conditions are quite restrictive. Condition (a) forbids very natural functions such as, equal:α,α->bool, cf., [Han89]. Condition (b) forbids, for instance, to have ground facts in a program, such as, append([a],[b], [a,c]), when the type of append is, [list(α), list(α),list(α)].

In this paper abstract interpretation methods are described that avoid completely or partially run-time type-checking for typed logic programs that do not meet both conditions (a) and (b) above. Type declarations associate types to every node of a term t. The type of the root of t is called the **global type** of t, whereas the types of any other node of t is an **inside type** of t.

First, programs that satisfy only condition (a) are considered, these programs are called **preserving**. It is easy to see that for preserving programs the run-time type-checking that must be done is much less than that of the general case: when unifying two atoms A and B, as a type-checking it suffices, in fact, to pairwise unify their global types. However, no type-checking at all would be better! This is possible for an important class of the preserving programs, those that are **type-recurrent**. Intuitively, a program is type-recurrent when the set of the types of the atoms appearing in its (typed) SLD-resolutions is finite. This definition is inspired by the notion of recurrent programs of [Bez89]. It is important to understand that whereas recurrent programs always terminate (when starting from a bounded goal, [Bez89]), type-recurrent programs do not necessarily terminate.

For a preserving type-recurrent program there is a compile-time test that, when satisfied, guarantees that the program analyzed is type-correct. This static test is realized by a particularly simple abstract interpretation that uses the set of all type substitutions as abstract domain and the abstract unification is just the normal unification applied to the types.

When one considers unrestricted programs, i.e., niether condition (a) nor (b) holds, it seems impossible to avoid run-time type-checking of the global and the inside types of the arguments of the atoms produced in the SLD-derivations. However, also in this case abstract interpretation can be of help. Using an appropriate abstract domain, it is possible to synthesize the following information about any given program P: for any atom A of P that is selected during an SLD-derivation, what variables of A may be instantiated to terms containing only type-preserving symbols. These terms do not require type-checking in the inside-types and thus, based on this information, at run-time some type-checking can be avoided.

Several abstract interpretation frameworks have been proposed in the past few years. The most successful are [MSo89,Bru88,KKa90]. The goal of a framework is to facilitate the task of describing a specific abstract interpretation application: one has only to define the abstract domain, the abstract unification and to show that domain and unification meet some natural conditions (relating them to the concrete domain and unification). For describing our abstract interpretation applications we have chosen to rely on the framework of [KKa90], that is based on the OLDT-resolution of [TSa86], refined according to the ideas introduced in [CoF91].

The paper is organized as follows. Section 2 contains all definitions concerning types. Section 3 is devoted to the description of the general abstract interpretation framework that will be the basis for the definition of the abstract interpretation applications presented in Sections 4 and 5. A short conclusion closes the paper.

2. Types

The notions related to logic programming are assumed to be known, see for instance [Llo87]. Let Pred and Func be disjoint ranked alphabets of predicate and functor symbols. Var is the infinite set of variable symbols. All considered programs use only symbols from these alphabets. If t is a term, an atom or a clause, Var(t) is the set of the variables of t.

For the definitions about types we follow [MO'K84], for a semantical approach see [Han89]. Let Tcons and Tvar be disjoint alphabets of type constructors and type variables, respectively. The variables of Tvar are $\alpha, \beta, \ldots$ whereas the usual variables of Var are $x, y, z, \ldots$. The set of types corresponding to Tcons and Tvar is defined by the following grammar:

TYPE::=Tvar | Tcons(TYPE*)

The elements of TYPE are generally denoted with $\rho, \tau, \ldots$ and are called types. As usual, $\rho \in$ TYPE is a monotype if it does not contain any type variable, otherwise it is a polytype.

Example. Assume that Tcons contains Char and List, then *Char, List(α)* and *List(Char)* are types in TYPE; the 2nd one is a polytype whereas the remaining two are monotypes.

Types are terms and thus one can extend to them the notions of substitution and unification and one can also define the relation 'more general' between them.

A type substitution is a set $\gamma=\{\alpha 1/\sigma 1,...,\alpha n/\sigma n\}$ where $\alpha i \in$ Tvar and $\sigma i \in$ TYPE. Only idempotent type substitutions are considered. We take advantage of the similarity between normal and type substitutions and define some terminology that applies to both of them. For a (type) substitution γ, $D(\gamma)=\{x \mid x\gamma \neq x\}$ and $R(\gamma)=\cup\{Var(x\gamma) \mid x \in D(\gamma)\}$. Subst and Tsubst denote, respectively, the set of (renaming equivalence classes of) idempotent substitutions and type substitutions, respectively. For σ in Subst or Tsubst, $eq(\sigma)=\{x=t \mid x/t\in\sigma\}$ and the **restriction** of σ to the set of variables V in Var, is $\pi(\sigma,V)=\{x/t \mid x/t\in\sigma$ and $x\in V\}$.

A term (type) $\sigma 1$ is **more general** than $\sigma 2$, denoted $\sigma 1 \geq \sigma 2$, when there exists a (type) substitution γ, such that $\sigma 1\gamma=\sigma 2$. The relation $\geq$ is easily extended to (type substitutions). Type renamings are defined in the obvious way. When $\sigma 1 \geq \sigma 2$ and $\sigma 1 \leq \sigma 2$, then $\sigma 1$ and $\sigma 2$ are equal modulo renamings, denoted with $\sigma 1 \approx \sigma 2$.

For two terms (types) $\sigma 1$ and $\sigma 2$, assume that there is a solution of $\sigma 1=\sigma 2$, and, in this case let γ be a mgu of it, then $\sigma 1\gamma$ is a **most general common instance** of $\sigma 1$ and $\sigma 2$.

A type declaration of a functor symbol f$\in$ Func, of arity k$\geq$0, is DEF(f)=[$\sigma 1,...,\sigma\kappa$->$\sigma 0$], with $\sigma i \in$ TYPE. Similarly, a type declaration of a predicate symbol p$\in$ Pred of arity k, is DEF(p)=[$\sigma 1,...,\sigma\kappa$].

The type declaration DEF(f)=[$\sigma 1,...,\sigma\kappa$->$\sigma 0$] is **type-preserving**, [Han89], when all type variables of $\sigma 1,...,\sigma\kappa$ also appear in $\sigma 0$.

From now on with DEF we denote a set of type definitions for the symbols in Func and Pred. The definition of a functor or predicate symbol q is DEF(q). When all definitions in DEF are type preserving, we say that DEF is type preserving.

Obviously, the type declaration of functor and predicate symbols determine the class of the well typed terms containing those symbols. Let us examine precisely how this works. Let us consider any term t such that all its nodes have an element of TYPE attached to them, with the only requirement that multiple occurrences of a variable have the same type. Such a term is called a typed term and is denoted with t:τ, where τ is the type of the root of t (τ is also the global type of the term t). The set of typed terms (wrt DEF) is denoted **TT**. A typed term t:τ is **well typed** w.r.t. DEF, if it satisfies the following points, [Han89]:

(i) if t=x then t:τ is well typed

(ii) if t=c, c$\in$ Func, then t:τ is well typed if there is a type substitution γ such that DEF(c)$\gamma \approx \tau$.

(iii) if t=f(t1,..,tk) and the types of t1, ..,tk are $\tau 1,...,\tau k$, then t:τ is well typed if there is a type substitution γ such that DEF(f)$\gamma \approx [\tau 1,..,\tau k$->$\tau]$.

A typed atom p(t1:$\tau 1$,...,tk:τk) is well typed (w.r.t. DEF) if each ti:τi is a well typed term and if there is a type substitution γ such that DEF(p)$\gamma \approx [\tau 1,..\tau k]$. A well typed clause and a well typed program can be defined similarly (the only problem are the multiple occurrences of variables). The unification operation can be easily extended to well typed atoms. The type-checking can be done by unifying simultaneously the terms and their types. Obviously, this typed unification, when successful, computes two substitutions: a normal one and a type one. It may fail either because of a usual mismatch or because of a type mismatch. See [Han89] for more details. In what follows, when considering SLD-derivations of typed programs we assume that typed unification is used.

In [Han89] it is shown that this typed SLD-resolution has a model theoretic counterpart and that the desirable results of soundness and completeness hold between the two semantics.

Let P be a well typed logic program, P satisfies the **genericity condition**, [MO'K84] see also [Han89], when each clause of P meets the following condition: let p(t1:τ1,....,tk:τk) be the head of the clause, then DEF(p)≈[τ1,..,τk],

Example. Let TYPE = {*Nat, Bool, Char*, α,β,...}, Func = {succ, bigger}. Let also, DEF(succ)=[α->α] and DEF(bigger)=[α,α->Bool]. DEF(succ) is type-preserving whereas DEF(bigger) is not. The typed term, succ:α(x:α) is well typed. Similarly, succ:*Nat*(2:*Nat*) is also well typed. It is more interesting to notice that succ:*Nat*(x:*Nat*) is also well typed. The term, bigger:*Bool*(succ:α(x:α),y:α) is also well typed of type *Bool*. The term, bigger:*Bool*(succ:*Nat*(x:*Nat*),y:*Char*) is clearly not well typed.

It is not difficult to see that the following result is true.

<u>Lemma 2.1</u>. Let DEF contain only type-preserving definitions. Consider two well typed atoms A=p(t1:τ1,....,tk:τk) and B=p(s1:σ1,. ..,sk:σk), such that they have neither variables nor type variables in common. If there is a solution of E={τ1=σ1,....τk=σk} then the typed unification of A and B will surely not fail because of a type error (but, it may fail because of a normal mismatch). Moreover, if γ is an mgu of E and ν is an mgu of {t1=s1,..,tk=sk}, then p(t1ν:τ1γ,...,tkν:τkγ) is a most general common instance of A and B.

<u>Theorem 2.1</u>. [MO'K84]. Assume that DEF is type-preserving. Let P be a typed logic program and G a typed goal that satisfy the genericity condition. P and G are type-correct, i.e., no SLD-derivation of PU{G} fails because of a type error.

3. A general framework for abstract interpretation

This section contains a brief description of a general scheme for describing techniques based on the abstract interpretation paradigm, for more details see [CoF91]. The scheme owns much to the notion of OLDT-resolution of [TSa86] and to the use of this notion for abstract interpretation made by [KKa90].

<u>Definition 3.1</u>. A **computation system** is a 4-tuple C=(C,≤,+,π), where each component is as follows:
- C is a subset of the first-order formula on some fixed alphabet some of whose symbols may have a fixed interpretation. In particular, the predicate symbol "=" is interpreted as the identity. All equations x=y with x and y in Var are in C.
- ≤ is a partial order on C with minimal element ⊥C,
- + is a function of type: C²-> C, that is called composition and must be associative,
- π is a function of type C,2^Var->C, that is called projection.
 To prevent possible ambiguities we will sometimes add subscripts to + and π, obtaining +C and πC.

<u>Definition 3.2</u>. A **program** for a computation system C, also called a C-program, is a set of formulas of this form:
p(x1,..,xk):- c, q1(y(1,1),...,y(1,k1)),...,qm(y(m,1),...,y(m,km)),
where, m≥0, c∈C, and if x=[x1,..,xk] and yi=[y(i,1),..,y(i,ki)] then all variables in x and yi are distinct.
Similarly, a C-goal is a pair, [(q1(y1),...,qm(ym)), c], with c∈C and m≥0. If m=0 then the goal is **empty**.

<u>Definition 3.3</u>. Let $C=(C,\leq,+,\pi)$. Given a C-program P and a C-goal G, a **computation** of $P\cup\{G\}$ is a sequence of C-goals, $G0,G1,...,Gr$, where $G0=G$ and, for $i\in[1,r]$, if $Gi-1=[(A1,..,An),c]$ then $Gi=[(B1,..,Bm,A2,..,An),c']$, such that the following holds: there is a renamed clause $H:-c1,B1,..,Bm$, of P, where $A1=p(x1,..,xk)$, $H=p(x1',..,xk')$, and $c'=c+c1+\{xi=xi'$ /$i\in[1,k]\}$.

The computation is **complete** if Gr is an empty goal. Note that c' can be $\bot_C$, this corresponds to the unification failure.

The reader wondering about the use of the projection function of computation systems, must be patient till the end of this section: the projection function is useful for "tabled" computations. Roughly, π restrict the values of C to a fixed set of variables.

It should be clear that the above definitions generalize the usual definitions of logic program, goal, and SLD-derivation and also the corresponding notions for typed logic programs. More precisely, let $S=(Subst,\leq_S,mgu,\pi_S)$ be as follows: Subst is the set of renaming equivalence classes of substitutions, $\leq_S$ is the usual 'more general' partial order between elements of Subst, π_S is the restriction of a substitution to a fixed set of variables, and mgu is the unification in the following sense: $mgu(\sigma1,\sigma2)$, where $\sigma1$ and $\sigma2 \in Subst$, is (the renaming equivalence class of) a mgu of the set of equations $eq(\sigma1)\cup eq(\sigma2)$. Notice that Subst is not a set of first-order formula as it is required for computation-systems. However, it is easy to see that Subst can be easily represented by such a set. The same is true for TTS in the next example.

It should be easy to see how any logic program can be transformed into an S-program: every clause $p(t):-q1(r1),..,qm(rm)$, becomes $p(x):-s,q1(y1),..,qm(ym)$, where x and yi are tuples of new variables and s contains all the equations corresponding to the argument replacements, i.e., if xj has replaced the argument tj, then s contains xj=tj. Let P and G be a logic program and a goal and P' and G' be the corresponding S-program and S-goal constructed as explained above.
(#) It is easy to see that for any SLD-derivation w of $P\cup\{G\}$ there is a corresponding computation w' of $P'\cup\{G'\}$ such that for each resolvent $<-A1,..,Ak$ of w, w' has a resolvent $[(A1',..,Ak'),c]$, such that $Ai\approx Ai'c$, $i\in[1,k]$, and viceversa.

Obviously, also typed SLD-derivations can be described as computations of a computation system very similar to S. Let $TTS=(TTS,\leq_{TTS}, mgu_{TTS}, \pi_{TTS})$ be defined as follows: TTS is the set of renaming equivalence classes of typed term substitutions, where a typed term substitution is a set $\gamma=\{x1/t1:\tau1,...,xn/tn:\tau n\}$, i.e., the variables xi are assigned a well typed term of TT. The composition mgu_{TTS}, the relation $\leq_{TTS}$ and the projection π_{TTS} are obvious extensions of mgu, $\leq_S$ and π_S. Following the above ideas for (normal) logic programs, one can construct a TTS-program P' corresponding to any given typed logic program P such that the computations of P' are equivalent to those of P as in (#) above.

Let us now tackle the notion of a computation system simulating another one.

<u>Definition 3.4</u>. Given two computation systems $C1=(C1,\leq_1,+_1,\pi_1)$ and $C2=(C2,\leq_2,+_2,\pi_2)$. **C1 simulates C2**, if the following two conditions hold:
(1) There is a monotonic function $\gamma: C1->2^{C2}$ such that for any c in C2 there is at least one c' in C1 such that $c\in\gamma(c')$.
(2) For any c1 and c1' of C1, and c2 and c2' of C2, such that $c2\in\gamma(c1)$ and $c2'\in\gamma(c1')$, then $c2+_2c2'\in\gamma(c1+_1c1')$.

<u>Lemma 3.1</u>. Let C1 and C2 be as in the above definition. If C1 simulates C2, then for any C2-program P and any C2-goal G, there is a C1-program P' and a C1-goal G', such that for any computation w of $P\cup\{G\}$, there is a computation w' of $P'\cup\{G'\}$ such that for every goal $[(A1,..,Ak),c]$ of w, w' contains a goal $[(A1,..,Ak),c']$, where $c\in\gamma(c')$.

Notice that, in general, the program P' and the goal G' of the above lemma are not unique. Often there are *best* P' and G': when C1 is such that every subset of C1 contains a least element.

Obviously, the fact that C1 simulates C2 does not imply that all the computations of C1 are finite. At the contrary, from the above observations it follows that there must be at least as many divergent C1's computations as there are divergent C2's computations.
Intuitively, there may be divergent computations of C that are **intrinsically finite**: during the computation a finite number of (renamed) values from C are infinitely repeated in the resolvents. At the contrary, a computation is **genuinely infinite**, when it computes an infinite set of distinct values from C.
Let us be more precise, let w be a computation. The values of C contained in the goals of w concern in general all the variables of w. Thus, since new variables are introduced at each step, always new values are produced. How can a computation be intrinsically finite then? The definition of 'the values computed in w' that is needed is as follows: for a resolvent $[(A1...Ak),c]$ of w, $[A1,\pi_C(c,Var(A1))]$ is the corresponding **value computed in w**. The computation w is intrinsically finite when the set of the values computed in w modulo variable renaming is finite. A C-program P and a C-goal G are **intrinsically finite** when all computations of $P \cup \{G\}$ are intrinsically finite.

Example. The program p(X):-p(X) is an intrinsically finite program that still has an infinite computation. The only value computed by this program with goal :-p(X) is the set of atoms that are renamings of p(X). The program p(X):-p(f(X)) insted is genuinely infinite, because it computes: p(X), p(f(X)), p(ff(X)) and so on.

If P and G are intrinsically finite, then all the values (modulo renaming) that are computed in the computations of $P \cup \{G\}$, can be computed in finite time by using a tabulation method. The notion of tabulation has been introduced in [TSa86] and used for abstract interpretation in [KKa90]. In this paper we will not redefine this notion. The reader is referred to the above mentione references and to [CoF91]. A method similar to tabulation is also studied in [BAK 89] under the name of equality loop-check.

For computation systems corresponding to abstract interpretation applications, it is in general the case that every program and goal is intrinsically finite. In fact, in general, in such computation systems the set of possible values on a finite set of variables (intuitively the variables of the program to analyse) is finite. This will be the case only for the second application described in this paper, cf. Section 5. Unfortunately, for the first application, cf. Section 4, the situation is different: the domain of values of the computation system is Tsubst and, obviously, there are infinite type substitutions on a finite domain! Because of this reason an extra condition (type recurrency) on the programs and goals will be necessary in order to guarantee intrinsic finiteness.

Obviously one wants that the tabled computations compute the same values (in the sense explained above) as the normal computation. Such a result is shown in [KKa90] only for the standard computation system S, but we want to consider any computation system C. More precisely, the following correctness result must be shown:

Correctness result: for any computation system C and any computation w:[G,c],...,[G1,c1], of C, there exists a tabled computation w':[G,c],...,[G1',c1'], such that w and w' compute the same set of values (modulo variable renaming).
One can show that the above correctness result holds for all computation systems C that satisfiy the following **condition(X):** let $C=(C,\leq,+,\pi)$

Consider the computation w:[G0,c0], [G1,c1]...,[Gk,ck], of C, where G0=A,A1..An, G1= B1...Bm, A2..An, the clause used in the step G0->G1 is cs: H:-c',B1..Bm, and Gk=B1'..Bp',A2..An. Let also c0'=π(c0,Var(A)), the following two points hold:

(i) π(c1,Var(cs))=π(c0'+c'+{A=H},Var(cs))

(ii) let ck'=π(ck,Var(cs)); if c=c0'+ ck'+{A=H}, then π(ck,Var(A))=π(c,Var(A)).

Condition (i) means that in order to compute c1 on the variables of cs it suffices to use c0' (instead of c0) and, similarly, condition (ii) means that c0' and ck' suffice for computing ck on Var(A).

Let us summarize all the notions introduced in this section pointing out how they define a general framework for abstract interpretation. Let us assume that one wants to define an abstract interpretation application in which a certain information I must be synthesized from any given program. This can be done in four steps:

(i) A computation system C=(C,$\le$,+,π) must be defined such that the elements of C represent the information I plus eventually some extra information that is useful for computing I , + will be an operation simulating on C the unification and $\le$ and π will be appropriately defined for C.

(ii) One must show that C simulates the concrete computation system from which he/she intends to synthesize the information I, for instance, in our applications we consider first the TTS and secondly the S computation-system, cf., section 4 and 5.

(iii) One must show that C satisfies condition(X) and thus that the computations of C can be safely tabled.

(iv) Finally, one has to prove that the computations of C are intrinsically finite for all programs or only for a subset of them. This implies the finiteness of the tabled computations.

If all these conditions are satisfied then the tabled computation realizes the desired application for all programs (or for the subset of point (iv)). This framework will be used in the next sections for defining 2 applications that are concerned with the static type-checking of typed logic programs.

4. The first application

This section describes how an abstract interpretation application can function as a test of well typedness for typed logic programs where all type definitions are type-preserving. More precisely, the application concerns only a subclass of these programs, the type-recurrent ones. This class will be defined in (iv) below. For defining this application we will follow the 4 points specified at the end of the previous section.

Point(i). Let T=(Tsubst, $\le_T$,mgu$_T$,π_T) be as follows: $\le_T$ is the 'more general' relation among type substitutions, mgu$_T$ is the type unification, and π_T is the restriction of type substitutions to a fixed set of type variables.

Point(ii) T simulates TTS. Two facts must be shown, cf. Def.3.4.

(1)Let γ:Tsubst->2^{TTS}, be defined as follows: for any $\beta\in$ Tsubst, let $\gamma(\beta)$={σ / $\sigma\in$TTS, and $\forall X\in D(\beta)$ the global type of $X\sigma$ is more specific than $X\beta$}. It is easy to see that γ is monotonic.

(2) That for any $\sigma1$ and $\sigma2\in$ TTS, such that $\sigma1\in\gamma(\tau1)$ and $\sigma2\in\gamma(\tau2)$ mgu$_{TTS}$($\sigma1$,$\sigma2$)$\in\gamma$(mgu$_T$($\tau1$,$\tau2$)) can be shown as follows: if mgu$_{TTS}$($\sigma1$,$\sigma2$)=$\perp_{TTS}$, then the result is trivial, otherwise, it suffices to observe that, as for normal terms, unifying terms with more specific types one obtains a term with a more specific type.

The following result is an immediate consequence of Lemma 3.1.

Lemma 4.1. For any TTS-program P and TTS-goal G, there exist T-program P' and T-goal G' such that if no computation of P'∪{G'} produces the value ⊥_T, then P and G are type-correct, i.e., no type error occurs in any computation of P∪{G}.

Point(iii). That T satisfies condition(X) can be seen intuitively as follows: assume the notation of condition (X), since the usual standardization apart is assumed, the variables of c0 that are not in A cannot have any effect on the values of c1=c0+c'+{A=H} on Var(cs); the same holds for condition (ii).

Point(iv). Let us consider for any finite set of variables V, the subset of Tsubst that contains all the type substitutions whose domain is included in V. This set is in general infinite. This implies that there may be computations of T that are genuinely divergent. Hence, we can only hope to find a subclass of T-programs and of T-goals such that their computations are intrinsically finite. In practice we need to find conditions on the T-programs and T-goals that guarantee the termination of their tabled computations. To this end it is very useful that T-programs are so similar to (normal) logic programs, because this allows us to use for T-programs results already existing about the termination of logic programs. In [Bez89] a class of logic programs and goals is defined such that all their computations terminate. The definition are as follows.
A level mapping is a function | |: HB->Nat, where HB is the Herbrand basis of our first order language. A logic program P is **recurrent** wrt a level mapping | | if for each ground instance A:-B1...,Bk of every clause of P, it is true that |A|>|Bi| for each i∈[1,k].
For an atom A, possibly with variables, [A] is the set of all ground instances of A. A goal <-A1...Ak is **bounded** wrt | | if there is an m≥1 such that for each Ai' in [Ai] |Ai'|<m.

Theorem 4.1. [Bez89]. Let P be a recurrent program and G a bounded goal wrt | |. All SLD-derivations of P∪{G} are finite.

This definition can be improved in our case because we want tabled computations to be finite and not just (not tabled) SLD-derivations. The extension is as follows. For A and B in HB, A depends on B, denoted by A<-B, if there is a ground instance of a clause of P such that A is the head and B is in the body. A level mapping is c-t (for controllable by tabulation) if there is no infinite sequence A1<-A2<-.... containing an infinite number of different atoms of HB and such that |A1|=|A2|=.....
A C-program P is **c-t recurrent** wrt a c-t level mapping | |, if for each ground instance A:-B1...Bk of each clause of P, |A|≥|Bi|, for each i∈[1,k]. The definition of goal ct-bounded is similar.

The difference between this definition and the previous one of [Bez89] is that the equality of the level value between |A| and |Bi| is now acceptable. This fact, by the requirement on c-t level mappings, allows infinite (not tabled) computations that are intrinsically finite. For instance, the program p(X):-p(X) is c-t-recurrent. Thus, these computations will become finite when tabulation is introduced. The following result is an easy extension of the previous theorem.

Theorem 4.2. Let P be a c-t recurrent C-program and G a c-t bounded C-goal. All tabled computations of P∪{G} are finite.

A TTS-program P is **type-recurrent** if there is a T-program P' corresponding to it as il Lemma 4.1, such that P' is c-t recurrent wrt some c-t level mapping. **Type-bounded** TTS-goals are defined similarly. The following result is an immediate consequence of Lemma 4.1 and Theorem 4.2.

Corollary 4.1. Let P be a TTS-program and G a TTS-goal that are type recurrent and type-bounded, respectively. Let P' and G' be the corresponding T-program and T-goal. The tabled computation of $P' \cup \{G'\}$ is finite. Moreover, if during this computation no resolvent $[F, \perp_T]$ is produced, then, P and G are type-correct.

Notice that the fact that a TTS-program is type-recurrent does by no mean imply that also its (normal) computation is finite.

5. The second application

The second application has the aim of partially avoiding run-time type-check in the case of not type preserving type definitions. As usual, TYPE is the set of types and DEF is the set of type definitions for the symbols of the alphabet considered. For this application the types of the analyzed program are not important (we are only concerned with the functors that are not type preserving). Thus, the S computation system is taken as the concrete semantics.

Let A be an atom of S-program P. A will in general be selected in several different computations, instantiated by different substitutions. Let the activation set of A, ACT(A), be the set of all these substitutions. We want to compute for A a set of abstract values that represents some information about ACT(A). In this specific application, we are interested in representing the following information: for each $\sigma \in$ ACT(G),

(a) which are the variables of A that are not instantiated in σ to a term containing a not type preserving functor, such variables are called safe in σ,

(b) the covering between the variables of A, defined as follows: let S1 and S2 be subsets of Var(A), S1 covers S2 in σ if, $\cup \{ Var(X\sigma) / X \in S1 \} \supseteq \cup \{ Var(Y\sigma) / Y \in S2 \}$.

(c) the **independence** of the variables of A, defined as follows: X and Y in Var(A) are independent in σ if, $Var(X\sigma) \cap Var(Y\sigma) = \emptyset$.

Notice that the covering information (b) contains also groundness information: X is ground in σ, if it is covered by the empty set. Since only information (a) is related to type-checking, one may wonder why points (b) and (c) are there. The reason is as follows: the independence is needed in order to guarantee the correctness of the computation of (a), whereas the covering is there (only) to improve the quality of this computation.

Example Let A=p(X,Y) be the selected atom and $\sigma = \{X/f(Z)\ Y/Z\}$ be the current substitution. Let also p(W1, g(W2)) be the head of the clause against which A must be unified; g is a not type preserving functor, called **unsafe** from now on. After the unification both X and Y will be instantiated to a term containing g (they become unsafe too), but we can discover it only if we know that X and Y are not independent in σ. Thus independence is necessary for the correctness.

The above example shows that unsafeness propagates via not-independence (called sharing). From this, it is easy to understand that groundness information is very useful for limiting the sharing : if a variable is ground it cannot share with any other variable! Finally, that covering is necessary for improving the quality of the groundness information is well known, [MSo89,CFi91].

Let us now describe the computation system for computing the information (a)-(c), above,. The four points given at the end of Section3 are followed.

Point(i). Let $D=(D, \leq_D, +_D, \pi_D)$ be as in (1)-(4) below:

(1) Each element d of D is a triple d=(U,C,SH), where U is a subset of Var, C contains propositional formula $\wedge S1 \rightarrow \wedge S2$, where S1 and S2 are subsets of Var, and SH contains pairs (X,Y), where X and Y are in Var. U,C, and SH stand for unsafe, covering and sharing, respectively. Let also GR(C)={X / $\emptyset \rightarrow X \in$ C}. GR(C) contains

the variables that are surely ground in d. The following properties are satisfied by each element of D:

(a) SH is transitive, symmetric and reflexive and for each $(X,Y) \in SH$ neither X nor Y are in GR(C).

(b) For each X->Y in C, there is (X,Y) in SH,

(c) C is closed under semantic implication, i.e., if $Cl = \wedge S1 -> \wedge S2$ then $\wedge S1 -> \wedge S2 \in C$.

(2) Let for $i \in [1,2]$ $di=(Ui,Ci,SHi)$, $d1 \leq_D d2$ if: $U2 \supseteq U1$, $SH2 \supseteq SH1$, $C1 \supseteq C2$.

(3) The D-unification $+_D$ is as follows: let d1 and d2 be as above and $\sigma = \{X1=X1',..,Xk=Xk'\}$; $d1 +_D d2 +_D \sigma = (U,C,SH)$ as follows: let $U'=U1 \cup U2$, $C'=C1 \cup C2 \cup \{Xi->Xi', Xi'->Xi / i \in [1,k]\}$ and $SH'=SH1 \cup SH2 \cup \{(Xi,Xi'),(Xi',Xi) / i \in [1,k]\}$;

(a) Let SS be the transitive, reflexive and symmetric closure of SH',

(b) C is the closure of C' under semantic implication. GRS (S stands for safe) is the following subset of GR(C): let $K=C'-\{\emptyset -> \wedge S / \exists X \in S$ s.t. $X \in U'\}$ and let K^+ be the closure of K under semantic implication, then $GRS=GR(K^+)$.

(c) Let $SHU=\{(X,Y) / (X,Y) \in SS$ and X and Y $\notin GRS\}$. U is the minimal set containing U' and such that if $X \in U$ and $(X,Y) \in SHU$ then $Y \in U$.

(d) $SH=\{(X,Y) / (X,Y) \in SS$ and X and Y $\notin GR(C)\}$.

Some explanations are probably useful. The set GRS contains the variables that are ground in C and that are surely safe. Because of this fact, a sharing pair involving an element of GRS can surely not propagate unsafeness. Thus all these pairs are eliminated from SH, obtaining SHU. This shows once more that groundness plays a role in limiting sharing and thus also in limiting the propagation of unsafeness.

(4) The projection π_D is as follows: for $d=(U,C,SH)$, $\pi_D(d,V)=(U',C',SH')$, where, $U'=\{X / X \in U \cap V\}$, $C'=\{\wedge S1 -> \wedge S2 / \wedge S1 -> \wedge S2 \in C$, and $V \supseteq S1 \cup S2\}$, $SH'=\{(X,Y) / (X,Y) \in SH$ and X and $Y \in V\}$.

Point(ii). D simulates S. Two facts must be shown, cf. Def 3.4.

(1) For $d=(U,C,SH)$ let $\gamma(d)$ contain all the substitutions σ such that:

(a) for each X such that $X\sigma$ contains an unsafe symbol, $X \in U$,

(b) for all pairs of variables X and Y that share in σ, $(X,Y) \in SH$,

(c) for every $\wedge S1 -> \wedge S2 \in C$, S1 covers S2 in σ.

It is easy to see that γ is monotonic wrt $\leq_D$ and that for each substitution σ there is an element d of D such that $\sigma \in \gamma(d)$.

(2) We have to show that: for every $\sigma1 \in \gamma(d1)$, $\sigma2 \in \gamma(d2)$, and $mgu(\sigma1,\sigma2) \in \gamma(d1 +_D d2)$.

The proof is omitted because of lack of space. A similar proof can be found in [CFi91]. The main point is to show that, for the computation of U it suffices to use SHU instead of the whole SH.

Point(iii) The proof of this point is not difficult, but quite technical. Therefore it is omitted for the shortness sake.

Point(iv). All D-programs and goals are intrinsically finite because for any finite set of variables V the set of values of D on V is obviously finite. This is generally the case in abstract interpretation applications.

Conclusion

Two abstract interpretation applications have been defined. The first one works as a compile-time finite test of the correctly-typed property for a subclass of the type preserving logic programs.

The second application detects the arguments of the atoms of resolvents that are surely safe, i.e., that do not contain not type-preserving functors. When such safe terms are unified, the type-checking is necessary only for their global types. These two applications can also work together: for programs that pass the test of the first application no type-checking at all is needed when safe terms are unified.

An interesting problem that still needs some study is whether classes of programs could be defined that are larger than the class of type-recurrent programs and that still enjoy the property of having finite tabled computations. To this end the work of [APe90] and [BCF90] about program termination could be useful.

References

[APe90] K.Apt & D.Pedreschi; Studies in pure Prolog: termination. Proc. Symp. on Computational Logic, Lec. Notes in AI, n.1, Springer Verlag,1990.

[Bak57] J.Bakus et al. ; The FORTRAN automatic coding system. Western Joint Computer Conf. (1957), 188-198.

[BAK89] R.N.Bol, K.A.Apt, J.W.Klop; An analysis of loop checking mechanisms for logic programs. CWI, R. CS-R8942, 1989.

[BCF90] A.Bossi, N.Cocco & M.Fabris; Proving termination of logic programs by exploiting term properties. Dep. of Math., Univ. of Padova, R.21, 1990. Accepted to the TAPSOFT Conf., Brighton, 1991.

[Bez89] M.Bezem; Characterizing termination of logic programs with level mappings. Proc. North American conf. on Logic Programming, Cleveland, eds. L.Lusk & R.Overbeek, MIT Press, (1989), 69-80.

[Bru88] M.Bruijhooghe; A practical framework for the abstract interpretation of logic programs. To appear in the J. of Logic Programming.

[CFi 91] A.Cortesi & G.Filè; Abstract Interpretation: an abstract domain for groundness, sharing, freeness and compoundness analysis. Dep. of Math.,University of Padova, R.4-1991; also to appear in the Proc. of the ACM SIGPLAN Symposium on partial evaluation and semantics based program manipulation, New Haven, 1991.

[CoF91] P.Codognet & G.Filè; Coomputations, abstractions and constraints. In preparation.

[Han89] M.Hanus; Horn clause programs with polymorphic types: semantics and resolution. Proc. of TAPSOFT 89, LNCS 352, 225-240.

[KKa90] T.Kanamori & T.Kawamura; Abstract interpretation based on OLDT-resolution. ICOT TR, 1990.

[Llo87] J.Lloyd; Foundations of logic programming. Springer Verlag, 2nd edition, 1987.

[Mil78] R.Milner; A theory of type polymorphism in programming. J.Comput. System Sci. 17, (1978), 348-375.

[MO'K84] A.Mycroft & R.O'Keefe; A polymorphic type system for Prolog. Artificial intelligence 23, (1984), 295-307.

[MSo89] K.Merriot & H.Sondergaard; A tutorial on abstract interpretation of logic program. Tutorial of the North American conf. on Logic Programming, Cleveland, 1989.

[Nau65] P.Naur; Checking of operand types in Algol compilers. BIT 5, (1965), 151-163.

[TSa86] H.Tamaki & T.Sato; OLD resolution with tabulation. Proc. of the 3rd Conf. on Logic programming, London, LNCS 225, (1986), 84-98.

A Technique for Recursive Invariance Detection and Selective Program Specialization

F. Giannotti[1]
CNUCE-CNR
Via Santa Maria 36,
56100 Pisa, Italy
fosca@gmsun.cnuce.cnr.it

M. Hermenegildo[2]
Univ. Politécnica de Madrid (UPM)
Facultad de Informática
28660-Boadilla del Monte, Madrid-Spain
herme@fi.upm.es *or* herme@cs.utexas.edu

Abstract

This paper presents a technique for achieving a class of optimizations related to the reduction of checks within cycles. The technique uses both Program Transformation and Abstract Interpretation. After a first pass of an abstract interpreter which detects simple invariants, program transformation is used to build a hypothetical situation that simplifies some predicates that should be executed within the cycle. This transformation implements the heuristic hypothesis that once conditional tests hold they may continue doing so recursively. Specialized versions of predicates are generated to detect and exploit those cases in which the invariance may hold. Abstract interpretation is then used again to verify the truth of such hypotheses and confirm the proposed simplification. This allows optimizations that go beyond those possible with only one pass of the abstract interpreter over the original program, as is normally the case. It also allows selective program specialization using a standard abstract interpreter not specifically designed for this purpose, thus simplifying the design of this already complex module of the compiler. In the paper, a class of programs amenable to such optimization is presented, along with some examples and an evaluation of the proposed techniques in some application areas such as floundering detection and reducing run-time tests in automatic logic program parallelization. The analysis of the examples presented has been performed automatically by an implementation of the technique using existing abstract interpretation and program transformation tools.

Keywords: Logic Programming, Abstract Interpretation, Program Transformation, Program Specialization, Parallel Logic Programming, Cycle Invariant Detection, Compile-time Optimization.

1 Introduction

This paper presents a technique for achieving a class of optimizations related to the reduction of conditionals within cycles by performing selective program specialization. Much work has been done on techniques such as partial evaluation which allow safe program transformations [1, 9, 25, 24, 5, 2]. Much work has also been done in abstract interpretation, which has been used to achieve several types of high level optimizations: mode inference analysis [8, 21], efficient backtracking [6], garbage collection [19], aliasing and sharing analysis [16, 22, 23] type inferencing [4, 20], etc. It has also been proposed to use the results from abstract interpretation to produce specialized versions of predicates for different run-time instantiation situations [10, 11]. One important issue in program specialization is when to create specialized versions of predicates and how to select the different versions. In [26] and [17] an elegant framework is proposed for abstract interpretation-based program specialization. An automaton is generated which allows run-time selection of the appropriate version. Specialization is controlled by a "wish-list" of optimizations which would be generated by the compiler. Although quite powerful, the framework as described requires the construction of a complex, specialized abstract interpreter with knowledge of the specialization process being done and capable of understanding the "wish-list" from the compiler.

---

[1] This author is supported by C.N.R. - Italian Research Council.
[2] This author is supported in part by ESPRIT projects "PEPMA" and "PRINCE," and CICYT project 361.208.

This paper presents a technique (developed independently of [26] and [17]) which attempts similar results but by quite different means. We assume the existence of an abstract interpreter. We also assume that this interpreter uses a domain that is adequate for the type of optimizations that the compiler performs. Based on these assumptions, and rather than asking the compiler for a "wish-list" of desired optimizations, we develop an abstract domain related notion ("abstract executability") which will guide the process of specialization and invariant detection. Also, rather than modifying the abstract interpreter to be aware of the specialization process we leave the interpreter unmodified.[3] Rather, we propose to perform the program specialization and simplification steps externally to the interpreter while still achieving our objective of extracting repetitive run-time tests to the outermost possible level. This is achieved by using program transformation to build a hypothetical situation that would reduce the predicates to be executed in the cycle according to their abstract executability and then using the abstract interpreter again to verify the truth of the hypothesis and (possibly) confirm the proposed simplification.

Consider the following conditional:

$$p(X) \leftarrow q(X), cond(test(X), p(X), r(X)).$$

Program transformation can be used to build a hypothetical situation that reduces the predicates to be executed in the cycle. For example, we can hypothesize that once the test in the conditional succeeds, it will always succeed. A correct program transformation under this hypothesis would be

$$p(X) \leftarrow q(X), cond(test(X), p1(X), r(X))$$
$$p1(X) \leftarrow q(X), p1(X).$$

Of course, this transformation is legal if we are sure that $test(X)$ will remain true in all the recursive calls of p in the *then branch*. If that is the case the relevance of the obtained optimization will depend on the complexity of $test(X)$ and on the number of nested recursive calls in the *then branch*. The interesting issue is whether the abstract interpretater can derive if $test(X)$ will be true in all the recursive calls. This depends on the capabilities of the abstract interpreter and precisely these capabilities can be used as a guideline for when to perform the transformation. I.e. given an abstract interpreter, a class of predicates that can be executed directly (reduced to true or false) on the information generated by the abstract interpreter can be identified. Then, rather than blindly performing hypothesis and transformations this class is used to select only potentially useful transformations. The abstract interpreter, run for a second time on the transformed program to verify the truth of the hypothesis formulated, has then a chance of being successful in its task. Our conviction is that such classes of predicates can be easily found for each abstract interpreter.

The idea of leaving the abstract interpreter unmodified is motivated by the consideration that the interpreter is probably already a quite complex module which may be quite difficult to modify and that therefore there is practical advantage in using this module as is. This appears to be the case with most current implementations. In addition, this allows the use of several different abstract interpreters with only minor modifications to the rest of the system. Our description, thus, will be quite independent from the abstract interpreter, which will be considered as a "black box."

The paper is organized as follows: the following section (section 2) recalls the basic ideas of Abstract Interpretation and introduces the concept of and-or graph to represent the result of an abstract interpretation process. Section 2 presents a class of predicates that may be executed at compile-time by using the information collected by a generic abstract interpreter. In sections 3 and 4 the and-or graph representation is exploited to describe the basic program transformation and optimization techniques proposed, based on the concept of abstract executability. The possibility of performing these optimizations using abstract interpretation and program transformation occurred to us while considering their implementation in the context of the abstract interpreter of the &-Prolog system [13].[4] Section 5 is dedicated to some examples illustrating the applicability of such techniques

---

[3] Including any other type of program specialization and optimisation that it may be doing.
[4] Although the techniques that will be proposed are of general applicability, examples from the &-Prolog system will

to several programs, including optimizing the automatic parallelization process in the &-Prolog system.
Finally, section 6 presents our conclusions.

Some knowledge of Prolog and Abstract Interpretation is assumed.

2 Abstract Interpretation of Logic Programs

Abstract interpretation is a useful technique for performing global analysis of a program in order to compute at compile-time characteristics of the terms to which the variables in that program will be bound at run-time for a given class of queries. The interesting aspect of abstract interpretation vs. classical types of compile-time analyses is that it offers a well founded framework which can be instanciated to produce a rich variety of types of analysis with guaranteed correctness with respect to a particular semantics [3, 4, 7, 8, 16, 18, 21, 23].

In abstract interpretation a program is "executed" using *abstract substitutions* instead of actual substitutions. An abstract substitution is a finite representation of a, possibly infinite, set of actual substitutions in the concrete domain. The set of all possible terms that a variable can be bound to in an abstract substitution represents an "abstract domain" which is usually a complete lattice or cpo which is ascending chain finite (such finiteness required, in principle, for termination of fixpoint computation).

Abstract substitutions and sets of concrete substitutions are related via a pair of functions referred to as the *abstraction* (α) and *concretization* (γ) functions. In addition, each primitive operation u of the language (unification being a notable example) is abstracted to an operation u' over the abstract domain. Soundness of the analysis requires that each concrete operation u be related to its corresponding abstract operation u' as follows: for every x in the concrete computational domain, $u(x)$ is "contained in" $\gamma(u'(\alpha(x)))$.

The input to the abstract interpreter is a set of clauses (the program) and set of "query forms." In its minimal form (least burden on the programmer) the query forms can be simply the names of the predicates which can appear in user queries (i.e., the program's "entry points"). In order to increase the precision of the analysis, query forms can also include a description of the set of abstract (or concrete) substitutions allowable for each entry point.

The goal of the abstract interpreter is then to compute in abstract form the set of substitutions which can occur at all points of all the clauses that would be used while answering all possible queries which are concretizations of the given query forms. Different names distinguish abstract substitutions depending on the point in a clause to which they correspond: *abstract call substitution* and the *abstract success substitution* are meant for the literal level while the terms *abstract entry substitution* and *abstract exit substitution* refer to the clause level.

A general mechanism for representing the output of the abstract interpretation process is the *and-or graph* associated with a logic program for a given entry point. It is a finite representation of a usually infinite *and-or tree* adorned with the *abstract substitutions*. It is worth noting that, even if the notion of and-or graph is typical of the *top-down* approaches to abstract interpretation, it also encompasses the *bottom-up* approaches which increasingly emphasize recording the *call patterns* of a program. Thus or-nodes are decorated with the abstract call and success substitutions for the associated predicate, whereas and-nodes are decorated with the abstract entry and exit substitutions for the associated clause. Such nodes of an and-or graph are defined by the following equations:

$$and\_node ::= A \times 2^{Sub} \times 2^{Sub} \times or\_node^*$$
$$or\_node ::= rec\_call \text{ of } A \mid A \times 2^{Sub} \times 2^{Sub} \times and\_node^*$$

where A is the set of all atoms and Sub is the set of concrete substitutions, $\mid$ denotes discriminated union, rec_call is the constructor for recursive calls and "*" stands for sequence construction.

---

be used throughout the paper for motivational purposes.

Intuitively, an and-node represents a clause in the program, and it is composed by an atom that is the head of the clause, the entry and the exit substitutions for the clause, and the list of children or-nodes each corresponding to a literal in the body of the clause.

An or-node represents a literal in the body of a clause; if the literal is a descendent of a literal that calls the same predicate then the construction of the and-or tree is suspended and the or-node refers back to the ancestor or-node. Otherwise, the or-node is composed by a literal, the call and the success substitutions for the literal, and the list of children and-nodes each corresponding to a clause for the predicate of the literal.

The various actual abstract interpreters differ in the way the finite and-or graph is constructed, i.e. when they decide to prune recursion and whether they allow different instances of the same clause. These options affect the accuracy of the computed abstract substitutions for the *recursive cliques*, i.e. the path on the graph that connects a predicate with one of its recursive calls, if there are.

Note that if the abstract interpreter itself already performs a certain degree of program specialization (as is sometimes the case) the abstract and-or tree represents the transformed program. In this case we assume that the specialized predicates have been renamed appropriately.

3 Abstract Executability and Program Transformation

In this section the concept of abstract executability is defined. As mentioned before, the recognition of potentially abstractly executable goals will be the guiding heuristic in guiding program especialization to obtain that optimizations desired. For example, sometimes Prolog programs (and very often the programs resulting from the transformations performed in the first stages of a Prolog compiler) mix logic predicates with meta-predicates which deal with program variables characteristics or types. Examples of such predicates are $var(X)$, $nonvar(X)$, $number(X)$, $list(X)$, $integer(X)$, $ground(X)$, $atomic(X)$, $independent(X,Y)$, etc. Some can be based on others: for example $independent(X,Y)$, which explores the two terms to which X and Y are bound to check that they do not have any common variables, can be written in terms of $var/1$ and $==/2$. Such predicates can be part of the original program or they may have been introduced by compilation stages performing tasks such as type-checking, indexing, program parallelization, etc. A characteristic of these predicates is that they seldom modify the current substitution. They are mostly used as "tests" to have an effect on the control of the program, rather than on the logic (which would obviously not be appropriate, given their generally extralogical nature).

Another important characteristic of such predicates is that they can potentially be executed on practical abstract domains, that is, they can be reduced to true or false by simply operating on the abstract representation of the possible bindings of their arguments. Consider for instance the abstract domain consisting of the three elements $\{int, free, any\}$. These elements respectively correspond to the set of all integers, the set of all unbound variables, and the set of all terms. An abstract substitution is then defined as a mapping from program variables to elements of the abstract domain. Assume a correct abstract interpreter whose estimates of integer type and freeness are conservative in the sense that it infers these values only when it is possible to guarantee that all possible substitutions bind respectively to integers and unbound variables. Consider the following clause containing the predicate $ground(X)$:

$$p(X,Y) \leftarrow q(Y), ground(X), r(X,Y).$$

Assume now that such an interpreter infers the call substitution for $ground(X)$ to be $\{Y/free, X/int\}$. This means, that if S is the set of all possible concrete substitutions corresponding to the program point just before $ground(X)$, then $\forall \theta_i \in S$, $X/n \in \theta_i$ where n is an integer. Since given any term t $integer(t) \rightarrow ground(t)$, then, and knowing that $ground/1$ doesn't modify its arguments, $\forall \theta_i \in S$, $ground(X) \equiv ground(\gamma(int)) \equiv true$. Therefore we can "execute on the abstract domain" the $ground(X)$ literal and reduce it to true. Note that this also trivially holds in the case where the abstract domain directly captures the information required by the literal for executability. This would be the case in the previous example had the abstract domain been for example $\{ground, free, any\}$ and the call substitution for $ground(X)$ been $\{Y/free, X/ground\}$.

We call the characteristic of a predicate sketched above "abstract executability." In general the condition for *abstract executability* according to an abstract domain D equipped with the concretization function γ and the abstraction function α is the following:

Definition (Abstractly executable goal): Given an abstract call substitution λ and a literal A, A is said to be abstractly executable and succeed (resp. fail) if $\forall\theta \in \gamma(\lambda)$ the concrete evaluation of $A\theta$ succeeds with empty concrete answer substitution (resp. finitely fails).

Trivially, we can state that a literal which is abstractly executable and succeeds (rep. fails) can be substituted by *true* (resp. *false*).

In other words, the (potentially) abstractly executable predicates behave as recognizers of the abstract values. In fact, if a predicate *testT* is a recognizer for the abstract value T then $\forall t \in \gamma(T)$ $testT(t) = true$. Now, if the abstract call substitution for an atom $testT(X)$ contains X/T then $testT(T)$ is a tautology and we are authorized to reduce it to true.

For instance the literal $ground(X)$ can be abstractly reduced to *true* w.r.t. the abstract call substitution $\{X/ground\}$, it can be abstractly reduced to *false* w.r.t. the abstract call substitution $\{X/free\}$ while it cannot be abstractly executed w.r.t. the abstract call substitution $\{X/any\}$.

Let us remark that the notion of abstract executability is only meaningful with respect to the final outcome of the abstract interpreter since only the final abstract substitutions represent *all* the possible concrete substitutions. Nevertheless, it is a design option of an abstract interpreter to let the abstractly executable predicates affect the abstract interpretation itself. In fact if recognizers of the abstract values are considered built-ins then an abstract counterpart already exists. As an example, if the abstract interpreter exploits the semantics of the built-in ground, it will include the binding $X/ground$ in the exit substitution of a literal $ground(X)$. In other words, the abstractly executable predicates play two different roles. At the end of the abstract interpretation process, they can be simplified by examining their final abstract call substitutions. During the abstract interpretation process they are clearly useful, as any other builtin, for enriching the information being inferred through the knowledge of their abstract exit substitutions.

The possibility of abstractly executing predicates reducing them to true or false allows for a further analysis phase aimed at performing simplifications of the final and-or graph and, consequently, of the program itself. This is done using standard program transformation and partial evaluation techniques. It is not the purpose of this paper to give an overview of such techniques. Rather, a simple (and non-exhaustive) repertoire of simplifications is proposed, described by the following rewriting rules on the and-or graph (head unification is assumed to be expressed in the form of unification goals at the beginning of the body, i.e. clauses are in the "normal form" of [3]):

- Case 1 – true in and-node: • Case 2 – false in and-node:

 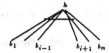 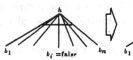

note that if the literals to the left of the literal reduced to *false* are pure, then they can also be eliminated.[5] In general, all literals to the right of the rightmost impure literal can be eliminated.

---

[5] Ignoring infinite computations. This transformation in fact has the arguably beneficial effect of augmenting the finite failure set of the program.

- Case 3 – Simplifying an or-node:
  - Case 4 – Collapsing an or-node:

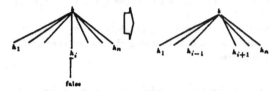

Note that, since no unification appears in the head, the call from g to h cannot produce any bindings as is essentially just parameter passing.

Further optimization of the resulting programs can be done using well known program transformation and partial evaluation techniques.

As an example, let's consider the if-then-else construct of Prolog. First, note that, provided $test(X)$ has no side-effects, is sufficiently instantiated (or $nottest$ can be safely defined), and modulo additional computation time,[6]

$$cond(X) \leftarrow (test(X) \rightarrow then(X); else(X)).$$

is essentially an (efficient) abbreviation of:

$$cond(X) \leftarrow test(X), then(X).$$
$$cond(X) \leftarrow nottest(X), else(X).$$

The associated and-or graph is labeled "a)" in the following figure:

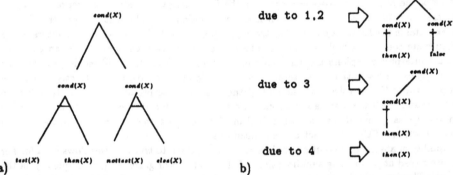

Assuming that $test(X)$ can be abstractly reduced to $true$ and hence that $nottest(X)$ can be reduced to $false$, the sequence of simplifications in figure "b)" above takes place. As a consequence, the original call to $cond(X)$ in the and-or graph can be replaced with the call to $then(X)$.

An interesting application of the conditional simplification is during a test for *floundering* in the evaluation of negated goals (or, better stated, *warning* against floundering). A conditional goal like $(ground(X) \rightarrow not\ p(X); error, abort)$ warns against incorrectness with respect to the negation as failure rule. If ground(X) is simplified according to the above observation, correctness is preserved. The optimization is relevant because the cost of a groundness check is proportional to the size of the term being checked.

As a final remark, notice that the correctness of the proposed simplifications is a straightforward consequence of the correctness of the associated Abstract Interpretation framework (w.r.t. the semantics that the abstract interpreter is derived from).

---

[6]For conciseness, the actual parameter passing has been abstracted out, as in the rest of this section, into a single variable X. The extension of the transformation to handle actual arguments is trivial and is illustrated in the examples shown in section 5.

4 A transformation for detecting deeper invariants

Although the direct applicability of the simplification criteria introduced in the previous section is in practice pretty rare, they do suggest a further possibility: that further optimization is possible in the case of recursive predicates. The idea is to transform the program into an equivalent form which gives more chances to obtain a simplification by highlighting possible invariant predicates which can then be extracted out of cycles. Consider the following recursive clause for a predicate p:

$$p(X) \leftarrow f(X), (test(X) \rightarrow p(X); q(X)).$$

and assume that the abstract interpreter cannot reduce $test(X)$ to either $true$ or $false$. It is still possible that during the concrete interpretation, the atom $test(X)$ after its first evaluation to true will also succeed in all the subsequent recursive calls through the $then$ $branch$. This consideration brings us to define a transformation of the clause which separates the first iteration from the subsequent ones:

$$p(X) \leftarrow f(X), (test(X) \rightarrow p1(X); q(X)).$$
$$p1(X) \leftarrow f(X), (test(X) \rightarrow p1(X); q(X)).$$

The above transformation introduces a copy of the predicate p with a new name, and the recursive calls refer to new predicate $p1$. Picture "c)" below represents the transformation at and-or graph level hiding at this stage the abstract substitutions:

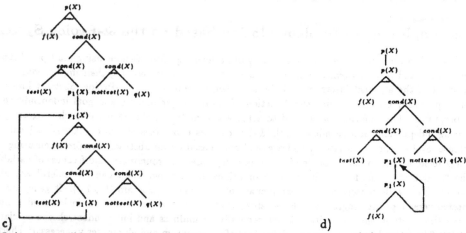

c)   d)

It is now possible to run the abstract interpreter to check whether $test(X)$ can be simplified to $true$ in $p1$ exploiting the fact the the abstract interpreter reached $p1$ under the hypothesis $test(X) = true$ in p, i.e. the initial call substitution of $p1$ has been extended with the new binding $\{X/T\}$ if $test/1$ is the recognizer for T. Note that if $test(X)$ is a builtin "understood" by the abstract interpretation framework, the extension of abstract substitution is for free by the application of the abstract interpreter. Now, if when running again the abstract interpreter, the and-or tree contains again the binding $\{X/T\}$ as abstract call substitution for $p1$, the basic simplification for the conditional in the clause for $p1$ applies yielding the final and-or graph of figure "d)" above, corresponding to the following optimized code:

$$p(X) \leftarrow f(X), (test(X) \rightarrow p1(X); q(X)).$$
$$p1(X) \leftarrow f(X), p1(X).$$

The above optimization corresponds to the detection of the invariant $test(X)$ for the recursive predicate $p1$; the abstract interpreter actually proved that $test(X)$ is invariant for $p1$: in fact $test(X)$ is $true$ in the preconditions for $p1$ and the abstract interpreter derives that it is $true$ in all the derivations from $p1$. Thus, in the final code the invariant has been extracted out of the cycle: $test(X)$, if *true*, will be executed only the first time; otherwise, the control remains to *p* (executing the else branch) with the open possibility of capturing possible invariance later on.

The generalization of the ideas presented above hinges on the and-or graph definition. We designate the critical portions of the graph, i.e. recursive cliques containing abstractly executable predicates. The transformation then consists in the duplication of such clique, and, after the simplification process, rebuilding a minimal logic program from the transformed graph. The full algorithm is given in [12]. The overall process includes the following phases: Abstract Interpretation (decorates the and-or graph with the computed abstract substitutions), Simplification, Transformation (subgraphs containing executable predicates within cycles are duplicated), second Abstract Interpretation (aimed at confirming the invariants), Simplification, and Final Transformation (new parts of the graph that have not been affected by simplifications are eliminated and the final program is generated).

It should be mentioned that although a minimum of two passes of the abstract interpreter is needed to achieve the desired result additional iterations might improve the program further. This cycle could be repeated until fixpoint. However, two passes appear to be a good compromise between precision and efficiency.

A final consideration worth noting is that the correctness of the process relies on two observations: a) the transformations are equivalence preserving since they perform copies of the graph, connecting them by proper renaming; b) the simplifications on the transformed graph are correct provided that the Abstract Interpreter is sound in the sense of section 2.

5 Examples on an Implementation based on the &-Prolog System

As mentioned in the introduction, the possibility of performing optimizations using multiple abstract interpretation and program transformation occurred to us while considering their implementation in the context of the abstract interpreter of the &-Prolog system [13]. In &-Prolog, abstract interpretation is primarily concerned with the detection of argument groundness and goal independence in logic programs. This information is used to achieve automatic parallelization of such programs, by exploiting independent and-parallelism[14]. &-Prolog uses two layers of static analysis: a local one (at clause level — the "annotator") and a global one (based on an abstract interpretation using the sharing/freeness domain [23], capable of inferring independence, groundness, and freeness of variables — the "analyzer"). The former determines conditions under which goals can be "safely" executed in parallel (i.e. while preserving the search space of the original program [14, 15]). It generates an annotated version of the program which contains parallel conjunctions, sometimes inside conditionals. The run-time tests in these conditionals are generally groundness and independent checks on clause variables. The latter layer analyses the global flow of the program and eliminates unnecessary checks. This is an important efficiency issue since these tests can be quite expensive. Although the abstract interpreter has proven quite powerful, there are situations where such checks can be eliminated only in particular cases. Indeed, we have often discovered further opportunities for optimization of conditionals inside cycles by performing program transformation and running the abstract interpreter a second time on the transformed programs. In order to illustrate the practical application of the techniques presented in the previous sections several experiments performed with this system are presented.

An interesting practical example, in the context of program parallelization, is the multiplication of matrices of integers. This program is sometimes used as a benchmark for evaluating the performance of parallel logic programming languages. The following is a part of the program relevant to our example:

```
main:- read(M), read(V), multiply(M,V,Result), write(Result).

multiply([],_,[]).
multiply([VO|Rest], V1, [Result|Others]):-
        multiply(Rest, V1, Others), vmul(VO,V1,Result).
```

```
vmul([],[],0).
vmul([H1|T1], [H2|T2], Result):-
        vmul(T1,T2, Newresult), scalar_mult(H1,H2,Product),
        add(Product,Newresult,Result).
```

The tests introduced by the Annotator on the procedure multiply are simplified by the first pass of the abstract interpreter resulting in the following program (note that the abstract interpreter has inferred that Result and Others are free variables):

```
multiply([],_,[]).
multiply([VO|Rest],V1,[Result|Others]) :-
      ( ground(V1), indep([[VO,Rest]]) ->
        multiply(Rest,V1,Others) & vmul(VO,V1,Result)
      ; multiply(Rest,V1,Others),  vmul(VO,V1,Result) ).
```

Where & means that two goals can be executed in parallel and ground(V1), indep([[VO,Rest]]) are dynamic tests for groundness and independence that express the sufficient condition under which the two goals can be parallelized.

A relevant hypothesis is that the test in the conditional will be an invariant in the recursive loop. Another hypothesis is built to detect whether there is an invariant in the else branch. The transformed program is the following:

```
multiply([],_,[]).
multiply([VO|Rest],V1,[Result|Others]) :-
        ( ground(V1), indep([[VO,Rest]]) ->
        multiply_then(Rest,V1,Others) & vmul(VO,V1,Result)
      ; multiply_else(Rest,V1,Others),  vmul(VO,V1,Result) ).

multiply_then([],_,[]).
multiply_then([VO|Rest],V1,[Result|Others]) :-
        ( ground(V1), indep([[VO,Rest]]) ->
        multiply_then(Rest,V1,Others) & vmul(VO,V1,Result)
      ; multiply(Rest,V1,Others),        vmul(VO,V1,Result) ).

multiply_else([],_,[]).
multiply_else([VO|Rest],V1,[Result|Others]) :-
      ( ground(V1), indep([[VO,Rest]]) ->
        multiply(Rest,V1,Others) & vmul(VO,V1,Result)
      ; multiply_else(Rest,V1,Others), vmul(VO,V1,Result) ).
```

The second pass of the abstract interpreter then determines that V1 is always ground in multiply_then. With this information we see that the multiply_else transformation was not successful in determining an invariant, and that the multiply_then transformation was partially successful: the first half of the test (ground(V1)) is found to be an invariant. The program is therefore rewritten as:

```
multiply([],_,[]).
multiply([VO|Rest],V1,[Result|Others]) :-
      ( ground(V1), indep([[VO,Rest]]) ->
        multiply_then(Rest,V1,Others) & vmul(VO,V1,Result)
      ; multiply(Rest,V1,Others),        vmul(VO,V1,Result) ).

multiply_then([],_,[]).
multiply_then([VO|Rest],V1,[Result|Others]) :-
```

```
( indep([[VO,Rest]]) ->
  multiply_then(Rest,V1,Others) & vmul(VO,V1,Result)
; multiply(Rest,V1,Others),       vmul(VO,V1,Result) ).
```

The repeated tests for ground(V1) are eliminated once the test succeeds in the first iteration. Note that in the normal use of the multiply predicate the remaining test checks for the independence of two free variables. This is a quite inexpensive check, the resulting program thus being quite efficient.

A different example which is unrelated to parallelization and takes advantage of the fact that "nonvar" is executable on the sharing/freeness abstract domain is the following:

```
map_add1(_,[]).
map_add1(X,[Y|TY]):-
     ( nonvar(X) -> Y is X+1, map_add1(X,TY)
                 ;  X is Y-1, map_add1(X,TY) ).
```

The first abstract interpretation pass cannot determine which side of the conditional will be taken. Again the hypothesis is that once a branch is taken the condition will hold. The specialized version of the program for testing the hypothesis is:

```
map_add1(_,[]).
map_add1(X,[Y|TY]):-
     ( nonvar(X)-> Y is X+1, map_add1_nvX(X,TY)
                ;  X is Y-1, map_add1(X,TY) ).

map_add1_nvX(_,[]).
map_add1_nvX(X,[Y|TY]):-
     ( nonvar(X)-> Y is X+1, map_add1_nvX(X,TY)
                ;  X is Y-1, map_add1(X,TY) ).
```

The abstract interpreter can now determine that X is always non-var in the body of map_add1_nvX and generates the simplified version:

```
map_add1(_,[]).
map_add1(X,[Y|TY]):-
     ( nonvar(X)-> Y is X+1, map_add1_nvX(X,TY)
                ;  X is Y-1, map_add1(X,TY) ).

map_add1_nvX(_,[]).
map_add1_nvX(X,[Y|TY]):-
     Y is X+1, map_add1_nvX(X,TY).
```

which dynamically and inexpensively traps the invariant avoiding run-time checking.

6 Conclusions

The paper has presented a technique for achieving a class of optimizations related to the reduction of conditionals within cycles. The technique is somewhat analogous to cycle invariant detection and removal in traditional compilers. The application of the techniques results in a specialized program and code which detects when it is legal to use the specialized version. A important feature of the technique presented is that it does not require any changes to an existing abstract interpreter to achieve a quite useful form of program specialization. Rather, the technique makes use of repeated applications (generally two) of the abstract interpretater, with program transformations interspersed. The transformations are used to build a hypothetical situation that could bring to the simplification of some predicates that couldn't be simplified by the first pass of the abstract interpreter. The abstract interpreter is the used to verify the truth of such hypothesis and possibly confirm the proposed simplification.

The technique was applied to a series of prototypical benchmarks, showing its usefulness. We have also applied these techniques to other programs, such as the Boyer and Moore theorem prover. The obtained performance figures are appealing enough to motivate further investigation, in particular regarding how to extend the class of predicates that can be executed at abstract interpretation-time and also which other types of control constructs are amenable to similar transformations.

Acknowledgements: We would like to thank Vitor Santos-Costa of the University of Bristol and the anonymous referees for their useful comments on previous drafts of this paper.

References

[1] K. Benkerimi and J. Lloyd. A partial evaluation procedure for logic programs. In *Proceedings of the North American Conference on Logic Programming*, pages 343–358. MIT Press, October 1990.

[2] D. Bjorner, A.P. Ershov, and N.D. Jones, editors. *Partial Evaluation and Mixed Computation – Proceedings of the Gammel Avernaes Workshop*. Noth-Holland, October 1987.

[3] M. Bruynooghe. A Framework for the Abstract Interpretation of Logic Programs. Technical Report CW62, Department of Computer Science, Katholieke Universiteit Leuven, October 1987.

[4] M. Bruynooghe and G. Janssens. An Instance of Abstract Interpretation Integrating Type and Mode Inference. In *5th Int. Conf. and Symp. on Logic Prog.*, pages 669–683. MIT Press, August 1988.

[5] M Bugliesi and F. Russo. Partial evaluation in prolog: Some improvements about cut. In *1989 North American Conference on Logic Programming*. MIT Press, October 1989.

[6] J.-H. Chang and Alvin M. Despain. Semi-Intelligent Backtracking of Prolog Based on Static Data Dependency Analysis. In *International Symposium on Logic Programming*, pages 10–22. IEEE Computer Society, July 1985.

[7] P. Cousot and R. Cousot. Abstract Interpretation: A Unified Lattice Model for Static Analysis of Programs by Construction or Approximation of Fixpoints. In *Conf. Rec. 4th Acm Symp. on Prin. of Programming Languages*, pages 238–252, 1977.

[8] S. K. Debray and D. S. Warren. Detection and Optimization of Functional Computations in Prolog. In *Third International Conference on Logic Programming*, number 225 in Lecture Notes in Computer Science, pages 490–505. Imperial College, Springer-Verlag, July 1986.

[9] J. Gallagher. Transforming logic programs by specializing interpreters. In *Proc. of the 7th. European Conference on Artificial Intelligence*, 1986.

[10] J. Gallagher and M. Bruynooghe. The derivation of an algorithm for program specialization. In *1990 International Conference on Logic Programming*, pages 732–746. MIT Press, June 1990.

[11] J. Gallagher, M. Codish, and E. Shapiro. Specialisation of prolog and fcp programs using abstract interpretation. *New Generation Computing*, 6:159–186, 1988.

[12] F. Giannotti and M. Hermenegildo. A Technique for Recursive Invariance Detection and Selective Program Specialization. Technical report, U. of Madrid (UPM), Facultad Informatica UPM, 28660-Boadilla del Monte, Madrid-Spain, 1991.

[13] M. Hermenegildo and K. Greene. &-Prolog and its Performance: Exploiting Independent And-Parallelism. In *1990 International Conference on Logic Programming*, pages 253–268. MIT Press, June 1990.

[14] M. Hermenegildo and F. Rossi. On the Correctness and Efficiency of Independent And-Parallelism in Logic Programs. In *1989 North American Conference on Logic Programming*, pages 369–390. MIT Press, October 1989.

[15] M. Hermenegildo and F. Rossi. Non-Strict Independent And-Parallelism. In *1990 International Conference on Logic Programming*, pages 237–252. MIT Press, June 1990.

[16] D. Jacobs and A. Langen. Accurate and Efficient Approximation of Variable Aliasing in Logic Programs. In *1989 North American Conference on Logic Programming*. MIT Press, October 1989.

[17] D. Jacobs, A. Langen, and W. Winsborough. Multiple specialization of logic programs with run-time tests. In *1990 International Conference on Logic Programming*, pages 718–731. MIT Press, June 1990.

[18] N. Jones and H. Sondergaard. A semantics-based framework for the abstract interpretation of prolog. In *Abstract Interpretation of Declarative Languages*, chapter 6, pages 124–142. Ellis-Horwood, 1987.

[19] H. Mannila and E. Ukkonen. Flow Analaysis of Prolog Programs. In *4th IEEE Symposium on Logic Programming*. IEEE Computer Society, September 1987.

[20] A. Marien, G. Janssens, A. Mulkers, and M. Bruynooghe. The impact of abstract interpretation: an experiment in code generation. In *Sixth International Conference on Logic Programming*, pages 33–47. MIT Press, June 1989.

[21] C.S. Mellish. Abstract Interpretation of Prolog Programs. In *Third International Conference on Logic Programming*, number 225 in Lecture Notes in Computer Science, pages 463–475. Imperial College, Springer-Verlag, July 1986.

[22] K. Muthukumar and M. Hermenegildo. Determination of Variable Dependence Information at Compile-Time Through Abstract Interpretation. In *1989 North American Conference on Logic Programming*. MIT Press, October 1989.

[23] K. Muthukumar and M. Hermenegildo. Combined Determination of Sharing and Freeness of Program Variables Through Abstract Interpretation. In *1991 International Conference on Logic Programming*. MIT Press, June 1991.

[24] D. Sahlin. The mixtus approach to the automatic evaluation of full prolog. In *Proceedings of the North American Conference on Logic Programming*, pages 377–398. MIT Press, October 1990.

[25] P. Sestoft. A bibliography on partial evaluation and mixed computation. In *Proceedings of the Workshop on Partial Evaluation and Mixed Computation*. North-Holland, October 1987.

[26] W. Winsborough. Path-dependent reachability analysis for multiple specialization. In *1989 North American Conference on Logic Programming*. MIT Press, October 1989.

Dynamic Detection of Determinism in Functional Logic Languages

Rita Loogen and Stephan Winkler
RWTH Aachen, Lehrstuhl für Informatik II
Ahornstraße 55, W-5100 Aachen, Germany
e-mail: {rita,winkler}@zeus.informatik.rwth-aachen.de

Abstract

Programs in functional logic languages usually have to satisfy a *nonambiguity condition*, that semantically ensures completeness of conditional narrowing and pragmatically ensures that the defined (non-boolean) functions are deterministic and do not yield different result values for the same argument tuples. The nonambiguity condition allows the dynamic detection of determinism in implementations of functional logic languages. In this paper we show how to achieve this and what can be gained by this optimization.

1 Introduction

Functional logic languages are extensions of functional languages with principles derived from logic programming [Reddy 87]. While their syntax almost looks like the syntax of conventional functional languages, their operational semantics is based on *narrowing*, an evaluation mechanism that uses unification instead of pattern matching for parameter passing. Narrowing is a natural extension of reduction to incorporate unification. It means applying the minimal substitution to an expression in order to make it reducible, and then to reduce it.

In general, functional logic programs have to satisfy a *nonambiguity condition*, that semantically ensures completeness of narrowing [Bosco et al. 89b], [Kaplan 84] and pragmatically ensures that the defined (non-boolean valued) functions are deterministic and do not yield various different result values for the same argument tuples.

The aim of this paper is to show how to make use of the nonambiguity condition in order to perform a *dynamic detection of determinism* in implementations of functional logic languages. The techniques that we present are independent from the implementation method. Especially for the implementation of narrowing, a lot of machine models have been proposed in the literature. A coarse classification distinguishes machines based on the Warren Abstract Machine (WAM) [Warren 83], see e.g. [Balboni et al. 89, Bosco et al. 89a], [Hanus 90], [Mück 90], and extensions of functional reduction machines [Kuchen et al. 90, Moreno et al. 90], [Loogen 91], [Chakravarty, Lock 91]. In this paper we focus on the stack narrowing machine of [Loogen 91]. It is however no problem to incorporate a corresponding determinism check in the other machine models.

The effect of the determinism check can be compared to a *dynamic and safe version of the cut operator* known from Prolog. It enables a dynamic reduction of the computation tree, whenever it is safe to do so, due to the nonambiguity of programs. It is important to note that the boolean valued functions corresponding to Prolog predicates always satisfy the nonambiguity condition. Thus, the dynamic detection of determinism applies also to Prolog and logic programs. The main application is however the optimized handling of purely functional computations in functional logic languages.

The paper is organized as follows. In Section 2, we describe the syntax and operational semantics of the functional logic language BABEL [Moreno,Rodríguez 88,89]. This section also contains a discussion of the nonambiguity condition. Section 3 contains an overview of the main components of the stack narrowing machine of [Loogen 91] and the compilation of BABEL programs into code for this machine. We explain the detection of determinism in two stages. In Section 4 the simple case of function definitions without guards, which correspond e.g. to unconditional function definitions and Prolog facts, is handled, while the general case is discussed in Section 5. In both sections we first show under which conditions a rule application is deterministic, i.e. no other rule for the corresponding function symbol needs to be considered. This is formally proved using the nonambiguity condition. Then we show how to check these conditions in the narrowing machine. Sections 6 and 7 finally contain some conclusions and a discussion of related work, respectively.

2 The Functional Logic Language BABEL

In this section, we introduce the first order weakly typed subset of the functional logic language BABEL [Moreno,Rodríguez 88,89]. BABEL is a higher order polymorphically typed functional logic language based on a constructor discipline. Its operational semantics is based on narrowing.

2.1 Syntactic Domains

Let $DC = \bigcup_{n \in \mathbb{N}} DC^n$ and $FS = \bigcup_{n \in \mathbb{N}} FS^n$ be ranked alphabets of *constructors* and *function symbols*, respectively. We assume the nullary constructors 'true' and 'false' to be predefined. Predefined function symbols are the boolean operators and the equality operator. In the following, letters $c, d, e \ldots$ are used for constructors and the letters $f, g, h \ldots$ for function symbols.

The following syntactic domains are distinguished:

- *Variables* $X, Y, Z \ldots \in Var$

- *Terms* $s, t \ldots \in Term$:

$$t ::= \begin{array}{ll} X & \text{\% Variable} \\ | \quad c(t_1, \ldots, t_n) & \text{\% } c \in DC^n, n \geq 0 \end{array}$$

- *Expressions* $M, N \ldots \in Exp$:

$$M ::= \begin{array}{ll} X & \text{\% Variable} \\ | \quad c(M_1, \ldots, M_n) & \text{\% } c \in DC^n, n \geq 0 \\ | \quad f(M_1, \ldots, M_n) & \text{\% } f \in FS^n, n \geq 0 \\ | \quad (B \to M) & \text{\% guarded expression} \\ | \quad (B \to M_1 \Box M_2) & \text{\% conditional expression} \end{array}$$

$B \to M$ and $B \to M_1 \Box M_2$ are intended to mean "if B then M else not defined" and "if B then M_1 else M_2", respectively.

2.2 Functional Logic Programs

A *BABEL program* consists of a finite set of defining rules for the not predefined function symbols in *FS*. Let $f \in FS^m$. Each *defining rule for* f must have the form:

$$\underbrace{f(t_1, \ldots, t_n)}_{\text{lhs}} := \underbrace{\underbrace{\{B \to\}}_{\text{optional guard}} \quad \underbrace{M}_{\text{body}}}_{\text{rhs}} .$$

and satisfy the following conditions:

1. *Term Pattern:* $t_i \in Term$.

2. *Left Linearity:* $f(t_1, \ldots, t_m)$ does not contain multiple variable occurrences.

3. *Restrictions on free variables:* Variables occurring in the right hand side (rhs), but not in the left hand side (lhs), are called *free*. Occurrences of free variables are allowed in the guard, but not in the body.

4. *Nonambiguity:* Given any two rules for the same function symbol f:

$$f(t_1, \ldots, t_n) := \{B \to\}M. \text{ and } f(s_1, \ldots, s_n) := \{C \to\}N.$$

one of the three following cases must hold:

(a) *No superposition:* $f(t_1, \ldots, t_n)$ and $f(s_1, \ldots, s_n)$ are not unifiable.

(b) *Fusion of bodies*: $f(t_1, \ldots, t_n)$ and $f(s_1, \ldots, s_n)$ have a most general unifier (m.g.u.) σ such that $M\sigma$ and $N\sigma$ are identical. ($M\sigma$ denotes the expression M where all variables X have been replaced by $\sigma(X)$. $M\sigma$ is called an *instance* of M.)

(c) *Incompatibility of guards*: $f(t_1, \ldots, t_n)$ and $f(s_1, \ldots, s_n)$ have a m.g.u. σ such that the boolean expressions $B\sigma$ and $C\sigma$ are incompatible.

In [Moreno, Rodríguez 89] the notion of incompatibility has been formalized in such a way that

(i) it is decidable whether two expressions B and B' are incompatible, and

(ii) if B and B' are incompatible first order expressions, there is no Herbrand algebra $\mathcal{A}$ such that, for some substitution ρ over $\mathcal{A}$, $[B]_{\mathcal{A}}\rho = [B']_{\mathcal{A}}\rho = true_{\mathcal{A}}$.

The formal definition of BABEL's declarative semantics is out of the scope of this paper. For our purpose it is only important that the notion of incompatibility of expressions is chosen in such a way that incompatible expressions cannot simultaneously be evaluated to true.

2.1 Example: Let $even(X)$ be true, if X represents an even number and $odd(X)$ be true, if X represents an odd number. Consider the program rules

$$R_1 = \left\{ \begin{array}{llll} f(X) := & even(X) & \to M_1. \\ f(X) := & not(even(X)) & \to M_2. \end{array} \right\} \text{ and } R_2 = \left\{ \begin{array}{llll} f(X) := & even(X) & \to M_1. \\ f(X) := & odd(X) & \to M_2. \end{array} \right\}.$$

Then, the guards of R_1 are incompatible, while the guards of R_2 are not, because incompatibility is only a decidable syntactic approximation of unsatisfiability.

In the next sections we will discuss the consequences of the nonambiguity condition in detail and show how this condition can be used to optimize implementations of functional logic and logic languages by a dynamic detection of deterministic (parts of) computations.

Guarded rules allow a simple translation of Prolog clauses into BABEL. The body of the clause becomes the guard of the BABEL rule whose body is identical to true. To ensure left linearity the guard must be extended by appropriate equality conditions. Note that the nonambiguity condition is trivially satisfied by logic programs, because the bodies of the rules are always identically true and thus condition 4.(b) (fusion of bodies) applies.

2.2 Example:

The Prolog rules	*father* (bob, john).	*male* (X) :- *father* (X, Y).	correspond to
the BABEL rules	*father* (bob, john) := true.	*male* (X) := *father* $(X, Y) \to$ true.	

If necessary, the following rules for the predefined function symbols are implicitly added to a program:

- Rules for the boolean operations:

not false	:=	true.
not true	:=	false.

false , Y	:=	false.
true , Y	:=	Y.

false ; Y	:=	Y.
true ; Y	:=	true.

- Rules for the equality operator '=':

$(c = c)$	:= true.	% $c \in DC^0$,
$(c(X_1, \ldots, X_n) = c(Y_1, \ldots, Y_n))$	:= $(X_1 = Y_1), \ldots, (X_n = Y_n)$.	% $c \in DC^n$, $n > 0$
$(c(X_1, \ldots, X_n) = d(Y_1, \ldots, Y_m))$	:= false.	% $c \in DC^n$, $d \in DC^m$, % $c \neq d$ or $n \neq m$

2.3 Operational Semantics

The operational semantics of BABEL is based on the following *narrowing rule*, which describes how to apply a BABEL rule to an expression through unification:

Let $f(t_1, \ldots, t_m) := R.$ be some variant of a rule in the program and $f(M_1, \ldots, M_m)$ an expression which shares no variables with the rule variant.

If $\theta \cup \sigma : Var \to Exp$ is a most general unifier with $t_i\theta = M_i\sigma$ for $1 \leq i \leq m$, then:

$$f(M_1, \ldots, M_m) \longrightarrow_\sigma R\theta.$$

To evaluate guarded and conditional expressions, one needs the following special narrowing rules:
$$(\text{true} \to M) \longrightarrow_\varepsilon M \qquad\qquad (\text{true} \to M_1 \square M_2) \longrightarrow_\varepsilon M_1 \qquad\qquad (\text{false} \to M_1 \square M_2) \longrightarrow_\varepsilon M_2,$$
where ε denotes the empty substitution, i.e. $\varepsilon(X) = X$ for all $X \in Var$.

The *one step narrowing relation* $\Longrightarrow_\sigma \subseteq Exp \times Exp$ with $\sigma : Var \to Exp$ is inductively defined by:

- If $M_i \longrightarrow_\sigma N_i$ for $i \in \{1, \ldots, n\}$ and $\phi \in DC \cup FS$, then

$$\phi(M_1, \ldots, M_i, \ldots, M_n) \Longrightarrow_\sigma \phi(M_1\sigma, \ldots, N_i, \ldots, M_n\sigma)$$

- $B \longrightarrow_\sigma B'$ implies

$$(B \to M) \Longrightarrow_\sigma (B' \to M\sigma) \qquad\qquad \text{and} \qquad\qquad (B \to M_1 \square M_2) \Longrightarrow_\sigma (B' \to M_1\sigma \square M_2\sigma)$$

The execution of several computation steps is given by the transitive, reflexive closure of the narrowing relation with composition of the substitutions, $\overset{\bullet}{\Longrightarrow}_\sigma$.

Narrowing of a BABEL expression M may lead to the following outcomes:

- *Success:* $M \overset{\bullet}{\Longrightarrow}_\sigma t$ with $t \in Term$,

- *Failure:* $M \overset{\bullet}{\Longrightarrow}_\sigma N$, N is not further narrowable and $N \notin Term$,

- *Nontermination.*

For simplicity we consider in the following sections the *leftmost innermost* narrowing strategy. This corresponds to the evaluation strategy of Prolog. Using the program transformation described in [Moreno et al. 90], our techniques can at once be incorporated in *lazy narrowing* implementations.

2.3 Example: Considering the Prolog-like rules of Example 2.2, we get the following innermost narrowing sequence for the expression $male(X)$.

$$male(X) \;\Longrightarrow_\varepsilon\; (father\,(X, Y) \to true.) \;\Longrightarrow_{[X/bob, Y/john]}\; (true \to true.) \;\Longrightarrow_\varepsilon\; true$$

In [Moreno, Rodríguez 88] soundness and completeness results have been stated, which relate the narrowing semantics to a declarative semantics for first order BABEL. A proof of these results can be found in [Moreno, Rodríguez 89]. The soundness result implies that it is not possible to narrow incompatible expressions simultaneously to true.

2.4 Lemma: Let B and B' be incompatible expressions. If $B \overset{\bullet}{\Longrightarrow}_\sigma true$ for some substitution σ, there exists no narrowing sequence $B'\sigma \overset{\bullet}{\Longrightarrow}_{\sigma'} true$, where σ' is any substitution.

Proof: The soundness of the narrowing semantics implies that $[\![B\sigma]\!]_{\mathcal{A}\rho} = true_{\mathcal{A}}$ for any Herbrand model $\mathcal{A}$ and any environment ρ. Due to property (ii) of the incompatibility notion we have that $[\![B'\sigma]\!]_{\mathcal{A}\rho} \neq true_{\mathcal{A}}$ for any Herbrand model $\mathcal{A}$ and any environment ρ. Using again the soundness of the narrowing semantics we can deduce that the existence of a narrowing sequence $B'\sigma \overset{\bullet}{\Longrightarrow}_{\sigma'} true$ contradicts this statement. This completes the proof. $\qquad\square$

3 Implementation of Functional Logic Languages

We will explain the dynamic detection of determinism for a simple stack-based narrowing machine that has been introduced in [Loogen 91]. The presented optimizing modifications are however independent from the chosen implementation technique and can analogously be incorporated into other machines which are capable of doing narrowing or SLD-resolution. In order to perform backtracking such machines will work with components, which correspond to the trail and the choice points known from Warren's Prolog Engine [Warren 83]. The modifications that are necessary concern mainly these components and the representation of variables in the machine.

In this section we shortly describe the structure and the behaviour of the stack narrowing machine. We concentrate on the components that are involved in the optimizations.

To evaluate an expression, the narrowing machine tries to reduce it to a normalized form. This means that the left hand sides of rules for the defined and predefined function symbols are unified with appropriate subexpressions, which are then replaced by the corresponding instances of the rule's right hand sides. The machine tries the program rules in their textual ordering and evaluates arguments from left to right; it backtracks when a failure or a user request for alternative solutions occurs.

3.1 Components of a Narrowing Machine

The *store* of the narrowing machine contains the following components:

- a *program store* and an *instruction pointer*
  The program store contains the translation of the program rules into abstract machine code. This component remains unchanged during the evaluation of programs. The instruction pointer points at the instruction in the program store that has to be executed next.

- *graph* $G : Adr \dashrightarrow GNodes$
  The graph or heap is necessary for the representation of terms, i.e. variables and structured terms. The set *GNodes* of graph nodes contains the following types of nodes:

  - *variable nodes:* $\langle \text{VAR}, a \rangle$ with $a \in Adr \cup \{?\}$ ($\langle \text{VAR}, ? \rangle$ represents an unbound variable)
  - *constructor nodes:* $\langle \text{CONSTR}, c, a_1 : \ldots : a_m \rangle$ with $c \in DC$ and $a_i \in Adr$ $(1 \leq i \leq m)$

- *(environment) stack*
  The environment stack is the central component of the machine. It is used to store the *environments* of function calls and *choice points* which keep track of possible alternative computations. Thus, environments contain the control information for forward computations while choice points control backward computations, i.e. backtracking. The environment stack is accessed via two pointers:

  - the *environment pointer ep* $\in \mathbb{N}$ indicates the topmost environment on the stack;
  - the *backtrack pointer bp* $\in \mathbb{N}$ indicates the topmost choice point.

  The *environments* of function calls contain locations for the local variables of the function rules, the arguments of the function call, the saved pointer to the previous environment block and the return address of the function call, i.e. the program address at which the computation has to be continued after a successful termination.

  When the environment block is created, the local variable positions are initialized with the symbol ? to indicate that a binding has not yet occurred. During unification the ? will be overwritten by the pointer to the graph node representing the expression to which the local variable is bound. The arguments are represented by pointers to their graph representation.

  Choice points contain the information that is necessary for restoring a previous state of the machine, i.e. the depth of the data stack up to which this has to be deleted on backtracking and, if necessary, a number of saved data stack entries, the length of the trail, the saved backtrack pointer, i.e. the pointer to the previous choice point and finally the backtrack address, which indicates the code address of the next alternative rule. Resetting the trail means unbinding the variables whose addresses have been noted in the trail.

  The environment stack of the narrowing machine in general contains a mixture of choice points and environments. The top of the stack is always indicated by the maximum of the environment pointer and the backtrack pointer.

- *trail tr* $\in Adr^*$
  The trail is used to mark variable bindings that have to be undone in case of backtracking.

- *data stack d* $\in Adr^*$
  The data stack is used for all accesses to the graph. Its entries are graph addresses.

  Instead of a data stack the WAM uses sets of registers. Our modifications, however, do not concern the organisation of the data access.

3.2 Compilation of Functional Logic Programs

After grouping the rules of BABEL programs according to the function symbols, a program has the general form:

$$\mathcal{P} = \{ (f^{(j)}(t_{i1}^{(j)}, \ldots, t_{im}^{(j)}) = \text{rhs}_i^{(j)}. \mid 1 \leq i \leq r_j) \mid 1 \leq j \leq k \}$$

Let f be a function symbol with arity m and maximally k local variables. Then:

$$proctrans((f(t_{i1}, \ldots, t_{im}) := rhs_i. \mid 1 \leq i \leq r, r \geq 2))$$

$:=$ TRY_ME_ELSE $rule_2$
      $ruletrans(ft_{11} \ldots t_{1m} := rhs_1., k)$

$rule_2:$ RETRY_ME_ELSE $rule_3$
      $ruletrans(ft_{21} \ldots t_{2m} := rhs_2., k)$

$rule_3:$ $\vdots$

$rule_r:$ TRUST_ME_ELSE_FAIL
      $ruletrans(ft_{r1} \ldots t_{rm} := rhs_r., k)$

$ruletrans(f(t_1, \ldots, t_m) := B \to M., k)$

$:=$ $unifytrans(ft_1 \ldots t_m, k)$
      $guardtrans(B)$
      JMP_TRUE body
      BACKTRACK

body: $bodytrans(M, m + k)$
      RET $m + k$

Figure 1: Compilation schemes *proctrans* and *ruletrans*

The code generated for such a program consists of the code for the various procedures (groups of rules for the same function symbol), which will be produced using the scheme *proctrans* given in figure 1. The code for a procedure is executed whenever an application of the corresponding function symbol has to be evaluated. A CALL instruction stores an environment block on top of the environment stack and writes the address of the code for the procedure into the instruction pointer. The defining rules of a function symbol are tested in their textual ordering. Before the first rule is tried, the instruction 'TRY_ME_ELSE $rule_2$' puts a choice point on top of the environment stack to keep note of the alternative rules. This choice point always contains the code address of the next rule, which is reset by the 'RETRY_ME_ELSE label' instruction if a rule is not applicable and the next rule is tried. The choice point is removed by the TRUST_ME_ELSE_FAIL instruction, when the last rule is tried.

If there exists only a single rule for the symbol f, the code for the procedure corresponds to the code produced for this rule by the *ruletrans* scheme. The translation of each rule consists of code for the unification of the arguments of the function application with the terms on the left hand side of the rule and code for the evaluation of the right hand side. After evaluation of the guard, the instruction JMP_TRUE tests whether the guard has been evaluated to true and if so, jumps to the code for the evaluation of the body. Otherwise backtracking is started by the BACKTRACK command. After a successful evaluation, the RET command gives the control back to the calling procedure. If the environment block of the procedure is on top of the stack, it can be removed by the RET instruction. This behaviour corresponds to the stack implementations of ordinary procedure oriented programming languages. If a choice point is on top of the stack, the environment must not be deleted, because it may be needed again in case of backtracking.

For the explanation of our modifications it is not necessary to go further into the details of the translation schemes for the unification, guard and body evaluation and the machine instructions that are used in these schemes. The interested reader is referred to [Loogen 91].

4   Determinism Detection for Rules without Guards

In this section we take a closer look at the nonambiguity condition of functional logic programs and show how this condition can be used to detect determinism. We first consider the special case of rules without guards which is rather common. Most pure function definitions and Prolog facts have this form.

4.1   The Language Level

In the case of rules without guards the nonambiguity condition shrinks to the form:

Given any two rules for the same function symbol f:

$$f(t_1, \ldots, t_m) := M. \text{ and } f(s_1, \ldots, s_m) := N.,$$

if $f(t_1, \ldots, t_m)$ and $f(s_1, \ldots, s_m)$ have a m.g.u. σ, then $M\sigma$ and $N\sigma$ are identical.

and we can state the following simple lemma.

4.1 Lemma: Let $\qquad\qquad (R_1)\ f(t_1,\ldots,t_n) := M.$ and $(R_2)\ f(s_1,\ldots,s_n) := N.$
be any two program rules for the same function symbol f and let $f(a_1,\ldots,a_n)$ be an expression with $a_i \in Term\ (1 \le i \le n)$. W.l.o.g. the rules and the expression do not have common variables.

If $f(t_1,\ldots,t_n)\theta = f(a_1,\ldots,a_n)$ for some substitution θ, i.e. R_1 is applicable without that variables in $f(a_1,\ldots,a_n)$ are bound, then R_2 does not need to be considered, because either R_2 is not applicable or the result of applying R_2 yields an instance of the application of R_1.

Proof: Assume that also R_2 is applicable, i.e. there exists a substitution θ' with

$$f(s_1,\ldots,s_n)\theta' = f(a_1,\ldots,a_n)\theta'.$$

As R_1 and R_2 do not have common variables and as R_1 is applicable with substitution θ, this implies that

$$f(s_1,\ldots,s_n)\theta\theta' = f(t_1,\ldots,t_n)\theta\theta'.$$

Thus, the left hand sides of R_1 and R_2 are unifiable and the nonambiguity condition tells us that in this case $M\theta\theta' = N\theta\theta'$. As θ binds only variables of R_1, we get $N\theta\theta' = N\theta'$, which shows that $N\theta'$ is an instance of $M\theta$. $\qquad\square$

This lemma means that we need not check applicability of other rules whenever no variables in the expression that is being evaluated are bound during the unification with the left hand side of a rule.

4.2 The Implementation Level

The detection of this situation is very simple in our narrowing machine. After a successful unification, we simply have to check whether variables have been bound. Each variable binding is noted in the trail and the length of the trail is noted in the choice point before the unification phase, if there are alternative rules. Thus, we have to compare the length of the trail after the unification with the length noted in the choice point on top of the stack. If the trail has not changed, we are allowed to delete the choice point on top of the environment stack, because Lemma 4.1 ensures that no other rules need to be considered for the current function application. Of course we assume here that one does not note bindings of local rule variables in the trail. Otherwise, one has to use a flag within the choice point that is set whenever a global variable is bound.

The specification of the narrowing machine in [Loogen 91] contains already a machine instruction, called POP, that checks whether an expression variable has been bound during unification and if so, removes the choice point on top of the stack. In the code for a function with more than one rule the POP instruction is placed after the code for the unification.

$$ruletrans(f(t_1,\ldots,t_m) := M., k) := \quad unifytrans(f(t_1,\ldots,t_m), k)$$
$$\text{POP}$$
$$bodytrans(M, m + k)$$
$$\text{RET } m + k$$

In the code for the last rule of a function and if there is only one rule, the POP instruction is superfluous, because the choice point has already been deleted by the TRUST_ME_ELSE_FAIL instruction or no choice point has been created. It should therefore be omitted in these cases.

This special treatment of BABEL functions without guards improves purely or partially deterministic computations. Its effect can be compared to the cut operator that can be used in Prolog to express determinism, because it also cuts off parts of the computation tree. In contrast to the Prolog cut, which has to be placed by the programmer and often restricts the bidirectional use of programs, the instruction POP performs a *dynamic cut*, which is *safe*, because no solutions are lost, and which does not restrict the usability of the program.

4.2 Example: The following BABEL rules define a binary function that determines, given a number n and a list l, the prefix of l consisting of n elements.

$$prefix(s(N), [Y \mid Ys]) \quad := \quad [Y \mid prefix(N, Ys)].$$
$$prefix(0, Ys) \qquad\qquad := \quad [\,].$$

Whenever this function is called with ground arguments or with arguments, which have the pattern of the left hand side of the first rule, our optimization applies, i.e. the second rule is not tried. When the first solution is found, one immediately recognizes that no further solutions exist, because there is no more choice point on the stack.

The second advantage of our dynamic detection of determinism is that the rules can nevertheless be used to compute the inverse function or the graph of *prefix*. Expressions of the form $prefix(N, [a, b, c]) = [a, b]$, $prefix(N, Ys) = [a, b, c]$ or $prefix(N, Ys)$ can still be solved appropriately. For the first two goal expression the representation of the solutions depends on the implementation of the equality rules, which is out of the scope of this paper.

Please note that this detection and optimized treatment of functional computations is not possible in Prolog and logic languages. In these languages functions are represented by predicates, which describe the function graphs. Even in queries which correspond to purely functional calls, i.e. the function arguments are ground terms and the function value is represented by an unbound variable, this result variable will in general be bound during unification and the POP instruction would not be applicable.

4.2 Example (continued): In Prolog the following rules define the same function as in the previous example

$$prefix(s(N), [Y \mid Ys], [Y \mid Zs]) \ :\text{-} \ prefix(N, Ys, Zs).$$
$$prefix(0, Ys, []).$$

The additional third parameter represents the result value. Deterministic calls of these rules have the form $prefix(n, l, Z)$ where n and l are ground terms representing a number and a list, respectively. As the result parameter will be bound during the unification phase, our optimization is not directly applicable.

The effect of the POP instruction can be achieved by placing a cut on the right hand side of the first rule. This will however restrict the applicability of the rules, as e.g., using the rules with the cut, queries of the form $prefix(N, [a, b, c], Z)$ will only yield one solution instead of four solutions without the cut.

In order to optimize the handling of predicates with functional character in Prolog or logic languages one could distinguish between output, input and general variables using mode declarations and use this information to detect determinism, but the effort seems to be higher than for functional logic languages.

In most Prolog systems the above example will be handled in an optimized way by the use of *indexing* that has for this example a similar effect as our POP instruction. In general, indexing is however only used for the first argument. Swapping the first two arguments of *prefix* makes indexing useless, but does not affect the dynamic detection of determinism in BABEL.

The simple case of BABEL rules without guards is given in most purely functional definitions. For Prolog clauses, which correspond to boolean valued BABEL functions, this special situation is only present in facts. Nevertheless, our detection of determinism reduces in special cases the computation tree by avoiding unnecessary backtracking and sometimes avoiding the generation of redundant solutions, when there is more than one possibility to derive an (intermediate) solution, which can especially happen in dynamically modified procedures.

In the next section we will consider the general case of guarded rules which may contain free variables in the guard. In the general case the detection of determinism is slightly more involved, but may lead to important reductions of the computation tree especially for logic programs.

5 Determinism Detection for Rules with Guards

5.1 The Language Level

A guarded rule is applicable when the unification of its left hand side with the expression is successful and when the guard of the right hand side of the rule can be evaluated to true. If during the unification and during the guard evaluation no variables of the evaluated expression are bound, no other rule needs to be tested. Lemma 4.1 can be generalized to the following theorem.

5.1 Theorem: Let $\quad (R_1)\ f(t_1,\ldots,t_n) := B \to M.$ and $(R_2)\ f(s_1,\ldots,s_n) := C \to N.$
be any two program rules for the same function symbol f and let $f(a_1,\ldots,a_n)$ be an expression with
$a_i \in Term\ (1 \le i \le n)$. W.l.o.g. the rules and the expression do not have common variables.

If $f(t_1,\ldots,t_n)\theta = f(a_1,\ldots,a_n)$ for some substitution θ and $B\theta \Rightarrow_\sigma^* true$, where σ binds only free
variables in B, i.e. R_1 is applicable without that variables in $f(a_1,\ldots,a_n)$ are bound, then R_2 does
not need to be considered, because either R_2 is not applicable or the result of applying R_2 yields an
instance of the application of R_1.

Proof: Again we assume that R_2 is also applicable, i.e. there exists a substitution θ' with

$$f(s_1,\ldots,s_n)\theta' = f(a_1,\ldots,a_n)\theta' \text{ and } C\theta' \Rightarrow_{\sigma'}^* true.$$

As in the proof of Lemma 4.1 we deduce that $\theta\theta'$ is a unifier of the left hand sides of R_1 and R_2. The
general nonambiguity condition tells us now that either

1. $M\theta\theta' = N\theta\theta'$ (fusion of bodies) or
2. $B\theta\theta'$ and $C\theta\theta'$ are incompatible. (Note that instances of incompatible expressions are also incompatible.)

As $B\theta \Rightarrow_\sigma^* true$ and σ binds only free guard variables, the same narrowing steps are possible for an
instance of B that binds only non-free variables. Thus, we also have that $B\theta\theta' \Rightarrow_\sigma^* true$. As we
have assumed that additionally $C\theta' \Rightarrow_{\sigma'}^* true$ and $C\theta\theta'\sigma = C\theta'$ (θ and σ bind only variables of R_1),
Lemma 2.4 implies that $B\theta\theta'$ and $C\theta\theta'$ are *not* incompatible. Consequently, the bodies of R_1 and R_2
must be identical with respect to $\theta\theta'$, i.e. case 1. (fusion of bodies) is true. To complete the proof we
now show that the application of R_2 yields an instance of the application of R_1:

R_1 yields the expression $M\theta\sigma$ that equals $M\theta$, because σ binds only free variables of B, which are not
allowed to occur in M.

R_2 yields the expression $N\theta'\sigma'$ that equals $N\theta\theta'\sigma'$, because θ does not bind variables in R_2. Because
of the fusion bodies, this is identical to $M\theta\theta'\sigma'$ and thus is an instance of $M\theta$. $\quad\square$

This theorem shows that, whenever a rule is applicable without that variables of the expression, which
is currently evaluated, are bound, then no other rules for the corresponding function symbol need to be
considered and no alternative guard evaluations are necessary.

As Prolog or logic programs are trivially nonambiguous, the theorem is also applicable. Whenever only
free variables of the right hand side of a rule have been bound during the application of a rule, no other rule
for the corresponding function symbol needs to be considered and no reevaluation of the right hand side of
the rule is necessary. This corresponds to a dynamically added cut operator at the rightmost position in a
rule.

5.2 The Implementation Level

In order to check whether no other rules need to be considered, one has to detect whether after the unification
and a successful guard evaluation only local variables and free guard variables have been bound. A first
approach is to place a POP instruction immediately before the code for the body evaluation, which checks
whether the length of the trail noted in the choice point corresponding to the active environment block
(which is not necessarily on top of the stack, because more environments and choice points may have been
pushed on the stack during the guard evaluation) equals the current length of the trail. If this is the case, no
global variable has been bound, and all the choice points and environments on top of the active environment
can be removed from the stack. Unfortunately, this simple approach applies only for guarded rules without
free guard variables.

When a guard contains free variables, the detection whether only the free guard variables have been
bound during the evaluation of the guard is not possible in the current version of the narrowing machine.
The guard variables always appear within parameters of predicates. When these predicates are called,
the local guard variables become global variables that cannot be distinguished from other global variables.
Particularly, their bindings will be trailed during the predicate evaluation. Consequently, the trail will grow
in most guard evaluations, even if only free guard variables have been bound.

Consequently, the following modifications are necessary in our narrowing machine:

— *Distinct Graph Representation of Guard Variables*

In order to be able to check whether only guard variables have been bound, one has to distinguish the representation of variables and guard variables in the machine.

Furthermore, nested guard evaluations make it necessary to distinguish the free guard variables of different guards. This is achieved by adding a pointer to the environment of the function call that initiated the guard evaluation, to graph representation of the guard variable.

Thus, we have chosen to represent the guard variables by the following type of nodes:

$$(\text{GUARDVAR}, a, ep) \text{ with } a \in Adr \cup \{?\} \text{ and } ep \in \mathbb{N}.$$

Such guard variables are generated by a newly introduced instruction 'INIT_GUARD_VARS nr', which is placed immediately before the code for the guard evaluation. The parameter is the number of free guard variables within the environment of the function call.

— *Extension of Environments*

The detection of determinism also makes sense for functions with only one defining rule, because it may allow to remove the choice points and environments of the guard evaluation corresponding to this rule. In order to be able to handle such cases, where no choice point is generated that saves the length of the trail, we have to note the length of the trail additionally in the environment block of a function call.

— *Extension of the Trail and the POP Instruction*

To simplify the test whether the variables noted in the trail after a guard evaluation are guard variables or not, we note additional information on the bound variables in the trail.

We define $tr \in ((\text{VAR} \times Adr) \cup (\text{GUARDVAR} \times Adr \times \mathbb{N}))^*$.

The tags VAR/GUARDVAR indicate the type of the bound variable and for guard variables we additionally note the environment pointer that is stored in the node of the guard variable.

The extended trail allows a simple check whether only guard variables, whose corresponding environment block lies above or is equal to the block of the current function call, i.e. the saved environment pointer is greater or equal to the current environment pointer, have been bound during a guard evaluation. Note that guard variables of previous function calls (saved environment pointer is less than the actual environment pointer) have to be considered as global variables. The POP instruction is modified appropriately. It removes all environments and choice points on top of the actual environment, when the check of the trail is successful. If a global variable has been bound, the environment stack remains unchanged.

Consequently, the following code is produced for a rule. The additional parameter of *ruletrans* gives the number of free guard variables within the rule. We assume that the guard variables are placed after the other local variables within the environment block of the function call.

$$ruletrans(f(t_1, \ldots, t_m) := B \to M., k, gv) \; := \quad \begin{aligned} &unifytrans\,(ft_1 \ldots t_m, k) \\ &\text{INIT\_GUARD\_VARS } gv \\ &guardtrans\,(B) \\ &\text{JMP\_TRUE body} \\ &\text{BACKTRACK} \\ body: \quad &\text{POP} \\ &bodytrans\,(M, m + k) \\ &\text{RET } m + k \end{aligned}$$

We conclude this section considering a small example.

5.2 Example: Consider a simple library data base in Prolog:

user (john).	*has_book* (john, b15).	*active* (X) :- *has_book* (X, B).
user (mary).	*has_book* (john, b11).	
user (bob).	*has_book* (john, b5).	
$\vdots$	$\vdots$	
	has_book (mary, b107).	
	$\vdots$	

Without our detection of determinism the query '*user*(X), *active*(X)' will yield the solution '$X = $ john' as many times as the user john has a book, because the subquery *has_book*(john,B) succeeds several times, binding the free guard variable B to the different books. With our dynamic detection of determinism such redundant solutions are avoided.

6 Results and Conclusions

The dynamic detection of determinism has been incorporated in the implementation of our stack narrowing machine. It shows good results, *avoiding unnecessary global backtracking* for deterministic computations after the first result has been found, and in general when local determinism can be detected. The speedup of computations depends of course on the grade of determinism that is present in the various example programs. In most of our examples the runtime could be increased by 10 to 40 % of the runtime without the determinism check. The additional test that is performed by the POP instruction does not slow down computations too much. The *early deletion of choice points* increases the space efficiency of the machine. [Winkler 91] contains a detailed discussion of examples and run time results.

The execution of purely functional programs is the same as in reduction machines. As soon as an applicable rule has been found, the POP instruction detects the determinism and deletes the choice point on top of the stack. Note that reduction machines, which do not presume a pattern matching compiler, also have to work with choice points noting the address of the next alternative rule during pattern matching. The only difference to the narrowing machine is the size of choice points and the check that has to be performed to detect the determinism.

The detection of local determinism corresponds to a *dynamic cut* of the computation tree that is *safe*, because it is only done, when no solutions are lost. Note that the modifications of the narrowing machine could also be incorporated in the Warren Abstract Machine [Warren 83].

One could argue that the same or even better backtracking behaviour can be achieved by *intelligent backtracking* schemes. This is true, if backtracking is initialized by a failure. If, however, backtracking is started, because the user asks for more solutions, our method is advantageous, because we delete choice points as soon as determinism is detected. Intelligent backtracking will go back to the topmost choice point, which may be superfluous, while our approach will avoid such superfluous computations. Incorporation of an intelligent backtracking scheme in the narrowing machine will of course lead to further improvements in the execution of programs.

7 Related Work

During the last years a lot of proposals have been made for the implementation of narrowing, see e.g. [Balboni et al. 89, Bosco et al. 89a], [Kuchen et al. 90, Moreno et al. 90], [Mück 90], [Hanus 90,91], [Chakravarty, Lock 91]. Except of [Kuchen et al. 90, Moreno et al. 90] and [Chakravarty, Lock 91], where unification and backtracking have been incorporated in a functional reduction machine, all these approaches are based on the Warren Abstract Machine.

In [Loogen 91] it has been shown, how to extend a stack reduction machine for functional languages in order to be able to perform narrowing. This approach led to the question whether it would be possible to detect purely functional computations in the extended machine. This paper gives a positive answer to this question and presents a general method for the dynamic detection of local determinism, that can also be useful for purely logic programs.

Except of [Hanus 90,91] the other proposals do not contain a special treatment of deterministic computations, but it seems that it is straightforward to integrate our techniques into the various machine models.

[Hanus 90,91] presents an extension of Warren's Abstract Machine which is able to evaluate Horn clause logic programs with equality. A preprocessor transforms programs into two groups of conditional equations, for narrowing steps and rewriting steps, respectively. Accordingly, the abstract machine is capable of performing narrowing and rewriting, where rewriting is supported by a separate mechanism. Before each narrowing step the machine tries to rewrite expressions. If no rewriting is possible, narrowing is done. In this way, no narrowing is necessary for the execution of purely functional programs. If however narrowing cannot be avoided, there will be an overhead of first trying to perform rewriting. In contrast, our machine always does narrowing, where narrowing is viewed as a generalization of reduction, which is achieved by replacing pattern matching by unification. After the applicability of a rule is proved, it is checked whether pattern matching or unification has been done, i.e. whether reduction or rewriting instead of narrowing can be used to evaluate the expression. No overhead is necessary to switch between rewriting and narrowing.

References

[Balboni et al. 89] G.P.Balboni, P.G.Bosco, C.Cecchi, R.Melen, C.Moiso, G.Sofi: *Implementation of a Parallel Logic Plus Functional Language*, in: P.Treleaven (ed.), Parallel Computers: Object Oriented, Functional and Logic, Wiley 1989.

[Bellia, Levi 86] M. Bellia, G. Levi: *The Relation between Logic and Functional Languages*, Journal of Logic Programming, Vol.3, 1986, 217–236.

[Bosco et al. 89a] P.G.Bosco, C.Cecchi, C.Moiso: *An extension of WAM for K-LEAF: A WAM-based compilation of conditional narrowing*, Int. Conf. on Logic Programming, Lisboa, 1989.

[Bosco et al. 89b] P.G.Bosco, C.Cecchi, E.Giovannetti, C.Moiso, C.Palamidessi: *Using resolution for a sound and efficient integration of logic and functional programming*, in: J.de Bakker (ed.): *Languages for Parallel Architectures: Design, Semantics, Implementation Models*, Wiley 1989.

[Chakravarty, Lock 91] M.M.T.Chakravarty, H.C.R.Lock: *The Implementation of Lazy Narrowing*, Symp. on Prog. Language Impl. and Logic Prog. (PLILP) 1991, LNCS, Springer Verlag 1991.

[DeGroot, Lindstrom 86] D.DeGroot, G.Lindstrom (eds.): *Logic Programming: Functions, Relations, Equations*, Prentice Hall 1986.

[Hanus 90] M.Hanus: *Compiling Logic Programs with Equality*, Workshop on Programming Language Implementation and Logic Programming (PLILP) 1990, LNCS 456, Springer Verlag 1990.

[Hanus 91] M.Hanus: *Efficient Implementation of Narrowing and Rewriting*, Draft Paper, University of Dortmund 1991.

[Kaplan 84] S.Kaplan: *Fair conditional term rewriting systems: unification, termination and confluence*, Technical Report no. 194, University of Orsay 1984.

[Kuchen et al. 90] H.Kuchen, R.Loogen, J.J. Moreno-Navarro, M.Rodríguez-Artalejo: *Graph-based Implementation of a Functional Logic Language*, European Symp. on Prog. (ESOP) 1990, LNCS 432, Springer Verlag 1990.

[Loogen 91] R.Loogen: *From Reduction Machines to Narrowing Machines*, Coll. on Combining Paradigms for Software Development, TAPSOFT 1991, LNCS 494, Springer Verlag 1991.

[Moreno, Rodríguez 88] J.J.Moreno-Navarro, M.Rodríguez-Artalejo: *BABEL: A functional and logic programming language based on constructor discipline and narrowing*, Conference on Algebraic and Logic Programming 1988, LNCS 343, Springer Verlag 1989.

[Moreno, Rodríguez 89] J.J.Moreno-Navarro, M.Rodríguez-Artalejo: *Logic Programming with Functions and Predicates: The Language BABEL*, Technical Report DIA/89/3, Universidad Complutense, Madrid 1989, to appear in the Journal of Logic Programming.

[Moreno et al. 90] J.J.Moreno-Navarro, H.Kuchen, R.Loogen, M.Rodríguez-Artalejo: *Lazy Narrowing in a Graph Machine*, Conf. on Algebraic and Logic Prog. (ALP) 1990, LNCS 463, Springer Verlag 1990.

[Mück 90] A.Mück: *Compilation of Narrowing*, Workshop on Programming Language Implementation and Logic Programming (PLILP) 1990, LNCS 456, Springer Verlag 1990.

[Reddy 87] U.S.Reddy: *Functional Logic Languages, Part I*, Workshop on Graph Reduction, LNCS 279, Springer Verlag 1987, 401–425.

[Warren 83] D.H.D.Warren: *An Abstract Prolog Instruction Set*, Technical Note 309, SRI International, Menlo Park, California, October 1983.

[Winkler 91] St. Winkler: *Development of a Program System for the Functional Logic Language BABEL by Implementing a Program Environment, a Compiler and a Stack Based Abstract Machine*, Diploma Thesis, RWTH Aachen, June 1991 (in german).

UNFOLDING - DEFINITION - FOLDING, IN THIS ORDER, FOR AVOIDING UNNECESSARY VARIABLES IN LOGIC PROGRAMS

Maurizio Proietti
IASI-CNR
Viale Manzoni 30
00185 Roma, Italy
proietti@irmiasi.rm.cnr.it

Alberto Pettorossi
Electronics Department
University of Rome II
00173 Roma, Italy
adp@irmiasi.rm.cnr.it

ABSTRACT

We take a syntactically based approach to the automatic improvement of performances of logic programs by using the unfold / fold transformation technique. A cause of program inefficiency is often the presence of variables which are *unnecessary*, in the sense that they force computations of redundant values or multiple visits of data structures.

We propose a strategy which automatically transforms initial program versions into new efficient versions by avoiding unnecessary variables. Our strategy is an extension of the one which was introduced in an earlier paper [Proietti-Pettorossi 90]. It is based on the syntactical characterization of the unnecessary variables and the introduction of a composite transformation rule made out of unfolding-definition-folding steps, in this order. The strategy consists in the repeated application of this composite rule to each clause with unnecessary variables. It avoids the search for *eureka* definitions which is often required by other techniques proposed in the literature. We define a class of programs for which our transformation strategy is successful and we propose a variant of that strategy which uses the so-called Generalization Rule. This variant is *always* terminating, but, in general, *not* all unnecessary variables are eliminated.

1. INTRODUCTION

When writing programs one uses variables for storing input and output data. They can be considered as *necessary* variables, because they are needed for expressing the meaning of programs. Often one also uses variables which are *unnecessary*, in the sense that they are not required for describing the input-output relation. Unnecessary variables are used by the programmer because they often allow for a more transparent way of writing programs or an easier proof of their correctness.

It is the case that many strategies for program transformation, which are given in the literature, are successful and improve program efficiency, precisely because they eliminate unnecessary variables, and thus, they avoid the construction of intermediate values and multiple traversals of data structures.

In this paper we propose some syntactically based techniques for avoiding unnecessary variables, which can be considered as a development of the tupling strategy and the composition strategy (also called loop-fusion, when used for merging consecutive loops).

For a presentation of those strategies in the case of functional programs the reader may refer to [Burstall-Darlington 77, Pettorossi 77, Feather 86, Wadler 88], while in the case of imperative and logic programs he may refer to [Paige-Koenig 82, CIP 84], and [Pettorossi-

This work has been partially supported by the "Progetto Finalizzato Sistemi Informatici e Calcolo Parallelo" of the CNR and MPI 40% (Italy).

Proietti 87, Debray 88], respectively.

In this paper we restrict our attention to definite logic programs (that is, logic programs without negation) [Lloyd 87], and we define two kinds of unnecessary variables, namely, the *existential variables* and the *multiple variables*, and through various examples we show that those variables may cause inefficiency. We then propose an automatic program transformation strategy for driving the use of the unfold/fold rules [Burstall-Darlington 77, Tamaki-Sato 84]. That strategy eliminates unnecessary variables in a large class of programs, which will be formally specified later on.

Let us start off by giving the definition of the existential and multiple variables of a clause.
DEFINITION 1. Given a clause C the *existential variables* of C are the variables which occur in the body of C and *not* in its head, and the *multiple variables* of C are the variables which occur *more than once* in the body of C.
Existential and multiple variables are collectively called *unnecessary variables*. ∎

In the practice of logic programming, multiple variables are often used for constructing programs out of various predicates which act on the same data structures, while existential variables are used only for storing intermediate results.

The following example shows a typical case where existential and multiple variables are used, and it demonstrates that they may cause inefficiency. In this example we introduce some concepts which will be important for the formal presentation of our transformation techniques to be given in the later sections.

Example 1. (*Rotate_Prune.*) Let us consider the following three predicates over pairs of binary trees labelled by natural numbers:
i) rotate(Tree1,U) which holds iff there is a sequence of trees $<T_1,T_2,...,T_{m-1},T_m>$ such that: T_1=Tree1, T_m=U, and for i=1,...,m-1, T_{i+1} is obtained from T_i by interchanging the left and right subtrees of a node. (rotate is the nondeterministic version of the flip function given in [Wadler 88].)
ii) prune(U,Tree2) which holds iff Tree2 can be obtained from U by substituting a leaf labeled by 0 to each subtree whose root has label 0.
iii) rotate_prune(Tree1,Tree2) which holds iff there exists a binary tree U such that both rotate(Tree1,U) and prune(U,Tree2) hold.

A logic program, called Rotate_Prune defining the predicate rotate_prune is the following:
1. rotate_prune(Tree1,Tree2) ← rotate(Tree1,U), prune(U,Tree2).
2. rotate(tip(N),tip(N)).
3. rotate(tree(L,N,R),tree(L1,N,R1)) ← rotate(L,L1), rotate(R,R1).
4. rotate(tree(L,N,R),tree(R1,N,L1)) ← rotate(L,L1), rotate(R,R1).
5. prune(tip(N),tip(N)).
6. prune(tree(L,0,R),tip(0)).
7. prune(tree(L,s(N),R),tree(L1,s(N),R1)) ← prune(L,L1), prune(R,R1).
In the body of clause 1, the variable U is both existential and multiple. The construction of its binding is *not* needed for the final output, as we will show by transforming the given program into one without unnecessary variables.

Our proposed program transformation strategy consists in performing on any clause with unnecessary variables the following sequence of steps: an unfolding step, followed by some definition steps, followed by some folding steps.

Those steps eliminate all unnecessary variables from any given clause, at the expenses of possibly introducing definitions with unnecessary variables. However, we may repeat the unfolding-definitions-foldings steps w.r.t. the newly introduced definitions, in the hope that the process will terminate and all unnecessary variables will be eliminated.

Let us now apply the proposed strategy to clause 1 with the unnecessary variable U.

Unfolding. We unfold the atom rotate(Tree1,U) in clause 1, and we get three clauses:

8. rotate_prune(tip(N), T2) ← prune(tip(N), T2).
9. rotate_prune(tree(L,N,R), T2) ← rotate(L, L1), rotate(R, R1), prune(tree(L1,N,R1), T2).
10. rotate_prune(tree(L,N,R), T2) ← rotate(L, L1), rotate(R, R1), prune(tree(R1,N,L1), T2).

Clause 8 does not contain any unnecessary variable.

Definitions. In clause 9 we detect the presence of the unnecessary variables L1 and R1. We then use the definition rule which introduces the new predicate new1 by means of a new clause whose body is equal to set of atoms with the unnecessary variables L1 and R1 (that is, the whole body of clause 9), and we get:

11. new1(L, N, R, T2) ← rotate(L, L1), rotate(R, R1), prune(tree(L1,N,R1), T2).

Foldings. We now fold clauses 9 and 10 using the newly introduced clause 11, and we get:

12. rotate_prune(tree(L, N, R), T2) ← new1(L, N, R, T2).
13. rotate_prune(tree(L, N, R), T2) ← new1(R, N, L, T2).

We can now replace clause 1 by clauses 8, 11, 12, and 13, without changing the set of answer substitutions which are computed for the predicates in the initial program version [Kawamura-Kanamori 88]. In the new program version the definition for the predicate rotate_prune consists of clauses 8, 12, and 13, which do not have unnecessary variables.

On the contrary, some unnecessary variables occur in clause 11, which defines the predicate new1. For eliminating those variables we repeat on clause 11 the unfolding-definitions-foldings process which we have performed above on clause 1.

Unfolding. We unfold the atom prune in clause 11, and we get the following clauses:

14. new1(L, 0, R, tip(0)) ← rotate(L, L1), rotate(R, R1).
15. new1(L, s(N), R, tree(L2,s(N),R2)) ← rotate(L, L1), prune(L1, L2),
    rotate(R, R1), prune(R1, R2).

Definitions. In clause 14 there is the unnecessary variable L1, which occurs in the atom rotate(L,L1). Thus, we define the new predicate new2 by introducing the following clause:

16. new2(T) ← rotate(T, T1).

This clause is sufficient to eliminate also the unnecessary variable R1 from clause 14.

As far as clause 15 is concerned, we do not need to introduce any new definition, because the initial clause 1 allows us to perform the folding steps which are required for eliminating the unnecessary variables.

Foldings. By folding clause 14 using clause 16, we get:

17. new1(L, 0, R, tip(0)) ← new2(L), new2(R).

By folding clause 15 using clause 1, we get:

18. new1(L, s(N), R, tree(L2,s(N),R2)) ← rotate_prune(L, L2), rotate_prune(R, R2).

We replace clause 11 by clauses 16, 17, and 18. Finally, we have to eliminate the unnecessary variables from the definition of the newly introduced predicate new2. By performing again the unfolding-definitions-foldings process described above we get:

19. new2(tip(N)).
20. new2(tree(L,N,R)) ← new2(L), new2(R).

No new definition has been introduced for eliminating the unnecessary variables from the clauses which define the predicate new2. Thus, the current program version consists of the clauses 8, 12, 13, 17, 18, 19, and 20, which do *not* contain any unnecessary variable.

As an extra transformation process, we can simplify the derived program by unfolding the atom prune(tip(N), T2) in clause 8 and by unfolding also the occurrences of the predicate new1 in clauses 12 and 13. By doing so we get our final program for the predicate rotate_prune:

rotate_prune(tip(N), tip(N)).
rotate_prune(tree(L,0,R), tip(0)) ← new2(L), new2(R).
rotate_prune(tree(L,s(N),R), tree(L2,s(N),R2)) ← rotate_prune(L,L2), rotate_prune(R,R2).
rotate_prune(tree(L,s(N),R), tree(R2,s(N),L2)) ← rotate_prune(L, L2), rotate_prune(R, R2).

This program does not construct any intermediate tree to be passed from the computation of rotate to the computation of prune. Our computer experiments using a Prolog interpreter confirm the expected improvements of program performances. If there is no pruning, the final version requires about 30% less time and 50% less memory, while if there is pruning, the final version is drastically better because it does not rotate subtrees which will be pruned.

We do not formalize in this paper the relationship between the elimination of unnecessary variables and the increase of program efficiency. Indeed, this is true in most cases, but it depends on the values of the input data. ∎

The rest of the paper is structured as follows. In Section 2 we describe in more formal terms the procedure used in the above example for eliminating the unnecessary variables. That procedure may not terminate, and this depends also on the rule which is applied for selecting the atom to be unfolded.

In Section 3 we present a class of programs and a selection rule for which the procedure for avoiding unnecessary variables terminates, and thus, *all* unnecessary variables are eliminated.

In Section 4 we show how to extend the applicability of our technique to *arbitrary* programs, that is, programs which do not belong to the class defined in Section 3, and we propose a general procedure for the *partial* elimination of the unnecessary variables.

In Section 5 we show that the proposed transformation techniques can easily be enhanced by exploiting the functionality of some predicates in the initial program.

2. A PROCEDURE FOR ELIMINATING UNNECESSARY VARIABLES

In this section we present the basic procedure for eliminating the unnecessary variables from a given program and we study its properties. A variant of this procedure will be presented in Section 4.

We need some preliminary definitions.

DEFINITION 2. (*Linking Variables.*) Given a clause C of the form: $H \leftarrow A_1,..., A_n$, and a subset B of the atoms $\{A_1,..., A_n\}$, the *linking* variables of B in C are the variables occurring both in B and in the set $\{H, A_1,..., A_n\} - B$. ∎

Given a term t, we denote by vars(t) the set of variables occurring in t. The same notation will be adopted for the variables occurring in atoms, sets of atoms, and clauses.

DEFINITION 3. (*Blocks of a clause.*) Consider a set B of atoms. We define a binary relation ↓ over B as follows: given two atoms A_1 and A_2 in B, we have that:

$A_1 \downarrow A_2$ iff $vars(A_1) \cap vars(A_2) \neq \emptyset$.

Given a clause C, let ⇓ denote the reflexive and transitive closure of the relation ↓ over the body B of C. We denote by Part(C) the *partition* of B into *blocks* w.r.t. the equivalence relation ⇓. ∎

For instance, the partition of the body of clause 14: new1(L, 0, R, tip(0)) ← rotate(L, L1), rotate(R, R1) consists of the two blocks: {rotate(L, L1)} and {rotate(R, R1)}.

DEFINITION 4. (*Faithful variant.*) A block S1 in the body of a clause C1 is a *faithful variant* of a block S2 in the body of a clause C2 iff there exists a renaming substitution ρ, such that:
i)   S1 = S2ρ (that is, S1 is a variant of S2), and
ii) $\forall X \in vars(S1)$. X is a linking variable of S1 in C1 iff Xρ is a linking variable of S2 in C2. ∎

Notice that each variable occurring in a block of atoms of a clause is *either* an existential variable for that clause *or* it is a linking variable for that block, but *not* both. Thus, Definition 4 does not change if we replace 'linking variable' by 'existential variable'.

DEFINITION 5. (*Failing Clauses.*) We say that a clause in a given program P is a *failing* clause iff its body contains an atom which is *not* unifiable with the head of any other clause in P. ■

DEFINITION 6. (*Definition Clauses.*) Let P be a program and C a non-failing clause in P. The clause C is said to be a *definition* clause of P iff it has the form $p(X_1,...,X_n) \leftarrow A_1,...,A_m$ where p is a predicate symbol not occurring elsewhere in P and $X_1,...,X_n$ are the distinct variables occurring in $\{A_1,...,A_m\}$. ■

PROCEDURE 1. (*Elimination of Unnecessary Variables,* or *Elimination Procedure* for short.)
Input: <P,C> where P is a program and C is a definition clause of P.
Output: a set TransfC of clauses without unnecessary variables such that for each goal containing only predicates in P, the sets of answer substitutions computed by the programs P and $(P - \{C\}) \cup$ TransfC are equal.
Let InDefs and OutDefs be sets of clauses.
InDefs := {C}; OutDefs := { }; TransfC := { };
while there exists a clause D in InDefs *do*
  begin
    (*Unfolding*) Unfold *an* atom in the body of D and collect in U_D the set of derived non-failing clauses;
    (*Definitions*) For each clause E in U_D and for each block $B \in$ Part(E) such that:
      i) B contains at least one unnecessary variable, and
      ii) B is *not* a faithful variant of the body of a clause in InDefs $\cup$ OutDefs,
      add to InDefs the new clause: $newp(X_1,...,X_n) \leftarrow$ B, where newp is a fresh predicate symbol and $X_1,...,X_n$ are the linking variables of B;
    (*Foldings*) For each clause E in U_D fold all blocks of Part(E) using clauses in InDefs $\cup$ OutDefs and collect the resulting clauses in a set F_D;
    InDefs := InDefs − {D}; OutDefs := OutDefs $\cup$ {D}; TransfC := TransfC $\cup F_D$
  end ■

Some extra simplification steps can be performed at the end of the Elimination Procedure on the program $(P - \{C\}) \cup$ TransfC. In particular, we may unfold non-recursive and terminating atoms.

REMARK. The condition that the input clause for the Elimination Procedure should be a definition clause, is motivated by the fact that we want to use that clause to perform folding steps (see [Tamaki-Sato 84, Kawamura-Kanamori 88] for this condition on the applicability of the folding rule). However, that requirement is not restrictive, as we now indicate.

Suppose that C is a clause in a program P, and the body B of C contains some unnecessary variables. We introduce a definition clause C1 whose body is B and the head is $newp(X_1,...,X_n)$, where newp is a fresh predicate symbol and $X_1,...,X_n$ are the linking variables of B in C.

We then apply the Elimination Procedure to the program P and the clause C1. If it terminates we can eliminate the unnecessary variables of C by folding its body using C1. ■

The reader may easily verify that the Elimination Procedure is partially correct. However, it can be shown that the problem of deciding whether or not that procedure terminates for a given program and a given clause is not solvable.

Notice also that the Elimination Procedure is non-deterministic if we do not specify the function for selecting the atom to be unfolded, and unfortunately, it is often the case that the termination of the procedure is affected by the choice of that selection function.

When we specify a selection function S, we say that the Elimination Procedure is performed *via S*.

3. A CLASS OF PROGRAMS FOR WHICH THE ELIMINATION OF UNNECESSARY VARIABLES SUCCEEDS.

In this section we address the crucial problem of finding a class of <program, clause> pairs for which the Elimination Procedure terminates via a suitable selection function. For defining this class we first need the notions of *non-ascending* programs and *tree-like* programs. We then introduce a syntactic operation on programs, called *split*, and we present the main result of this section which ensures the termination of the Elimination Procedure if the program can be split into a non-ascending and a tree-like program.

DEFINITION 7. (*Depth of a variable in a term.*) Let X be a variable and t be a term with at least one occurrence of X. The *depth* of X in t, denoted by depth(X,t), is defined as follows:
- if t = X then depth(X,t) = 0, and
- if f is an n-ary function symbol and $t = f(t_1,...,t_n)$ then
  depth(X,t) = max{depth(X,t_i) | X occurs in t_i and i=1, ..., n}+1. ∎
DEFINITION 8. If t and u are terms we say that t is *smaller* than u, and we write t ≤* u, iff for each variable X in vars(t) ∩ vars(u) we have that: depth(X,t) ≤ depth(X,u). ∎

The definitions of depth and ≤* relation can be extended from terms to atoms, by considering the predicate symbols as terms constructors. Notice that the ≤* relation is *not* a partial ordering relation, because it is neither antisymmetric nor transitive.

DEFINITION 9. (*Linear Expressions.*) A term, or an atom, or a set of atoms is *linear* iff each variable occurs in it at most once. A clause is said to be *right linear* iff its body is linear. ∎
DEFINITION 10. (*Non-Ascending Programs.*) A clause of the form: $H \leftarrow A_1,...,A_n$, is said to be *non-ascending* iff i) it is a right linear clause and ii) for all i=1,...,n we have that: $A_i \leq^* H$.
A program is said to be non-ascending iff all its clauses are non-ascending. ∎

Example 2. The clause rotate(tree(L,N,R),tree(R1,N,L1)) ← rotate(L,L1), rotate(R,R1) is non-ascending. The clause rev([H|T],Acc,R) ← rev(T,[H|Acc],R) is *not* non-ascending, because the depth of the variable Acc in the body is greater than the one in the head.
The clause rotate_prune(Tree1,Tree2) ← rotate(Tree1,U), prune(U,Tree2) is *not* non-ascending because it is *not* right linear (indeed, the variable U occurs twice in the body). ∎

DEFINITION 11. (*Tree-like Programs.*) A clause C is said to be *tree-like* iff *either*
i) C has the form: $p_0(t_0) \leftarrow p_1(t_1),..., p_n(t_n)$, where $t_0,t_1,...,t_n$ are linear terms and, for i=1,...,n, t_i is a subterm of t_0, *or*
ii) C can be obtained from a tree-like clause $B_0 \leftarrow B_1,...,B_n$ by replacing some B_i, i=0,...,n, by a nullary predicate symbol.
A set of atoms is tree-like iff it is the body of a tree-like clause. A program is tree-like iff it consists of tree-like clauses only. ∎
Example 3. The following clauses are tree-like clauses:
    btree(tip(N)).
    btree(tree(L,N,R)) ← btree(L), btree(R).
    p ← btree(L), btree(R). ∎

The classes of non-ascending and tree-like programs are not large enough to include many programs one writes in practice. However, as already mentioned, it is often possible to decompose programs into pairs of tree-like and non-ascending programs and in that case we are guaranteed that the Elimination Procedure terminates, as Theorem 14 below shows.

DEFINITION 12. (*Split.*) Let P be a program and π a set of pairs $<p, i>$ which is a total function from the predicate symbols in P to the natural numbers, such that for each predicate symbol p, i is not greater than the arity of p.

i) Let $<p, i>$ be a pair in π. The *split* of the atom $p(t_1,...,t_n)$ w.r.t. π is the pair of atoms $<L,R>$ defined as follows:

if n=0 (that is, p is a nullary predicate) then $<p, p>$,

if i=0 then $<p, p(t_1,...,t_n)>$,

if n=1 and i=1 then $<p(t_1), p>$, and

if n>1 and i>0 then $<p(t_i), p(t_1,...,t_{i-1},t_{i+1},...,t_n)>$.

ii) The split of a set of atoms $A_1,...,A_m$ w.r.t. π is the pair $<\{L_1,...,L_m\}, \{R_1,...,R_m\}>$, where $<L_i, R_i>$ is the split of A_i w.r.t. π, for i=1,...,m.

iii) The split of the clause $A_0 \leftarrow A_1,...,A_m$ w.r.t. π is $<L_0 \leftarrow L_1,...,L_m, R_0 \leftarrow R_1,...,R_m>$, where $<L_i, R_i>$ is the split of A_i w.r.t. π, for i=0,...,m.

iv) The split of the program $\{C_1,...,C_k\}$ w.r.t. π is the pair of programs: $<\{C_{1L},...,C_{kL}\}, \{C_{1R},...,C_{kR}\}>$, where $<C_{iL}, C_{iR}>$ is the split of C_i w.r.t. π, for i=1,...,k.

We say that the split of P w.r.t. π is *disjoint* iff for every clause $C_i=H_i \leftarrow Body_i$, i=1,...,k, we have that: $vars(C_{iL}) \cap vars(C_{iR}) \subseteq vars(H_i) - vars(Body_i)$. ∎

Example 4. Let us consider the program Rotate_Prune of Example 1 and let π be $\{<rotate\_prune, 0>, <rotate, 2>, <prune, 1>\}$. The splits of clauses 1 and 3 w.r.t. π are the pairs $<1.L, 1.R>$ and $<3.L, 3.R>$, whose components are the following clauses:

1.L rotate_prune ← rotate(U), prune(U).

1.R rotate_prune(Tree1,Tree2) ← rotate(Tree1), prune(Tree2).

3.L rotate(tree(L1,N,R1)) ← rotate(L1), rotate(R1).

3.R rotate(tree(L,N,R)) ← rotate(L), rotate(R).

Notice that the split of clause 3 is disjoint, because $vars(3.L) \cap vars(3.R) = \{N\}$, and N does not occur in the body of the clause. ∎

We now introduce a class of suitable selection functions to be applied during the execution of the Elimination Procedure.

DEFINITION 13. (*Synchronized Descent Rules.*) Given a set G of atoms, an atom M in G is said to be a *maximal* atom iff for each atom A in G we have that $A \leq^* M$. A selection function S is said to be a *Synchronized Descent Rule* (or SDR for short) iff for each clause C whose body B has a maximal atom, S selects a maximal atom of B. ∎

Notice that the maximal of a set of atoms may not exist and, if it exists, need not be unique.

We are now ready to state the main result of this section.

THEOREM 14. Let P be a program and C a definition clause in P. Let P − {C} be non-ascending, and π be a function from the predicates of P to the natural numbers, such that:

i) the split $<B_L, B_R>$ of the body of C w.r.t. π is disjoint, B_L is tree-like, and B_R is linear, and

ii) the split $<Q, R>$ of P − {C} w.r.t. π is disjoint, Q is a tree-like program, and R is a non-ascending program.

Then the Elimination Procedure via any SDR terminates for the input $<P, C>$. ∎

The reader may verify that the Rotate_Prune program satisfies the hypotheses of the above theorem and, indeed, the derivation presented in Example 1 is just an application of the Elimination Procedure via an SDR.

4. APPLICATION OF THE TRANSFORMATION TECHNIQUE TO UNRESTRICTED PROGRAMS

In this section we will show the use of our transformation techniques for classes of programs which are *not* included in the ones described in the previous section. We introduce the notion of *marked* programs, and we present a variant of the Elimination Procedure to be applied on those programs.

The concept of marked programs is related to the one of *blazed* terms in the so-called deforestation technique of [Wadler 88]. However, in Wadler's approach 'blazing' is assigned on the basis of the type information, while here we use an untyped logic language.

We will also introduce a new transformation rule, called *Generalization + Equality Introduction*, and we will show the use of that rule for extending the applicability of our transformation techniques to *all* programs (but, in that case, only a *partial* elimination of the unnecessary variables is guaranteed).

DEFINITION 15. A *marked program* is a program in which the predicate symbols are divided into two disjoint sets: the relevant ones and the non-relevant ones. We assume that if a relevant predicate occurs in the body of a clause then the head predicate of that clause is relevant as well.

A clause of a marked program is called marked clause. An atom whose predicate is relevant is called relevant atom. The *relevant portion* of a marked clause C is the clause obtained from C by deleting the non-relevant atoms. The relevant portion of a marked program is the set of the relevant portions of the program clauses.   ∎

Given a marked clause C, we denote by PartR(C) the partition of the set of the relevant atoms of the body of C with respect to the equivalence relation ⇓ introduced in Section 2.

We now give the version of the Elimination Procedure to be applied on marked programs.

PROCEDURE 2. (*Extended Elimination Procedure.*)

Input: <P,C> where P is a marked program and C is a marked definition clause of P. We assume that all atoms in C are relevant.

Output: a set TransfC of marked clauses such that: i) for each goal containing only predicates in P, the sets of answer substitutions computed by the programs P and $(P - \{C\}) \cup$ TransfC are equal, ii) the relevant portion of TransfC does *not* contain any multiple variables, and iii) the existential variables occurring in relevant atoms of TransfC are *only* the ones which are in common with non-relevant atoms.

Let InDefs and OutDefs be sets of marked clauses.

InDefs := $\{C\}$; OutDefs := $\{\}$; TransfC := $\{\}$;

while there exists a clause D in InDefs *do*

  begin

      (*Unfolding*) Unfold *a* relevant atom in the body of D and collect in U_D the set of derived non-failing marked clauses;

      (*Definitions*) For each clause E in U_D and for each block $B \in$ PartR(E) such that:

        i) B contains either a multiple variable or an existential variable which is *not* a linking variable, and

        ii) B is *not* a faithful variant of the body of a clause in InDefs $\cup$ OutDefs,

        add to InDefs the new marked clause: $newp(X_1,...,X_n) \leftarrow B$, where newp is a fresh predicate symbol which is assumed to be relevant, and $\{X_1,...,X_n\}$ is the set of linking variables of B;

      (*Foldings*) For each clause E in U_D fold all blocks of PartR(E) using clauses in InDefs $\cup$ OutDefs and collect the resulting marked clauses in a set F_D;

      InDefs := InDefs $- \{D\}$; OutDefs := OutDefs $\cup \{D\}$; TransfC := TransfC $\cup F_D$

  end   ∎

Notice that the linking variables of a block in PartR(C) may *not* coincide with the non-existential variables of that block (and this is why we must refer to the notion of linking variables).

It is easy to check the partial correctness of the Extended Elimination Procedure. As far as its termination is concerned, we can state the following obvious extension of Theorem 14.

THEOREM 16. Let P be a marked program and C a marked definition clause in P. If the relevant portions of P and C satisfy the hypotheses of Theorem 14 then the Extended Elimination Procedure via any SDR terminates for the input <P, C>. ■

In the above theorem by 'an SDR' we mean a function which selects a maximal atom (if any) in the set of the relevant atoms of the body of a clause.

We are now faced with the problem of choosing the predicates to be considered as relevant. This choice can be done according to the observation that we may eliminate more unnecessary variables if the relevant portion we consider is larger or more instantiated. For this reason we now introduce an additional transformation rule, called *Generalization + Equality Introduction Rule*. This rule allows us to obtain, for any given marked program P, an equivalent marked program Q with the *maximal* relevant portion which subsumes the relevant portion of P and satisfies the hypotheses of Theorem 14 (see the *Generalization Strategy* below).

DEFINITION 17. Given a clause of the form: $H \leftarrow A_1, ..., A_n$ the *Generalization + Equality Introduction Rule* (or *Generalization Rule*, for short) allows us to derive the new clause:

$$GenH \leftarrow eq(X_1,t_1), ..., eq(X_r,t_r), GenA_1, ..., GenA_n.$$

where: $(GenH, GenA_1, ..., GenA_n) \theta = (H, A_1, ..., A_n)$ and $\theta = \{X_1 / t_1, ..., X_r / t_r\}$ is a substitution which is not a variable renaming. ■

As usual we assume that the equality predicate is defined by the unit clause $eq(X,X)$.
Since the equalities introduced by an application of the above rule can be simplified by unfolding, and the unfolding rule preserves the set of computed answer substitutions, we have that also the Generalization Rule preserves the set of computed answer substitutions.

Now, let us recall that a clause $H \leftarrow A_1, ..., A_m$ *subsumes* a clause $K \leftarrow B_1, ..., B_n$ iff there exists a substitution θ such that $H\theta = K$ and $\{A_1\theta, ..., A_m\theta\} \subseteq \{B_1, ..., B_n\}$. Subsumption is a partial order on clauses and it can be extended to a pre-order on programs as follows. We say that a program P1 subsumes a program P2 iff for each clause C1 in P1 there exists a clause C2 in P2 such that C1 subsumes C2.

We apply the Generalization Rule according to the following *Generalization Strategy*:
i) we assume that all equalities introduced by the generalization rule are non-relevant, and then
ii) we repeatedly apply the generalization rule starting from a given marked program P so that the resulting final marked program Q satisfies the following conditions:
α) the relevant portion of Q subsumes the relevant portion of P,
β) the relevant portion of Q satisfies the hypotheses of Theorem 14, and
γ) the relevant portion of Q is a maximal (w.r.t. the subsumption pre-order) among all programs which satisfy conditions α and β, that is, for every program R, if R subsumes the relevant portion of P and the relevant portion of Q subsumes R, then R does *not* satisfy the hypotheses of Theorem 14.

Example 6. (Sum and Size of a Binary Tree.) Let us consider a binary tree T whose nodes are labelled by natural numbers. In order to compute the sum of the numbers in T and the size of T we can use the following program Sum_Size which traverses twice the tree T:

1. sum_size(T,S,N) ← sumtree(T,S), size(T,N).
2. sumtree(T,S) ← sumtree1(T,0,S).
3. sumtree1(tip(W),A,S) ← plus(W,A,S).
4. sumtree1(tree(L,W,R),A,S) ← plus(W,A,A1), sumtree1(L,A1,SL), sumtree1(R,SL,S).
5. size(T,N) ← size1(T,0,N).
6. size1(tip(W),A,s(A)).
7. size1(tree(L,W,R),A,N) ← size1(L,s(A),NL), size1(R,NL,N).

8. plus(N,0,N).

9. plus(M,s(N),s(S)) ← plus(M,N,S).

We would like to avoid the multiple traversal of the tree by eliminating the multiple variable T from clause 1. We know that this is possible if the pair <Sum_Size, clause 1> satisfies the hypotheses of Theorem 16. Unfortunately, in the program made out of clauses 2,3,...,9 clauses 4 and 7 are *not* non-ascending. Thus, we apply the Generalization Strategy and we transform clauses 4 and 7 into the following clauses 4g and 7g, respectively:

4g. sumtree1(tree(L,W,R),A,S) ← eq(A1,A2), eq(SL1,SL), plus(A,W,A1),
$$\text{sumtree1(L,A2,SL), sumtree1(R,SL1,S).}$$

7g. size1(tree(L,W,R),A,N) ← eq(A,A1), eq(NL1,NL), size1(L,s(A1),NL), size1(R,NL1,N).

Now, if we consider the program Sum_Size1 made out of clauses 1, 2, 3, 4g, 5, 6, 7g, 8, and 9, and we assume that all predicates different from equalities are relevant, then the relevant portions of Sum_Size1 and clause 1 satisfy the hypotheses of Theorem 14, with $\pi = \{<\text{sum\_size},1>, <\text{sumtree},1>, <\text{size},1>, <\text{sumtree1},1>, <\text{plus},1>, <\text{size1},1>\}$. (The reader may easily verify that any program which subsumes Sum_Size and it is subsumed by the relevant portion of Sum_Size1 does *not* satisfy the hypotheses of Theorem 14.)

Thus, Theorem 16 ensures that the Extended Elimination Procedure terminates for the input <Sum_Size1, clause 1>, and indeed, after some extra simplification steps, we get the following final program:

sum_size(tip(W),W,s(0)).

sum_size(tree(L,W,R),S,N) ← new(L,W,SL,s(0),NL), new(R,SL,S,NL,N).

new(tip(W),A1,S,A2,s(A2)) ← plus(A1,W,S).

new(tree(L,W,R),A1,S,A2,N) ← plus(A1,W,A3), new(L,A3,SL,s(A2),NL),
$$\text{new(R,SL,S,NL,N).}$$

In this final program there are still some unnecessary variables (they are SL, NL, and A3), but it is more efficient than the original program because it avoids multiple traversals of the binary trees to be visited. Computer experiments using a Prolog interpreter demonstrate that the derived program is about 20% more efficient in time and 30% more efficient in space. ∎

The above example shows that our techniques can be used not only for avoiding the construction of intermediate data structures, as shown by the Rotate_Prune example, but also for avoiding multiple traversals of data structures.

5. ENHANCING THE TRANSFORMATION TECHNIQUES BY EXPLOITING FUNCTIONALITY.

We would like to remark that the transformation techniques described so far do not use any information about the properties which may hold for some predicates occurring in the programs. In this section we will show how to exploit the information about the functionality of some predicates, while applying our techniques. We assume that the functionality information is given to us either by an oracle (because functionality is undecidable, in general) or by some known methods (see, for instance [Debray-Warren 89]) which work for particular classes of programs.

We say that a predicate $p(X_1,...,X_h,Y_1,...,Y_k)$ defined by a program P is *functional w.r.t. the arguments* $X_1,...,X_h$ iff for every goal ← $p(t_1,...,t_h,u_1,...,u_k)$, with $t_1,...,t_h$ ground terms, there exists a unique ground answer substitution computed by P.

The knowledge that some predicates are functional may be used, during the program transformation process, according to the following rule, which is an extension of the one in [Tamaki-Sato 84].

DEFINITION 18. (*Functionality Rule.*) Let P be a program and $p(X_1,...,X_h,Y_1,...,Y_k)$ be a predicate occurring in P. Suppose that p is functional w.r.t. the arguments $X_1,...,X_h$, and C is a clause of the form $H \leftarrow p(t_1,...,t_h,u_1,...,u_k), p(t_1,...,t_h,v_1,...,v_k), A_1,...,A_n$. If there exists a most general unifier θ between $u_1,...,u_k$ and $v_1,...,v_k$ then C can be replaced by the clause:

$(H \leftarrow p(t_1,...,t_h,u_1,...,u_k), A_1,...,A_n) \theta$

otherwise C can be deleted. ∎

It can be shown that the Functionality Rule preserves the set of computed answer substitutions.

The only modification we make to the Extended Elimination Procedure is the following: after each unfolding step we apply the Functionality Rule whenever possible. This variant of Procedure 2 which can be used for eliminating unnecessary variables from functional predicates will be called *Functional Elimination Procedure*. The enhancement of our techniques is based on the following result.

THEOREM 19. Let P be a marked program where each relevant predicate p is functional w.r.t. the argument in position f_p. Let π be the function mapping each predicate p to the position f_p, and let C be a definition clause in P. Suppose that:

i) the split $<B_L,B_R>$ of the body of C w.r.t. π is disjoint, B_L is tree-like, and B_R is linear, and

ii) the split $<Q, R>$ of P $-$ {C} w.r.t. π is disjoint, Q is a tree-like program, and R is a non-ascending program.

Then the Functional Elimination Procedure via any SDR terminates for the input $<P, C>$. ∎

Example 7. (Fibonacci Numbers.) The initial program is:

1. fib(0,s(0)).
2. fib(s(0),s(0)).
3. fib(s(s(N)),F) ← fib(s(N),F1), fib(N,F2), plus(F1,F2,F).

Let us assume that fib is a relevant predicate and plus is not. Clause 3, with the unnecessary variable N, is not a definition clause and it contains the non-relevant predicate plus. Thus, we introduce a new definition clause whose body consists of the relevant atoms of the body of clause 3, that is:

4. new1(N,F1,F2) ← fib(s(N),F1), fib(N,F2).

By applying the Functional Elimination Procedure and then performing some simplification steps we get the following program, which corresponds to the *linear* evaluation of the Fibonacci function:

1. fib(0,s(0)).
2. fib(s(0),s(0)).
3'. fib(s(s(N)),F) ← new1(N,F1,F2), plus(F1,F2,F).
5. new1(0,s(0),s(0)).
6. new1(s(N),F1,F2) ← new1(s(N),F2,F21), plus(F2,F21,F1). ∎

6. CONCLUSIONS

We have presented a transformational approach to the automatic improvement of logic programs. In particular we have introduced some versions of a strategy, encoded in the Elimination, Extended Elimination, and Functional Elimination Procedures, for guiding the application of the Unfold / Fold rules, and improving program efficiency. That strategy is an extension of the one presented in [Proietti-Pettorossi 90]. It allows us to eliminate all unnecessary variables from a class of programs defined by induction on arbitrary data structures, which satisfy *non-linear* domain equations (like trees of various kinds), while the previous strategy was successful only in the case of data structures satisfying *linear* domain equations (like natural numbers and lists).

The transformation technique we have proposed is algorithmic, and the consequent improvements of program efficiency are due to the fact that by eliminating unnecessary variables one often avoids redundant computations.

Our proposed technique also extends previous ones developed by other authors, and in particular the loop-fusion method of [Debray 88], because it does *not* require a fixed recursive pattern for the initial program versions.

The application of our techniques achieve the advantages which can usually be obtained in the case of functional programs by applying the composition (or deforestation [Wadler 88]) and tupling strategies.

7. ACKNOWLEDGEMENTS

We would like to thank our colleagues of the Compulog Project for their stimulating conversations and their interest in our work. The University of Rome Tor Vergata and the IASI Institute of the National Research Council of Italy gave us the required research facilities.

8. REFERENCES

[Burstall-Darlington 77] Burstall, R.M. and Darlington, J.: "A Transformation System for Developing Recursive Programs", JACM, Vol. 24, No. 1, January 1977, pp. 44-67.

[CIP 84] CIP Language Group: "The Munich Project CIP" Lecture Notes in Computer Science, 1984.

[Debray 88] Debray, S.K.: "Unfold/Fold Transformations and Loop Optimization of Logic Programs", Proc. SIGPLAN 88, Conference on Programming Language Design and Implementation, Atlanta, Georgia, 1988.

[Debray-Warren 89] Debray, S.K. and Warren, D.S.: "Functional Computations in Logic Programs", ACM TOPLAS, 11 (3), 1989, pp. 451-481.

[Feather 86] Feather, M.S.: "A Survey and Classification of Some Program Transformation Techniques", Proc. TC2 IFIP Working Conference on Program Specification and Transformation, Bad Tölz, Germany, 1986.

[Hogger 81] Hogger, C.J.: "Derivation of Logic Programs", JACM, No. 28, 2, 1981, pp. 372-392.

[Kawamura-Kanamori 88] Kawamura, T. and Kanamori, T.: "Preservation of Stronger Equivalence in Unfold/Fold Logic Program Transformation", Proc. Int. Conf. on Fifth Generation Computer Systems, Tokyo, 1988, pp. 413-422.

[Lloyd 87] Lloyd, J.W.: "Foundations of Logic Programming", Springer-Verlag, Berlin, Heidelberg, New York, Tokyo, 2nd edition, 1987.

[Paige-Koenig 82] Paige, R. and Koenig, S.: "Finite Differencing of Computable Expressions" ACM TOPLAS 4 (3) July 1982, pp. 402-454.

[Pettorossi 77] Pettorossi, A.: "Transformation of Programs and Use of Tupling Strategy", Proc. Informatica 77, Bled, Yugoslavia, 1977, pp. 3 103, 1-6.

[Pettorossi-Proietti 87] Pettorossi, A. and Proietti, M.: "Importing and Exporting Information in Program Development", Proc. IFIP TC2 Workshop on Partial Evaluation and Mixed Computation, Gammel Avernaes, Denmark, North Holland 1987, pp. 405-425.

[Proietti-Pettorossi 90] Proietti, M. and Pettorossi, A.: "Synthesis of Eureka Predicates for Developing Logic Programs", Proc. ESOP 90, Copenhagen, 1990, Lecture Notes in Computer Science No. 432, pp. 306-325.

[Tamaki-Sato 84] Tamaki, H. and Sato, T.: "Unfold/Fold Transformation of Logic Programs", Proc. 2nd International Conference on Logic Programming, Uppsala, Sweden, 1984.

[Wadler 88] Wadler, P. L.: "Deforestation: Transforming Programs to Eliminate Trees", Proc. ESOP 88, Nancy, France, 1988, Lecture Notes in Computer Science 300, pp. 344-358.

Efficient Integration of Simplification into Prolog[*]

P.H. Cheong   L. Fribourg

LIENS, URA 1327 du CNRS
45, rue d'Ulm, 75005 Paris, FRANCE
e-mail: {cheong,fribourg}@dmi.ens.fr

Abstract

SLOG is a language integrating logic and functional programming that is based on innermost narrowing and rewriting. Although innermost narrowing can be simulated by Prolog, the same is not true for rewriting. This has motivated a weaker notion of rewriting, called simplification, in the proposal of "Prolog with Simplification". In this paper, we address the issue of efficiently integrating simplification into Prolog.

1   Introduction

Many approaches on the amalgamation of logic programming (LP) and functional programming (FP) are narrowing-based [BL86]. If theoretical results pertaining to strategy refinements and their completeness [BGM87, Fri85, Höl88, LPB+87, Red85, You89] are relatively abundant, efficient implementations of narrowing are only beginning to appear [BCM89, Han90, KLMR90, Müc90].

It is well-known that innermost narrowing possesses a simple and yet efficient implementation in Prolog, by means of the flattening technique. In practice, innermost narrowing is seldom used alone, but often in conjunction with rewriting, to reinforce the priority of determinate computations over nondeterminate ones. This direction was first taken in SLOG [Fri85]—a language that integrates LP and FP—and, more recently, ALF [Han90].

As rewriting is not entirely compatible with the flattening technique, new abstract machines like the A-WAM [Han90] have been specially designed for innermost narrowing and rewriting. A simpler approach has however been taken in "Prolog with Simplification" [Fri88]. The idea there was to use a weaker notion of rewriting, called simplification, that is more suited to the flattening technique, and hence more suited to a portable implementation in Prolog.

In this paper, we address the issue of efficiently integrating simplification into Prolog. We present a simple meta-interpreter for Prolog with Simplification which, with respect to the one previously given in [Fri88], incorporates two basic improvements. First, clauses

---

[*]This work has been partially supported by ESPRIT project BRA 3020.

are represented in a way that reflects the simplification mechanism (cf. section 3). Second, matching calls are opened up using techniques of partial evaluation (cf. section 4).

2 Prolog with Simplification

2.1 Syntax

We consider a first order language that distinguishes between (defined) functions and (primitive) constructors. We suppose that equality = is the only predicate. (Other predicates can however be interpreted as syntactical sugar for truth-value functions.)

Clauses and goals are defined as usual, except for the following restrictions—adopted from SLOG—on the clause heads and the body literals:

- The lhs of a clause head begins with a function symbol and contains no other function symbol.

- The rhs of a body literal is a ground constructor term.

Note that the second condition does not incur a loss in generality, since all body literals like $s = t$, where t is not necessarily a ground constructor term, can be preprocessed as $equal(s,t)$—which stands for "$equal(s,t) = true$"—where $equal$ is defined by the clause $equal(X,X)$.

Example 1 The following are some examples of clauses and goals:

```
reverse([])=[].
reverse([X|Xs])=append(reverse(Xs),[X]).
append([],Y)=Y.
append([X|Xs],Y)=[X|append(Xs,Y)].
:- append(append(A,B),C)=[].
```

A correct answer substitution for the goal is {A/[],B/[],C/[]}.

2.2 Operational semantics

The flat form

The actual interpreter works on flat clauses and flat goals, which only involve literals of the form $f(s) = t$, where s and t are some constructor terms. The flat form can be obtained by iteratively applying the following rules [BGM87, vEY87]:

$$\frac{f(s) = t :- \ldots}{f(s) = t[u \leftarrow x] :- \ldots, t/u = x} \quad \text{if} \left\{ \begin{array}{l} t/u \text{ begins with a function} \\ x \text{ is a fresh variable} \end{array} \right.$$

$$\frac{\ldots, f(s) = t, \ldots}{\ldots, f(s[u \leftarrow x]) = t, s/u = x, \ldots} \quad \text{if} \left\{ \begin{array}{l} s/u \text{ begins with a function} \\ x \text{ is a fresh variable} \end{array} \right.$$

where t/u denotes the subterm of t at occurrence u; $t[u \leftarrow t']$ denotes the term obtained from t by replacing the subterm at occurrence u by t'.

Example 2 The clauses and goals in example 1 are flattened as:

```
reverse([])=[].
reverse([X|Xs])=Y :- reverse(Xs)=Z,append(Z,[X])=Y.
append([],Y)=Y.
append([X|Xs],Y)=[X|Z] :- append(Xs,Y)=Z.
:- append(A,B)=D,append(D,C)=[].
```

In what follows, we suppose that all clauses and goals are in flat form. A *deterministic* clause is a clause with a lhs that is not unifiable with the lhs of other clause heads. A literal is said to be *simplifiable* by a deterministic clause if the lhs of clause head matches the lhs of the literal. A goal is said to be *simplified* if it does not contain any simplifiable literal.

Refutation

Prolog with Simplification interprets the flat form by leftmost *SLD-resolution* and *Simplification*.

Example 3 The solution of example 2 can be found by innermost narrowing or, alternatively, by the following leftmost SLD-resolution:

```
:- append(A,B)=D,append(D,C)=[].
           | A/[],B/D
:- append(D,C)=[].
           | D/[],C/[]
         true
```

Leftmost SLD-resolution is equivalent to innermost narrowing [Der83, vEY87], but what is simplification? Intuitively, it is SLD-resolution restricted to literals which are "sufficiently instantiated". The idea definitely has a close connection to rewriting, since (1) SLD-resolution is equivalent to narrowing and (2) rewriting is also a restriction of narrowing to terms that are sufficiently instantiated.

Simplification is defined as follows:

One-step simplification:

$$\frac{:\text{-} L_1,\ldots,L_{i-1},L_i,L_{i+1},\ldots,L_n}{:\text{-} (L_1,\ldots,L_{i-1},B_1,\ldots,B_m,L_{i+1},\ldots,L_n)\sigma}$$

if L_i is simplifiable by $A :\text{-} B_1,\ldots,B_m$ and $\sigma = mgu(A,L_i)$. Note that σ may be undefined (due to unification failure), in which case the goal is determinately replaced by "fail", which provokes immediate backtracking. Otherwise (σ is defined), the goal is determinately replaced by the resolvent.

Example 4 Consider the previous SLD-derivation which, upon backtracking, generates the unsolvable goal :- append(A1,B)=D1,append([X1|D1],C)=[]. Although that goal causes an infinite loop in Prolog, it can be simplified to "fail", since the rightmost literal is simplifiable (by the second append clause) and there is unification failure.

Simplification is however a weaker notion than rewriting, and the following example illustrates the difference.

Example 5 Consider the following (flat) definition for multiplication:

```
0*Y=0.
s(X)*Y=Z :- X*Y=W,X+W=Z.
0+Y=Y.
s(X)+Y=s(Z) :- X+Y=Z.
```

The goal `:- A+B=C,0*C=0` can't be simplified into `true` (but into `:- A+B=C`), whereas reduction on `0*(A+B)=0` directly yields `true`.

In general, simplification is weaker than rewriting when there are rewrite rules that delete variables on the lhs (as above) or when there are conditional rewrite rules (our notion of deterministic clauses is too weak to take into account such rules).

Remark It can be shown that Prolog with Simplification is complete for first-order logic with equality with respect to the solutions built over constructors (see [Fri85]) if every function is completely defined by the program over constructors.

In what follows, SLD-resolution shall always be intended as leftmost SLD-resolution. We shall terminate this section by presenting some general ideas of a meta-interpreter for Prolog with Simplification.

2.3 Meta-interpreter

Difference-lists

The interpreter will be manipulating goals, sometimes taking them apart and sometimes joining them back together. A necessary operation is of course the concatenation operation, and that leads us to use the well-known difference-lists' technique here.

The difference-list representation of a clause

$$A :- B_1, \ldots, B_n$$

is

$$\text{clause}(A, [B_1, \ldots, B_n|U] - U)$$

where U is some distinct variable.

For instance, the difference-list definition of append is:

```
clause(append([],Y)=Y,U-U).
clause(append([X|Xs],Y)=[X|Z],[append(Xs,Y)=Z|U]-U).
```

The difference-list representation of a goal can be defined similarly. For example, $[B_1, \ldots, B_n|U] - U$ stands for the goal $:- B_1, \ldots, B_n$.

The interpreter therefore works on difference-lists and (leftmost) SLD-resolution is implemented as:

```
resolve(X-Y,X-Y) :- X==Y,!.          % empty goal
resolve([X|Y]-Z,U-Z) :- clause(X,U-Y). % nonempty goal
```

The toplevel

To emphasize the priority of determinate computations over nondeterminate ones, goals are always simplified before any SLD-resolution step. This strategy is depicted in the following toplevel loop:

```
prolog_with_simp(X-Y) :- X==Y,!.          % empty goal
prolog_with_simp(X) :-                     % nonempty goal
      simplify(X,Y),                       % simplify first
      resolve(Y,Z),                        % then resolve
      prolog_with_simp(Z).
```

where `simplify(X,Y)` means that the goal Y is a simplified form of X.

It remains to discuss about the implementation of `simplify`.

3   Simplification

Goals are simplified in zero or more steps, by application of the (one-step) simplification rule, until the simplified form is obtained. The essential point here is that simplification can follow a left-to-right strategy.

Given two literals L and L', we say that L' is *functionally dependent* on L (written $L \prec L'$, in keeping with the notation in K-LEAF [LPB$^+$87]) if the rhs of L shares variables with the lhs of L'. A goal :- $L_1, \ldots, L_n$ is said to be *ordered* if $\prec$ is acyclic on the set $\{L_1, \ldots, L_n\}$ and $i < j \Rightarrow L_j \not\prec L_i$.

Proposition 1 All initial goals (that is, those obtained by flattening) are ordered.

Proposition 2 Ordering is preserved by SLD-resolution and simplification, hence all dynamically generated goals are also ordered.

Proposition 3 Consider an ordered goal :- $L_1, \ldots, L_n$. Simplification on some literal L_i will not instantiate the lhs of L_j, for all $j \leq i$.

According to the above propositions, the application of the simplification rule on a literal will not affect the "simplification status" of literals to its left (since their lhs will not be instantiated). This justifies using a left-to-right strategy:

```
simplify(X-Y,X-Y) :- X==Y,!.
simplify([X|Y]-Z,U-W) :-
      simp_1(X,U-V),           % simplify the leftmost literal first
      simplify(Y-Z,V-W).       % then followed by the others
```

where `simp_1` stands for the simplification of a literal.

Whether a literal $f(s_1, \ldots, s_n) = s$ is simplifiable can be tested by matching the input arguments s_i against the input parameters t_i of some deterministic clause head $f(t_1, \ldots, t_n) = t$. If matching is successful, then the simplification process can proceed to simplify the clause body (from-left-to-right), with variable-bindings that are computed, as usual, by unification. However, it is only necessary to unify s against t, since the s_i has already been matched against the t_i.

The desired behavior can be simulated by writing every deterministic clause

$$f(t_1,\ldots,t_n) = t \text{ :- } l_1,\ldots,l_m$$

as

```
simp_l(f(X1,...,Xn)=X,U1-U) :-
    match(X1,t1),                % match X1 against t1
    ...
    match(Xn,tn),                % match Xn against tn
    !,                           % literal is simplifiable
    unify(X,t),                  % syntactic unification
    simp_l(11,U1-U2),            % simplify leftmost literal first
    simp_l(12,U2-U3),            % followed by the second
    ...
    simp_l(lm,Um-U).             % followed by the last
```

where $X1,\ldots,Xn,X,U1,\ldots,Um,U$ are distinct variables. Note that the cut (!) expresses the determinate nature of simplification.

The above clauses take into account the case where the given literal is simplifiable. So it is sufficient to complete that definition by:

```
    simp_l(X,[X|Y]-Y).          % should be last simp_l clause
```

for the case where the given literal is already simplified.

Remark Although we have defined simplification primarily in terms of deterministic clauses, inductive clauses and closed-world-assumption clauses (CWA-clauses) [Fri88] like:

```
equal([X|Y],[Z|T]) :- equal(X,Z),equal(Y,T).  % inductive clause
equal([],[X|Y]) :- fail.                        % CWA-clause
```

can also be declared. These clauses can be represented as simp_l clauses, in exactly the same way as the representation of deterministic clauses.

4 Simulation of Matching

To implement matching, or one-way unification, in Prolog, it is necessary to forbid the instantiation of the term to be matched. It is clear that such an "impure" feature can only be simulated via some impure primitives of Prolog. The idea of our implementation is very simple: to decompose matching calls into syntactic unification and some common (impure) Prolog primitives (i.e., nonvar/1 and ==/2). The following expansion rules are used to implement matching:

- "match(X,c)", where c is a constant, expands to "X==c".

- "match(X,c(t1,...,tn))" expands to "nonvar(X),unify(X,c(Y1,...,Yn)), match(Y1,t1),...,match(Yn,tn)", where the Yi's are distinct variables.

- "match(X,Y)" expands to

  - "unify(X,Y)", if the occurrence of the variable Y is the first (leftmost) in the clause. In this case, X and Y can be immediately unified at compile-time.
  - "X==Y", if the occurrence of Y is not the first.

Let's illustrate with an example. Suppose

```
append([],Y)=Y.
append([X|Xs],Y)=[X|Z] :- append(Xs,Y)=Z.
```

are deterministic clauses. The corresponding simp_1 clauses are:

```
simp_1(append(A,B)=C,U-U) :- match(A,[]),
                             match(B,Y),!,
                             unify(C,Y).
simp_1(append(A,B)=C,U-V) :- match(A,[X|Xs]),
                             match(B,Y),!,
                             unify(C,[X|Z]),
                             simp_1(append(Xs,Y)=Z,U-V).
```

In the above clauses, to match A against a constant [] can be simulated by the call A==[]; to match A against a cons-structure [X|Xs] can be simulated by the call "nonvar(A), unify(A,[X|Xs])"; and finally, to match B against Y can be simulated by the (compile-time) unification of Y and B. We thus obtain the following simp_1 clauses:

```
simp_1(append(A,B)=C,U-U) :- A==[],!,unify(C,B).
simp_1(append(A,B)=C,U-V) :- nonvar(A),unify(A,[X|Xs]),!,
                             unify(C,[X|Z]),
                             simp_1(append(Xs,B)=Z,U-V).
```

5   An example: permutation sort

We define permutation sort by the following flat definition:

```
clause(psort(X,Y),[permute(X,Y),ordered(Y)|U]-U).
clause(permute([],[]),U-U).
clause(permute([X|Xs],[Y|Z]),[delete([X|Xs],Y,T),permute(T,Z)|U]-U).
clause(delete([X|Xs],X,Xs),U-U).
clause(delete([X|Xs],Y,[X|Z]),[delete(Xs,Y,Z)|U]-U).
clause(ordered([]),U-U).
clause(ordered([X|Xs]),[min(X,Xs),ordered(Xs)|U]-U).
clause(min(X,[]),U-U).
clause(min(X,[Y|Z]),[le(X,Y),min(X,Z)|U]-U).
clause(le(0,Y),U-U).
clause(le(s(X),s(Y)),[le(X,Y)|U]-U).
```

Only the clauses for delete are not deterministic. This gives rise to the following definition for simp_1.

```
simp_1(psort(X,Y),U-V) :-
    simp_1(permute(X,Y),U-W),
    simp_1(ordered(Y),W-V).
simp_1(permute(X,Y),U-U) :- X==[],Y==[],!.
simp_1(permute(E,F),U-V) :-
    nonvar(E),unify(E,[X|Xs]),
    nonvar(F),unify(F,[Y|Z]),!,
    simp_1(delete([X|Xs],Y,T),U-W),
    simp_1(permute(T,Z),W-V).
simp_1(ordered(X),U-U) :- X==[],!.
simp_1(ordered(E),U-V) :-
    nonvar(E),unify(E,[X|Xs]),!,
    simp_1(min(X,Xs),U-W),
    simp_1(ordered(Xs),W-V).
simp_1(min(X,Y),U-U) :- Y==[],!.
simp_1(min(X,E),U-V) :-
    nonvar(E),unify(E,[Y|Z]),!,
    simp_1(le(X,Y),U-W),
    simp_1(min(X,Z),W-V).
simp_1(le(X,Y),U-U) :- X==0,!.
simp_1(le(E,F),U-V) :-
    nonvar(E),unify(E,s(X)),
    nonvar(F),unify(F,s(Y)),!,
    simp_1(le(X,Y),U-V).
simp_1(X,[X|Y]-Y).
```

6 Final Remarks

Following the above ideas, we have implemented a meta-interpreter in Sicstus Prolog and conducted some preliminary experiments (cf. figure 1 for some benchmarks) to demonstrate its feasibility.

In the first two examples (rev and revI), the interpreter is about four times slower than Prolog. The difference stems from the interpretive overhead incurred by our approach, but that is more than compensated by impressive speedups in the case of generate-and-test programs (psort and queens). (The definition for queens is given in appendix A.)

Theoretically, it is possible to construct a new machine that minimizes the overhead due to simplification, but such an approach may introduce other types of drawbacks as well. One the one hand, portability is lost and, on the other hand—since much more effort has gone into the development of standard WAM's—the new machine would very likely be slower than the current or future breed of Prolog compilers. It is for this reason that our approach can compensate for its simplicity by taking advantage of a fast Prolog compiler; in fact benchmarks of figure 2 shows that it is only slightly slower that the

	Prolog with Simplification	Prolog
rev: 100 elements	680	180
rev: 500 elements	16620	4540
revI: 30 elements	740	179
revI: 60 elements	4820	1399
psort: 8 elements	4750	37590
psort: 10 elements	30059	3417000
8-queens	3020	7460
10-queens	3980	103600

rev : naive reverse
revI : naive reverse in inverted mode
psort : permutation sort of a list already sorted in inverse order

Figure 1: Benchmarks in milliseconds on SUN 3

	Prolog with Simplification	A-WAM
rev: 30 elements	30	19
revI: 30 elements	350	210
psort: 8 elements	2450	1500

Figure 2: Comparison with A-WAM on SUN 4

more complex approach taken in the A-WAM. (A similar observation also applies to the implementation of lazy narrowing in Prolog [Che90], compared to other specialized machines.)

The difference with A-WAM is that simplification does not simulate full rewriting, either because it can't take into account the deletion of lhs variables, or because our notion of deterministic clauses is too weak to consider conditional term rewriting. The first restriction does not seem very serious, but the second restriction may limit some practical applications; for instance, conditional term rewriting with the quicksort program is *a priori* not possible. However, rules that are conditional on simple tests (as in the case of quicksort) can actually be considered in a straightforward way (see appendix B).

Acknowledgements The authors would like to thank the anonymous referees for their helpful comments. The second author was visiting the IDA department at the University of Linköping, Sweden, while the final version of this paper is written, and he expresses hereby his warmest thanks to Prof. Maluszynski and the LOGPRO group for the nice work environment they have been providing all along his stay.

References

[BCM89] P.G. Bosco, C. Cecchi, and C. Moiso. An extension of WAM for K-LEAF. In *Proceedings of the 6th International Conference on Logic Programming, Lisboa*, pages 318–333, 1989.

[BGM87] P.G. Bosco, E. Giovannetti, and G. Moiso. Refined strategies for semantic unification. In *TAPSOFT'87, LNCS 150*, pages 276–290, 1987.

[BL86] M. Bellia and G. Levi. The relation between logic and functional languages. *Journal of Logic Programming*, 3:217–236, 1986.

[Che90] P.H. Cheong. Compiling lazy narrowing into prolog. Technical Report 25, LIENS, 1990. To appear in *Journal of New Generation Computing*.

[Der83] P. Deransart. An operational semantics of prolog programs. In *Programmation Logique, Perros-Guirec, CNET-lannion*, 1983.

[Fri85] L. Fribourg. SLOG: A logic programming language interpreter based on clausal superposition and rewriting. In *Proceedings of the IEEE Symposium on Logic Programming, Boston*, pages 172–184, 1985.

[Fri88] L. Fribourg. Prolog with simplification. In K. Fuchi and M. Nivat, editors, *Programming of future generation computers*, pages 161–183. Elsevier Science Publishers B.V. (North Holland), 1988.

[Han90] M. Hanus. Compiling logic programs with equality. In *PLILP'90, LNCS 456*, pages 387–401, 1990.

[Höl88] S. Hölldobler. From paramodulation to narrowing. In *Proceedings of the 5th International Conference on Logic Programming, Seattle*, pages 327–342, 1988.

[KLMR90] H. Kuchen, R. Loogen, J.J. Moreno, and M. Rodríguez. Graph-based implementation of a functional logic language. In *ESOP' 90, LNCS 432*, pages 279–290, 1990.

[LPB+87] G. Levi, C. Palamidessi, P.G. Bosco, E. Giovannetti, and C. Moiso. A complete semantic characterization of K-LEAF, a logic language with partial functions. In *Proceedings of the IEEE Symposium on Logic Programming*, pages 318–327, 1987.

[Müc90] A. Mück. The compilation of narrowing. In *PLILP'90, Springer LNCS 456*, 1990.

[Red85] U.S. Reddy. Narrowing as the operational semantics of functional languages. In *Proceedings of the IEEE Symposium on Logic Programming, Boston*, pages 138–151, 1985.

[vEY87] M.H. van Emdem and K. Yukawa. Logic programming with equations. *Journal of Logic Programming*, 4:265–288, 1987.

[You89]   Y.H. You. Enumerating outer narrowing derivations for constructor-based term rewriting systems. *Journal of Symbolic Computation*, 7:319–343, 1989.

Appendix A

We adapt from [Fri88] the following formulation of the k-queens problem:

```
clause(queens(X,Y),[permute(X,Y),safe(Y)|U]-U).
clause(permute([],[]),U-U).
clause(permute([X|Xs],[Y|Z]),[delete([X|Xs],Y,T),permute(T,Z)|U]-U).
clause(delete([X|Xs],X,Xs),U-U).
clause(delete([X|Xs],Y,[X|Z]),[delete(Xs,Y,Z)|U]-U).
clause(safe([]),U-U).
clause(safe([X|Y]),[nodiag(X,Y,1),safe(Y)|U]-U).
clause(nodiag(X,[],N),U-U).
clause(nodiag(X,[Y|Z],N),[noattack(X,Y,N),nodiag(X,Z,N+1)|U]-U).
clause(noattack(X,Y,N),
        [prolog_eval(N =\= X-Y),prolog_eval(N =\= Y-X)|U]-U).
clause(prolog_eval(P),U-U) :- call(P).
```

where prolog_eval is used to indicate a call to a Prolog primitive. All clauses are deterministic, except those for delete. This yields the following definition for simp_1:

```
simp_1(queens(X,Y),[permute(X,Y),safe(Y)|U]-U) :- !.
simp_1(permute(X,Y),U-U) :- X==[],Y==[],!.
simp_1(permute(E,F),U-V) :-
        nonvar(E),unify(E,[X|Xs]),
        nonvar(F),unify(F,[Y|Z]),!,
        simp_1(delete([X|Xs],Y,T),U-W),
        simp_1(permute(T,Z),W-V).
simp_1(safe(A),U-U) :- A==[],!.
simp_1(safe(A),U-V) :-
        nonvar(A),unify(A,[X|Y]),!,
        simp_1(nodiag(X,Y,1),U-W),
        simp_1(safe(Y),W-V).
simp_1(nodiag(X,A,N),U-U) :- A==[],!.
simp_1(nodiag(X,A,N),U-V) :-
        nonvar(A),unify(A,[Y|Z]),!,
        simp_1(noattack(X,Y,N),U-W),
        simp_1(nodiag(X,Z,N+1),W-V).
simp_1(noattack(X,Y,N),U-V) :-
        simp_1(prolog_eval(N =\= X-Y),U-W),
        simp_1(prolog_eval(N =\= Y-X),W-V).
simp_1(prolog_eval(P),U-U) :- call(P),!.
simp_1(X,[X|U]-U).
```

Note that the simplification mechanism treats prolog_eval(P) in a different manner. If call(P) is successfully executed by Prolog, then simplification succeeds with the new bindings of variables. Otherwise, prolog_eval(P) is considered to be simplified (which is taken care of by the last simp_1 clause).

Appendix B

Quicksort can be defined by the following flat clauses:

```
clause(quicksort([])=[],U-U).
clause(quicksort([X|Xs])=Ys,[partition(Xs,X)=(Littles,Bigs),
                             quicksort(Littles,Ls),
                             quicksort(Bigs,Bs),
                             append(Ls,[X|Bs])=Ys|U]-U).
clause(partition([],Y)=([],[]),U-U).
clause(partition([X|Xs],Y)=([X|Ls],Bs),
        [prolog_eval(X=<Y),partition(Xs,Y)=(Ls,Bs)|U]-U).
clause(partition([X|Xs],Y)=(Ls,[X|Bs]),
        [prolog_eval(X>Y),partition(Xs,Y)=(Ls,Bs)|U]-U).
```

The last two clauses for partition are not deterministic in the sense of section 2.2, although the arithmetic tests in these clauses allow to discriminate between the two. It is immediate to consider these clauses as deterministic, provided we use the arithmetic tests as a sort of guards:

```
simp_1(partition(A,Y)=B,U-V) :-
     nonvar(A),unify(A,[X|Xs]),
     call(X=<Y),!,
     unify(B,([X|Ls],Bs)),
     simp_1(partition(Xs,Y)=(Ls,Bs),U-V).
simp_1(partition(A,Y)=B,U-V) :-
     nonvar(A),unify(A,[X|Xs]),
     call(X>Y),!,
     unify(B,(Ls,[X|Bs])),
     simp_1(partition(Xs,Y)=(Ls,Bs),U-V).
```

Note that this method is correct if the guards are mutually exclusive and they consist of simple (non recursive) tests.

Lazy Evaluation in Logic*

Sergio Antoy

Portland State University
Department of Computer Science
and
Center for Software Quality Research
Portland, OR 97207

Abstract

We generalize, simplify, and optimize the technique for doing lazy evaluation in logic programming proposed in [10]. Our approach is based on term rewriting and a strategy for the design of rewrite rules proposed in [1]. We describe a novel transformation scheme from a rewriting system $\mathcal{R}$ to a logic program $\mathcal{H}$, characterized by the fact that a reduction step is performed by $\mathcal{H}$ only when needed. Improvements over [10] include: our approach can be applied to any constructor-based data type, the soundness and completeness conditions of the reduce rules are automatically provided for deterministic computations, the complexity of a logic computation is bounded by a simple function of the reduction being performed, and backtracking never occurs.

1 Introduction

Lazy evaluation is a computational strategy which ensures that expressions are evaluated only when needed. This strategy is appealing because it makes efficient use of resources and, more interesting, it supports an expressive programming style which includes infinite structures. [7] regards lazy evaluation as a fundamental attribute of modern functional programming languages. [10] presents a technique for doing lazy evaluation in logic through a set of motivating and illustrative examples.

In this note we discuss an approach to lazy evaluation in logic which generalizes, simplifies, and optimizes the technique presented in [10]. Our starting point is a rewriting system and a strategy for the design of rewrite rules introduced in [1]. We define a

---

*This material is based upon work supported by the National Science Foundation Grant No. CCR-8908565.

transformation scheme from a rewriting system to a logic program. In our approach a reduction step is performed only when needed and the complexity of a logic computation is a simple function of a reduction sequence.

We compare our method briefly with *vEM* [15] and more extensively with the method presented in [10]. The Appendix contains a significant example. The proofs of our results, omitted because of limitations of space, are in [2].

2 Preliminaries and Notations

We consider a term rewriting system $\mathcal{R}$ with a many-sorted signature partitioned into a set $\mathcal{C}$ of *constructors* and a set $\mathcal{D}$ of *(defined) operations*. For each sort s, we assume an arbitrary, but fixed ordering, called *standard*, among the constructors of s. The rules of $\mathcal{R}$ are characterized below.

$\mathcal{X}$ denotes a set of sorted variables and variables are denoted by upper case letters. $\square$ denotes a nullary, sort-overloaded symbol called *place* distinct from constants and variables. For any signature Σ, $T(\Sigma)$ is the set of terms built over Σ. Any term referred to in this note type checks. The leading symbol or principal functor of a term t is called *root* of t. The elements of $T(\mathcal{C})$, $T(\mathcal{C} \cup \mathcal{X})$, $T(\mathcal{C} \cup \mathcal{D})$, and $T(\mathcal{C} \cup \mathcal{D} \cup \mathcal{X} \cup \{\square\})$ are respectively called *values*, *constructor terms*, *ground terms*, and *templates*. A term t of the form $f(x_1, \ldots, x_n)$, where f is an operation and the arguments are constructor terms is called *f-rooted constructor term*. The rules in $\mathcal{R}$ are characterized by having an operation-rooted constructor term as the left side.

An *occurrence* is denoted by a string of dot-separated integers or occurrences, i.e. the symbol "$\cdot$" is overloaded. For terms t and t', occurrence p, and variable X, t/p denotes the subterm of t at p; $t[p \leftarrow t']$ denotes occurrence substitution; and $\{t'/X\}t$ denotes variable substitution. If α is a reduction sequence, then $|\alpha|$ denotes the number of reduction steps in α; if t is a term, then $|t|$ denotes the number of occurrences in t. More specific notation will be introduced where needed.

3 Translation

This section describes the transformation of a rewriting system into a set of Horn clauses. The algorithm is based on a representation of the rules which, in addition to the rules themselves, stores also certain "decisions" made during the design of the rules.

3.1 Definitional Trees

The defining rules of an operation f are represented by a tree-like structure formally defined below. *Rule*, *var*, and *ind* are uninterpreted functions. This structure is a by-product of a definitional process, called *binary choice strategy*, described in [1]. Roughly speaking, a *rule* node of a *pdt* is a leaf representing a rule, while *var* and *ind* nodes are branches respectively representing "variable" and "inductive" choices.

Definition 1 d is a *partial definitional tree*, or *pdt*, if and only if one of the following cases holds:

$d = rule(l, r)$, where l is a template in which there are no occurrences of places and r is a term whose variables are in l, too.

$d = var(t, o, v, d')$, where t is a template, o is the occurrence of a place in t, v is a fresh variable, and d' is a *pdt* such that the template in its root is $t[o \leftarrow v]$.

$d = ind(t, o, d_1, \ldots, d_k)$, where t is a template, o is the occurrence of a place in t, the sort of t/o has constructors $c_1, \ldots, c_k$ in standard ordering, and for all i, $1 \leq i \leq k$, d_i is a *pdt* such that the template in its root is $t[o \leftarrow c_i(\Box, \ldots, \Box)]$.

Definition 2 A *definitional tree* is a pair $\langle f, d \rangle$ where f is an operation and d is a *pdt* such that the template in d is $f(\Box, \ldots, \Box)$.

If $\langle f, d \rangle$ is a definitional tree, with some abuse of terminology, we call d a definitional tree of f. Note that templates do not need explicit representation in a definitional tree. However, we make them explicit if this simplifies the discussion. For example, consider an operation *append* for concatenating lists. The constructors of the sort *list* are (in standard ordering) [] and the mixfix symbol [...|...]. The rules of *append* are:

$$append([], Y) \rightarrow Y$$
$$append([E|X], Y) \rightarrow [E|append(X, Y)] \tag{1}$$

Some design of the operation *append* using the binary choice procedure generates the following definitional tree:

$$var(2, Y, ind(1, rule(Y), var(1 \cdot 1, E, var(1 \cdot 2, X, rule([E|append(X, Y)]))))) \tag{2}$$

Figure 1, which shows the same tree in pictorial representation, gives an intuitive idea of the definitional process. Throughout Section 5, unless otherwise stated, we assume that the term rewriting systems we discuss are designed using the binary choice strategy and consequently every operation has a definitional tree.

3.2 Translation Algorithm

The algorithm described below is the key component of the transformation of a rewriting system $\mathcal{R}$ into a logic program $\mathcal{H}_\mathcal{R}$ which evaluates lazily the ground terms of $\mathcal{R}$. The algorithm takes a definitional tree of some operation and generates a set of clauses containing only one binary predicate denoted by the infix symbol $\Rightarrow$. Clauses are denoted using Edinburgh Prolog syntax.

Procedure *Translate(d)*.

    Input: d: a *pdt*,
    Output: $\mathcal{H}_d$: a set of Horn clauses.

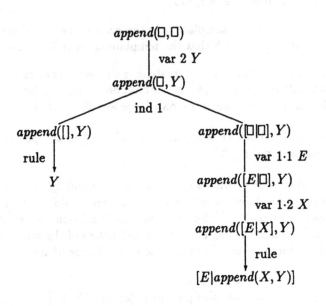

Figure 1: Pictorial representation of a definitional tree of the operation *append* defined in the text. The templates are explicitly represented and the information stored in a node has been spatially rearranged to emphasize the definitional process.

Execute either of the following cases depending on d:

1. $d = rule(l, r)$: execute either of the following cases depending on r:

    1.1. r is constructor-rooted: output $l \Rightarrow r$.

    1.2. r is a operation-rooted: output $l \Rightarrow X :- r \Rightarrow X$, where X is a fresh variable.

    1.3. r is a variable:

    1.3.1. for each constructor c of the same sort of r output $\{c(X_1, \ldots, X_k)/r\}$ $(l \Rightarrow r)$, where for all i, $1 \le i \le k$, X_i is a fresh variable, and

    1.3.2. output $l \Rightarrow X :- r \Rightarrow X$, where X is a fresh variable.

2. $d = var(t, o, v, d')$: execute $Translate(d')$.

3. $d = ind(t, o, d_1, \ldots, d_k)$:

    3.1. for all i, $1 \le i \le k$, execute $Translate(d_i)$, and

    3.2. instantiate all places of t with fresh variables and output $t[o \leftarrow Z] \Rightarrow X :- Z \Rightarrow Y, t[o \leftarrow Y] \Rightarrow X$, where X, Y, and Z are fresh variables too.

End *Translate*.

For example, the clauses generated from the definitional tree of Figure 1 are:

```
append([],[]) => [].
append([],[A|B]) => [A|B].
append([],A) => B :- A => B.
append([A|B],C) => [A|append(B,C)].
append(A,B) => C :- A => D, append(D,B) => C.
```
(3)

We execute the procedure *Translate* once for each operation f of $\mathcal{R}$—input a definitional tree of f. The output constitutes the core of $\mathcal{H}_{\mathcal{R}}$. A few additional clauses needed to complete $\mathcal{H}_{\mathcal{R}}$ are described in Section 5.

4   Analysis of $\Rightarrow$

The analysis of the relation $\Rightarrow$ requires the introduction of two new concepts: *state of a computation* and *λ-reduction*.

4.1   State of Computation

A state of a computation is a tree which generalizes *proof trees* as defined, e.g., in [14]. We need this generalization to reason about non-terminating computations. Goal clauses are denoted by prefixing the symbol ":−" to a list of positive literals.

Definition 3 Let :− $\mathcal{N}$ be a goal clause, for some positive literal $\mathcal{N}$, and $\mathcal{L}$ be a logic program. If there exists a clause ξ in $\mathcal{L}$ such that :− $\mathcal{N}$ and ξ resolve with resolvent :− $\mathcal{N}_1, \ldots, \mathcal{N}_k$, for some $k \geq 0$, we call *expansion* of :− $\mathcal{N}$ by ξ the ordered tree $\mathcal{T}$ which has :− $\mathcal{N}$ as root and :− $\mathcal{N}_1, \ldots,$:− $\mathcal{N}_k$ as leaves. Note that if $k = 0$, then $\mathcal{T}$ has just one leaf, the empty clause.

Definition 4 Let $\mathcal{L}$ be a logic program and $\mathcal{N}$ a positive literal. An ordered tree $\mathcal{T}$ is *state of a computation* of :− $\mathcal{N}$ in $\mathcal{L}$ if either $\mathcal{T}$ consists of the single node :− $\mathcal{N}$, or $\mathcal{T}$ is obtained by expanding the leftmost leaf that is different from the empty clause of some state of a computation of :− $\mathcal{N}$ in $\mathcal{L}$.

The nodes of a state of a computation are labeled according to their expansion state as follows. A node is *empty* if it is the empty clause, *unexpanded* if it is a leaf different from the empty clause, and *expanded* if it is not a leaf. Furthermore, a node $\mathcal{N}$ is *solved* if the subtree rooted at $\mathcal{N}$ is finite and all its leaves are empty, otherwise it is *unsolved*. A proof tree is a state of a computation whose root is solved and represents a computation successfully terminated. If $\mathcal{T}$ is a state of a computation, then $|\mathcal{T}|$ denotes the number of non-empty nodes in $\mathcal{T}$.

4.2   λ-reductions

A computation of :− $t \Rightarrow u$ in $\mathcal{H}_{\mathcal{R}}$, where t and u are respectively a ground term and a term in $\mathcal{R}$, computes a reduction sequence of t in $\mathcal{R}$. This reduction sequence, crucial to our approach, is called λ-reduction. An operational definition of this concept is obtained via a procedure executing a traversal of a state of a computation.

The *classification* of a clause is a labeling of the clauses of $\mathcal{H}_R$ resulting from the procedure *Translate*. We call *n-clause* a clause generated by the execution of instruction n of the procedure *Translate* and *n-goal* (or *n-node*) any goal of a state of a computation expanded by an n-clause.

Procedure $Reduce(\mathcal{T}, \mu, o, \lambda)$

    Input: $\mathcal{T}$: a state of a computation,

                μ: a reduction sequence in $\mathcal{R}$,

                o: an occurrence.

    Output: λ: a reduction sequence in $\mathcal{R}$.

Let $\mathcal{N}$ denote the root of $\mathcal{T}$ and let t denote the term reduced by μ.

If $\mathcal{N}$ is empty, then abort; else $\mathcal{N}$ must be equal to $:- t_l \Rightarrow t_r$, for some terms t_l and t_r.

If $\mathcal{N}$ is unexpanded, then abort; else $\mathcal{N}$ has been expanded by some clause ξ of $\mathcal{H}_R$ with match σ.

Execute either of the following cases depending on the classification of $\mathcal{N}$.

1. $\mathcal{N}$ is either a 1.1-node or a 1.3.1-node: Then ξ must be equal to $l \Rightarrow r$, for some terms l and r.
   Set λ to extend μ with the reduction step $t'[o \leftarrow \sigma(r)]$.

2. $\mathcal{N}$ is either a 1.2-node or a 1.3.2-node: Then ξ must be equal to $l \Rightarrow X :- r \Rightarrow X$, for some terms l and r, and $\mathcal{N}$ must have only one child, say $\mathcal{T}'$.
   Set μ' to extend μ with the reduction step $t'[o \leftarrow \sigma(r)]$ and execute $Reduce(\mathcal{T}', \mu', o, \lambda)$

3. $\mathcal{N}$ is a 3.2-node: Then ξ must be equal to $t_Z \Rightarrow X :- Z \Rightarrow Y, t_Y \Rightarrow X$, for some terms t_Z and t_Y, and let p denote the occurrence of Z in t_Z. Also, $\mathcal{N}$ must have two children, say $\mathcal{T}'$ and $\mathcal{T}''$.
   Execute $Reduce(\mathcal{T}', \mu, o{\cdot}p, \lambda')$ and execute $Reduce(\mathcal{T}'', \lambda', o, \lambda)$.

End *Reduce*.

Reduce terminates normally on a state of a computation $\mathcal{T}$ if and only if $\mathcal{T}$ is a proof tree, i.e. its root goal is solved.

Definition 5 Let $\mathcal{T}$ be a proof tree of $:- t \Rightarrow u$ in $\mathcal{H}_R$, where t and u are respectively a ground term and a term in $\mathcal{R}$. The λ-*reduction* of t computed by $\mathcal{T}$ is the value of λ computed by $Reduce(\mathcal{T}, t \xrightarrow{*} t, \Lambda, \lambda)$.

4.3 Properties of $\Rightarrow$

The relation $\Rightarrow$ is a formalization and generalization of the concept of *useful simplification* suggested in [10]. Moreover, any clause selected in $\mathcal{H}_R$ during a computation never needs to be reconsidered, and any reduction computed by the execution of $\mathcal{H}_R$ is a necessary one.

If $t \Rightarrow t'$, for some ground terms t and t', then let T denote a proof tree of $:- t \Rightarrow t'$ and λ the λ-reduction of t computed by T. The following conditions hold.

1. t is operation-rooted;

2. λ reduces t to t';

3. t' is the only constructor-rooted term in λ;

4. $|T| < 2|\lambda|$.

The *completeness* and *parsimony* properties [1] provided by the binary choice strategy imply that for any ground operation-rooted *constructor* term t there exists *exactly* one rule $l \to r$ of $\mathcal{R}$ such that $t = \sigma(l)$, for some substitution σ. This property is extended by the procedure *Translate* to any ground operation-rooted (*arbitrary*) term in the following sense.

If t is a ground operation-rooted term and T a variable, then the following conditions hold:

1. *(existence)* there exists a clause in $\mathcal{H}_\mathcal{R}$ whose head matches $t \Rightarrow T$;

2. *(uniqueness)* no clause in $\mathcal{H}_\mathcal{R}$ witnessing condition 1, except the first in the order of generation of the procedure Translate, satisfies the call $t \Rightarrow T$.

This implies that a goal of the form $:- t \Rightarrow T$, where t is operation-rooted and T is a variable, never fails in $\mathcal{H}_\mathcal{R}$, though it may not succeeds. In particular, for such goals, backtracking is unnecessary and proof trees are unique.

Rewriting systems designed with the binary choice strategy are *left-linear* and *non-ambiguous* and in such systems, any term t has a set of redexes that must be reduced to compute the normal form of t [8]. The crucial issue is the definition of the *descendants* of a redex s in a term t after a reduction step is performed on t. [8] gives a formal, quite technical definition of this concept. [9] offers a less formal, more intuitive formulation. Suppose we have a reduction sequence $t \overset{*}{\to} t'$ and a subterm s of t. The descendants of s in t' are computed as follows. Underline the root of s and perform the reduction sequence $t \overset{*}{\to} t'$. The descendants of s in t' are the subterms of t' which have an underlined root. Then, a redex s in a term t is *needed* if in every reduction of t to normal form a descendant of s is reduced. We also say that a redex s of t is *maximal* if, for any other redex r of t, s is not a proper subterm of r.

If $t \Rightarrow t'$, for some ground terms t and t', then any *redex* reduced in a λ-reduction of t is *maximal* and *needed*.

5 Evaluation

[8] proves that if a term t has a normal form, then repeated rewriting of needed redexes leads to that normal form. Thus, we can use $\Rightarrow$ to attempt computing normal forms. For this task we introduce a new binary relation on terms, denoted by the infix symbol $\twoheadrightarrow$. Intuitively, $\twoheadrightarrow$ is a driver which guides the application of $\Rightarrow$ to appropriate subterms of a term t until the normal form of t is reached.

Definition 6 For all ground terms t and t', $t \twoheadrightarrow t'$ if and only if one of the following conditions holds:

1. $t' = c(t'_1, \ldots, t'_k)$, if $t = c(t_1, \ldots, t_k)$ for some constructor c and terms $t_1, \ldots, t_k$, for some $k \geq 0$, and for all i, $1 \leq i \leq k$, $t_i \twoheadrightarrow t'_i$.

2. $t \Rightarrow t''$ and $t'' \twoheadrightarrow t'$, for some term t''.

The term t of condition 2 of the above definition is operation-rooted, thus the cases of the definition are mutually exclusive. Intuitively, if $t \twoheadrightarrow t'$, then t' is computed from t by "skipping over" constructors and applying $\Rightarrow$ to operation-rooted terms.

The logic implementation of the relation $\twoheadrightarrow$ requires one clause to satisfy condition 2 and, for each constructor in $\mathcal{R}$, one clause to satisfy condition 1. For example, the following clauses must be added to those presented in display 3.

$$\begin{array}{ll} \texttt{[] ->> [].} & \\ \texttt{[A|B] ->> [C|D] :- A ->> C, B ->> D.} & \text{(4)} \\ \texttt{A ->> B :- A => C, C ->> B.} & \end{array}$$

From now on, we assume that $\mathcal{H}_\mathcal{R}$ defines the relation $\Rightarrow$ as well as $\twoheadrightarrow$. Many of the properties we have discussed in the previous section are extended with little modification. We call $\Rightarrow$-*goal* a goal whose predicate is $\Rightarrow$ and $\twoheadrightarrow$-*goal* a goal whose predicate is $\twoheadrightarrow$. We extend the definition of λ-reduction of t by enabling the procedure *Reduce* to handle $\twoheadrightarrow$-nodes.

We classify $\twoheadrightarrow$-nodes according to the cases of Definition 6. If $\mathcal{N}$ is a $\twoheadrightarrow$-node, then, for some terms t and t', $\mathcal{N} = :- t \twoheadrightarrow t'$. If t is constructor-rooted, we say that $\mathcal{N}$ is a *1-node*, otherwise t is operation-rooted and we call it a *2-node*. The case statement of the procedure *Reduce* is extended with the following two new entries.

Extension to the procedure *Reduce*

4. $\mathcal{N}$ is a 1-node: Then $\mathcal{N}$ must have has k children, say $\mathcal{T}_1, \ldots, \mathcal{T}_k$, for some $k \geq 0$. Let $\mu_0 = \mu$, for i from 1 to k execute $Reduce(\mathcal{T}_i, \mu_{i-1}, o \cdot i, \mu_i)$, and let $\lambda = \mu_k$.

5. $\mathcal{N}$ is a 2-node: Then $\mathcal{N}$ must have two children, say $\mathcal{T}'$ and $\mathcal{T}''$. Execute $Reduce(\mathcal{T}', \mu, o, \mu')$ and execute $Reduce(\mathcal{T}'', \mu', o, \lambda)$.

The extended procedure *Reduce* extends the definition of λ-reduction to states of a computation which include $\twoheadrightarrow$-goals. The basic properties of the relation $\Rightarrow$ discussed in the previous section are extended as follows.

If $t \twoheadrightarrow t'$, *for some ground terms* t *and* t', *then let* $\mathcal{T}$ *denote a proof tree of* $:- t \twoheadrightarrow t'$ *and* λ *the* λ-reduction *of* t *computed by* $\mathcal{T}$. *The following conditions hold.*

1. λ *reduces* t *to* t';

2. t' *is a value;*

3. $|\mathcal{T}| \leq 2|\lambda| + |t'|$;

4. Any redex reduced in λ is maximal and needed.

The existence and uniqueness properties of $\Rightarrow$ hold for $\twoheadrightarrow$ too.

If t is a ground operation-rooted term and T a variable, then the following conditions hold.

  1. *(existence) there exists a clause in $\mathcal{H}_\mathcal{R}$ whose head matches $t \twoheadrightarrow T$;*

  2. *(uniqueness) no clause in $\mathcal{H}_\mathcal{R}$ witnessing condition 1, except the first, satisfies the call $t \twoheadrightarrow T$.*

Our main result states that if a term has a normal form in $\mathcal{R}$, then $\mathcal{H}_\mathcal{R}$ computes it.

For any ground term t and value v, if $t \xrightarrow{*} v$, then $t \twoheadrightarrow v$.

The λ-reduction of a term t may reduce several times a redex s if there are several distinct occurrences of s in t. We do not address the problem of retaining the value computed during the first evaluation of s and using it when another occurrence of s must be evaluated. [10] suggests a solution to this problem. This solution can be incorporated in our approach as well.

6 Extensions

The binary choice procedure generates the set of terms used as left sides of the rewrite rules defining some operation. A generalization of the strategy allows us to significantly extend the class of rewriting systems to which our transformation can be applied. For each left side l we consider a set of right sides $\{r_1, \ldots, r_k\}$, for some $k \geq 0$, rather than a single right side. We obtain the set of rules $\{l \rightarrow r_1, \ldots, l \rightarrow r_k\}$. Definitional trees accommodate this change easily if we replace the term in a *rule* node with a set of terms. We call *ruleset* node a node so defined. A *ruleset* node represents the set of rewrite rules sharing a common left side.

The translation of rewrite rules into Horn clauses is minimally affected by this generalization. Only one line of the procedure *Translate* needs to be changed. The change concerns branch 1 of the main case statement, which is reformulated as follows:

  1. $d = ruleset(l, \{r_1, \ldots, r_k\})$: for all j, $1 \leq j \leq k$, set r to r_j and
     execute either of the following cases (1.1 through 1.3) depending on r:

The uniqueness and existence of a match for a goal of the form $:- t \twoheadrightarrow T$ is not preserved by the generalization. but this is an obvious and desired consequence of the generalization. We conjecture that $\xrightarrow{*}$ and $\twoheadrightarrow$ still define the same relation. The problem in proving this claim is that the notion of needed redex is defined only for *regular* [8] (*orthogonal* [9]) systems. The rewriting systems we are discussing violate a key assumption of regularity, i.e. they are *ambiguous* according to [8] and *inconsistent* according to [11].

A more important consideration is that without the unique match condition satisfied by both $\Rightarrow$ and $\twoheadrightarrow$, the choice of a reduction rule, rather than only a redex, becomes critical, thus we need a concept stronger than *needed redex*. It seems that substantial preliminary work is required to determine sufficient conditions for a computationally efficient solution. An interim solution lies in proving that the *soundness* and *completeness* conditions defined in [10] hold for $\mathcal{H}_\mathcal{R}$.

7 Related work and concluding remarks

The implementation of term rewriting systems in Prolog has received considerable attention. Early work, summarized in [13], was directed toward prototyping algebraic specifications through their direct implementation and/or through the use of Prolog-supported automatic theorem provers. More recent efforts, [4] and [6], aim at the integration of functional and/or equational programs into logic programs. An extensive analysis and comparison appears in [3]. [15] presents two approaches to this problem, one of which, referred to as *vEM*, is concerned with a class of systems largely contained in the one we consider—any operation discussed in the *vEM* examples of [15] has a definitional tree. With respect to *vEM* we do not need *Noetherianity*, nor *strong canonicity*, nor do we place any restriction on or require any preprocessing of the terms to be reduced. As far as efficiency is concerned, the size of a *vEM* proof tree for the evaluation of a term t is at least equal to the length of the reduction sequence it computes, and the preprocessing of t implies the traversal of t. Thus, in the worst case, our method is slower than *vEM* by a factor less than 2. On the other hand, *vEM* cannot handle some of our cases, for example those involving infinite structures, see the Appendix, and there are cases accepted by both methods where *vEM* is either slower than our method by a factor arbitrarily large or, more likely, will eventually run out of available resources.

The most interesting comparison is with [10] which shares our goals. We overcome a number of limitations in this approach. In [10], a logic program must satisfy the *soundness* and *completeness* conditions whose proofs are non-trivial even for simple examples. We require these proofs only for non-deterministic computations and conjecture that for some of these they might be unnecessary anyway. Narain's method makes the operator "[...|...]" strict in its first argument, and in some cases, similar to *vEM*, this makes Narain's approach slower than ours by a factor arbitrarily large or not usable at all. Since Narain's presentation does contain efficiency measures and his method is not related to rewriting, our measure of performance based on the size of a term and the length of its reduction sequence cannot be easily compared. However, a simple experiment points out some differences. Let t be $append([a, b], [c, d])$, where a, b, c, and d are irreducible atoms. Both Narain's method and ours build a proof tree with 15 nodes for the evaluation of t. During the construction of the proof tree with Narain's method, there are 10 failures and 6 instances of backtracking. Our method guarantees that neither failures nor backtracking ever take place.

Appendix

The following example is adapted from [5, p.52], see also [12, p.140] and [10, p.262]. The operations *sieve*, *filter*, and *show* are incomplete in the sense of [1]. To apply the procedure *Translate* and insure the soundness and completeness conditions of [10], we could complete these operations with arbitrary rules. The clauses generated from these arbitrary rules are unnecessary in the logic program and can be dropped, since the prime numbers are infinite and the list of primes used by the computation never becomes empty. In reality, we use a refinement of the procedure *Translate* which accepts definitional trees with "empty" *rule* nodes. We use the Prolog implementation of numbers, thus matching a term with an occurrence of *succ* (successor) needs some adjustment and the predicate *factor* is defined, using built-in predicates, only in the logic program.

$$primes \rightarrow sieve(ints\_from(2))$$
$$sieve([A|B]) \rightarrow [A|sieve(filter(B, A))]$$
$$filter([A|B], C) \rightarrow if(factor(C, A), filter(B, C), [A|filter(B, C)])$$
$$ints\_from(A) \rightarrow [A|ints\_from(succ(A))]$$
$$show(0, [A|B]) \rightarrow []$$
$$show(succ(A), [B|C]) \rightarrow [B|show(A, C)]$$
$$if(false, A, B) \rightarrow B$$
$$if(true, A, B) \rightarrow A$$

The procedure *Translate* generates from the above rules the following clauses.

```
primes => A :- sieve(ints_from(2)) => A .
sieve([A|B]) => [A|sieve(filter(B,A))] .
sieve(A) => B :- A => C , sieve(C) => B .
filter([A|B],C) => D :- if(factor(C,A),filter(B,C),[A|filter(B,C)]) => D .
filter(A,B) => C :- A => D , filter(D,B) => C .
ints_from(A) => [A|ints_from(B)] :- B is A+1 .
show(0,[A|B]) => [] .
show(A,[B|C]) => [B|show(D,C)] :- A > 0 , D is A-1 .
show(A,[B|C]) => D :- A => E , show(E,[B|C]) => D .
show(A,B) => C :- B => D , show(A,D) => C .
if(false,A,[]) => [] .
if(false,A,[B|C]) => [B|C] .
if(false,A,B) => C :- B => C .
if(true,[],A) => [] .
if(true,[A|B],C) => [A|B] .
if(true,A,B) => C :- A => C .
if(A,B,C) => D :- A => E , if(E,B,C) => D .
```

The remaining clauses of the program follow. *factor* is defined using built-in predicates. In the underlying implementation of integers, every number is a normal form, thus $x \twoheadrightarrow x$, for any integer x.

```
factor(A,B) => true :- 0 is B mod A , ! .
factor(_,_) => false .
[] ->> [] .
[A|B] ->> [C|D] :- A ->> C , B ->> D .
A ->> A :- integer(A) .
A ->> B :- A => C , C ->> B .
```

For example, the call $show(10, primes) \twoheadrightarrow T$ instantiates T to [2,3,5,7,11,13,17,19,23,29].

References

[1] Sergio Antoy. Design strategies for rewrite rules. In *CTRS'90*, Montreal, Canada, June 10-14 1990.

[2] Sergio Antoy. Lazy evaluation in logic. Technical Report 90-17, Rev. 1, Portland State University, Portland, OR, March 1990.

[3] L. G. Bouma and H. R. Walters. Implementing algebraic specifications. In J. A. Bergstra, J. Heering, and P. Klint, editors, *Algebraic Specifications*, chapter 5. Addison-Wesley, Wokingham, England, 1989.

[4] Doug DeGroot and Gary Lindstrom, editors. *Logic Programming: Functions, Relations, and Equations*, Englewood Cliffs, NJ, 1986. Prentice-Hall.

[5] Joseph A. Goguen and Timothy Winkler. Introducing OBJ3. Technical Report SRI-CSL-88-9, SRI International, Menlo Park, CA, 1988.

[6] S. Hölldobler. *Foundation of Equational Logic Programming*. Springer-Verlag, Berlin, 1989. Lect. Notes in Artificial Intelligence, Vol. 353.

[7] Paul Hudak. Conception, evolution, and application of functional programming languages. *Computing Surveys*, 21:359–411, 1989.

[8] Gérard Huet and Jean-Jacques Lévy. Call by need computations in non-ambiguous linear term rewriting systems. Technical Report 359, INRIA, Le Chesnay, France, 1979.

[9] Jan Willem Klop and Aart Middledorp. Sequentiality in orthogonal term rewriting systems. Technical Report CS-R8932, Stichting Mathematisch Centrum, Amsterdam, 1989.

[10] Sanjai Narain. A technique for doing lazy evaluation in logic. *The Journal of Logic Programming*, 3:259–276, 1986.

[11] Michael J. O'Donnell. Computing in systems described by equations. Springer-Verlag, 1977. Lect. Notes in Comp. Sci., Vol. 58.

[12] Michael J. O'Donnell. *Equational Logic as a Programming Language*. MIT Press, 1985.

[13] H. Petzsch. Automatic prototyping of algebraic specifications using Prolog. In *Recent Trends in Data Type Specification*, pages 207–223. 3rd Workshop on Theory and Applications of Abstract Data Types, Springer-Verlag, 1985.

[14] Leon Sterling and Ehud Shapiro. *The Art of Prolog*. The MIT Press, Cambridge, MA, 1986.

[15] Maarten H. van Emden and Keitaro Yukawa. Logic programming with equations. *The Journal of Logic Programming*, 4:265–288, 1987.

S-SLD-resolution — An Operational Semantics for Logic Programs with External Procedures

Johan Boye

Dept. of Computer and Information Science
University of Linköping
S-58183 Linköping, Sweden
e-mail:johbo@ida.liu.se

Abstract

This paper presents a new operational semantics for logic programs with external procedures, introduced in [BM88]. A new resolution procedure *S-SLD-resolution* is defined, in which each step of computation is characterized by a goal and a set of equational constraints, whose satisfiability cannot be decided with the information at hand. This approach improves the completeness of the resulting system, since further computation may result in the information needed to solve some earlier unsolved constraints. We also state a sufficient condition to distinguish a class of programs where no unsolved constraints will remain at the end of computation.

1 Introduction

We will address the problem of defining an operational semantics for logic programs with external procedures, a formalism presented in [BM88, Bon89]. In this approach the external procedures, which can be written in any language, are regarded as "black boxes" that reduce ground terms. Under this assumption an external procedure implicitly defines an infinite set Eq of ground equational axioms. This allows us to consider a logic program with external procedures as an equational logic program [JLM84]. However, since the equational axioms are not explicitly given, it is not always possible to compute a complete set of Eq-unifiers of two terms. Thus unification in this formalism becomes inherently incomplete.

[BM88] introduces an incomplete unification algorithm *S-unify*, which has three possible outcomes: success, fail and "don't know". The "don't know" case occurs when S-unify encounters equations whose solvability cannot be determined, because the external procedures will not reduce non-ground terms. An operational semantics based on *pr-resolution* is developed in [Bon90], which basically is SLD-resolution with Robinson unification exchanged for S-unification. When a "don't know" answer is returned from S-unify, the resolution process is cancelled.

In this paper we will present an extension of S-unification that, instead of answering "don't know", simply returns the set of equational constraints which cannot be solved with the information at hand. This algorithm is then used to define a new resolution procedure, *S-SLD-resolution*, where every step of the computation is characterized by a goal and a set of equational constraints. Thus S-SLD-resolution formalizes the natural idea of delaying constraints in the computational process. This approach resembles the one taken by CLP-languages [JL87], and increases the completeness of the resulting system since further computation may decide the solvability of the constraints. To guarantee that no unsolved constraints remain at the end of computation, we also give a sufficient condition for programs. The condition is defined by using analogies with attribute grammars, presented in [DM85, DM91].

2 Preliminaries

We let our external procedures be represented by certain defined or *external* symbols, and a term built from an external symbol is called a *functional call*. A functional call with ground arguments may be reduced to a ground term by invocation of the corresponding procedure. A variable occurring in a functional call is said to be *in the scope* of the external functor. A term built entirely from constructors and variables is called a *structure*. We have three types of relational symbols: the internal, the external and $\doteq$, the equality symbol, which will be considered neither internal nor external (the symbol $=$ will be used to denote syntactical identity). An *amalgamated* program is a finite set of definite clauses, where the top symbol of the head of each clause is an internal relational symbol.

3 S-unification

S-unification, presented in [BM88], is a unification algorithm solving equations modulo Eq, i.e. finding a *complete set* of Eq-unifiers to the equations. As motivated in the introduction, this algorithm has three possible outcomes when trying to unify two terms t_1 and t_2:

- *success* and preferably a complete set of unifiers;

- *fail* (if the algorithm finds that t_1 and t_2 are not unifiable);

- *don't know* (if the algorithm is unable to determine whether t_1 and t_2 are unifiable or not).

The core of S-unify is an algorithm which transforms a set U of equations to *normal form*. We first define this notion:

Definition 1 Let N be a finite set of equations. N is in *normal form* iff:

- $N = \{x_1 \doteq s_1, \ldots, x_n \doteq s_n, r_1 \doteq d_1, \ldots, r_m \doteq d_m\}$, where $n, m \geq 0$,

- each x_i is a variable which occurs only once in N,

- each d_i is a nonground functional call which occurs only once in N,

- all s_i and r_j are structures.

If, in addition, $r_1 \ldots r_m$ are all distinct variables and there exists a permutation $\wp$ on $\{1 \ldots m\}$ such that $1 \leq i \leq j \leq m$ implies $r_{\wp i} \notin var(d_{\wp j})$, then N is in *solved form*. In this case the substitution σ_N is defined by:

$$\sigma_N = \{x_1/s_1, \ldots, x_n/s_n\}\{r_{\wp 1}/d_{\wp 1}\} \cdots \{r_{\wp m}/d_{\wp m}\} \qquad \square$$

Example 1 Let s and 0 be constructors and let f be an external functor. Then the sets:

$$N_1 = \{x \doteq s(y),\ y \doteq f(y),\ 0 \doteq f(s(y))\} \text{ and}$$
$$N_2 = \{x \doteq s(y),\ y \doteq f(s(w)),\ w \doteq f(z)\}$$

are both in normal form. In addition, N_2 is in solved form, and:

$$\sigma_{N_2} = \{x/s(f(s(f(z)))),\ y/f(s(f(z))),\ w/f(z)\} \qquad \square$$

The idea is that if the normal form algorithm returns a set of equations N that is in solved form, then $\{\sigma_N\}$ is a complete set of Eq-unifiers for N. If the result is failure then $\emptyset$ is a complete set of Eq-unifers for N. If the result is a set N in normal form but not in solved form, it is impossible to determine a complete set of Eq-unifiers for N.

Algorithm 1 (Normal form algorithm)
Initially, let $U_0 = U$ and $i = 0$. Repeatedly do the following: Select any equation $u \doteq t$ in U_i such that one of rule 1 to 9 applies. If no such equation exists then stop with U_i as result. Otherwise perform the corresponding action, that is; stop with failure or construct U_{i+1} from U_i. Then increment i by 1.

1. $u = t$. Remove $u \doteq t$.

2. u is a variable which occurs elsewhere in U_i and t is a structure distinct from u. If u occurs in t then stop with failure, otherwise replace all other occurrences of u by t (i.e. leave $u \doteq t$ unchanged).

3. u is a nonvariable and t is a variable. Replace $u \doteq t$ by $t \doteq u$.

4. $u = c_1(u_1 \ldots u_n)$ and $t = c_2(t_1 \ldots t_n)$ where c_1 and c_2 are constructors. If $c_1 \neq c_2$ then stop with failure, otherwise replace $u \doteq t$ by $u_1 \doteq t_1 \ldots u_n \doteq t_n$.

5. u and t are both functional calls. Let z be a variable not in $var(U_i) \cup var(U)$. Replace $u \doteq t$ by the two equations $z \doteq u$ and $z \doteq t$.

6. u or t has a functional call d as proper subterm. Let z be a variable not in $var(U_i) \cup var(U)$. Replace all occurrences of d in U_i by z, then add the equation $z \doteq d$.

7. u is a functional call and t is a structure. Replace $u \doteq t$ by $t \doteq u$.

8. u is a structure and t is a functional call which occurs elsewhere in U_i. Replace all other occurrences of t by u (i.e. leave $u \doteq t$ unchanged).

9. t is a ground functional call. Replace $u \doteq t$ by $u \doteq s$, where $t \doteq s$ is in Eq. $\qquad \square$

Lemma 1 For each input set U of equations, algorithm 1 terminates and returns failure or a set U' of equations in normal form. If U' is in solved form, $\{\sigma_{U'}\}$ is a complete set of Eq-unifiers for U. If the result is failure, the set of Eq-unifiers to U is $\emptyset$.

The rules in algorithm 1 can be applied in any order, the results obtained will still be equivalent in the following sense:

Lemma 2 Let U' and U'' be the resulting sets obtained from algorithm 1 by applying the rules in two different orders. Then U' is in solved form iff U'' is in solved form, in which case $\sigma_{U'}$ and $\sigma_{U''}$ are equal up to renaming of variables.

Proofs of the two lemmas can be found in [Bon89]. The version of S-unify presented in [Bon90] can now be outlined:

Algorithm 2 (S-unify1)
Use the normal form algorithm to transform U. Then choose, depending on the result of the transformation, the appropriate alternative below:

1. If the result is failure, then terminate and output $\emptyset$.

2. If the result is a set N in solved form, then terminate and output $\{\sigma_N\}$.

3. If the result is a set in normal form which is not in solved form, then terminate and output "don't know". $\qquad\qquad\Box$

4   The problem of operational semantics

The classical way of drawing conclusions from a logic program is via SLD-resolution. This would mean that we fix a computation rule, which in each resolution step chooses an atom in the goal to be refuted by the program. However, the don't-know case of S-unify complicates matters; how shall we proceed if the selected goal atom yields "don't know" when S-unified with some head in the program? Of course we could decide that it is not possible to determine whether the goal is refutable or not, and let our resolution procedure answer "don't know". But this would be giving up too easily, as shown by the following example:

Example 2 Consider the program:

$$p(Z, Z) \leftarrow$$
$$q(a) \leftarrow$$

and the goal $\leftarrow p(f(X), X), q(X)$. If we had fixed a left-to-right computation rule, we would get the equation set $\{X \doteq Z, Z \doteq f(X)\}$ and a "don't-know" answer from S-unify. But if we used a right-to-left computation rule, we would get the substitution $\{X/a\}$ and the new goal:

$$\leftarrow p(f(a), a)$$

Now we would get the equation set $\{Z \doteq f(a), Z \doteq a\}$, which would result in success or fail, depending on the definition of f. □

The example shows that if we define our resolution procedure in this way, then our results will be dependent on the computation rule! This suggests that we can do better than this "eager" strategy. Of course we would like to employ a strategy as "lazy" as possible, i.e. avoid answering "don't know" until we have exhausted every other possibility.

The reason to a set of equations' not ending up in solved form can be attributed to certain kinds of equations, the *constraints*, whose solvability is impossible to determine by means of S-unification:

Definition 2 Let N be a set of equations in normal form. The *constraints* occurring in N are all equations $r \doteq d$ such that d is a functional call and one of the following cases apply:

1. r is a structure which is not a variable.

2. r is a variable occurring in d.

3. r is a variable occurring in another equation $r \doteq x$ (or $x \doteq r$) where x is a structure or a variable occurring in d. □

The basic problem with all of these three cases is the insufficient instantiation of variables occurring in the functional calls. As shown by example 2 the solvability of the constraints becomes determinable if the variables occurring therein become instantiated to ground terms. The idea of *delaying* constraints comes to mind, as suggested by the following example:

Example 3 Let f,g and h be external functors in:

$$p(X, f(Y, Z)) \leftarrow q(Y, g(X, Y), Z), r(h(X, Y, Z), Z, Y)$$
$$q(X, X, a) \leftarrow$$
$$r(X, X, b) \leftarrow$$

Calling the program with $\leftarrow p(a, W)$ would result in the substitution $\{X/a, W/f(Y, Z)\}$ and the new goal:

$$\leftarrow q(Y, g(a, Y), Z), r(h(a, Y, Z), Z, Y)$$

Trying to refute the two atoms would result in the equations $\{Y \doteq X_0, X_0 \doteq g(a, Y), Z \doteq a\}$ and $\{X_0 \doteq h(a, Y, Z), X_0 \doteq Z, Y \doteq b\}$ respectively, neither of which is in solved form. How should we proceed?

Let us suppose for a moment that there exists a Y such that $Y \doteq g(a, Y)$. Then instead of delaying the first atom, we consider it refuted, yielding the (partial) substitution $\{Z/a\}$. Then we get the new goal:

$$\leftarrow r(h(a, Y, a), a, Y)$$

Now Y gets instantiated to b and we get the equations $X_0 \doteq a$ and $X_0 \doteq h(a, b, a)$, which we can determine to be true or false. If $h(a, b, a)$ reduces to something else than a, we can safely fail, but what happens otherwise? We have yielded the empty goal, but our reasoning was based upon the assumption that $Y \doteq g(a, Y)$. Well, during the course of the refutation Y has got instantiated to b, so our constraint has been transformed into $b \doteq g(a, b)$. If $g(a, b)$ reduces to something other than b, then again we can safely fail, otherwise we have a success case. □

Example 3 suggests a "very lazy" strategy, where we hope to the very last minute that all the important variables (i.e. those in the scope of external functors) will be instantiated to ground terms. We will formalize this line of thought in the next section. There are some dangers related to this strategy; unsolved constraints may be returned as an answer at the end of computation, or the computation may loop cumulating unsatisfiable constraints. It is however possible to develop static checks that guarantee that, given a program and a goal, neither of these cases will occur during the course of computation.

5 S-SLD-resolution

In order to handle the constraints in an appropriate way, we start by making a slight extension of the S-unification algorithm. As input it takes a finite set U of equations and a (possibly empty) set C of constraints.

Algorithm 3 (S-unify2)
Let $U' = U \cup C$. Use the normal form algorithm to transform U'. Then choose, depending on the result of the transformation, the appropriate alternative below:

- If the result is failure, then terminate and output $\emptyset$.

- If the result is a set N in normal form, then:

  1. Let C' be the set of all constraints occurring in N.
  2. Let $N' := N - C'$. N' will now be in solved form, and we therefore construct $\sigma_{N'}$.
  3. Terminate and output $\{\sigma_{N'}\}$ and C' as result. □

The correctness of the algorithm follows directly from lemma 1. If the algorithm returns $\emptyset$, this means that the unification has failed, taking the earlier accumulated constraints into account.

Example 4 Let $sub1$ and $add1$ be external functors with the interpretation $sub1(X) = X - 1$ and $add1(X) = X + 1$. Assume that we call the redefined S-unification algorithm with the set

$$U = \{X \doteq sub1(Y), 1 \doteq add1(Z), W \doteq add1(Z)\}$$

of equations and the set $C = \{2 \doteq add1(W)\}$ of constraints. First $U \cup C$ is submitted to the normal form algorithm:

$$U_0 = \{X \doteq sub1(Y), 1 \doteq add1(Z), W \doteq add1(Z), 2 \doteq add1(W)\}$$
$$U_1 = \{X \doteq sub1(Y), 1 \doteq add1(Z), W \doteq 1, 2 \doteq add1(W)\} \qquad \text{by } 8,$$
$$U_2 = \{X \doteq sub1(Y), 1 \doteq add1(Z), W \doteq 1, 2 \doteq add1(1)\} \qquad \text{by } 2,$$
$$U_3 = \{X \doteq sub1(Y), 1 \doteq add1(Z), W \doteq 1, 2 \doteq 2\} \qquad \text{by } 9,$$
$$U_4 = \{X \doteq sub1(Y), 1 \doteq add1(Z), W \doteq 1\} \qquad \text{by } 1$$

U_4 is in normal form and thus returned as the result of the normal form algorithm.

1. $C' := \{1 \doteq add1(Z)\}$

2. $N' := N - C' = \{X \doteq sub1(Y), W \doteq 1\}$
   $\sigma_{N'} := \{X/sub1(Y), W/1\}$

3. The result of the unification is thus $\{\sigma_{N'}\} = \{\{X/sub1(Y), W/1\}\}$ and
   $C' = \{1 \doteq add1(Z)\}$ □

We next define a new resolution procedure, *S-SLD-resolution*, which uses the redefined S-unification algorithm. For a given program and a given goal, the S-SLD-resolution procedure builds a *S-SLD-tree*, whose branches are of four different types: success, failure, infinite and *don't-know*-branches.

Let P be a program, $H \leftarrow B_1, \ldots, B_n$ a (renamed) clause in P, $\mathcal{R}$ a computation rule and C a set of constraints, initially set to $\emptyset$. A *S-SLD-derivation* is a sequence $\langle G_0, \epsilon, \emptyset; G_1, \theta_1, C_1; \ldots \rangle$ defined as follows: Let $G_i = \leftarrow A_1, \ldots, A_m$ be a goal in a derivation and C_i the corresponding constraint set, and assume that $\mathcal{R}(G_i) = A_j$. Then G_{i+1} and C_{i+1} is obtained by:

1. Call S-unify2(H, A_j, C_i). Suppose the result is θ_{i+1} and C_{i+1}.

2. $G_{i+1} := \leftarrow (A_1, \ldots, A_{j-1}, B_1, \ldots, B_n, A_{j+1}, \ldots, A_m)\theta_{i+1}$.

The notion of *S-SLD-tree* follows from the algorithm above: a tree is said to be a S-SLD-tree for a goal G_0 and a program P iff it satisfies the following conditions:

- $\langle G_0, \emptyset \rangle$ is the root of the tree.

- $\langle G_{i+1}, C_{i+1} \rangle$ is a child of $\langle G_i, C_i \rangle$ iff G_{i+1} is obtained from G_i and C_{i+1} is obtained from C_i via the algorithm above.

A *success* branch of a S-SLD-tree is a branch which ends with the leaf $\langle \Box, \emptyset \rangle$. A *failure* branch is a branch ending with the leaf $\langle G, C \rangle$, where $G \neq \Box$. A *don't-know* branch is a branch ending with the leaf $\langle \Box, C \rangle$, where $C \neq \emptyset$. An *infinite* branch, finally, is a branch without a leaf.

S-SLD-resolution is sound and independent of computation rule in the sense of theorem 1 below. In the proof of the theorem we use the following idea, introduced in [DM91]: Every finite branch B in the S-SLD-tree gives rise to a *proof tree skeleton*, which could be seen as a derivation tree of the underlying context-free grammar of the program. Each node in B has an associated set of equations, corresponding to the unification made at that node. The union U of all these sets is the associated set of equations to the proof tree skeleton. If U has a solution then B is a success branch, in which case the skeleton can be *decorated* into a *proof tree*.

Theorem 1 (Independence of computation rule)

Let P be a program, G a goal and $\mathcal{R}$ a computation rule. Suppose there exists an S-SLD-derivation $S = \langle G, \epsilon, \emptyset, G_1, \theta_1, C_1, \ldots, \Box, \theta_n, C_n \rangle$ via $\mathcal{R}$. Then for every other computation rule $\mathcal{R}'$ there exists another S-SLD-derivation $S' = \langle G, \epsilon, \emptyset, G_1', \theta_1', C_1', \ldots, \Box, \theta_n', C_n' \rangle$ via $\mathcal{R}'$. Moreover, if σ and σ' are the respective computed answer substitutions then $G\sigma$ is a variant of $G\sigma'$, and C_n is a variant of C_n'.

Proof sketch : Let T be the proof tree skeleton induced by S, with the corresponding set of equations U. The process of refuting G can abstractly be viewed as trying to find a solution to U. Every computation rule would obviously yield a S-SLD-sequence S' which has T as proof tree skeleton.

Let U_i and U'_i be the sets of equations associated with the unification in step i in S and S', respectively. Then $U = \bigcup_{i=1...n} U_i = \bigcup_{i=1...n} U'_i$. According to lemma 2, the order in which we apply the rules of algorithm 1 to U has no importance, the result will still be equivalent. Since the same set of equations will be treated in both cases, the result follows.

Definition 3 A program P is *operationally complete* for a goal G if the S-SLD-tree for P and G via some computation rule contains no don't-know branches.

When P is operationally complete for G it is easily seen, using theorem 5.5 in [Höl89], that every correct answer substitution for $Eq \cup P \cup G$ is an instance of some σ associated with a success branch in the S-SLD-tree.

We would now like to characterize classes of programs which are operationally complete. The general problem of determining if a given program is operationally complete for a given goal, is undecidable. This can be proved by reducing it to the halting problem of Turing machines, in the same fashion as was done with the occur-check problem in [DM85]. This means that the characterization must be given as a sufficient condition. We will define our condition in terms of *predicate annotations*, used in [DM85, DFT89, DM91] for relating logic programs and attribute grammars.

6 A class of operationally complete programs

Some predicates have the property that some of their arguments are always supposed to have a value when the predicate is called (*inherited* arguments), while others are supposed to have a value when the computation of the call is successfully completed (*synthesized* arguments). This inspires to the following definition:

Definition 4 A *direction-assignment* (in short *d-assignment*) μ for a program P is a mapping from arguments into the set $\{\downarrow, \uparrow\}$, where $\downarrow$ is read "inherited" and $\uparrow$ "synthesized". If μ is a total function, μ is said to be a total d-assignment, otherwise a partial d-assignment. For each clause, the inherited argument positions of the head and the synthesized argument positions of the body are called *input* positions. Analogously, the synthesized argument positions of the head and the inherited argument positions of the body are called *output* positions.

Definition 5 For each clause C, we define a *dependency relation* D_C as follows: $p_i D_C q_j$ iff p_i is an input position, q_j is an output position and p_i and q_j have at least one common variable in C. Let C' be an instance of a clause C. The compound dependency graph for a proof tree skeleton t is obtained by pasting together the dependency graphs $D_{C'}$ of the C':s used in t. If this graph contains cycles it is said to be *circular*, otherwise *non-circular*. We denote the set of argument positions occurring in t by T. The compound dependency graph could be seen as the graph of a relation defined on T, whose reflexive and transitive

closure will be denoted by R_t. If the graph is non-circular, then R_t is a partial ordering.
□

Definition 6 The d-assignment is said to be *non-circular* iff the relations R_t for all derivation trees are partial orderings. A d-assignment is said to be *simple* if all variables occurring in output positions in a clause also occur in some of the input positions in the same clause.
□

As pointed out earlier, the problem about the constraints is that the variables occurring therein are not instantiated enough. If all these variables were to become ground, the correctness of each pending constraint could be determined. We therefore devote our attention towards finding a condition which guarantees that the variables occurring in the terms containing external functors always become ground. A very crude way of achieving this is of course to make sure that *all* variables become ground, i.e. every possible proof tree is ground.

It has been proved for ordinary logic programs that if there exists a simple and non-circular d-assignment μ for a program P, then every proof tree will be completely ground if the inherited positions (following μ) of the original goal are ground ([DM85, DM91]). This is based on the following observation (which is true also for amalgamated programs):

Observation 1 If all the variables occurring at minimal elements of R_t are instantiated to ground terms then every variable occurring in t will be instantiated to a ground term (if the program is simple and non-circular).

For ordinary logic programs, the fact that the inherited arguments of the original goal are ground would guarantee that the input positions of the root of every proof tree will be ground. However, this is not the case if we include external functors. Consider for instance the following example.

Example 5 Let f be an external functor and P the program:

$$p(f(X,Y), X, Y) \leftarrow$$

with the d-assignment $p : \downarrow\uparrow\uparrow$. Clearly, the program is both simple and non-circular. The goal $\leftarrow p(3, X, Y)$ yields the equation $3 \doteq f(X,Y)$, but that's it! The output positions remain non-ground.
□

Obviously, we must impose some further restriction on our amalgamated programs. In the following, all programs considered will be simple and non-circular unless otherwise stated.

As shown by the example above, a constraint produced by a S-unification at a minimal element w.r.t. R_t, prevents the instantiation of these argument instances. We would therefore like to find a class of programs for which this will never happen. The only minimal elements w.r.t. R_t for some t are, apart from the input positions of the root, the elements where an input position has unified with a variable-free output position. We must thus demand that:

- No external functors occur in the input positions of any head that unifies with an atom in the original goal.

- No external functors occur in those input positions of a clause that unify with variable-free output positions of the head of some other clause.

An (again rather austere) way of making sure that the requirements above will be met is to ban completely the existence of external functors in *any* input position. We can thus formulate a condition for a program P:

Condition 1 There exists a d-assignment μ for P that is simple and non-circular. Moreover, P does not have any external functors in any input position (following μ).

Theorem 2 Let P be a program fulfilling condition 1 and G any goal with ground inherited positions. Then every proof tree constructed from P and G via some computation rule $\mathcal{R}$ will be completely ground (and thus the S-SLD-tree will have no don't-know branches).

Proof sketch : Let t be some proof tree constructed from P and G. At minimal elements w.r.t. $\mathcal{R}_t$, a unification is always made between a ground term and term not containing external functors. Thus every variable occurring at these positions will be instantiated to a ground term. Since P is simple and non-circular, the result follows. □

7 Conclusions

We have proposed a new operational semantics for logic programs with external procedures. The idea stems from viewing S-unification as a constraint satisfaction problem. The constraints whose satisfiability cannot be determined with the information at hand, are delayed in the computation. The notion of *S-SLD-resolution* provides a formalization of this idea, allowing us to state correctness of our operational semantics and study its properties.

The formalization gives a perspective of our work in relation to logic programming and constraint logic programming. In particular we exploit well-known relations between logic programs and attribute grammars to develop a sufficient condition for the operational completeness of amalgamated programs. The condition can be checked statically, but a compile-time check algorithm will contain two steps with a worst-case exponentiality: generating simple d-assignments and testing for non-circularity. However, the non-circularity test can be replaced by a test for strong non-circularity, which can be carried out in polynomial time [DJL88, DM85].

The future work includes:

- development of an efficient implementation of S-unification. A first step in this direction has already been made [KK91].

- refinement of the sufficient condition by relating dependency analysis techniques to abstract interpretation [Bru87, JS87, Mel87, Nil90].

- investigation of the problem of looping accumulating constraints.

References

[BM88]   Bonnier, S. and J. Małuszyński. Towards a clean amalgamation of logic programs with external procedures. In *5th Int. conf. and symp. on logic programming*, pp. 311-326, MIT Press, 1988.

[Bon89]   Bonnier, S. *Horn clause logic with external procedures: towards a theoretical framework*, Licenciate thesis, University of Linköping, Sweden, 1989.

[Bon90]   Bonnier, S. *Horn clause logic with external procedures: towards a formal framework*, Research report, University of Linköping, Sweden, 1990.

[Bru87]   Bruynooghe M. *A framework for the abstract interpretation of logic programs.* Report CW62, Katholieke Universiteit, Leuven, 1987.

[DFT89]   Deransart, P., G. Ferrand and M. Téguia. *Une nouvelle construction de grammaires attribuées associées à un programme logique et application au problème du test d'occurrence*, Research report 89-3, Laboratoire d'Informatique, University of Orléans, France, 1989.

[DJL88]   Deransart, P., M. Jourdan and B. Lorho. *Attribute grammars. Definitions, systems and bibliography*, LNCS 323, Springer-Verlag, 1988.

[DM85]   Deransart, P. and J. Małuszyński. Relating logic programs and attribute grammars. *Journal of logic programming*, 2(2), pp. 119-156, 1985.

[DM91]   Deransart, P. and J. Małuszyński. *A grammatical view on logic programming*, MIT Press. To appear.

[Höl89]   Hölldobler, S. *Foundations of equational logic programming*, Lecture notes in artificial intelligence 353, Springer Verlag, 1989.

[JL87]   Jaffar, J. and J-L. Lassez. Constraint logic programming. In *14th ACM POPL Conf.*, ACM, 1987.

[JLM84]   Jaffar, J., J-L. Lassez and M. Maher. A theory of complete logic programs with equality. *Journal of logic programming*, 1(3), pp. 211-223, 1984.

[JS87]   Jones, N.D. and H. Søndergaard. A semantics-based framework for the abstract interpretation of Prolog. In Abramsky and Hankin editors, *Abstract interpretation of declarative languages*, pp. 123-142, Ellis Horwood, 1987.

[KK91]   Kågedal, A and F. Kluźniak. *Enriching Prolog with S-unification*, Research report LiTH-IDA-R-91-12, University of Linköping, Sweden, 1991.

[Mel87]   Mellish C. Abstract interpretation of Prolog programs. In Abramsky and Hankin editors, *Abstract interpretation of declarative languages*, pp. 181-198, Ellis Horwood, 1987.

[Nil90]   Nilsson, U. Systematic semantic approximations of logic programs. In Deransart and Małuszyński editors, *Proc. PLILP'90*, pp. 293-306, Springer Verlag, 1990.

Operational Semantics of
Constraint Logic Programming over Finite Domains

Pascal Van Hentenryck
Brown University, Box 1910,
Providence, RI 02912
Email: pvh@cs.brown.edu

Yves Deville
University of Namur, 21 rue Grandgagnage
B-5000 Namur (Belgium)
Email: yde@info.fundp.ac.be

Abstract

Although the Constraint Logic Programming (CLP) theory [7] is an elegant generalization of the LP theory, it has some difficulties in capturing some operational aspects of actual CLP languages (e.g. [8, 6, 3, 12, 18, 19]). A difficulty comes from the intentional incompleteness of some constraint-solvers. Some constraints are simply delayed until they can be decided by the constraint-solver. Others are approximated, providing an active pruning of the search space without being actually decided by the constraint-solver.

This paper presents an extension of the *Ask & Tell* framework [14] in order to give a simple and precise operational semantics to (a class of) CLP languages with an incomplete constraint-solver. The basic idea is to identify a subset of the constraints (the basic constraints) for which there exists an efficient and complete constraint-solver. Non-basic constraints are handled through two new combinators, relaxed ask and relaxed tell, that are in fact relaxations of the standard ask and tell.

The extended framework is instantiated to CLP on finite domains, say CLP(F) [16, 6]. Arc-consistency is shown to be an efficient and complete constraint-solver for basic constraints. We also present how non-basic constraints can be approximated in CLP(F). The resulting semantics precisely describes the operational semantics of the language, enables the programmer to reason easily about the correctness and efficiency of his programs, and clarifies the links of CLP(F) with the CLP and *Ask & Tell* theories.

It is believed that the approach can be used as well to endow other CLP languages such as BNR-Prolog [12], CLP(Σ^*) [19], and parts of Trilogy [18] with a precise operational semantics.

1 Introduction

Constraint Logic Programming (CLP) is a generalization of Logic Programming (LP) where unification, the basic operation of LP languages, is replaced by the more general concept of constraint-solving over a computation domain.

Syntactically, a CLP program can be seen as a finite set of clauses of the form

$$H \leftarrow c_1 \wedge \ldots \wedge c_n \, \Diamond \, B_1 \wedge \ldots \wedge B_m$$

where $H, B_1, \ldots, B_m$ are atoms and $c_1, \ldots, c_n$ are constraints. A goal in a CLP language is simply a clause without head.

The declarative semantics of a CLP language can be defined either in terms of logical consequences or in an algebraic way. An answer to a CLP goal is no longer a substitution but rather a conjunction of constraints $c_1, \ldots, c_n$ such that

$$P, T \models (\forall)\,(c_1 \wedge \ldots \wedge c_n \Rightarrow G) \text{ (logical version)}$$
$$P \models_S (\forall)\,(c_1 \wedge \ldots \wedge c_n \Rightarrow G) \text{ (algebraic version)}$$

where P is a program, S is a structure, T is the theory axiomatizing S, and $(\forall)\,(F)$ represents the universal closure of F. The rest of the presentation can be read from a logic or algebraic point of view and we use the notation $\mathcal{D} \models$ to denote that fact.

The operational semantics of CLP amounts to replacing unification by constraint-solving. It might be defined by considering configurations of the form $\langle G, \sigma \rangle$ where G is a conjunction of atoms and σ is a consistent conjunction of constraints. Now the only transition that needs to be defined between configurations is the following:

$$\frac{\begin{array}{l} H \leftarrow c_1 \wedge \ldots \wedge c_n \diamond B_1 \wedge \ldots \wedge B_m \in P \\ \mathcal{D} \models (\exists)(\sigma \wedge c_1 \wedge \ldots \wedge c_n \wedge H = A) \end{array}}{\langle A \wedge G, \sigma \rangle \longmapsto \langle B_1 \wedge \ldots \wedge B_m \wedge G, \sigma \wedge c_1 \wedge \ldots \wedge c_n \wedge H = A \rangle}$$

In the above transition rule, A is the selected atom (it can be any atom in the goal since the order in a conjunction is irrelevant) and $(\exists)(F)$ represents the existential closure of F.

The CLP theory [7] imposes a number of restrictions on $\mathcal{S}$, $\mathcal{T}$, and their relationships to establish equivalences between the semantics. Also the CLP language should be embedded with a *complete* constraint-solver, which means that, given a conjunction of constraints σ, the constraint-solver should return *true* if $\mathcal{D} \models (\exists)(\sigma)$ and *false* otherwise (i.e $\mathcal{D} \models (\forall)(\neg\sigma)$).

Although the CLP theory is an elegant generalization of the LP theory, it has some difficulties in capturing all operational aspects of actual CLP languages such as [8, 6, 3, 12, 18, 19]. One difficulty comes from the possibility offered by these languages to state constraints that cannot be decided upon by the constraint-solver. This is the case for instance of non-linear constraints over rational and real numbers which are simply delayed until they become linear. Another difficulty comes from the fact that, for some classes of problems, an intentionally incomplete but efficient constraint-solver might well turn out to be more valuable from a programming standpoint than a complete constraint-solver. This is justified by the tradeoff, that appears in many combinatorial problems, between the time spent in pruning and searching.

The present paper originates from an attempt to define in a precise and simple way the operational semantics of CLP over finite domains, say CLP(F) [16, 6], which suffers for both difficulties mentioned above. CLP(F) was designed with the goal of tackling a large class of combinatorial optimization problems with a short development time and an efficiency competitive with procedural languages. Typical applications for which CLP(F) is successful include sequencing and scheduling, graph coloring and time-table scheduling, as well as various assignments problems. The basic idea behind CLP(F) is to associate to variables a domain which is a finite set of values. Constraints in CLP(F) include (possibly non-linear) equations, inequalities, and disequations. It is simple to see that a complete constraint-solver exists for the above constraints but would necessarily require exponential time (unless P = NP). Moreover many of the abovementioned applications require different solutions and use various kinds of constraints, heuristics, and problem features. Hence it is unlikely that a complete constraint-solver be adequate for all of them. Fortunately most of them basically share the same pruning techniques and the idea behind CLP(F) was precisely to provide the language with those techniques[1]. However CLP(F) does not inherit directly its operational semantics from the CLP framework since, on the one hand, some constraints are only used when certain conditions are satisfied (e.g. disequations) and, on the other hand, some constraints are used to prune the search space although the constraint-solver cannot, in general, decide their satisfiability (e.g. inequalities). Previous approaches to define the operational semantics of CLP(F) were not based on the CLP framework but rather were given in terms of inference rules [15].

The new operational semantics presented here is based on the *Ask & Tell* framework [14] which generalizes the CLP framework by adding the concept of constraint entailment (i.e. ask) to the concept of constraint-solving (i.e. tell). Ask constraints directly account for the first difficulty in CLP languages: they might be used to restrict the context in which a constraint is executed. However to account for the incompleteness of the constraint-solver, we need to generalize the framework[2]. The basic idea behind the semantics presented here is to split the set of primitive constraints into two sets: (1) *basic* constraints for which there exists an efficient and complete constraint-solver; (2) *non-basic* constraints which are only approximated.

---

[1] Note that for other types of problems a complete constraint-solver might be more appropriate [1].

[2] In fact Saraswat considers his framework as parametrized on the combinators as well so that we are actually instantiating his framework.

The basic constraints are handled following the standard CLP theory and are precisely those constraints that can be returned as an answer to a goal. Non-basic constraints are handled through two new combinators, relaxed ask and relaxed tell. Contrary to ask (resp. tell), relaxed ask (resp. relaxed tell) check entailment (resp. consistency) not wrt the accumulated constraints but rather wrt a relaxation of them. Moreover relaxed tell enables to deduce new basic constraints approximating the non-basic one. Formally, relaxed ask and relaxed tell are not mere combinators but rather they define a family of combinators parametrized by a relaxation and an approximation function. The relaxation function specifies the relaxation of the accumulated constraints while the approximation specifies how to infer new basic constraints.

Describing a CLP language in the extended framework (i.e. with relaxed ask and tell) amounts to specifying:

- the basic constraints and their associated complete constraint-solver;

- the non-basic constraints and their associated (1) relaxation functions, (2) approximation functions, and (3) constraint-solvers that are complete for the conjunction of a non-basic constraint and the relaxation of a conjunction of basic constraints.

The extended framework is then instantiated to CLP(F). We identify the subset of basic constraints and show that arc-consistency [9] provides an efficient and complete constraint-solver for them. We also define the relaxation and approximation functions used in CLP(F) and suggest the non-basic constraint-solvers.

The contributions of the paper are twofold.

1. It gives a precise and simple operational semantics to CLP(F), characterizing what is computed (e.g. what is an answer) and how the computation is achieved (e.g. what is the pruning at some computation point). The semantics allows the programmer to formally reason about the correctness and the efficiency of his programs. It also clarifies the relationships between CLP(F) on the one hand, and the CLP and *Ask&Tell* frameworks on the other hand.

2. It proposes an extended framework that can possibly be instantiated to endow languages such as BNR-Prolog [12], CLP(Σ^*) [19], and parts of Trilogy [18] with a precise operational semantics.

The rest of this paper is organized in the following way. The next section defines the operational semantics of basic CLP. Section 3 instantiates basic CLP to finite domains. Section 4 describes the new combinators for non-basic constraints, relaxed ask and relaxed tell. Section 5 instantiates relaxed ask and relaxed tell to CLP(F). Section 6 contains the conclusion of this paper and directions for future research.

2 Basic CLP

In this section, we define the syntax and operational semantics of the class of languages we consider for Constraint Logic Programming. We use a structural operational semantics [13] similar to the one used in [14]. There is an important difference however due to the various possible interpretations of nondeterminism. Saraswat's semantics, as well as other semantics for concurrent logic programming languages, describes all possible executions of a program on a given goal. Hence an actual implementation might actually fail (resp. succeed) for some elements of the success (resp. failure) set. Our semantics captures a single execution and hence an element of the success set might never fail in an actual implementation. This difference does not show up in the transition rules but rather in the definition of the success, failure, flounder, and divergence sets.

2.1 Syntax

The abstract syntax of the basic language can be defined by the following (incomplete) grammar:

$$P ::= H \leftarrow B \mid P.P$$
$$B ::= A \mid ask(c) \rightarrow B \mid tell(c) \mid B \& B \mid \exists x\, B \mid true$$

In words, a program is a non-empty set of clauses. A clause is composed of a head and a body. The head is an atom whose arguments are all distinct variables. The body is either an atom, an implication $ask(c) \rightarrow B$ where $ask(c)$ is an ask on constraint c, a tell on constraint c, a conjunction of two bodies, an existential construction $\exists x\, B$ where B is a body with variable x free, or $true$. For completeness, some semantic rules should also be added, for instance the rule stating that any variable in a clause is either existentially quantified or appears in the head.

The concrete syntax can be the one of any existing CLP language suitably enhanced to include the implication construct. There is no difficulty in translating any of these concrete syntax to the abstract one.

2.2 Basic Constraints

Basic constraints are split into two sets: basic tell-constraints, simply referred to as basic constraints, and basic ask-constraints. The constraint-solver for basic CLP should be complete for consistency of basic constraints and entailment of basic ask-constraints.

Definition 1 A basic constraint-solver is *complete* iff it can decide (1) consistency of c and σ (i.e. $\mathcal{D} \models (\exists)(\sigma \wedge c)$) and (2) entailment of c' wrt σ (i.e. $\mathcal{D} \models (\forall)\,(\sigma \Rightarrow c')$) where c is a basic constraint, c' a basic ask-constraint, and σ a conjunction of basic constraints.

CLP languages usually maintain constraints in a reduced form to obtain an incremental behaviour necessary to achieve efficiency[3].

Definition 2 A *reduction* function is a total function red which, given a consistent conjunction of basic constraints σ, returns a conjunction of basic constraints, such that $\mathcal{D} \models \sigma \Leftrightarrow red(\sigma)$.

In the following, we assume the existence of a reduction function red for basic constraints and denote by CCR the set of consistent conjunctions of constraints in reduced form (i.e. the codomain of red). The actual choice of the reduction function depends upon the computation domain.

2.3 Configurations

The configurations in the transition system are of the form $\langle B, \sigma \rangle$ where B is a body and σ a conjunction of basic constraints in reduced form. Informally B represents what remains to be executed while σ represents the constraints accumulated so far. Successful computations end up with a conjunction of basic constraints in reduced form. Computations may also flounder when an ask on constraint c cannot be decided upon (i.e. neither c nor $\neg c$ is entailed by the accumulated constraints). We use the terminal *flounder* to capture that behaviour.
Terminal configurations are thus described by $T = \{\sigma \mid \sigma \in CCR\} \cup \{flounder\}$.
Configurations are described by $\Gamma = \{\langle B, \sigma \rangle \mid B$ is a body and $\sigma \in CCR\} \cup T$.
A transition $\gamma \longmapsto \gamma'$ can be read as "configuration γ nondeterministically reduces to γ'".

2.4 Transition Rules

Goals are assumed to be resolved against clauses from some program and constraints are assumed to be checked for consistency and entailment in a given structure or theory $\mathcal{D}$. Since the structure (or theory) never changes, we simply omit it and assume that it is clear from the context. As the program changes, we make use of an indexed transition system and use $P \vdash \gamma \longmapsto \gamma'$ to denote that the transition $\gamma \longmapsto \gamma'$ takes place in the context of program P. When the transition does not depend on program P, we simply drop the prefix $P \vdash$. In the following, σ denotes a conjunction of basic constraints in reduced form, γ a configuration, B and G bodies, P a program, and c a basic constraint. These are possibly subscripted.

True: The atom $true$ simply leads to a terminal configuration. $\langle true, \sigma \rangle \longmapsto \sigma$

---

[3]Most operational semantics do not include this function as it plays no fundamental role in their description. As we will see, it plays an important role in showing how efficiently the relaxation function can be computed in CLP(F).

Tell: The tell operation on a basic constraint is successful if the constraint is consistent with the accumulated constraints.

$$\frac{\mathcal{D} \models (\exists)(\sigma \wedge c)}{\langle tell(c), \sigma \rangle \longmapsto red(\sigma \wedge c)}$$

Implication: An implication $ask(c) \rightarrow B$ never fails. If the basic ask-constraint c is entailed by the accumulated constraints, it reduces to the body B. If $\neg c$ is entailed by the accumulated constraints, the implication terminates successfully. Otherwise, the implication flounders.

$$\frac{\mathcal{D} \models (\forall)(\sigma \Rightarrow c)}{\langle ask(c) \rightarrow G, \sigma \rangle \longmapsto \langle G, \sigma \rangle}$$

$$\frac{\mathcal{D} \models (\forall)(\sigma \Rightarrow \neg c)}{\langle ask(c) \rightarrow G, \sigma \rangle \longmapsto \sigma}$$

$$\frac{\mathcal{D} \models \neg(\forall)(\sigma \Rightarrow c) \quad \mathcal{D} \models \neg(\forall)(\sigma \Rightarrow \neg c)}{\langle ask(c) \rightarrow G, \sigma \rangle \longmapsto flounder}$$

Existential Quantification: An existential quantification can be removed by replacing the quantified variable by a brand new variable.

$$\frac{y \text{ is a brand new variable} \quad \langle G[x/y], \sigma \rangle \longmapsto \gamma}{\langle \exists x\, G, \sigma \rangle \longmapsto \gamma}$$

The fact that variable y be brand new can be formalized in various ways if necessary.

Conjunction: The semantics of conjunction is given here by the interleaving rule which is appropriate for CLP languages. If any of the goals in a conjunction can make a transition, the whole conjunction can make a transition as well and the accumulated constraints are updated accordingly. The conjunction flounders when both conjuncts flounder.

$$\frac{\langle G_1, \sigma \rangle \longmapsto \langle G_1', \sigma' \rangle}{\langle G_1 \,\&\, G_2, \sigma \rangle \longmapsto \langle G_1' \,\&\, G_2, \sigma' \rangle \quad \langle G_2 \,\&\, G_1, \sigma \rangle \longmapsto \langle G_2 \,\&\, G_1', \sigma' \rangle}$$

$$\frac{\langle G_1, \sigma \rangle \longmapsto \sigma'}{\langle G_1 \,\&\, G_2, \sigma \rangle \longmapsto \langle G_2, \sigma' \rangle \quad \langle G_2 \,\&\, G_1, \sigma \rangle \longmapsto \langle G_2, \sigma' \rangle}$$

$$\frac{\langle G_1, \sigma \rangle \longmapsto flounder \quad \langle G_2, \sigma \rangle \longmapsto flounder}{\langle G_1 \,\&\, G_2, \sigma \rangle \longmapsto flounder}$$

Procedure Call with one Clause: We now consider the solving of an atom $p(t_1, \ldots, t_n)$ wrt a clause $p(x_1, \ldots, x_n) \leftarrow B$. If $B[x_1/t_1, \ldots, x_n/t_n]$ can make a transition in the context of the accumulated constraints, then so can $p(t_1, \ldots, t_n)$ in the context of the clause and the constraints.

$$\frac{p(x_1, \ldots, x_n) \leftarrow B \;\vdash\; \langle B[x_1/t_1, \ldots, x_n/t_n], \sigma \rangle \longmapsto \gamma}{p(x_1, \ldots, x_n) \leftarrow B \;\vdash\; \langle p(t_1, \ldots, t_n), \sigma \rangle \longmapsto \gamma}$$

Procedure Call with several Clauses: If an atom can make a transition in program P_1, it can obviously make a transition in the program made up of P_1 and program P_2.

$$\frac{P_1 \vdash \gamma \longmapsto \gamma'}{\begin{array}{l} P_1.P_2 \vdash \gamma \longmapsto \gamma' \\ P_2.P_1 \vdash \gamma \longmapsto \gamma' \end{array}}$$

2.5 Operational Semantics

We now define the operational semantics of the language in terms of its success, floundering, divergence, and failure sets. Let $\stackrel{*}{\longmapsto}$ denote the reflexive and transitive closure of $\longmapsto$ and $initial(G, \sigma)$ the configuration $\langle G, red(\sigma) \rangle$. Also a configuration γ is said to diverge in program P if there exists an infinite sequence of transitions $P \vdash \gamma \longmapsto \gamma_1 \longmapsto \dots \longmapsto \gamma_i \longmapsto \dots$

The success, floundering, and divergence sets (not necessarily disjoint) can be defined in the following way.

$$
\begin{aligned}
SS[P] &= \{\langle G, \sigma \rangle \mid P \vdash initial(G, \sigma) \stackrel{*}{\longmapsto} \sigma'\} \\
FLS[P] &= \{\langle G, \sigma \rangle \mid P \vdash initial(G, \sigma) \stackrel{*}{\longmapsto} flounder\} \\
DS[P] &= \{\langle G, \sigma \rangle \mid initial(G, \sigma) \text{ diverges in P}\}
\end{aligned}
$$

The failure set can now be defined in terms of the above three sets.

$$FS[P] = \{\langle G, \sigma \rangle \mid \langle G, \sigma \rangle \notin SS[P] \cup FLS[P] \cup DS[P]\}$$

Another semantic definition can be given to capture the results of the computation.

$$RES[P]\langle G, \sigma \rangle = \{\sigma' \mid P \vdash initial(G, \sigma) \stackrel{*}{\longmapsto} \sigma'\}$$

3 Basic CLP(F)

3.1 Basic Constraints

In the following, x and y, possibly subscripted, denote variables and a, b, c, v, w, possibly subscripted, denote natural numbers.

Definition 3 The *basic* constraints of CLP(F) are either *domain* constraints (i.e. $x \in \{v_1, \dots, v_n\}$) or *arithmetic* constraints: (i.e. $ax \neq b$, $ax = b$, $ax = by + c$ ($a \neq 0 \neq b$), $ax \geq by + c$, $ax \leq by + c$).

The semantics of addition, multiplication, $=$, $\leq$, $\geq$, and $\neq$ is the usual one. Clearly, the negation of each basic constraint can be expressed as a conjunction or disjunction of basic constraints. Hence all the basic constraints can also be basic ask-constraints[4]. Note that the variables appearing in arithmetic constraints are expected to appear in some domain constraints. Every variable thus has a domain. This can be seen as an implicit ask-constraint and justifies the following definition.

Definition 4 A *system of constraints* S is a pair $\langle AC, DC \rangle$ where AC is a set of arithmetic constraints and DC is a set of domain constraints such that any variable occurring in an arithmetic constraint also occurs in some domain constraint of S. The set D_x is the *domain* of x in S (or in DC) iff the domain constraints of x in DC are $x \in D_1, \dots, x \in D_k$ and D_x is the intersection of the D_i's.

It follows that, provided that each variable has a domain, a conjunction of basic constraints can be represented by a system of constraints and vice versa.

We conclude this subsection by a number of conventions. If c is an arithmetic constraint with only one variable x, we say that c is *unary* and denote it as $c(x)$. Similarly, if c is an arithmetic constraint with two variables x and y, we say that c is *binary* and denote it $c(x, y)$. As usual, $c(x/v)$ and $c(x/v, y/w)$ denote the Boolean value obtained from $c(x)$ and $c(x, y)$ by replacing x and y by the values v and w respectively.

---

[4]Note that we may want to consider some of them as non-basic constraints for efficiency reasons.

3.2 Constraint-Solving

The constraint-solver of CLP(F), and hence of basic CLP(F), is based on consistency techniques, a paradigm emerging from AI research [9]. We start by defining a number of notions.

Definition 5 Let $c(x)$ be a unary constraint and D_x be the domain of x. Constraint $c(x)$ is said to be *node-consistent* wrt D_x if $c(x/v)$ holds for each value $v \in D_x$. Let $c(x, y)$ be a binary constraint and D_x, D_y be the domains of x, y. Constraint $c(x, y)$ is said to be *arc-consistent* wrt D_x, D_y if the following conditions hold: (1) $\forall v \in D_x \; \exists w \in D_y \; c(x/v, y/w)$ holds; (2) $\forall w \in D_y \; \exists v \in D_x \; c(x/v, y/w)$ holds.

We are in position to define a solved form for the constraints.

Definition 6 Let S be a system of constraints. S is in *solved form* iff any unary constraint $c(x)$ in S is node-consistent wrt the domain of x in S, and any binary constraint $c(x, y)$ in S is arc-consistent wrt the domains of x, y in S.

We now study a number of properties of systems of constraints in solved form.

Property 7 Let $c(x, y)$ be the binary constraint $ax \geq by + c$, arc-consistent wrt $D_x = \{v_1, \ldots, v_n\}, D_y = \{w_1, \ldots, w_m\}$. Assume also that $v_1 < \ldots < v_n$ and $w_1 < \ldots < w_m$. Then we have (1) $c(v_1, w_1)$ and $c(v_n, w_m)$ hold and (2) if $c(v_i, w_j)$ holds, then $c(v_{i+k}, w_{j-l})$ holds ($0 \leq k \leq n - i, \; 0 \leq l < j$).

Property 8 Let $c(x, y)$ be the binary constraint $ax \leq by + c$, arc-consistent wrt $D_x = \{v_1, \ldots, v_n\}, D_y = \{w_1, \ldots, w_m\}$. Assume also that $v_1 < \ldots < v_n$ and $w_1 < \ldots < w_m$. Then we have (1) $c(v_1, w_1)$ and $c(v_n, w_m)$ hold and (2) if $c(v_i, w_j)$ holds, then $c(v_{i-k}, w_{j+l})$ holds ($0 \leq k < i, \; 0 \leq l \leq m - j$).

Property 9 Let $c(x, y)$ be the binary constraint $ax = by + c$, arc-consistent wrt $D_x = \{v_1, \ldots, v_n\}, D_y = \{w_1, \ldots, w_m\}$. Assume also that $v_1 < \ldots < v_n$ and $w_1 < \ldots < w_m$. Then we have $n = m$ and $c(v_i, w_i)$ holds ($1 \leq i \leq n$).

The satisfiability of a system of constraints in solved form can be tested in a straightforward way.

Theorem 10 Let $S = \langle AC, DC \rangle$ be a system of constraints in solved form. S is satisfiable iff $\langle \emptyset, DC \rangle$ is satisfiable.

Proof It is clear that $\langle \emptyset, DC \rangle$ is not satisfiable iff the domain of some variable is empty in DC. If the domain of some variable is empty in DC, then S is not satisfiable. Otherwise, it is possible to construct a solution to S. By properties (1), (2), and (3), all binary constraints of S hold if we assign to each variable the smallest value in its domain. Moreover, because of node-consistency, the unary constraints also hold for such an assignment. $\square$

It is worth noting that, in a system of constraints in solved form, all the values within the domain of a variable do not necessarily belong to a solution. For instance, in the following system

$$\langle \{x \leq y, 3x \geq 2y + 1\}, \{x \in \{2, 4, 6\}, y \in \{2, 3, 6\}\} \rangle$$

there is no solution with the value 4 assigned to x although the system is in solved form.

It remains to show how to transform a system of constraints into an equivalent one in solved form. This is precisely the purpose of the node- and arc-consistency algorithms [9].

Algorithm 11 To transform the system of constraints S into a system in solved form S': first apply a node-consistency algorithm to the unary constraints of $S = \langle AC, DC \rangle$ to obtain $\langle AC, DC' \rangle$; then apply an arc-consistency algorithm to the binary constraints of $\langle AC, DC' \rangle$ to obtain $S' = \langle AC, DC'' \rangle$.

Theorem 12 Let S be a system of constraints. Algorithm 1 produces a system of constraints in solved form equivalent to S.

We now give a complete constraint-solver for the basic constraints. Given a system of constraints S, Algorithm 13 returns *true* if S is satisfiable and *false* otherwise.

Algorithm 13 To check the satisfiability of a system of constraints S: first apply Algorithm 11 to S to obtain $S' = \langle AC, DC \rangle$; now if the domain of some variable is empty in DC', return *false*; otherwise return *true*.

The complexity of Algorithms 11 and 13 is the complexity of arc-consistency algorithms. In [10], an arc-consistency algorithm is proposed whose complexity is $O(cd^2)$ where c is the number of binary constraints and d is the size of the largest domain. Given the form of the basic constraints, it is possible to design a specific arc-consistency algorithm whose complexity is $O(cd)$ [4] showing that basic constraints can be solved efficiently.

3.3 Reduced Form of the Basic Constraints

Definition 14 Let σ be a consistent conjunction of basic constraints. The reduced form of σ, denoted $red(\sigma)$, is obtained as follows:

1. Let S be the system of constraints associated with σ and $S' = \langle AC', DC' \rangle$ be its solved form.

2. Let DC'' be equivalent to DC' be with only one domain constraint per variable.

3. Let AC'' be AC' without (1) the unary constraints and (2) the binary constraints $c(x, y)$ satisfying $\forall v \in D_x \ \forall w \in D_y \ c(x/v, y/w)$ where D_x, D_y are the domains of x, y in S'.

4. $red(\sigma)$ is the conjunction of basic constraints in $\langle AC'', DC'' \rangle$.

The reduced form for the basic constraints is equivalent to our solved form since the unary constraints are node consistent (and hence implied by the domain constraints) and the binary constraints satisfying the given condition are also implied by the domain constraints.

3.4 Expressive Power

The basic constraints of CLP(F) have been chosen carefully in order to avoid an NP-Complete constraint-solving problem. For instance, allowing disequations with two variables leads to an NP-Complete problem (graph-coloring could then be expressed in a straightforward manner). Allowing equations and inequalities with three variables also leads to NP-complete problems. Finally, it should be noted that Algorithm 13 is not powerful enough to decide systems of constraints including constraints of the form $ax + by = c$.

The reason is that arc-consistency algorithms handle constraints locally while more global reasoning is necessary to check satisfiability of this class of constraints. As an example, the system

$$(\{x_1 + x_2 = 1, x_2 + x_3 = 1, x_3 + x_1 = 1\} , \{x_1 \in \{0, 1\}, x_2 \in \{0, 1\}, x_3 \in \{0, 1\}\})$$

is not satisfiable although it is arc-consistent. To verify the unsatisfiability, just add the three constraints to obtain $2x_1 + 2x_2 + 2x_3 = 3$. The left member is even while the right member is odd.

4 Non-Basic CLP

The purpose of this section is to provide a systematic way to describe the operational semantics of non-basic constraints that are approximated in terms of basic constraints. As mentioned previously, approximation is present in several CLP languages.

To capture that behaviour, we introduce two new combinators in the *Ask & Tell* framework, relaxed tell and relaxed ask. These combinators are parametrized by a relaxation and an approximation function such that we are in fact defining a family of combinators.

Before starting the more formal presentation, let us give an informal account to the approach. Let σ be the accumulated constraints and assume that we face a relaxed tell on a non-basic constraint c. Relaxed tell, instead of checking the consistency of $\sigma \wedge c$ which might be quite complex in general, only checks the consistency of a relaxation of $\sigma \wedge c$. Clearly if the relaxed problem is not satisfiable, the initial problem is not satisfiable either. Moreover the solutions of the relaxed problem cannot necessarily be expressed as a conjunction of basic constraints. Hence only an approximation of the solutions, in the form of a conjunction of basic constraints, can be added to the accumulated constraints. When the approximation is not equivalent to the initial problem, the non-basic constraint cannot be removed from the goal part of the configuration. Finally, if we face a relaxed ask, then entailment is not checked wrt σ, but rather wrt to its relaxation. As a consequence, relaxed ask could return undecided (i.e. flounder) while an ask (if implemented) would have returned *true* or *false*.

4.1 Relaxation and Approximation

Relaxed tell and relaxed ask require to associate, with each non-basic constraint, (1) a relaxation function; (2) an approximation function and (3) a complete constraint-solver.

Definition 15 A *relaxation* function is a total function $r : CCR \rightarrow CCR$ such that $\mathcal{D} \models (\forall)(\sigma \Rightarrow r(\sigma))$.

In other words, a relaxation function associates with each conjunction σ of basic constraints another conjunction of basic constraints that is implied by σ. Hence the relaxation of σ captures the solutions of σ and possibly some non-solutions.

We also would like to infer new basic constraints from a non-basic constraint. This is achieved through an approximation function.

Definition 16 Given a relaxation function r, an *approximation* function is a total function ap, which, given $\sigma \in CCR$ and a non-basic constraint c, returns a conjunction of basic constraints, such that $\mathcal{D} \models (\forall)((r(\sigma) \wedge c) \Rightarrow ap(\sigma, c))$.

This definition specifies that the approximation captures all solutions of $r(\sigma) \wedge c$ and possibly some non-solutions. In the following, we assume that a relaxation function and an approximation function have been defined for each constraint and denote them by the (overloaded) symbols r and ap respectively.

The non-basic constraint-solver has to satisfy a number of requirements.

Definition 17 A non-basic constraint-solver is *complete* iff it can decide (1) the consistency of c and $r(\sigma)$; (2) the entailment of c' by $r(\sigma)$ where c is a non-basic constraint, c' is a non-basic ask-constraint, and σ is a conjunction of basic constraints.

The following properties are straightforward but help understanding the transition rules.

Property 18 [Properties of relaxation and approximation]

1. $\mathcal{D} \models (\forall)((\sigma \wedge c) \Rightarrow (\sigma \wedge ap(\sigma, c)))$;
2. $\mathcal{D} \models \neg(\exists)(c \wedge r(\sigma))$ implies $\mathcal{D} \models \neg(\exists)(c \wedge \sigma)$;
3. $\mathcal{D} \models \neg(\exists)(ap(\sigma, c) \wedge \sigma)$ implies $\mathcal{D} \models \neg(\exists)(c \wedge \sigma)$;
4. $\mathcal{D} \models (\forall)(ap(\sigma, c) \Rightarrow c \wedge r(\sigma))$ implies $\mathcal{D} \models (\forall)((\sigma \wedge c) \Leftrightarrow (\sigma \wedge ap(\sigma, c)))$;
5. $\mathcal{D} \models (\exists)(ap(\sigma, c) \wedge \sigma)$ and $\mathcal{D} \models (\forall)(ap(\sigma, c) \Rightarrow c \wedge r(\sigma))$ implies $\mathcal{D} \models (\exists)(c \wedge \sigma)$;
6. $\mathcal{D} \models (\forall)(r(\sigma) \Rightarrow c)$ implies $\mathcal{D} \models (\forall)(\sigma \Rightarrow c)$.

4.2 Relaxed Tell

Termination: Termination of relaxed tell occurs when the approximation $ap(\sigma, c)$ is equivalent to $r(\sigma) \wedge c$ which means that $r(\sigma) \wedge c$ can be represented as a conjunction of basic constraints. Moreover if $ap(\sigma, c)$ is consistent with σ, we obtain a terminal configuration.

$$\frac{\mathcal{D} \models (\exists)(\sigma \wedge ap(\sigma, c)) \qquad \mathcal{D} \models (\forall)(ap(\sigma, c) \Rightarrow (r(\sigma) \wedge c))}{\langle relax\text{-}tell(c), \sigma \rangle \longmapsto red(\sigma \wedge ap(\sigma, c))}$$

Property 18.4 and 18.5 ensure respectively the equivalence of $\sigma \wedge ap(\sigma, c)$ and $\sigma \wedge c$ and the consistency of c and σ.

Pruning: Property 18.1 allows for the addition of new basic constraints (and hence pruning of the search space) provided that $r(\sigma) \wedge c$ and $ap(\sigma, c) \wedge \sigma$ be consistent, and $ap(\sigma, c)$ be not entailed by σ.

$$\frac{\begin{array}{l} \mathcal{D} \models (\exists)\,(r(\sigma) \wedge c) \\ \mathcal{D} \models (\exists)\,(\sigma \wedge ap(\sigma, c)) \\ \mathcal{D} \models \neg(\forall)\,(ap(\sigma, c) \Rightarrow (r(\sigma) \wedge c)) \\ \mathcal{D} \models \neg(\forall)\,(\sigma \Rightarrow ap(\sigma, c)) \end{array}}{\langle relax\text{-}tell(c), \sigma \rangle \longmapsto \langle relax\text{-}tell(c), red(\sigma \wedge ap(\sigma, c)) \rangle}$$

The third condition (non-termination) imposes to keep *relax-tell(c)* in the resulting configuration while the last condition (non-redundancy) avoids the infinite application of the rule. Progress is achieved here together with the conjunction rules because the search space is pruned by the new constraints.

Floundering: Floundering occurs when there is no termination and when the approximation is entailed by the accumulated constraints.

$$\frac{\begin{array}{l} \mathcal{D} \models (\exists)\,(r(\sigma) \wedge c) \\ \mathcal{D} \models \neg(\forall)\,(ap(\sigma, c) \Rightarrow (r(\sigma) \wedge c)) \\ \mathcal{D} \models (\forall)\,(\sigma \Rightarrow ap(\sigma, c)) \end{array}}{\langle relax\text{-}tell(c), \sigma \rangle \longmapsto flounder}$$

4.3 Relaxed Ask

Relaxed ask can be defined in a straightforward way by checking entailment of the non-basic ask-constraint by the relaxation of the accumulated constraints.

$$\frac{\mathcal{D} \models (\forall)\,(r(\sigma) \Rightarrow c)}{\langle relax\text{-}ask(c) \; \to \; G, \sigma \rangle \longmapsto \langle G, \sigma \rangle}$$

$$\frac{\mathcal{D} \models (\forall)\,(r(\sigma) \Rightarrow \neg c)}{\langle relax\text{-}ask(c) \; \to \; G, \sigma \rangle \longmapsto \sigma}$$

$$\frac{\begin{array}{l} \mathcal{D} \models \neg(\forall)\,(r(\sigma) \Rightarrow c) \\ \mathcal{D} \models \neg(\forall)\,(r(\sigma) \Rightarrow \neg c) \end{array}}{\langle relax\text{-}ask(c) \; \to \; G, \sigma \rangle \longmapsto flounder}$$

4.4 Relations with Ask and Tell

The soundness of the transition system can be proven by structural induction on the configurations. More interesting perhaps is the relationships between ask and tell and their relaxed versions in the case where *ask* and *tell* can be decided in the computation domain. Assume that P^* is the program P where all occurrences of relaxed ask and relaxed tell have been replaced by ask and tell. We state the following property without proof.

Property 19

- $SS(P) \subseteq SS(P^*)$

- $FS(P) \subseteq FS(P^*)$

5   Non-Basic CLP(F)

In Section 2, we have presented the basic constraints of CLP(F). These constraints can be handled by an efficient complete constraint-solver but are certainly not expressive enough to be the only constraints available in CLP(F). Moreover we have shown that apparently simple extensions to the basic constraints can lead to an NP-Complete constraint-solving problem. To complete the definition of CLP(F), it remains to specify the non-basic constraints and to define (1) the relaxations, (2) the approximations, and (3) the constraint-solvers.

Definition 20 Let σ be a conjunction of basic constraints in reduced form and $\langle AC, DC \rangle$ its associated system of constraints. The relaxation $r(\sigma)$ is the conjunction $x_1 \in D_1 \wedge \ldots \wedge x_n \in D_n$ such that $DC = \{x_1 \in D_1, \ldots, x_n \in D_n\}$.

In other words, the relaxation in CLP(F) simply ignores all constraints but the domain constraints. Note also that computing the relaxation does not induce any cost since the accumulated constraints are already in reduced form for incrementality purpose.

There are two approximation functions depending if full k-consistency is achieved (i.e. ap_1) or if the reasoning is only performed on the minimum and maximum values in the domains (i.e. ap_2). For some symbolic constraints, both approximations might be useful depending upon the problem at hand.

Definition 21 Let σ be a conjunction of basic constraints and c be a non-basic constraint with variables $x_1, \ldots, x_n$ such that $\mathcal{D} \models (\exists)(r(\sigma) \wedge c)$. Let $dom(r(\sigma), c)$ be the set of natural number tuples

$$\{\langle a_1, \ldots, a_n \rangle \mid \mathcal{D} \models ((r(\sigma) \wedge c)[x_1/a_1, \ldots, x_n/a_n])\}).$$

The approximation $ap_1(\sigma, c)$ is defined by $x_1 \in D_1 \wedge \ldots \wedge x_n \in D_n$, and the approximation $ap_2(\sigma, c)$ is defined by $x_1 \in \{min_1, \ldots, max_1\} \wedge \ldots \wedge x_n \in \{min_n, \ldots, max_n\}$, where D_i represents the projection of $dom(r(\sigma), c)$ along argument i, and min_i and max_i represent the minimum and maximum value in D_i.

Finally, the constraint-solvers for the non-basic constraints are once again based on consistency techniques. However they use the semantics of the constraints to come with an efficient implementation. For instance, inequalities and equations use a reasoning about variation intervals [5, 16]. There can be a large variety of (numeric and symbolic) constraints that might be considered as interesting primitive constraints so that we will not specify the constraint-solvers for them. Note only that the constraint-solvers can be made complete as only domain constraints (i.e relaxation of basic constraints) are considered in conjunction with a non-basic constraint. Also, in [17], we propose a new combinator that makes possible to define most interesting numerical and symbolic constraints from a small set of primitive constraints.

6 Conclusion

This paper has presented an extension to the *Ask & Tell* framework to capture, in a simple and precise way, the operational semantics of CLP languages where some constraints are approximated, inducing an incomplete constraint-solver. The extension consists of two new combinators, relaxed tell and relaxed ask, that are parametrized on a relaxation and approximation function. Constraint are divided into two sets, basic and non-basic constraints. Basic constraints (that can be decided efficiently) are handled as usual while non-basic constraints (that are approximated in terms of basic constraints) are handled through the new combinators.

The extended framework has been instantiated to CLP over finite domains. The basic constraints of CLP(F) have been identified and arc-consistency has been shown to be the basis of an efficient and complete constraint-solver. The relaxation and approximation functions for non-basic constraints in CLP(F) have also been described.

The contributions of this paper include (1) a precise and simple definition of the operational semantics of CLP(F) and (2) the definition of two new combinators that should allow for a precise operational semantics of several other CLP languages.

A structural operational semantics does not describe how to implement the language. What remains to be described for that purpose is how the constraint-solvers are implemented and under which conditions a non-basic constraint is reconsidered after floundering. An efficient arc-consistency algorithm for basic constraints is defined in [4]. The wakening of non-basic constraints is based on a generalisation of the techniques used for delay mechanisms (e.g. [11, 2]). Both issues are beyond the scope of this paper.

Future work includes the study of the properties of the above operational semantics as well as its relationships with the declarative and denotational semantics. Application of the approach to other CLP languages with an incomplete constraint-solver will also be considered.

Acknowledgments

Vijay Saraswat gave the initial motivation beyond this paper. His comments were instrumental in showing the value of the SOS semantics. Discussions with Baudouin Le Charlier and the CHIP team of ECRC are also gratefully acknowledged.

References

[1] W Buttner and al. A General Framework for Constraint Handling in Prolog. Technical report, Siemens Technical Report, 1988.

[2] A. Colmerauer. PROLOG II: Manuel de Reference et Modele Theorique. Technical report, GIA - Faculte de Sciences de Luminy, March 1982.

[3] A. Colmerauer. An Introduction to Prolog III. *CACM*, 28(4):412–418, 1990.

[4] Y. Deville and P. Van Hentenryck. An Efficient Arc-Consistency Algorithm for a Class of CSP Problems. Technical Report CS-90-36, CS Department, Brown University, 1990.

[5] M. Dincbas, H. Simonis, and P. Van Hentenryck. Extending Equation Solving and Constraint Handling in Logic Programming. In MCC, editor, *Colloquium on Resolution of Equations in Algebraic Structures (CREAS)*, Texas, May 1987.

[6] M. Dincbas, P. Van Hentenryck, H. Simonis, A. Aggoun, T. Graf, and F. Berthier. The Constraint Logic Programming Language CHIP. In *FGCS-88*, Tokyo, Japan, 1988.

[7] J. Jaffar and J-L. Lassez. Constraint Logic Programming. In *POPL-87*, Munich, FRG, 1987.

[8] J. Jaffar and S. Michaylov. Methodology and Implementation of a CLP System. In *ICLP-87*, Melbourne, Australia, May 1987.

[9] A.K. Mackworth. Consistency in Networks of Relations. *AI Journal*, 8(1):99–118, 1977.

[10] R. Mohr and T.C. Henderson. Arc and Path Consistency Revisited. *AI Journal*, 28:225–233, 1986.

[11] L. Naish. *Negation and Control in Prolog*. PhD thesis, University of Melbourne, Australia, 1985.

[12] W. Older and A. Vellino. Extending Prolog with Constraint Arithmetics on Real Intervals. In *Canadian Conference on Computer & Electrical Engineering*, Ottawa, 1990.

[13] G.D. Plotkin. A Structural Approach to Operational Semantics. Technical Report DAIMI FN-19, CS Department, University of Aarhus, 1981.

[14] V.A. Saraswat. *Concurrent Constraint Programming Languages*. PhD thesis, Carnegie-Mellon University, 1989.

[15] P. Van Hentenryck. A Framework for Consistency Techniques in Logic Programming. In *IJCAI-87*, Milan, Italy, August 1987.

[16] P. Van Hentenryck. *Constraint Satisfaction in Logic Programming*. Logic Programming Series, The MIT Press, Cambridge, MA, 1989.

[17] P. Van Hentenryck and Y. Deville. A new Logical Connective and its Application to Constraint Logic Programming. Technical Report CS-90-24, CS Department, Brown University, 1990.

[18] P. Voda. The Constraint Language Trilogy: Semantics and Computations. Technical report, Complete Logic Systems, North Vancouver, BC, Canada, 1988.

[19] C. Walinsky. CLP($\sum^*$). In *ICLP-89*, Lisbon, Portugal, 1989.

Constraints for Synchronizing Coarse-grained Sequential Logic Processes

Antonio Brogi, Maurizio Gabbrielli

Dipartimento di Informatica, Università di Pisa

Corso Italia 40, 56125 Pisa, Italy

e-mail: {brogi,gabbri}@dipisa.di.unipi.it

Abstract

We present a constraint based language designed to support the parallel execution of sequential logic theories. Each sequential computation is embedded in a (coarse-grained) process. According to the concurrent constraint paradigm, the parallel execution of processes is performed by using constraints for inter-process communication and synchronization. The operational semantics of the language is modelled according to the true concurrency approach.

1 Introduction

Recently, some efforts [2] have been devoted to realize concurrent systems where several logic theories (for instance Prolog programs) work in parallel by synchronizing via another shared theory. The granularity of processes (i.e. the theories) to be elaborated in parallel is coarse, mostly when compared with the fine-grained parallelism of concurrent logic languages [16]. Other proposals [1,9] define models for combining concurrent reactive and nondeterministic transformational languages, where a tight integration of don't know and don't care nondeterminism makes the computational model rather complex.

In this paper we propose a constraint based framework for a logic language which supports the parallel execution of sequential logic theories based on a smooth integration of the committed choice and the don't know nondeterminisms. Coarse-grained parallel processes synchronize and communicate via a shared set of constraints, as originally proposed in the concurrent constraint logic programming paradigm [14,15]. Differently from [14,15], each parallel process encapsulates a sequential theory, for example represented by a CLP (Constraint Logic Programming) [10] program. Several parallel processes synchronize via the shared state by checking the satisfiability of some constraints ("ask") in order to activate the theory, and by adding new constraints ("tell") resulting from the sequential derivation. The don't know nondeterminism is exploited within the sequential theories only and no distributed backtracking is needed. The operational semantics of the language can be modelled according to the true concurrent operational semantics approach [5,4].

A key issue is the neat separation between the sequential language and the synchronization primitives. The consequence of such a separation is twofold. On one hand, the sequential language can be any particular instance of the CLP framework as well as any other language or constraint solving system. It can also be any other programming language, if equipped with a suitable external interface defined in terms of constraints. The communication mechanism can thus be used for integrating different programming languages within a well-defined concurrent framework. On the other hand, from the implementation point of view, the communication issues can be addressed by exploiting standard tools (such as imperative concurrent languages or operating system primitives). The proposed framework can also be viewed as a rational redefinition of other existing proposals of multi-threaded logic languages. The remaining part of the paper is organized as follows. In the next section the formal definition of the language is given. Section 3 is devoted to discuss programming examples. Section 4 defines the operational semantics according to the true concurrency approach. Finally section 5 contains a discussion on the relation of our framework to other existing proposals and on some implementation issues.

2 Formal Definition

In this section we formally define the language $\mathcal{L}(C, S_C)$, which is parametric w.r.t. a constraint system C and a sequential language S_C, which is based on C too. The language $\mathcal{L}(C, S_C)$ allows to define separate S_C sequential computations, which are encapsulated into $\mathcal{L}(C, S_C)$ logic processes to be executed in parallel and which synchronize via a shared set C of constraints. Let us first introduce the definition of constraint system according to [15].

Definition 2.1 (Constraint system)
A first order constraint system is a quadruple $C = \langle \Sigma, A, Var, \Phi \rangle$ where Σ is a many sorted vocabulary with associated set of sorts S and ranking function ρ, A is a Σ-structure, Var is an S-sorted set of variables and Φ, the set of admissible constraints, is some non-empty subset of (Σ, Var)-formulas, closed under conjunction. □

As usual, the A-valuation is a mapping from Var to the domain of A. A constraint can be considered as the implicit definition of the possibly infinite set of its solutions, that is the valuations which satisfy the constraint. A widely used constraint system in logic programming is the Herbrand system which interprets the constraints on the free Σ-algebra.

The language S_C can be any sequential language which is defined on the constraint system C, that is any language whose semantics is defined by an input-output function on the domain of the constraint system C. Therefore we give the following general definition.

Definition 2.2 (S_C)
Let $C = \langle \Sigma, A, Var, \Phi \rangle$ be a constraint system where $\Pi_c \subseteq \Sigma$ is the set of constraint predicate symbols. Let Π_{st} be a set of many sorted predicate symbols with $\Pi_{st} \cap \Pi_c = \emptyset$. Then a sequential constraint language S_C on C, Π_{st} is any language such that:

1. *computations are activated by a pattern $\langle [p_1(\tilde{X}_1), \ldots, p_h(\tilde{X}_h)], \sigma \rangle$ where $p_i \in \Pi_{st}$ and σ is a set of constraints on C,*

2. *the result of a computation is a pair $\langle \Psi, \sigma' \rangle$ where $\Psi \in \{success, failure\}$ is a termination mode and σ' is a set of constraints on C.*

□

A CLP language [11,10], instantiated on the C domain, is definitely the most appropriate candidate for the S_C language of sequential theories, since the constraint based sequential computation of CLP naturally fits into the concurrent constraint framework. Sequential computations can exploit the internal don't know nondeterminism (implemented by back-tracking), which does not affect the (external) nondeterminism of the $\mathcal{L}(C, S_C)$ concurrent processes, ruled by the usual committed choice policy. According to the previous definition, CLP programs implementing sequential theories given an initial goal possibly terminate with a single answer (constraint). Multiple answers could be considered as well by allowing disjunction of constraints in the constraint system.

Even if it is the most appropriate, CLP is not the only kind of language that can be used in $\mathcal{L}(C, S_C)$ for the sequential part. Any sequential language whose input-output can be translated into a constraint form can be considered. Therefore we give the definition of $\mathcal{L}(C, S_C)$ parametrically w.r.t. a generic S_C language.

A generic $\mathcal{L}(C, S_C)$ process $p(X)$ is defined by a set of clauses of the form

$$p(X) :\text{-} \; ask : tell \rightarrow SeqGoal \; ; \; Tail$$

where *ask* and *tell* are constraints on C specifying the synchronization conditions for activating the sequential computations. *SeqGoal* is the initial goal for a S_C sequential theory. *Tail* contains the set of processes which are created after the termination of the sequential computation. Let us now give the formal definition of $\mathcal{L}(C, S_C)$.

Definition 2.3 (Syntax of $\mathcal{L}(C, S_C)$)
Let $C = \langle \Sigma, \mathcal{A}, Var, \Phi \rangle$ be a constraint system, where $\Pi_c \subseteq \Sigma$ is the set of constraint predicate symbols. Let Π_{st}, Π_{com} be sets of many sorted predicate symbols (with their signatures) such that $\Pi_{st} \cap \Pi_{com} = \emptyset$ and $(\Pi_{st} \cup \Pi_{com}) \cap \Pi_c = \emptyset$. Let S_C be any sequential language on C, Π_{st}. Then the formal syntax of $\mathcal{L}(C, S_C)$ is the following.

Clause	::=	$Head : - Agent$	Head	::=	$p(\tilde{X})$
Agent	::=	$nil \mid \exists \tilde{X}. ask : tell \rightarrow SeqGoal; Tail$	ask	::=	$c \in \Phi$
SeqGoal	::=	$nil \mid b(\tilde{X}) \mid b(\tilde{X}), SeqGoal$	tell	::=	$c \in \Phi$
Tail	::=	$nil \mid Head \mid Tail \parallel Tail$			

where $p \in \Pi_{com}$, $b \in \Pi_{st}$, $\rightarrow$ is the commit operator, $\parallel$ is the AND-parallel conjunction, which is assumed to be commutative. $\tilde{X}$ is a subset of Var . $\exists \tilde{X}$ denotes the existential quantification of the (possibly empty) set of variables $\tilde{X}$. Π_{com} is the collection of symbols for communication predicates, Π_{st} is the (disjoint) set of sequential theories predicate symbols. A top-level goal specifies the initial configuration of a $\mathcal{L}(C, S_C)$ system and is defined as follows:

$$TopGoal ::= Tail :: \sigma$$

where $\sigma \in \wp_f(\Phi)$, Φ denotes the set of all constraints (definition 2.1) and $\wp_f(\Phi)$ denotes the set of the finite parts of Φ. □

According to the previous definition, an agent represents a coarse-grained parallel process, which contains an *ask:tell* constraint, a sequential goal (*SeqGoal*) and a rewriting pattern (*Tail*) to activate other processes. Existential quantification is introduced to allow local variables. A commit operator ($\rightarrow$) follows the *ask:tell* guard.

A formal operational semantics for the $\mathcal{L}(C, S_C)$ language is given in section 4. Informally, a $\mathcal{L}(C, S_C)$ computation can be described as follows. According to the cc framework [15], at the communication processes level, the computation can be considered as the concurrent execution of agents which interact via a global set of constraints (named *store*) in order to synchronize and exchange the results of their computations. The basic actions performed by the agents, are asking for a constraint c being entailed by the store σ, and telling a constraint c to the store σ, if $c \cup \sigma$ is consistent. These basic computation steps allow to continue, suspend or fail the computation of an agent, depending on the result of the ask:tell evaluation. Such an evaluation is specified as follows.

Evaluation of ask:tell

Let us denote by $(\exists)(c)$ and $(\forall)$ the existential and the universal closure of c respectively. The evaluation of an ask constraint a in a store σ *succeeds* if $C \models (\forall)\sigma \Rightarrow a$ (σ entails a), *fails* if $C \models (\forall)\sigma \Rightarrow \neg a$ ($\sigma \wedge a$ is inconsistent), *suspends* if $C \models (\exists)(\sigma \wedge a)$ and $C \not\models (\forall)\sigma \Rightarrow a$ ($\sigma \wedge a$ is consistent but σ does not entail a).

The evaluation of a tell constraint t in the store σ either *succeeds* if $C \models (\exists)(\sigma \wedge t)$ or *fails* if $C \models (\forall)\sigma \Rightarrow \neg t$.

The evaluation of an $a : t$ (ask:tell) element is performed as an atomic action. It succeeds if the evaluations of both a and t succeed. It fails if the evaluation of either a or t fails and suspends otherwise. If $a : t$ succeeds, the store is updated, the new store being $\sigma' = \sigma \wedge t$. If the evaluation fails the store is unchanged.

The $\|$ operator has the usual meaning of parallel execution of agents, where the evaluation of $A_1 \| A_2$ succeeds iff the evaluation of both A_1 and A_2 succeeds, fails iff the evaluation of either A_1 or A_2 fails and suspends iff neither A_1 nor A_2 fail and A_i suspends, $i \in \{1,2\}$. The nondeterminism arising from clause selection is handled according to the committed choice rule.

Rule of commitment

Given a procedure call $p(\tilde{X})$ in a store σ, in order to continue the computation a clause $p(\tilde{Y}) : -Agent$ is nondeterministically selected among those whose guard (i.e. *ask : tell*) evaluation in the store $\sigma \cup \{\tilde{X} = \tilde{Y}\}$ can be successfully performed. If the evaluation of the guard of every clause for p suspends, then the computation of $p(\tilde{X})$ suspends (and can possibly be resumed later, if the store will provide further informations for the successful evaluation of the ask).

After the successful execution of the $ask : tell$ constraint in an agent $A = (ask : tell \rightarrow SeqGoal; Tail)$, the sequential computation starts according to the given sequential goal SeqGoal. This computation is performed on a local environment of the sequential theory. Different sequential computations can be performed in parallel by different agents provided that there has been a synchronization with the store. If the sequential computation for SeqGoal successfully terminates, its result is a constraint on local variables which has to be "projected" on the global variables determined by the initial activation pattern. The resulting constraint α is then added to the global store σ by an implicit tell operation. If either $\sigma \wedge \alpha$ is not consistent or the result of the sequential computation is a finite failure, the whole computation terminates with failure. Otherwise, if $\sigma \wedge \alpha$ is consistent (i.e. $C \models (\exists)(\sigma \wedge \alpha)$) the store becomes $\sigma \wedge \alpha$ and the (parallel) computation of the elements of the Tail is started. The element nil has the obvious operational meaning. A generic top-goal initial goal has the form $G :: \sigma$, where G is a Tail element and σ is a store. The computation terminates with *failure* when the evaluation of an agent finitely fails.

3 Programming Examples

Before introducing the formal operational semantics of $\mathcal{L}(\mathcal{C}, \mathcal{S}_\mathcal{C})$, let us discuss some programming examples. The main motivation for the definition of $\mathcal{L}(\mathcal{C}, \mathcal{S}_\mathcal{C})$ is to provide a general framework where sequential processes work in parallel and synchronize by means of constraints. The degree of parallelism in a program execution strictly depends on the granularity of the sequential computations. If the sequential processes are coarse-grained then the degree of parallelism is high since synchronization steps are rather unfrequent. On the other hand, if the grain size of sequential processes decreases then the number of synchronizations grows.

One of the best known application areas of A.I. is the so called *distributed problem solving*. Generally speaking, the task of solving a problem on some domain is approached by means of several agents. Knowledge is partitioned into several distinct agents which cooperate in accomplishing a task. The agents can work in parallel and, in many cases, each agent contains detailed knowledge on a particular (sub) domain. Given a general problem, the process of problem solving consists of separately solving sub-problems in parallel as well as of coordinating parallel activities (synchronization). A particularly suitable programming example for $\mathcal{L}(\mathcal{C}, \mathcal{S}_\mathcal{C})$ is the problem of solving systems of constraints, which can be tackled by solving several subsystems of constraints in parallel by specialized constraint solvers.

A large class of applications for $\mathcal{L}(\mathcal{C}, \mathcal{S}_\mathcal{C})$ can be inspired by blackboard systems [6]. The language Shared Prolog [2] is an example of a concurrent logic language based on the blackboard model of problem solving. Relations between $\mathcal{L}(\mathcal{C}, \mathcal{S}_\mathcal{C})$ and Shared Prolog are discussed in section 5. Here, the definition of a blackboard system given in [2] is adapted to $\mathcal{L}(\mathcal{C}, \mathcal{S}_\mathcal{C})$. In the blackboard model of problem solving, knowledge is divided into separate agents which work in parallel communicating (and synchronizing) via a shared blackboard. The blackboard is a shared data structure containing the current state of the solution. A special agent, called *control*, monitors the changes on the blackboard and decides which agent to start. Let us consider an example from a general class of blackboard systems where the blackboard data structure has a hierarchical structure. In particular, we discuss a problem of *data fusion*, originally presented in [3] and then programmed in Shared Prolog [2]. The problem of data fusion consists of merging data coming from different sensors and other information sources. Data are organized into a hierarchical structure, where input data belong to the lowest level of the hierarchy. Each agent inputs data at one level of the hierarchy, elaborates them, and produces higher level data as results.

Let us consider a simplified version of the system described in [3], which elaborates data coming from different sensors (e.g. radars) and tries to identify the objects moving in a given region of space. Suppose that the initial *TopGoal* of the system activates three processes in parallel:

```
data_acquisition(Plots, Rcs) || rcs_filter(Rcs, Rcsval, Status) ||
plot_filter(Plots, Altitude, Speed, Status')
```

The *data acquisition* process collects packets of plots coming from several sensors and packets of radar cross sections (rcs). The technical meaning of data is rather irrelevant here (plots denote tracks of detected objects and so on). The task of the *rcs_filter* process is to input a packet of rcs data (*Rcs*), to filter out measurements errors and to produce filtered rcs data (*Rcsval*). When the *rcs_filter* process is initialized (i.e. during its first activation), it dynamically activates a new process *rcs_expert*. Such a process will elaborate the data produced by the *rcs_filter* process and produce, in turn, higher level data. For simplicity of notation, in this example all the variables occurring in the *ask* part of a clause but not in the *head* are intended to be existentially quantified. Moreover, the *ask* constraints should be repeted in the *tell* part, too.

```
rcs_filter(Rcs, Rcsval, Status) :-
  Rcs=R.NextR, Rcsval=RV.NextRV, Status=initializing :
  Status'=initialized → rcs_filtering_procedure(R, RV);
  rcs_filter(NextR, NextRV, Status') ‖ rcs_expert(Rcsval, Rcs_hyp, Status)

rcs_filter(Rcs, Rcsval, Status) :-
  Rcs=R.NextR, Rcsval=RV.NextRV, Status=initialized :
  true : → rcs_filtering_procedure(R, RV);
  rcs_filter(NextR, NextRV, Status)
```

A committed-choice behaviour rules the selection of which definition of *rcs_filter* to choose depending on the internal status of the process itself.

The *plot_filter* process after filtering out measurement errors, derives data from the plots such as *Altitude* and *Speed*. The definition of the process consists of two alternative definitions as in the previous case.

```
plot_filter(Plots, Altitude, Speed, Status) :-
  Plots=P.NextP, Altitude=A.NextA, Speed=S.NextS, Status=initializing:
  Status'=initialized → filtering_procedure(P, A, S);
  plot_filter(NextP, NextA, NextS, Status') ‖
  altitude_expert(Altitude, Altitude_hyp) ‖ speed_expert(Speed, Speed_hyp)

plot_filter(Plots, Altitude, Speed, Status) :-
  Plots=P.NextP, Altitude=A.NextA, Speed=S.NextS, Status=initialized:
  true → filtering_procedure(P, A, S);
  plot_filter(NextP, NextA, NextS)
```

Notice that the filtering procedures are standard Prolog or CLP programs, while in real-time systems they might more realistically be C routines [3]. Since the task of the system is to dynamically monitor a region of space, packets of data continuously arrive and have to be elaborated. Correspondingly, the processes recursively activate themselves on streams of data at the end of each cycle. Other processes working at higher levels of abstraction have a similar structure. The *speed_expert* process, for instance, given the detected speed of an object, draws an hypothesis on the type of the object. Different knowledge bases and different selection criteria are used depending on the value of the speed.

```
speed_expert(Speed, Speed_hyp) :-
  Speed=S.NextS, S ≤ 40:
  Speed_hyp=Sh.NextSh → search_slow_obj(S, Sh);
  speed_expert(NextS, NextSh)
```

```
speed_expert(Speed, Speed_hyp) :-
  Speed=S.NextS, S > 40:
  Speed_hyp=Sh.NextSh → search_fast_obj(S, Sh);
  speed_expert(NextS, NextSh)
```

Another case of systems where coarse-grained processes synchronize are operating systems. A very common situation is to have a manager of some resources (e.g. files, printers, databases) and several processes running in parallel and sharing those resources. Accesses to shared resources are ruled by mutual exclusion protocols. The problem of solving mutual exclusion protocols in concurrent logic programming has been addressed by many authors (see for instance [16]) and can be rephrased according to the constraints paradigm (on the line of [14]). The major difference w.r.t. concurrent logic languages is that after a process has worked in mutual exclusion from others, it can perform a *purely* sequential computation.

4 Operational Semantics

In this section we define an operational semantics for the language. The first purpose is to provide a formal description of the operational behavior. The second aim is to discuss the properties of the language from the viewpoint of distribution. This is the reason why we have decided to define the operational semantics following the true concurrency approach, by using the constructive technique presented in [5], which extends the Plotkin's Structured Operational Style [12] to deal with distributed states. It would be also possible to define a fixpoint semantics based on an immediate consequence operator by generalizing the techniques used in [7] and [8]. Such a construction, however, would be based on interleaving and would not properly outline the above mentioned distribution properties.

According to [5], a $\mathcal{L}(\mathcal{C}, \mathcal{S}_\mathcal{C})$ system is decomposed into a set of configurations representing processes, possibly located in different places, which cooperate in accomplishing a task. A set of configurations represents the distributed state of the system. The behavior of a system is defined by a set of rewriting rules to be applied to sets of configurations to get computations. A rewriting rule specifies how a subset of the configurations composing a distributed state may evolve independently. A computation can be described by a sequence of distributed states (sets of configurations) and of rewriting rules, representing the evolution of system sub-parts.

Rewriting rules are adopted to describe the synchronization of parallel processes. The sequential $\mathcal{S}_\mathcal{C}$ derivations encapsulated into processes have to be modelled as well. To this purpose, we assume that a transition system describing $\mathcal{S}_\mathcal{C}$ computations is defined. Then, rewriting rules possibly resort to the transition system for $\mathcal{S}_\mathcal{C}$ when dealing with sequential computations. In other words, this corresponds to describing the semantics of $\mathcal{L}(\mathcal{C}, \mathcal{S}_\mathcal{C})$ by means of a hierarchy of two abstract machines (as in [2]). The lower level machine, defined by a transition system, describes the evolution of sequential computations. The higher level machine, defined by a set of rewriting rules, describes the behavior of $\mathcal{L}(\mathcal{C}, \mathcal{S}_\mathcal{C})$ programs.

Each sequential computation consists of the execution of a $\mathcal{S}_\mathcal{C}$ program for a given sequential goal *SeqGoal*. The derivation of *SeqGoal* can be either successful or failing. Given a store σ, the possible computations of *SeqGoal* can be described by the two following transitions, according to 2.2:

$$\langle SeqGoal, \sigma \rangle \mapsto_{\mathcal{S}_\mathcal{C}} \langle success, \sigma' \rangle \qquad \qquad \langle SeqGoal, \sigma \rangle \mapsto_{\mathcal{S}_\mathcal{C}} \langle failure, \sigma' \rangle$$

The first one means that given a store σ, $SeqGoal$ can be proved by computing the new store σ'. The second transition models the finite failure of the computation.

We are mostly interested in modelling the aspects of concurrency, so that the definition of the transition system for S_C is omitted here. Notice, however, that it can be easily defined (once S_C has been suitably specified) as it has been done in [13,4] for the Prolog language.

Given that we know how to model sequential derivations, we have to define how a system composed of a set of concurrent agents and a store behaves. A set of rewriting rules is introduced to describe how sets of configurations can evolve. The set of configurations for the $\mathcal{L}(C, S_C)$ language must describe all the possible states of the processes, that is of the store and of the agents. In order to describe a store σ, the configuration $\langle \sigma \rangle$ is adopted, so that a store is denoted by the set of constraints it contains. During a computation, an agent can be in one of the three following states:

1. waiting for starting a sequential computation,

2. performing a sequential computation,

3. waiting for adding to the store the results of its terminated sequential computation.

The first state is denoted by a configuration of the form $\langle Head \rangle$ where $Head$ is the name of the process waiting for synchronizing with the store. An agent which is currently performing a sequential computation is denoted by the pair $\langle (SeqGoal; Tail), \sigma \rangle$ where $SeqGoal$ is the sequential goal currently being executed, $Tail$ is the continuation of the process and σ is the local store of the process. Finally, an agent which is waiting for telling to the store the result of its succesfully terminated computation is denoted by the pair $\langle Tail, \sigma \rangle$, where $Tail$ is the tail recursion and σ is the local store, as updated by the sequential computation. Moreover, the special configuration $\langle failure \rangle$ represents a failure.

Definition 4.1 (Configurations)
The set of configurations is defined as follows:

$$\Gamma = \{\langle \sigma \rangle, \langle Head \rangle, \langle (SeqGoal; Tail), \sigma \rangle, \langle Tail, \sigma \rangle, \langle failure \rangle\}$$

where Head, SeqGoal and Tail are defined according to definition 2.3. Configurations are ranged over by γ. □

A set of configurations can syntactically be obtained from the definition of a $\mathcal{L}(C, S_C)$ goal.

Definition 4.2 (From $\mathcal{L}(C, S_C)$ goals to sets of configurations)
Let Θ be the set of all possible TopGoal and Tail (definition 2.3). Then the function
$dec : \Theta \to 2^\Gamma$ is defined by induction on the syntax of $\mathcal{L}(C, S_C)$ as follows:

$$
\begin{aligned}
dec(Tail :: \sigma) &= dec(Tail) \cup \{\langle \sigma \rangle\} \\
dec(Tail \parallel Tail') &= dec(Tail) \cup dec(Tail') \\
dec(Head) &= \{\langle Head \rangle\} \\
dec(nil) &= \emptyset
\end{aligned}
$$ □

Definition 4.3 (Evaluation of ask:tell)
Let Φ the set of all constraints (definition 2.1) and let Δ the set of all the ask:tell elements. Then the function $\nu : \wp_f(\Phi) \times \Delta \to \wp_f(\Phi)$, where $\wp_f(\Phi)$ denotes the set of the finite parts of Φ, defines the ask : tell evaluation rule as follows:

$$\nu(\sigma, \exists \tilde{Y}.ask : tell) \begin{cases} \sigma \wedge tell & if \; C \models (\forall)\sigma \Rightarrow \exists \tilde{Y}.ask \; and \; C \models (\exists)(\sigma \wedge ask \wedge tell) \\ fail & if \; C \models (\forall)\sigma \Rightarrow (\neg \exists \tilde{Y}.ask \wedge tell) \end{cases}$$

The function $\rho : \wp_f(\Phi \times SeqGoal) \to \wp_f(\Phi)$, given a store σ and a SeqGoal, returns a restriction of σ which depends on SeqGoal. □

In the previous definition, a specific definition of ρ depends on the specific sequential language. For example, $\rho(\sigma, SeqGoal)$ can be the restriction of σ to the variables in $SeqGoal$. More generally (by slightly modifying the syntax) the goal $SeqGoal$ can contain an explicit information on the restriction to be operated on σ. Let us now give the set of rewriting rules describing $\mathcal{L}(C, S_C)$.

Definition 4.4 (Rewriting Rules)
Let P be a $\mathcal{L}(C, S_C)$ program, the $\mathcal{L}(C, S_C)$ derivation relation $\Rightarrow_P$ is defined as the least relation satisfying the following rewriting rules.

(1) $$\frac{\exists \, (p(\tilde{X}) : -\exists Z.ask : tell \to SeqGoal; Tail) \in P \; s.t. \; \nu(\sigma \wedge (\tilde{X} = \tilde{Y}), \exists Z.ask : tell) = \sigma'}{\{\langle \sigma \rangle\} \cup \{\langle p(\tilde{Y}) \rangle\} \Rightarrow_P \{\langle \sigma' \rangle\} \cup \{\langle (SeqGoal; Tail), \sigma' \rangle\}}$$

(2) $$\frac{(SeqGoal, \sigma) \mapsto_{S_C} (success, \sigma')}{\{\langle (SeqGoal; Tail), \sigma \rangle\} \Rightarrow_P \{\langle Tail, \rho(\sigma', SeqGoal), \rangle\}}$$

(3) $$\frac{(SeqGoal, \sigma) \mapsto_{S_C} (failure, \sigma')}{\{\langle (SeqGoal; Tail), \sigma \rangle\} \Rightarrow_P \{\langle failure \rangle\}}$$

(4) $$\frac{\nu(\sigma, \{\} : \sigma') = \langle \sigma'' \rangle}{\{\langle \sigma \rangle\} \cup \{\langle Tail, \sigma' \rangle\} \Rightarrow_P \{\langle \sigma'' \rangle\} \cup dec(Tail)}$$

(5) $$\frac{\nu(\sigma, \{\} : \sigma') = fail}{\{\langle \sigma \rangle\} \cup \{\langle Tail, \sigma' \rangle\} \Rightarrow_P \{\langle failure \rangle\}}$$

(6) $$\frac{}{\{\gamma\} \cup \{\langle failure \rangle\} \Rightarrow_P \{\langle failure \rangle\}}$$ □

marks

As usual, we leave the parameter P implicit in the definition of the $\Rightarrow_P$ relation since the program does never change during the derivation. Rule 1 defines the activation of a process $p(\tilde{X})$. Rules 2 and 3 define the derivation of the result of a sequential computation, both resorting to the transition system for S_C. Rules 4 and 5 correspond to the *tell* operation of the local store which has been computed by a successful sequential computation. The tail recursion is defined by resorting to the definition of *dec* (4.2). Rule 6 makes every configuration evaporating in presence of a failure. □

Let us now define what a computation is, by following [4].

Definition 4.5 (Computation)
Let P be a $\mathcal{L}(C, S_C)$ program, and G be a TopGoal. The sequence of states and rewriting rules $\xi = S_0 \; R_0 \; S_1 \; R_1 \ldots$ is a computation iff:

- $S_0 = dec(G)$
- $R_j = \{I_{1,j} \Rightarrow_P I'_{1,j}, \ldots, I_{n,j} \Rightarrow_P I'_{n,j}\}$ *such that:*

- $I_{i,j} \Rightarrow_P I'_{i,j}$ belongs to the $\Rightarrow_P$ derivation relation $\forall i = 1, \ldots, n$
- $I_{1,j} \oplus \ldots \oplus I_{n,j} \subseteq S_j$
- $S_{j+1} = (S_j \backslash (I_{1,j} \oplus \ldots \oplus I_{n,j})) \oplus I'_{1,j} \oplus \ldots \oplus I'_{n,j}$

where $\oplus$ denotes the union of multisets.

A successful computation is a finite sequence $\xi = S_0\ R_0\ S_1\ R_1\ \ldots\ R_{n-1}\ S_n$ such that $S_n = \{\langle\sigma\rangle\}$ where σ is a store. The result of the computation is $\rho(\sigma, G)$, that is the projection of the store on the initial goal (definition 4.3).

Analogously, a failed computation is a finite sequence as above where: $S_n = \{\langle failure \rangle\}$ □

Remarks

A computation is modelled by a (possibly) infinite sequence of states and rewriting rules. At each step of the computation, the configuration denoting the global store represents the current global state of the computation. Each sequential computation which succeeds or finitely fails is denoted by one transition only (rule 2 and 3 of definition 4.4). Infinite sequential computations are interpreted as a local deadlock of the corresponding process, which is denoted by a configuration of the form $\langle\langle SeqGoal; Tail\rangle, \sigma\rangle$. Local deadlocks do not cause global deadlocks. A failure can arise either because a sequential computation finitely fails (rule 3) or because the store which has been locally computed by a sequential derivation is inconsistent with the global one (rule 5). According to rule 6, failures are propagated to all the other components of the system. Notice however, that other choices would have been possible, for instance a finitely failing sequential computation may be interpreted by the corresponding process remaining stuck. □

A true concurrent operational semantics shows the degree of (independent) parallelism which is possible in a given language, that is the number of transitions which can concurrently take place. In the above definition this happens for all the transitions related to the execution of the (sequential) components of different processes. The concurrent constraint framework is clearly a special case, where there are no sequential goals. In such a case, the above defined operational semantics shows that one transition only is possible at any given time. In other words, the true concurrent semantics would be very similar to an interleaving semantics. The relevant issue is therefore exactly in the internal (sequential) structure of processes. Note also that the number of transitions is controlled by the fact that only one process can communicate with the store. In other words, the unique shared set of constraints is a bound to concurrency. This can be improved, by splitting the store into a collection of separate stores, as already suggested in the case of Shared Prolog [2].

5 Comparisons with other Concurrent Languages and Implementation Issues

It is mandatory to compare $\mathcal{L}(\mathcal{C}, \mathcal{S}_C)$ with the family of concurrent logic languages [16] and, in particular, with concurrent constraint programming [14]. The main difference concerns the coarseness of the processes involved in a computation. In the other concurrent logic languages, processes are very fine-grained, that is they correspond to atomic goals. In our framework, AND-parallel processes are intended to be coarse-grained. The degree of parallelism derives from the size of sequential derivations. Indeed, since each parallel process encapsulates a sequential computation, if the sequential computations are short, the granularity of the processes is finer and the degree of parallelism decreases because of the many interactions with the store. The neat separation between the deduction and synchronization steps in $\mathcal{L}(\mathcal{C}, \mathcal{S_C})$ is reflected also by the operational semantics. In fact, the operational semantics of $\mathcal{L}(\mathcal{C}, \mathcal{S_C})$ can naturally be described by means of a distributed model (see section 4), while interleaving models only have been defined so far for the family of concurrent logic languages.

Another related work is Shared Prolog (SP) [2]. An SP system is composed of a set of parallel agents (theories) which are Prolog programs extended by communication patterns ruling the communication and the synchronization via a shared blackboard. The blackboard is a set of unit clauses (facts). The synchronization of the agents with the blackboard consists of adding and/or deleting facts to/from the blackboard. The computational model of SP is similar to that of $\mathcal{L}(\mathcal{C}, \mathcal{S_C})$. However, there are some major differences to be noticed. First, the synchronization model of $\mathcal{L}(\mathcal{C}, \mathcal{S_C})$ is based on constraints rather than on assert and retract operations. Moreover, the dynamic evolution of the shared store of constraints is *monotonic* while the shared blackboard of SP is intrinsically *nonmonotonic* as long as retract operations are permitted. This difference is reflected by the underlying semantic model, when compositionality is taken into account.

Let us now briefly discuss some issues related to the language implementation. The dynamic configuration of a $\mathcal{L}(\mathcal{C}, \mathcal{S_C})$ system consists of a parallel composition of a set of sequential processes and a store. Each sequential process can virtually be associated to a different processor capable of supporting the execution of $\mathcal{S_C}$ programs. Another process is in charge of managing the store. Actually, it will be required to support the standard basic functions of any constraint system, that is to embed some evaluation function working on constraints.

In addition to that, there are communication primitives supporting the exchange of messages between the various processes. Notice that such primitives can be defined as standard inter-process primitives, not being affected by the particular nature of the $\mathcal{L}(\mathcal{C}, \mathcal{S_C})$ language. There are three possible types of messages:

1. messages containing *ask* and *tell* constraints sent by some sequential process to the store manager as a request for starting an internal derivation,

2. messages sent by the store manager to a sequential process containing the authorization to start an internal derivation,

3. messages sent by a sequential process to the store manager.

It is worth noting that messages of type 2 contain - at least virtually - a copy of the current shared store to be passed to a sequential process. Indeed, in a real implementation it is not necessary to copy the whole store. It is sufficient that messages of type 2 contain only a copy of a subset of the store, roughly a suitable projection of the store on the variables occurring in the *Head* of the agent.

Finally, the tail recursion of a process possibly generates several new processes. Correspondingly, the implementation has to provide some primitives for the dynamic creation of processes.

References

[1] R. Bahgat and S. Gregory. Pandora: Non-deterministic Parallel Logic Programming. In G. Levi and M. Martelli, editors, *Proc. Sixth Int'l Conf. on Logic Programming*, pages 471–486. The MIT Press, Cambridge, Mass., 1989.

[2] A. Brogi and P. Ciancarini. The Concurrent Language Shared Prolog. *ACM Transactions on Programming Languages and Systems*, 1(1), 1991.

[3] A. Brogi, R. Filippi, M. Gaspari, and F. Turini. An Expert System for Data Fusion based on a Blackboard Architecture. In *Proc. of Eight Int'l Workshop on Expert Systems and their Applications*, pages 147–166, Avignon, 1988.

[4] A. Brogi and R. Gorrieri. A Distributed, Net Oriented Semantics for Delta Prolog. In J. Diaz and F. Orejas, editors, *Proc. of TAPSOFT-CAAP'89*, volume 351 of *Lecture Notes in Computer Science*, pages 637–654. The MIT Press, Cambridge, Mass., 1989.

[5] P. Degano and U. Montanari. Concurrent Histories: a Basis for Observing Distributed Systems. *Journal of Computer and System Science*, 34:442–461, 1987.

[6] R. Engelmore and T. Morgan. *Blackboard Systems*. Addison-Wesley, 1988.

[7] M. Gabbrielli and G. Levi. Unfolding and Fixpoint Semantics of Concurrent Constraint Programs. In H. Kirchner and W. Wechler, editors, *Proc. Second Int'l Conf. on Algebraic and Logic Programming*, volume 463 of *Lecture Notes in Computer Science*, pages 204–216. Springer-Verlag, Berlin, 1990. Extended version to appear in *Theoretical Computer Science*.

[8] M. Gabbrielli and G. Levi. Modeling Answer Constraints in Constraint Logic Programs. In K. Furukawa, editor, *Proc. Eighth Int'l Conf. on Logic Programming*. The MIT Press, Cambridge, Mass., 1991.

[9] S. Haridi and S. Janson. Kernel Andorra Prolog and its computation model. In D. H. D. Warren and P. Szeredi, editors, *Proc. Seventh Int'l Conf. on Logic Programming*, pages 31–48. The MIT Press, Cambridge, Mass., 1990.

[10] J. Jaffar and J.-L. Lassez. Constraint Logic Programming. In *Proc. Fourteenth Annual ACM Symp. on Principles of Programming Languages*, pages 111–119. ACM, 1987.

[11] J. Jaffar and J.-L. Lassez. Constraint Logic Programming. Technical report, Department of Computer Science, Monash University, June 1986.

[12] G. Plotkin. A structured approach to operational semantics. Technical Report DAIMI FN-19, Computer Science Department, Aarhus University, 1981.

[13] V. A. Saraswat. GHC: operational semantics, problems and relationship with CP($\downarrow$,$|$). In *IEEE Int'l Symp. on Logic Programming*, pages 347–358. IEEE, 1987.

[14] V. A. Saraswat. *Concurrent Constraint Programming Languages*. PhD thesis, Carnegie-Mellon University, January 1989.

[15] V. A. Saraswat and M. Rinard. Concurrent constraint programming. In *Proc. Seventeenth Annual ACM Symp. on Principles of Programming Languages*. ACM, 1990.

[16] E. Y. Shapiro. The family of concurrent logic programming languages. *ACM Computing Surveys*, 21(3):412–510, 1989.

EXTENDING EXECUTION TREES FOR
DEBUGGING AND ANIMATION
IN LOGIC PROGRAMMING

Diego Loyola*
ESLAI
Casilla de correo 1000, Correo central
Buenos Aires, Argentina

Debugging tools that give a global program execution image and direct access to -user selected- relevant information enable rapid error detection. In this work two Prolog debuggers that provide these facilities are presented. The concept of execution tree was extended in order to be used in the debugging and animation process.

The computation of a logic program can be seen as a sequence of nondeterministic goal reductions or as a proof tree. This tree has less information than the sequence of reductions as the clause and substitution used in each step of the derivation is unknown.

The key idea of this work is to associate to each node in the execution tree the derivation information used to solve it, and to obtain the trees corresponding to failed derivations in order to analyze why a goal can not be proved.

The enriched tree represents successful or failed derivations and has a node for each goal generated by the computation. Additional information is associated to the node: the clause used to solve the goal, variable bindings before and after the derivation (substitutions) and information on whether the node derivation was successful or not.

Program analysis is made by means of a flexible navigation through the extended tree. The system provides a visual feedback of program execution enabling access to detailed node information when required (zooming).

Two debugging alternatives are proposed:

a) The tree is stepwise displayed by levels, showing a node and its direct descendings. Each node has a mark (+ or -) indicating the exit or failure in its derivation. The first phase of the analysis includes the root node and its sons. The following steps focus on some of the sons or the parent, according to the user selection.

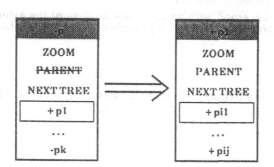

*   Current E-mail: ME1J@DLRVMKP.BITNET

b) The tree is shown graphically (using scrolling facilities) without any additional work from the user. Nodes that can not be proved are marked with a darker color than those that could be derived. The user can select any tree node in order to examine it.

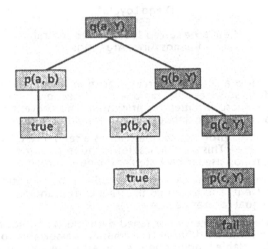

The first debugging alternative is incorporated to Arity Prolog using its meta-level reasoning facilities for computing the extended execution tree while a goal is solved.

The second one is part of a pure Prolog environment implemented in Turbo Prolog. The unification process is extended so that it now also returns the extended tree. The user interacts with a graphical display of this tree. The environment is composed of edition, consult, animation/debug and trace modules, that can be accessed through their own windows.

The extra information provided in each node and the possibility of skipping the derivation of correct or uninteresting parts of the execution allows the user to have a clear idea of the actual program behavior and directly examine suspicious nodes. This leads to fewer user interactions needed in order to detect an error.

Future work will be concerned with an automatic analysis of the extended execution tree looking for 'bug patterns' that identify the possible cause of the program failure.

References
[1] D. Loyola, Animacion y Debuggin en Prolog, Internal Report ESLAI 1988

A Narrowing-Based Theorem Prover

Ulrich Fraus

Bayerisches Forschungszentrum für Wissensbasierte Systeme,
Universität Passau, Postfach 2540, W-8390 Passau, Germany
fraus@forwiss.uni-passau.de

Heinrich Hußmann

Institut für Informatik, Technische Universität München,
Postfach 20 24 20, W-8000 München 2, Germany
hussmann@lan.informatik.tu-muenchen.dbp.de

Abstract

This work presents a theorem prover for inductive proofs within an equational theory which supports the verification of universally quantified equations. This system, called TIP, is based on a modification of the well-known *narrowing algorithm*. Particulars of the implementation are stated and practical experiences are summarized.

Equational axiomatic specifications are now widely accepted as a promising tool for the early phases of software development. The special style of equational specifications is interesting for the intermediate stage in which the first detailed formal description of the intended product is given. The main advantage of an axiomatic approach to software specification is formal reasoning: Axioms, together with a good calculus, allow computer-assisted verification of propositions about the specified piece of software. Within the software development process, such a tool helps, for instance, to check whether a first operational formulation of a system meets more informal requirements which have been formulated without any consideration for the operationality. It is the aim of the system presented here to contribute to such a verification of software in a very early phase of the development.

In our approach, we try to develop an alternative to the so-called "inductionless induction" method, which has been developed based on early work by Knuth/Bendix, Goguen, Huet/Hullot and others. The aim is here, to keep the proof method rather "natural", i.e. very similar to the method a human being uses when performing a proof with paper and pencil. A successful proof achieved by our system is understandable to everybody who knows the basic mathematical facts of equational reasoning and induction. Moreover, the situation in which a new (usually more general) lemma has to be invented looks more familiar to the user - therefore it is easier to use human intuition. And human intuition will be necessary anyway for mastering a non-trivial proof - regardless of the proof method that is used.

The TIP system (Term Induction Prover) presented in this paper, is restricted to so-called constructor-based specifications, i.e. every ground term has to be equivalent to a constructor ground term (sufficient completeness [Gu 75]). All equations have the form:

$$f(c_1, ..., c_n) \int t \qquad \text{where f is a function symbol, the } c_i \text{ are constructor terms (with variables) and t is an arbitrary term.}$$

The basic idea of our variant of term induction is to "misuse" the case analysis which is normally given in the left hand sides of the rules for an inductive proof. This idea allows us to make extensive use of the algorithms that are already available in an implementation of narrowing. Since narrowing enumerates all alternatives for a unification of a subterm of a goal with all patterns of the rules, it automatically generates an appropriate case analysis (when using a constructor-complete specification).

Besides, the TIP system allows the user to work on several theorems in parallel. So one can start to prove a necessary lemma or a generalization whenever one recognizes the need for it. After finishing such a subproof, this lemma can be applied immediately in other proofs.

A short glance at the proof algorithm of the TIP system:

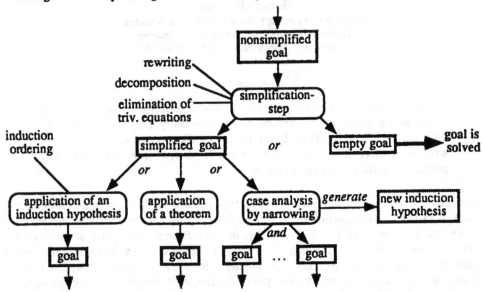

The behavior of the TIP system, in particular the degree of its automation, can be influenced by the user, if some system parameters are adjusted appropriately. Also it is possible to choose one of the four induction orderings, which are implemented in the system.

The theorem prover TIP is quite easy to use, because most parts of the tiresome work are done by the system automatically. So the user can concentrate on choosing the most promising proof steps. Easy proofs like the commutativity of the addition are found without any user interaction, but normally human intuition is needed to find the central idea(s) of a proof. All the standard examples for theorem proving can be done with the TIP system, too.

Up to now TIP can only prove unconditional goals, but we are working on an extention of the proof algorithm (with small changes) to conditional goals. Another planned improvement is a comfortable X-Window user interface.

References

[Fr, Hu 91]   U. Fraus, H. Hußmann: A Narrowing-Based Theorem Prover. Proceedings of the IMA Conference on "The Unified Comutation Laboratory", July 1990, University of Stirling (Scotland), Oxford University Press, to appear 1991.

[Gu 75]   J. V. Guttag: The specification and application to programming of abstract data types. Ph. D. thesis, University of Toronto, Toronto, 1975.

[Re 90]   U. S. Reddy: Term Rewriting Induction. CADE 10, Lecture Notes in Computer Science 249, 1990, 162-177.

[Wi e.a. 83]   M. Wirsing, P. Pepper, H. Partsch, W. Dosch, M. Broy: On hierarchies of abstract data types. Acta Informatica 20, 1983, 1-33.

The ALF System

Michael Hanus
Fachbereich Informatik, Universität Dortmund
W-4600 Dortmund 50, Germany
e-mail: michael@ls5.informatik.uni-dortmund.de

ALF (*Algebraic Logic Functional programming language*) is a language which combines functional and logic programming techniques. The foundation of ALF is Horn clause logic with equality which consists of predicates and Horn clauses for logic programming, and functions and equations for functional programming. Since ALF is a genuine integration of both programming paradigms, any functional expression can be used in a goal literal and arbitrary predicates can occur in conditions of equations. The operational semantics of ALF is based on the resolution rule to solve literals and narrowing to evaluate functional expressions. In order to reduce the number of possible narrowing steps, a leftmost-innermost basic narrowing strategy is used which can be efficiently implemented. Furthermore, terms are simplified by rewriting before a narrowing step is applied and also equations are rejected if the two sides have different constructors at the top. Rewriting and rejection can result in a large reduction of the search tree. Therefore this operational semantics is more efficient than Prolog's resolution strategy.

The ALF system is *an efficient implementation of the combination of resolution, narrowing, rewriting and rejection.* Similarly to Prolog, ALF uses a backtracking strategy corresponding to a depth-first search in the derivation tree. ALF programs are compiled into instructions of an abstract machine. The abstract machine is based on the Warren Abstract Machine (WAM) with several extensions to implement narrowing and rewriting. In the current implementation programs of this abstract machine are executed by an emulator written in C.

ALF has also a type and module concept which allows the definition of generic modules. A preprocessor checks the type consistence of the program and combines all needed modules into one flat-ALF program which is compiled into a compact bytecode representing an abstract machine program. The current implementation has the following properties:

- The machine code for pure logic programs without defined functions is identical to the code of the original WAM, i.e., for logic programs there is no overhead because of the functional part of the language.

- Functional programs where only ground terms have to be evaluated are executed by deterministic rewriting without any dynamic search for subterms positions where the next rewriting step can be applied. The compiler computes these positions and generates particular machine instructions. Therefore such programs are also efficiently executed.

- In mixed functional and logic programs argument terms are simplified by rewriting before narrowing is applied and therefore function calls with ground arguments are automatically evaluated by rewriting and not by narrowing. This is more efficient because rewriting is a deterministic process. Hence in most practical cases the combined rewriting/narrowing implementation is more efficient than an implementation of narrowing by flattening terms and applying SLD-resolution.

In order to get an impression of the current implementation, the following table contains a comparison of the ALF system with other functional languages.

Naive reverse for a list of 30 elements		
System:	Machine:	Time:
ALF	Sun4	19 msec
Standard-ML (Edinburgh)	Sun3	54 msec
CAML V 2-6.1	Sun4	28 msec
OBJ3	Sun3	5070 msec

It was mentioned above that rewriting and rejection can reduce the search space. A typical class of programs for this optimization are the generate-and-test programs. For instance, permutation sort is a program where a list is sorted by constructing a permutation of the list and checking whether the permuted list is a sorted one. The relational version of this program, which is a pure Prolog program, needs 38.4 seconds for sorting a list of 8 elements. The equivalent functional version of this program, which is executed by narrowing and rewriting, needs only 1.5 seconds for sorting the same list.

The current version of the ALF system, which is a first prototype implementation, is available on Sun-3 or Sun-4 machines under SunOS 4.0. The compiler is written in Prolog and the emulator of the abstract machine is written in C.

References:

1. M. Hanus: Compiling Logic Programs with Equality. *Proc. of the 2nd International Workshop on Programming Language Implementation and Logic Programming*, Linköping, 1990. Springer LNCS 456, pages 387–401

2. M. Hanus: Efficient Implementation of Narrowing and Rewriting. To appear in *Proc. Int. Workshop on Processing Declarative Knowledge*, Kaiserslautern, 1991 Springer Symbolic Computation Series.

3. M. Hanus, A. Schwab: ALF User's Manual. Fachbereich Informatik, Universität Dortmund, 1991

4. M. Hanus, A. Schwab: The Implementation of the Functional-Logic Language ALF. Fachbereich Informatik, Universität Dortmund, 1991

Experiences with Gentle:
Efficient Compiler Construction Based On Logic Programming

Jürgen Vollmer

GMD Forschungsstelle an der Universität Karlsruhe, Vincenz–Prießnitz–Straße 1, D–7500 Karlsruhe 1

email: vollmer@karlsruhe.gmd.dbp.de / vollmer@gmdka.uucp, Phone: +/49/721/6622-14

Abstract

Gentle [Schröer 89] is a compiler description language in the tradition of two level grammars [Koster 71] and logic programming [Warren 80]. It provides a common declarative notation for high level description of analysis, transformation, and synthesis. Imperative constructs like global variables and dynamic arrays, needed for efficient compiler construction, are introduced as well. A tool has been implemented to check the wellformedness of *Gentle* descriptions, and to generate very fast (generation speed 260.000 lines per minute) very efficient compilers (compilation speed nearly 90.000 line per minute on Dec 3100 workstation). The language and a supporting tool were designed and implemented by F.W. Schröer in 1989.

Logic Programming and Compiling

Using logic programming as a compiler–writing tool has a long tradition. [Warren 80] shows how *Prolog* may be used for this purpose and writes:

> To summarize, *Prolog* has the following advantages as a compiler–writing tool:
>
> 1. Less time and effort is required.
> 2. There is less likelihood of error.
> 3. The resulting implementation is easier to 'maintain' and modify.

The practicability of *Prolog* for compiler writing "depends on how efficiently *Prolog* itself is implemented". About the way this could be done [Warren 80] states:

> It is likely that most of the improvement will be attributable to a few relatively simple but heavily used procedures (e.g. lexical analysis, dictionary lookup), and so a mixed language approach may be an attractive possibility. An alternative view (which I favour) is to look for more sophisticated ways of compiling special types of *Prolog* procedure, guided probably by extra pragmatic information provided by the *Prolog* programmer.

The language *Gentle*, presented in this paper, uses both ways to improve the execution speed of the generated compiler. *Gentle* uses dynamic programming to overcome restrictions imposed by this approach. This technique is not discussed here.

Gentle

Compilation is often viewed as a process translating the source text into a sequence of intermediate languages, until the desired output is synthesized. These intermediate languages may be viewed as terms, and *Gentle* offers a simple and efficient way to transform these terms. These transformations are described in a declarative way using predicates. Besides the specification of terms and rules transforming them, the concrete syntax of the context free source language is declared using the same declarative notation. Due to the special nature of the task Horn logic as *Gentle*'s foundation is modified in several ways:

- *Gentle* is a typed language. Term type declarations are used to specify terms. Predicates have a typed signature. The type of local variables is derived from the context in which they occur. Global variables are declared explicitly together with their types. For example:

```
-- term declaration
EXPR = const  (INTEGER), var (IDENTIFIER),
       binary (OP, EXPR, EXPR).
OP =   plus, minus, mult, div.
-- external type declarations
'TYPE' IDENTIFIER.
-- context free grammar predicate signatures
'TOKEN' Identifier (-> IDENTIFIER).
'TOKEN' PLUS.
'NONTERM' Expression (-> EXPR).
-- term transformation predicate signatures
'ACTION' CodeExpr (EXPR -> REGISTER).
'CONDITION' get_meaning (IDENTIFIER->OBJECT).
-- local variables in a clause
CodeExpr (Expr -> ResultReg)
-- global variable / table declaration
'VAR' INT Level.
'TABLE' NODE_ATTRIBUTES Graph [NODE].
```

- The data flow inside the predicates is fixed. In a clause, the parameters left form the arrow -> are input parameters, those right of it are output parameters.

- The notion of a variable is more like that of functional languages.

- Several kinds of predicates are offered for different jobs during compilation. The context free grammar is specified using 'TOKEN' and 'NONTERM' predicates. A parser generator is used to generate a parser out of the context free grammar specified by the *nonterm* clauses.

Term transformation is specified by 'ACTION' and 'CONDITION' predicates. *Action* and *condition* predicates transform their input terms into output terms. Side effects (like writing to a file or a global variable) may be caused by them. *Action* predicates are used as an assertion a transformation must fulfill, while a *condition* predicate is used to test terms for the given condition.

```
-- a grammar clause
Expression (-> var (Id)) :
    Identifier (-> Id).
Expression (-> binary (plus, Left, Right)) :
    Expression (-> Left)
    PLUS
    Expression (-> Right).
-- an action clause
CodeExpr (const (N) -> ResultReg):
    GetNewReg (-> ResultReg)
    EmitCode (load_constant (N, ResultReg)).
-- a condition clause
Is8bit (N) :
    LessEqual (-128, N) LessEqual (N, 127).
```

- Backtracking is restricted, such that once a tail of a clause has been proven completely, all alternatives for that clause are discarded.

The price one must pay is that the *Gentle* proof procedure is not complete, but a more efficient implementation is possible. Our experience shows that the full power of the *Prolog* backtracking mechanism is not needed for compiler writing.

Compilers have to maintain global data, like symbol tables, or must deal with graphs, like basic block graphs, used for optimization. For an efficient implementation of these kinds of data, *Gentle* offers global variables and global tables, which are something like dynamic growing arrays in imperative languages.

```
-- creation of a new table entry
new (-> Node) : 'KEY' NODE Node.
-- read / write access of a table entry
Graph [Root] -> node (Info, Succ1, Succ2)
Graph [Root] <- node (Info, Succ1, Succ2)
```

Access to procedures implemented in other languages is possible. As a result *Gentle* is a mixture of declarative (backtracking, notation), functional (variables, fixed data flow) and imperative (global data, external procedures) features, which is well suited specifying compilers at a very high level of abstraction.

Measurements

Several projects are implemented using *Gentle*. Some of the important ones are:

- A program transformation tool for object oriented languages, which translates *Trellis* programs to *C++*. The specification consists of 6000 lines of *Gentle* program and external *C++* code for the lexical analyzer.

- A compiler for a subset of a *Pascal* like language, called *MiniLax* and with target processors M68k and VAX. The *Gentle* specification consists of 896 lines for the front-end and 534 lines for the code generator.
- In the *ESPRIT* project *Rex* a specification language for time critical, distributed systems was designed and is currently implemented.
- A compiler for a functional–logic language called *guarded term ML* is currently under development. As target processor a M68k is used.
- An industrial compiler for an object oriented language.
- *Gentle* itself is implemented with *Gentle*.

The followings table shows the generation speed of the *Gentle* tool and the compilation speed of the generated compiler. The times are measured on a *Dec station 3100 (MIPS processor)* using the *UNIX time* command (user time). The average of three runs is given. The generated *C* programs are compiled using the optimization capability of the *C* compiler.

For the *Generation* measurement, the *Gentle* tool generates itself. The specification consists of 3939 lines of *Gentle* program and 405 line *Gentle* library. Output are 4861 lines of *C* program, 1455 lines input for the parser generator and 30 lines auxiliary input to the scanner and parser generator. For the *Compilation* measurement, the generated *MiniLax* compiler compiles a 8055 line input program and generates a 58.718 lines of M68k assembler text.

	time (sec)	lines per minute
Generation	1,0	260.640
Compilation	5,4	89.499

References

[Koster 71] C.H.A Koster. Affix grammers. In J.E.L Peck, editor, *ALGOL 68 Implementation*, pages 95–109. North Holland, 1971.

[Schröer 89] F.W. Schröer. Gentle. In J. Grosch, F.W. Schröer, and W.M. Waite, editors, *Three Compiler Specifications*, GMD – Studien Nr. 166, pages 31–36. Gesellschaft für Mathematik und Datenverarbeitung, Forschungsstelle Karlsruhe, August 1989.

[Vollmer 91] Jürgen Vollmer. A tutorial on gentle. Arbeitsberichte der GMD Nr 508, GMD, German National Research Center for Computer Science, Vincenz–Prießnitz-Straße 1, D-7500 Karlsruhe-1, February 1991.

[Warren 80] David H.D. Warren. Logic programming and compiler writing. *Software–Practice and Experience*, 10:97–125, 1980.

The system FLR (Fast Laboratory for Recomposition)

Beate Baum, Peter Forbrig

Universität Rostock
FB Informatik
Albert - Einstein - Str. 21
2500 Rostock
Federal Republic of Germany

The system FLR /FoLä89/ developed at our department is a tool for generating text documents, programs in a programming language, formal specifications and grammars. FLR is implemented in Prolog.

FLR utilizes the generative aspect of Attribute Grammars. The knowledge of a problem domain is described by special Attribute Grammar rules. Each nonterminal contains two attributes at least. The first attribute called control character supports the generating process, the second attribute comments it. Different control characters for one nonterminal stand for different possibilities in generating software parts. The solution of a special problem is generated by selecting special rules out of the knowledge base.

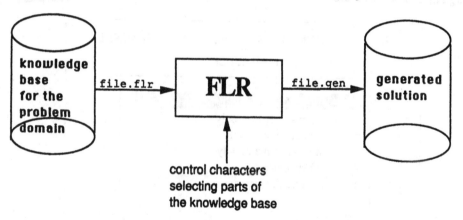

FLR provides functions for manipulating grammars and control characters in a menu - oriented interface. The grammars can be activated, modified and checked with regard to their properties, like reduceness, termination and attribute independencies. Also operations for handling files are available.

In our approach the system FLR is used for generating Attribute Grammars for compiler construction. This means, the FLR - grammar rules contain grammar parts of compiler descriptions for algorithmical languages in form of prolog clauses. So the compiler constructor has the possibility to select determined grammar parts for his applications. Some grammar parts, like scanning and the symbol table managment, are uniform for each application and can be reused always. The result of this generation is a concrete compiler grammar represented by a prolog program.

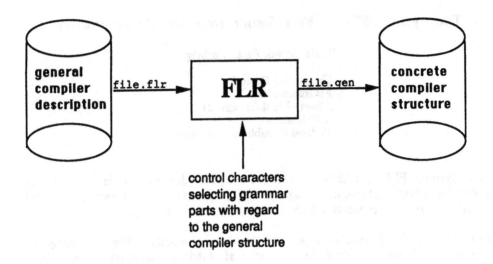

control characters
selecting grammar
parts with regard
to the general
compiler structure

In the following figure a window is shown demonstrating the process of selecting grammar parts. By deciding, for example for the control character w, the grammar rules describing the while-statement in algorithmical languages are generated.

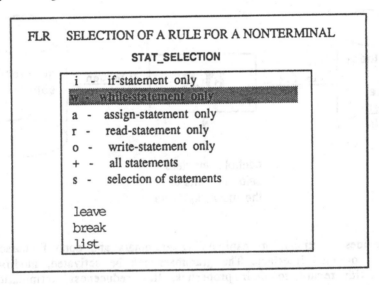

We try to extend our knowledge base and to use the system for further experiments.

/FoLä89/   Forbrig,P.; Lämmel,U.:
           Knowledge based program generation using attributed grammars
           in: bit'89
               Proceedings of the Berliner Informatik - Tage
               Informatik Informationen Reporte
               Akademie der Wissenschaften, Berlin, 1989

Multilanguage Interoperability

Giuseppe Attardi, Mauro Gaspari

Dipartimento di Informatica, Corso Italia 40, I-56125 Pisa, Italy

attardi@di.unipi.it, gaspari@di.unipi.it

1. Introduction

We present an approach to the interoperability of programming languages, based on a Common Runtime Support, and its implementation for Prolog, Common Lisp and C.

The interoperability of programming languages is more and more often required. As applications become more elaborate, the facilities required to build them grow in number and sophistication. Each facility is accessed through a specific package, quite complex itself. like in the cases of: modeling, simulation, graphics, hypertext, data bases, symbolic manipulation and so on. Reusability is a significant issue: once a package has been developed, tested and debugged, it is undesirable having to rewrite it in a different language just because the application is based in such other language. It is unlikely that all useful facilities be available in one language.

Several approaches have been proposed to the problem of combining code from different languages [Atkinson 89]: client/server model with remote procedure calls; foreign function call interfaces; common intermediate form on which all languages are translated.

The approach we propose achieves tightly coupled interoperability, i.e. procedures in one language can invoke procedures in another language and viceversa, and data can be shared or passed back and forth between procedures in different languages, without the overhead of transforming data representations. Interopreability is bidirectional, in the sense that not only languages with more sophisticated facilities (like memory management) can call procedures of less sophisticated languages, but also the opposite direction is supported, allowing for instance a C based application to call a package developed in Prolog or LISP.

2. Common Runtime Support

Our approach to interoperability is based on an intermediate support layer between the operating system and the high level programming language. This layer is called CRS (Common Runtime Support), and provides the essential functionalities which can be abstracted from modern high level languages. The CRS uses C as the common intermediate form for all languages. Such usage of C has been applied successfully to several programming languages, for instance Cedar, Common Lisp and Scheme at Xerox PARC [Atkinson 89], Modula3 and C at DEC SRC, Linda and also Prolog. Though convenient for portability, the use of C as intermediate form is not essential to the approach, and in fact the fundamental requirements are agreement on procedure call conventions and access to the facilities provided by the CRS (memory, I/O and processes) only through the functional interface provided by CRS. The CRS provides the following facilities:

- storage management, with dynamic memory allocation and *conservative* garbage collection
- symbol table management, including dynamic linking and loading, and image dumping
- support for debugging, allowing to debug mixed code programs through a single debugging tool
- multiple threads, to support concurrent programming constructs
- generic low level I/O, operating both on files and on network streams
- support for logic variables, unification and nondeterminism

The following diagram shows the CRS in relation with other elements of a programming environment.

The Lisp Abstract Machine as well as the Warren Abstract Machine, are provided as a library of C routines and data structures. The Common Lisp compiler translates Lisp code to C, which is then handed over to the native C compiler to produce executable code. Similarly Prolog is translated to C through an intermediate Lisp form.

3. Nondeterminism

We explored two alternative implementations of nondeterminism and backtracking: success continuations and failure continuations. In the technique of *success continuations*, each Prolog predicate is translated into a boolean function with as many arguments as the predicate plus an extra one, the continuation, which is a function of zero arguments, representing any remaining goals carried over from previous calls, i.e. the resolvent except the first goal. The continuation in general has to be a *closure*, since it may refer to some arguments of the predicate. To achieve interoperability with COMMON LISP, [Attardi 90] used Lisp closures as success continuations.

With the technique of *failure continuations*, the continuation argument is not needed: predicates are translated into functions which return either TRUE, or FALSE or SUCCESS. SUCCESS is returned when a failure continuation has been created and pushed onto the continuation stack, i.e. there are still other choices for the

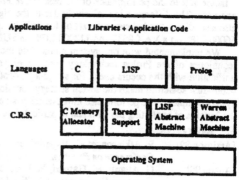

predicate, while TRUE is returned for the only (or last) choice in a predicate. Therefore determinate predicates return either TRUE or FALSE. Stack frames for predicates returning either TRUE or FALSE are immediately reclaimed, while stack frames are preserved for predicates returning SUCCESS.

With success continuations it is difficult to optimize the handling of the predicate cut. Consider a clause such as: a :- b, !, c. The predicate cut is executed in the continuation of b, i.e. two or more stack frames down from a. These frames may be reclaimed, since one will not return to them, however this is hard to do in a portable way. With failure continuations, the cut can just pop from the continuation stack all the elements since the invocation of a. Another optimization which is impossible with success continuations is discarding the last choice point when entering the last clause in a predicate. With failure continuations a similar effect is obtained by avoiding to issue a SUCCESS after the last body goal: any subsequent failure will skip this predicate to return to the immediately preceding failure continuation. Tail recursion optimization can be done for success continuations when the body consists of a single goal or all predicates involved in the body goals are known to be determinate. With failure continuations we can detect during execution determinate predicates and as a consequence perform tail-recursion optimization whenever appropriate. In [Attardi 90] we described techniques such as mode declaration and clause indexing which apply equally well to both approaches.

The CRS implements the mechanism for failure continuations through the following primitives:

SUCCEED   pushes a failure continuation into the failure stack and returns SUCCESS to the caller

FAIL   transfers control back to last failure continuation

FRAME_SKIP   preserves the stack frame referred by the last failure continuation

The approach of success continuations has an advantage in portability, since there are no machine dependencies in its implementation. Instead the FRAME_SKIP primitive is definitely machine dependent.

4. Related Work

Most of the work in the literature address the issue of integration between logic programming and functional programming by proposing new languages which combine the features of both approaches. Several authors have explored implementations of Prolog in a functional language or additions of logic programming to a functional language through the use of continuations, where "backtrack" continuations are either explicit in the user code, or hidden through macros. For the great part these solutions, though of theoretical interest, have lower practical impact. In fact, it is quite hard to develop efficient implementations of a combined logic/functional language (most papers leave unexplored the issue of efficiency or admit its difficulty), since the conflicting requirements of the two approaches (in terms of data mobility, memory management, control flow) invalidate the techniques that had been developed separately for each approach. A second reason is that any new language introduces a problem of backward compatibility: i.e. programs written previously will have to be rewritten or translated into the new language. The rate of acceptance of new languages is decreasing as the amount of code, libraries and programming tools for the language that one expects to be available is becoming significantly large.

Our approach is quite different, since it allows each language to *coexist* with any other. Applications libraries written in a language can be accessed readily from other languages. Predeveloped Prolog or Lisp programs can be embedded into full applications without any change. Backtracking is dealt through failure continuations, with a mechanism that allows to detect dynamically determinate predicates and to reclaim their stack frame immediately after return. Logic variables are implemented through locatives, and are first-class values which can be stored in variables or data structures. There is a single stack frame, for both C, Lisp and Prolog. A Common Runtime for the languages is an appealing solution, since it leaves programmers the freedom to use the most adequate language for each task, and also avoids giving some particular language a special role.

6. Conclusions

The current implementation of Prolog for the CRS is compatible with the specifications of the ISO working group on Prolog. Its performance, measured through the usual Prolog benchmarks, meets our goal to stay within a factor of 2 to the performance of the best native Prolog implementations, demonstrating the viability of the approach. The difference is exclusively due to the higher cost of memory allocation primitives which include also a small overhead of synchronization for multithread. Our benchmarks indicate for instance a performance of 56000 LIPS on a VAX 8530 and 110000 LIPS on SUN4/260.

We implemented with our system a complete theorem prover for first order predicate calculus using the techniques of the Prolog technology theorem prover (PTTP) by Mark Stickel. The theorem prover uses unification with the occurs check for soundness, depth-first iterative-deepening search instead of unbounded depth-first search to make the search strategy complete, and the model elimination rule to make the inference system complete. The performance improvements we achieved with respect to Stickel's implementation indicate the benefits of exploiting the CRS and in particular its support for logic variables and unification.

5. References

[Atkinson 89]   R. Atkinson, et al., Experiences creating a portable Cedar, *Proceedings of the SIGPLAN 89 Conference on Programming Language Design and Implementation*, 1989.

[Attardi 90]   G. Attardi, M. Gaspari, F. Saracco, Interoperability of AI lanaguages, *Proceedings of 9th European Conference on Artificial Intelligence*, 1990, 41–46.

Short overview of the CLPS system

Bruno LEGEARD - Emmanuel LEGROS

ENSMM Besançon - LAB
15, Impasse des St Martin
25000 BESANCON
Tel. 81 88 53 44
Fax. 81 88 65 02

Abstract

We propose an extension of Prolog to deal with set expressions using equality and membership operators and allowing restricted universal quantifiers in goals. Such extension allows a better expressive power, in particular when the language is used for a functional specification purpose. Those set expressions are treated as symbolic constraints and some mechanisms for constraints satisfaisability checking and solution values generating are provided. We describe, in the paper, the current implementation which is composed by three modules : preprocessor, satisfaction checker and solution values generator. Then, we demonstrate some promising results on the color problem comparing with standard Prolog and PROLOG III language.
In the future, we aim to extend our implementation to deal with the subset operator and the sequence data structure.

Key-Words

Logic Programming, Prolog, constraints, CLPS, set languages.

1- Overview of CLPS

We will, here, present the principal characteristics of CLPS.

1.1- Objects and operators

The basic objects of CLPS are atom and set. A set is an heriditary finite collection of coherent objects. That means that we accept sets of sets, but **at any level all sets are finite**.

Syntactically, a set is notified by brackets and {} denotes the empty set. The rest of notation of the so called Edimburgh Prolog are conserved. Set construction can be done in three way :

- by using the set construction function , noted 'l'. {E|S} means the adding of element E inside the set S. Differently speaking, E is an arbitrary element of the set {E|S}.
- by using the union operator - $\cup$ -, for example in a equality constraint - e.g. $S1 = \{E\} \cup S$, or in an expression as argument of a predicate : $pred1(\{E\} \cup S, ...)$.
- by explicit definition, for example {e,f,g} which is the set collecting the three characters e, f and g, or {3 .. 7} which is the collection of the integers 3,4,5,6 and 7.

Now, we have to consider the built-in operators. The ones are used to define set expressions, and the others to build set constraints. The first category covers union ($\cup$), intersection ($\cap$), cardinality(#) and n pow of a set (npow). Set expressions can appear as predicates arguments or inside a set constraints. The second category covers equality (=), membership ($\in$) and their negation. The set constraints appear only in the body of clauses. But,the set expressions can be used as arguments in the head of clauses.

1.2 Restricted Universal Quantifiers

CLPS allows an extension of Horn Clauses, which is a formula of the form : B :- L1, ..., Ln ; where Li can be either a set constraint, a restricted universal quantifier or a Prolog predicate (possibly using negation as failure).

A **set constraint** is an expression of the form F1 Op1 F2, where Fi is a set expression and Op1 a set constraint operator, member of $\{=, \neq, \in, \notin\}$.
A **set expression** is an atom, a set or has the form F1 Op2 F2 where Fi is a Set expression and Op2 a set expression operator, which takes values inside $\{\cup, \cap, \#, npow\}$.
A **restricted universal quantifier** is an expression of the form : $\forall X1 \in S1,..., \forall Xn \in Sn$: $(pred1(...),...,predk(...))$; which limit the scope of universal quantification of X1, ..., Xn to the predicates pred1 to predk. S1 ... Sn must be sets.

2 - General design.

The next general scheme presents the different stages required for obtaining the results of a set constraints problem. We clearly see observing this scheme, that the CLPS's environment consists of one preprocessor, one unit realizing consistency checking during the resolution process (in fact, classical prolog), and a unit generating combinations of values that must appear as solutions. The three phases of the solving process above mentioned, are each realized by one of the three units making up the CLPS interpreter. The first unit (the preprocessor) obtains the CLPS program to be executed, the second (Prolog & consistency checking) elaborates step by step the constraints system that will have to be verified by the results of the problem and finally, the last unit (results generator) solves the set constraints system elaborated during the previous stage.

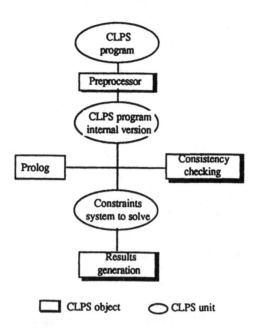

□ CLPS object   ◯ CLPS unit

3 - the color problem

To illustrate the foregoing, we present in this part a well known problem, solved with the CLPS interpreter. This example shows the two main advantages of constraints logic programming. Firstly, the possibility to express constraints directy in the discourse's domain increases significantly the expressiveness and secondly, the constraints solving strategy - we assign free variables only when all the constraints concerning them are known - which reduces the execution time by limiting useless backtracks (in opposition to standard Prolog strategy "Generate and Test").

<u>Standard Prolog program</u>

```
colouring([],_,S,S).
colouring([R|L],C,T,S):-
        member(Co,C),
        not incorrect([R,Co],T),
        colouring(L,C,[[R,Co]|T],S).
incorrect([R,C],S):-
        member([R1,C],S),
        adj(R1,R).
adj(R,R1):-
        adjacent(R,Eadj),
        member(R1,Eadj).
member(X,[X|_]).
member(X,[_|L]):-
        member(X,L).
```

<u>CLPS program</u>

```
colouring({},_,{}).
        colouring({R|L},C,{<R,Co>|S}):-
        Co in C,
        good_color(<R,C>,S),
        colouring(L,C,S).
good_color(<R,C>,S):-
        adjacent(R,Eadj),
        forall R1 in Eadj : (correct(<R,C>,R1,S)).
correct(<R,C>,R1,S):-
        <R1,C1> in S,
        C neg C1.
correct([R,C],R1,S):-
        <R1,_> nin S.
```

The CLPS system is meta-programmed on a Sun 4-60 station (with 8 Mega RAM) with the Delphia Prolog V 2.0.7 (compiled) language (standard Prolog and CLPS programs).

Lecture Notes in Computer Science

For information about Vols. 1–441
please contact your bookseller or Springer-Verlag

Vol. 486: J. van Leeuwen, N. Santoro (Eds.), Distributed Algorithms. Proceedings, 1990. VI, 433 pages. 1991.

Vol. 487: A. Bode (Ed.), Distributed Memory Computing. Proceedings, 1991. XI, 506 pages. 1991.

Vol. 488: R. V. Book (Ed.), Rewriting Techniques and Applications. Proceedings, 1991. VII, 458 pages. 1991.

Vol. 489: J. W. de Bakker, W. P. de Roever, G. Rozenberg (Eds.), Foundations of Object-Oriented Languages. Proceedings, 1990. VIII, 442 pages. 1991.

Vol. 490: J. A. Bergstra, L. M. G. Feljs (Eds.), Algebraic Methods 11: Theory, Tools and Applications. VI, 434 pages. 1991.

Vol. 491: A. Yonezawa, T. Ito (Eds.), Concurrency: Theory, Language, and Architecture. Proceedings, 1989. VIII, 339 pages. 1991.

Vol. 492: D. Sriram, R. Logcher, S. Fukuda (Eds.), Computer-Aided Cooperative Product Development. Proceedings, 1989 VII, 630 pages. 1991.

Vol. 493: S. Abramsky, T. S. E. Maibaum (Eds.), TAPSOFT '91. Volume 1. Proceedings, 1991. VIII, 455 pages. 1991.

Vol. 494: S. Abramsky, T. S. E. Maibaum (Eds.), TAPSOFT '91. Volume 2. Proceedings, 1991. VIII, 482 pages. 1991.

Vol. 495: 9. Thalheim, J. Demetrovics, H.-D. Gerhardt (Eds.), MFDBS '91. Proceedings, 1991. VI, 395 pages. 1991.

Vol. 496: H.-P. Schwefel, R. Männer (Eds.), Parallel Problem Solving from Nature. Proceedings, 1991. XI, 485 pages. 1991.

Vol. 497: F. Dehne, F. Fiala. W.W. Koczkodaj (Eds.), Advances in Computing and Intormation - ICCI '91 Proceedings, 1991. VIII, 745 pages. 1991.

Vol. 498: R. Andersen, J. A. Bubenko jr., A. Sølvberg (Eds.), Advanced Information Systems Engineering. Proceedings, 1991. VI, 579 pages. 1991.

Vol. 499: D. Christodoulakis (Ed.), Ada: The Choice for '92. Proceedings, 1991. VI, 411 pages. 1991.

Vol. 500: M. Held, On the Computational Geometry of Pocket Machining. XII, 179 pages. 1991.

Vol. 501: M. Bidoit, H.-J. Kreowski, P. Lescanne, F. Orejas, D. Sannella (Eds.), Algebraic System Specification and Development. VIII, 98 pages. 1991.

Vol. 502: J. Bārzdiņš , D. Bjørner (Eds.), Baltic Computer Science. X, 619 pages. 1991.

Vol. 503: P. America (Ed.), Parallel Database Systems. Proceedings, 1990. VIII, 433 pages. 1991.

Vol. 504: J. W. Schmidt, A. A. Stogny (Eds.), Next Generation Information System Technology. Proceedings, 1990. IX, 450 pages. 1991.

Vol. 505: E. H. L. Aarts, J. van Leeuwen, M. Rem (Eds.), PARLE '91. Parallel Architectures and Languages Europe, Volume I. Proceedings, 1991. XV, 423 pages. 1991.

Vol. 506: E. H. L. Aarts, J. van Leeuwen, M. Rem (Eds.), PARLE '91. Parallel Architectures and Languages Europe, Volume II. Proceedings, 1991. XV, 489 pages. 1991.

Vol. 507: N. A. Sherwani, E. de Doncker, J. A. Kapenga (Eds.), Computing in the 90's. Proceedings, 1989. XIII, 441 pages. 1991.

Vol. 508: S. Sakata (Ed.), Applied Algebra, Algebraic Algorithms and Error-Correcting Codes. Proceedings, 1990. IX, 390 pages. 1991.

Vol. 509: A. Endres, H. Weber (Eds.), Software Development Environments and CASE Technology. Proceedings, 1991. VIII, 286 pages. 1991.

Vol. 510: J. Leach Albert, B. Monien, M. Rodríguez (Eds.), Automata, Languages and Programming. Proceedings, 1991. XII, 763 pages. 1991.

Vol. 511: A. C. F. Colchester, D.J. Hawkes (Eds.), Information Processing in Medical Imaging. Proceedings, 1991. XI, 512 pages. 1991.

Vol. 512: P. America (Ed.), ECOOP '91. European Conference on Object-Oriented Programming. Proceedings, 1991. X, 396 pages. 1991.

Vol. 513: N. M. Mattos, An Approach to Knowledge Base Management. IX, 247 pages. 1991. (Subseries LNAI).

Vol. 514: G. Cohen, P. Charpin (Eds.), EUROCODE '90. Proceedings, 1990. XI, 392 pages. 1991.

Vol. 515: J. P. Martins, M. Reinfrank (Eds.), Truth Maintenance Systems. Proceedings, 1990. VII, 177 pages. 1991. (Subseries LNAI).

Vol. 516: S. Kaplan, M. Okada (Eds.), Conditional and Typed Rewriting Systems. Proceedings, 1990. IX, 461 pages. 1991.

Vol. 517: K. Nökel, Temporally Distributed Symptoms in Technical Diagnosis. IX, 164 pages. 1991. (Subseries LNAI).

Vol. 518: J. G. Williams, Instantiation Theory. VIII, 133 pages. 1991. (Subseries LNAI).

Vol. 519: F. Dehne, J.-R. Sack, N. Santoro (Eds.), Algorithms and Data Structures. Proceedings, 1991. X, 496 pages. 1991.

Vol. 520: A. Tarlecki (Ed.), Mathematical Foundations of Computer Science 1991. Proceedings, 1991. XI, 435 pages. 1991.

Vol. 521: B. Bouchon-Meunier, R. R. Yager, L. A. Zadek (Eds.), Uncertainty in Knowledge-Bases. Proceedings, 1990. X, 609 pages. 1991.

Vol. 522: J. Hertzberg (Ed.), European Workshop on Planning. Proceedings, 1991. VII, 121 pages. 1991. (Subseries LNAI).

Vol. 523: J. Hughes (Ed.), Functional Programming Languages and Computer Architecture. Proceedings, 1991. VIII, 666 pages. 1991.

Vol. 524: G. Rozenberg (Ed.), Advances in Petri Nets 1991. VIII, 572 pages. 1991.

Vol. 525: O. Günther, H.-J. Schek (Eds.), Large Spatial Databases. Proceedings, 1991. XI, 471 pages. 1991.

Vol. 526: T. Ito, A. R. Meyer (Eds.), Theoretical Aspects of Computer Software. Proceedings, 1991. X, 772 pages. 1991.

Vol. 527: J.C.M. Baeten, J. F. Groote (Eds.), CONCUR '91. Proceedings, 1991. VIII, 541 pages. 1991.

Vol. 528: J. Maluszyński, M. Wirsing (Eds.) Programming Language Implementation and Logic Programming. Proceedings, 1991. XI, 433 pages. 1991.